Exploring Medical Biotechnology – *In Vivo, In Vitro, In Silico*

This book on medical biotechnology offers a wide array of topics and cutting-edge research in the field featuring contributions from multiple authors, each specializing in their respective areas, making it highly valuable to science students and enthusiasts. The book provides comprehensive coverage on a diverse range of topics including sequence analysis, network pharmacology, drug discovery, CRISPR-Cas technology, precision medicine, neuroimaging biomarkers, therapeutics for neurodegenerative diseases, molecular pathogenesis of various diseases, plant-derived antioxidants, genetic approaches for disease diagnosis, cancer progression, immunotherapy, cancer biomarker identification, RNA Interference (RNAi) and nanotechnology in drug discovery and delivery, thus giving a holistic understanding of medical biotechnology.

Each chapter delves into the latest research and developments in its respective field and offers readers insights into cutting-edge advancements in medical biotechnology. This up-to-date information will be invaluable to science students and enthusiasts who want to stay at the forefront of the field.

The book adopts an interdisciplinary approach by incorporating elements of biology, genetics, computational techniques, nanotechnology and more, which enables readers to see the interconnectedness of different scientific disciplines and how they contribute to medical biotechnology.

The book emphasizes practical applications, such as drug discovery, disease treatment, gene editing and targeted drug delivery. This focus on real-world applications is useful to readers interested in applying biotechnological techniques in the medical field. The book also acknowledges the serious challenges faced in the field of medical biotechnology and offers potential solutions to overcome these challenges to foster innovative thinking.

As a reader, you will gain a comprehensive understanding of various aspects of medical biotechnology, equipping you with knowledge that can contribute to technological breakthroughs, advancements in medicine and the betterment of human life.

Confluence of Research in Biotechnology for Human Welfare
Series Editor: Susmita Mukherjee
(University of Engineering & Management, West Bengal, India)

Natural Products
Alternate Therapeutic as Quorum Sensing (QS) Inhibitors
Edited by Moupriya Nag, Dibyajit Lahiri, Jaideep Banerjee and Taniya Roy Chowdhury

Toxicity of Aquatic System and Remediation
The Contemporary Issues
Edited by Susmita Mukherjee, Sanket J. Joshi, Sonali Paul, and Rita Kundu

Exploring Medical Biotechnology – *In Vivo, In Vitro, In Silico*
Biotechnology from Labs to Clinics and Basic to Advanced
Edited by Pratik Talukder, Ritesh Tandon and Uttam Pal

For more information, please visit series page: https://www.routledge.com/Confluence-of-Research-in-Biotechnology-for-Human-Welfare/book-series/CRBHW

Exploring Medical Biotechnology – *In Vivo, In Vitro, In Silico*

Biotechnology from Labs to Clinics and Basic to Advanced

Edited by Pratik Talukder, Ritesh Tandon and Uttam Pal

CRC Press
Taylor & Francis Group
Boca Raton London New York

CRC Press is an imprint of the
Taylor & Francis Group, an **informa** business

First edition published 2024

by CRC Press
2385 NW Executive Center Drive, Suite 320, Boca Raton FL 33431

and by CRC Press
4 Park Square, Milton Park, Abingdon, Oxon, OX14 4RN

CRC Press is an imprint of Taylor & Francis Group, LLC

ISBN: 978-1-032-29555-8 (hbk)
ISBN: 978-1-032-29556-5 (pbk)
ISBN: 978-1-003-30213-1 (ebk)

DOI: 10.1201/9781003302131

Typeset in Times
by Apex CoVantage, LLC

Contents

Preface

The field of biomedical science has considerably affected our way of life in recent years. Beginning from lifesaving therapeutics to post-COVID immunization, biomedical methods have entered every part of our advanced occupied life. Hence, it is critical to broaden our insight into biomedical science to extend this new yet profoundly significant field of biotechnology. This book, "*Exploring Medical Biotechnology* – In Vivo, In Vitro, In Silico*: Biotechnology from Labs to Clinics and Basic to Advanced*", is about the new continuous exploration patterns and advances in the field of medical science to assist researchers and scholars. Recently, during the post-COVID situation, the significance of biomedical strategies have been expanding step-by-step for the advancement of mankind. This book features a combination of logical works from different yet interconnected exploration regions concerning both hypothetical and viable/specialized fields of medical biotechnology. Medical biotechnology is the utilization of living cells and cell materials to research and create drugs and other demonstrative items that help treat and forestall human sicknesses. Medical biotechnology incorporates innovative work utilized in clinical, rural and drug development, surgical methods with artificial intelligence and many more uses. Most clinical/medical biotechnologists work at universities or other sophisticated R&D labs that are generally outside the knowledge of general new aspirants of research. For these individuals, this book will be an eye opener.

Most work in this sector is associated with sophisticated, state-of-the-art techniques and develops bio-based electrical devices for various therapeutic analyses. Various modern therapeutic tests as a feature of clinical biotechnology explore modern facilities that often remain unknown to the masses. This book also plays an important role in public consciousness and pre-alerts for the need for one. Moreover, medical biotechnology has brought new ideas and applications to the field of food safety and associated health hazards. It includes the various multidisciplinary fields such as biomechanics and bio-electrical impulses that are crucial for the wellbeing of various patients. In addition, tissue engineering and regenerative medicine have also been incorporated and have made a parallel way alongside conventional therapeutic techniques. This book truly showcases an ensemble of scientific works from the diverse yet interconnected research areas in the theoretical and practical/technical fields of medical biotechnology.

Editor Biographies

Dr. Pratik Talukder obtained his Ph.D. in Science from the University of Calcutta, India. Currently, Dr. Talukder is working as Associate Professor in the Department of Biotechnology, University of Engineering and Management, Kolkata (UEMK). He is a university rank holder in his postgraduate degree. He is the recipient of the prestigious national postdoctoral fellowship of the Science and Engineering Research Board (SERB), Department of Science & Technology, Government of India. His background is in molecular biology, biophysics, bioinformatics and tissue culture. He has published more than 30 articles and reviewed articles for several journals of international repute. He is the assistant editor of the journal *Applied Biochemistry and Biotechnology*. He has also received research projects from the Government of India.

Dr. Uttam Pal obtained his Ph.D. in Science from Jadavpur University, India. He is a university rank holder in his postgraduate degree from the University of Calcutta and a recipient of an INSPIRE Fellowship. Dr. Pal served as a scientist working on translational research at the Technical Research Centre of S.N. Bose National Centre for Basic Sciences, India. His research interests are computational drug discovery, *in silico* design and the characterization of nano biomaterials for medical applications. He has developed binding specificity and drugability indices within the concept of molecular docking. He is a member of the Royal Society of Chemistry, London and several other scientific societies. He has published many research and review articles and reviewed articles for several journals. Additionally, he currently serves as a guest associate editor for the *Frontiers in Chemistry* and *Applied Biochemistry and Biotechnology* journals.

Dr. Ritesh Tandon is an Associate Professor of Microbiology and Immunology at the University of Mississippi Medical Center (UMMC). He contributes significantly to research, teaching and service. Dr. Tandon is recognized internationally for his research on herpes viruses. His research is supported by several extramural and intramural grants, and he serves on the American Heart Association (AHA) and the National Institutes of Health (NIH) grant review study sections. He is also a Fellow of the American Heart Association (FAHA). He directs the Graduate Virology Course at UMMC and team teaches three other microbiology courses. Dr. Tandon also holds a joint assistant professor position in the Department of Biomolecular Sciences at the University of Mississippi. He has published considerable research and reviewed articles in several journals of international repute.

Editor biographies

[illegible]

[illegible]

[illegible]

Contributors

Ayan Adhikary
University of Engineering and Management
Kolkata, India

Arpan Banerjee
University of Engineering and Management
Kolkata, India

Shreya Banerjee
Serampore College
Serampore, India

Sankha Banerjee
Presidency university
Kolkata, India

Maitrayee Banerjee Mukherjee
Krishnagar Government College
Krishnagar, India

Tista Basu
ICAR – Central Research Institute for Jute and Allied Fibers
Barrackpore, India

Kaushik Bera
CSIR – Indian Institute of Chemical Biology
Kolkata, India

Disha Bhattacharya
Guru Nanak Institute of Pharmaceutical Science and Technology
Panihati, India

Pritam Biswas
St. Xavier's College
Kolkata, India

Santanu Chakraborty
Presidency university
Kolkata, India

Srishti Chakraborty
University of Engineering and Management
Kolkata, India

Sounak Chanda
University of Engineering and Management
Kolkata, India

Nivedita Chatterjee
University of Engineering and Management
Kolkata, India

Amlan Das
National Institute of Biomedical Genomics
Kalyani, India

Anirban Das
University of Engineering and Management
Kolkata, India

Shreya Das
Jadavpur University
Kolkata, India

Suravi Datta
University of Engineering and Management
Kolkata, India

Chandrani Dey
Jadavpur University
Kolkata, India

Chiranjeeb Dey
Serampore College
Serampore, India

Anushka Dutta
University of Engineering and Management
Kolkata, India

Anushka Ghosh
University of Engineering and Management
Kolkata, India

Debabrata Ghosh Dastidar
Guru Nanak Institute of Pharmaceutical Science and Technology
Panihati, India

Shubham Kundu
CSIR – Indian Institute of Chemical Biology
Kolkata, India

Shimantika Maikap
University of Engineering and Management
Kolkata, India

Nakul C. Maiti
CSIR – Indian Institute of Chemical Biology
Kolkata, India

Anupam Maity
CSIR – Indian Institute of Chemical Biology
Kolkata, India

Pranga Prava Mandal
CSIR – Indian Institute of Chemical Biology
Kolkata, India

Meghna Mishra
University of Engineering and Management
Kolkata, India

Rajdip Misra
CSIR – Indian Institute of Chemical Biology
Kolkata, India

Animesh Mondal
Government General Degree College
Mangalkote, India

Arundhuti Mukherjee
University of Engineering and Management
Kolkata, India

Susmita Mukherjee
University of Engineering and Management
Kolkata, India

Nilay Nandi
University of Texas Southwestern Medical Center
Dallas, USA

Uttam Pal
CSIR – Indian Institute of Chemical Biology
Kolkata, India

Debashish Paul
Shiv Nadar University
Delhi-NCR, India

Sonali Paul
University of Engineering and Management
Kolkata, India

U. S. Poornima
Shiv Nadar University
Delhi-NCR, India

Sumit Kumar Pramanik
CSIR – Central Salt & Marine Chemicals Research Institute
Bhavnagar, India

Tatini Rakshit
Shiv Nadar University
Delhi-NCR, India

Rajarshi Ray
University of Engineering and Management
Kolkata, India

Kamalesh Roy
Institute of Genetic Engineering
Badu, India

Krishnendu Sarkar
Krishnagar Government College
Krishnagar, India

Mainak Sarkar
University of Engineering and Management
Kolkata, India

Suranjana Sarkar
Surendranath College
Kolkata, India

S. Saroj
Shiv Nadar University
Delhi-NCR, India

Arunima Sengupta
Jadavpur University
Kolkata, India

Shashank Shekhar
University of Texas Southwestern Medical Center
Dallas, USA

Abhilasha Singh
University of Engineering and Management
Kolkata, India

Baishakhi Sinha
University of Engineering and Management
Kolkata, India

Asmeeta Sircar
University of Engineering and Management
Kolkata, India

Pratik Talukder
University of Engineering and Management
Kolkata, India

Rajeswari Tiwari
CSIR – Central Salt & Marine Chemicals Research Institute
Bhavnagar, India

Ritesh Tandon
National Institute of Health
Bethesda, USA

Introduction

The most intriguing and complicated object of our world is life itself. Scientists have always had a question in their mind about how tiny molecules of DNA, proteins and enzymes perform the enormous task of regulating the entire physiology of a living organism. Advancement in the understanding of our own biology and the biology of the pathogens that we encounter is essential for technological breakthroughs for the sustenance and progress of our civilization and is one of the most important objective that one can accomplish during their lifetime. We are at the dawn of an efflorescence in medicinal biotechnology; this field provides an excellent basis for the application of modern physical, chemical and computational techniques. Hence, a three-pronged approach of *in vivo*, *in vitro*, *in silico* is the need of the hour. This book will play a pivotal role in the improvement of our understanding of the different facets of medical biotechnology. This book will diversify experiences and knowledge and enable one to learn and absorb new ideas and to use them to further the boundaries of modern research and make human life better. There are, and will be, many serious challenges in the field of medical biotechnology, and this book will help in finding their solutions.

The first chapter of this book by Animesh Mondal deals with sequence analysis to decipher protein functions in human health and disease. A comprehensive understanding of this chapter will help to decode the functional paradigm and folding behaviour of proteins in disease biology. The second chapter by Biswas et al. concerns network pharmacology as a potential tool to unlock the treasure trove of pharmaceuticals in natural products. The development of network pharmacology provides a way to potentially unlock and develop a scientific evidence-based library for the discovery of natural products for therapeutic applications. The third chapter by Misra et al. gives a comprehensive review of recent developments on drug discovery approaches that target dengue protease; in this chapter, an overview of the dengue viral life cycle, the origin of the disease and drug development approaches are presented with a specific focus on the recent research aimed at the antiviral discovery targeting viral NS2B/NS3 serine protease. The fourth chapter by Ghosh Dastidar et al. involves the modern computational tools for vaccine engineering. In this chapter, three major approaches for *in silico* vaccine engineering are discussed. The next chapter by Mukherjee et al. discusses the scope and significance of precision medicine; this recent concept is based on studying varying patient parameters such as their occupation, medical history and place of residence apart from genetic makeup. By analysing these data, patients are identified for clinical trials depending upon the probability of how well they might respond to a given therapy. The next two chapters address the latest CRISPR-Cas technology. The sixth chapter by Krishnendu Sarkar deals with the broader area of the use of the CRISPR-Cas system in medical biotechnology. The seventh chapter by Banerjee et al. focuses on the CRISPR-Cas9 molecular scissors. By using these techniques, genomic DNA sequences can be easily edited, thus manipulating their function. This gene therapy tool has a broad spectrum of applications, but this chapter focuses on different areas of disease biology including but not limited to sickle cell anemia, cancer, cystic fibrosis and Duchenne muscular dystrophy. The eighth chapter by Paul et al. describes an *in silico* drug development approach that uses tuberculosis as an example. The next six chapters (Chapters 9 to 14) concern the broad category of the role of medical biotechnology research in the field of disease treatment. The ninth chapter by Maitrayee Banerjee Mukherjee involves the use of neuroimaging biomarkers in the early detection of dementia and Alzheimer's disease; in this chapter, different neuroimaging approaches for the discovery of novel biomarkers are discussed along with the probable medical advantages and limitations. Chapter 10 by Kundu et al. focuses on therapeutics for neurodegenerative diseases. The eleventh chapter by Nandi et al. delineates the current progress in the development of novel biomarkers and therapeutic interventions to treat neurodegenerative diseases. Chapters 12, 13 and 14 by Talukder et al. address the molecular

DOI: 10.1201/9781003302131-1

pathogenesis, clinical manifestation and molecular orientation of three different but important diseases, namely, Huntington chorea, Leishmaniasis and Ebola. The next chapter (Chapter 15) by Basu et al. concerns the importance of plant-derived antioxidants in the field of medical biotechnology; this chapter focuses on the essential plant-derived antioxidants and their pivotal role in fighting many cardiovascular, neural, and diabetic diseases, as well as aging. Furthermore, the effects of antioxidant supplements and antioxidant rich foods from both dietary consumption and therapeutic aspects are discussed. The sixteenth chapter by Talukder et al. deals with the *in vitro* genetic approach to identify disease gene mutations; various types of mutations and uses of cytogenetic and molecular tools are explored for the genetic diagnosis of juvenile genetic illnesses. Chapter 17 by Talukder et al. discusses *in vitro* models for studying cancer progression. The eighteenth chapter by Mukherjee et al. focuses on immunotherapy; major applications and categories of immunotherapy, their history of development, the recent findings on tumour-infiltrating immune cells in human cancers and their connections with immunotherapy in addition to future perspectives and challenges are discussed. Saroj et al. in Chapter 19 outlines useful analytical tools to identify cancer biomarkers from cell-derived extracellular vesicles. In Chapter 20, Das et al. discuss the importance of RNA Interference (RNAi) in medical biotechnology. The final two chapters concern nanotechnology in drug discovery, delivery and therapy. The role of nano carriers for targeted drug delivery is detailed in Chapter 21 by Tiwari et al. The final chapter (Chapter 22) by Maity et al. concerns the application of nanomaterial in human health and disease. Accordingly, this book truly showcases an ensemble of scientific works from the diverse yet interconnected research areas on the theoretical, experimental and technical fields of medical biotechnology.

1 Sequence Analysis to Decipher Protein Functions in Health and Disease

*Animesh Mondal**

*Government General Degree College, Mangalkote, Purba Bardhaman 713132, India, animeshpresiz@gmail.com

1.1 INTRODUCTION

Throughout recent decades, an enormous amount of information has been received pertaining to the protein sequence of various organisms (van der Lee et al. 2014). This denotes the diversity in the sequential steps and in functional space. Currently, with the advancement of computational biology, protein segments that lack the annotation of molecular roles can be easily rendered from an analysis of sequential similarity through homology modeling. In 1952, Edsall conjectured that

> It is reasonable to hope that the endless variety of proteins found in nature and their extraordinary diverse and specific interactions with one another and with other substances can be explained on the basis of a few relatively rigid general patterns, simply by varying the nature and sequence of the side chains attached to the fundamental repeating pattern.
>
> (Edsall and Astbury 1953)

An illustration of many of the unwanted/uncharacterized proteins along with their segments could lead to the discovery of many novel attributions concerning the cellular function and crosstalk mechanisms. Moreover, regarding disease biology, this could shed new light by unveiling molecular details. Hence, structural deduction from the sequence of a protein segment is a key player in understanding protein functionality (van der Lee et al. 2014).

The conventional conformation function paradigm is viewed as a lock and key model where the distinctive biological tasks of a protein are controlled by specific amino acid sequences and where many proteins execute their function without a specific structure (Uversky 2019). These structurally independent proteins are known as intrinsically disordered proteins (IDPs) or protein hybrids, and they contain both intrinsically disordered protein regions (IDPRs) along with a domain that is in an ordered state (Uversky 2019). In silico analysis indicates that IDPs/IDPRs are frequently found in most living individuals and that it is well evident that an abundance of disorder becomes greater with the increase of a multiplicity of individual beings (Uversky 2019). The length of the disordered sequences may vary according to the organism. The portion of disordered sequences with a length >30 is the same in archaea and bacteria (Uversky 2019). However, the portion is quiet higher in eukaryotes. This estimation depends upon the increased role of IDPs in cellular signaling. Regarding the functional aspects of IDPs, the term "strangeness" echoes the proper orientation. This shows a paradigm shift from the classic structure function relationship of proteins where the folding of an amino acid sequence into a unique 3D structure is responsible for the execution of a particular function. The wide existence of IDPs/IDPRs in the majority of living organisms including viruses is a striking feature of these sequences (Uversky 2019). After establishing that IDPs/IDPRs comprise a new type of functional protein, it is now well known that these proteins/regions are highly prevalent in natural organisms (Uversky 2019). From a historical

DOI: 10.1201/9781003302131-2

perspective, the chronology of understanding regarding IDPs/IDPRs was very rapid. The initial collection of IDPRs comprised 67 disordered domains where longer than 7 residues were observed in 61 PDB proteins, which were categorized into sets of short, medium and long IDPRs bearing 7–21, 22–44 and 45 or more residues, respectively (Uversky 2016; Das et al. 2014). Through this progress, a set of 91 IDPs having an extended disorder was further determined (Uversky, Gillespie, and Fink 2000). Consequently, this list was also enriched with the addition of the experimentally validated IDPs of 157 proteins and then, of more than 200 proteins carrying not less than a 30-amino acid-long IDPR (Keith Dunker, Brown, and Obradovic 2002; Dunker et al. 2002). An interesting report declares that there are approximately 1,150 nonredundant proteins present in the record of IDPs that are well validated (DeForte and Uversky 2016). The application of bioinformatic tools to the Swiss Protein database captured IDPRs that have 40 consecutive amino acids existing in over 15,000 proteins, and a high degree of disorder was observed in more than 1,000 proteins. Further analysis showed that 31 genomes from 3 kingdoms (archaea, bacteria and eukaryotes) have disordered proteins. Eukaryotes possess a 52–67% disordered sequence of proteins. The association between intrinsic disorder and proteome size is exhibited for the 3,484 viral, archaeal, bacterial and eukaryotic species (Xue, Dunker, and Uversky 2012). This kind of natural abundance of disorder in protein was later reinforced by many extensive in silico investigations (Uversky 2019; Xue, Dunker, and Uversky 2012; Brown, Johnson, and Daughdrill 2010; Xue, Williams, Oldfield, Dunker, et al. 2010; Pal, Maity, et al. 2016; Pal, Roy, et al. 2016). Recent investigation through MobiDB-Lite prediction depicted that human proteomes harbor 43.6% disorder (Piovesan et al. 2021).

Although enzymes were initially not included in the list, recent advancements have established that enzymes also carry IDPRs of a similar length to nonenzymatic proteins. These IDPRs in enzymes have novel functions in addition to the parallel function of nonenzymatic IDPRs (DeForte and Uversky 2017). Based upon these observations, it is now well accepted that within the world of proteins, one can find disorder everywhere (Uversky 2019). These IDPs can engage in free interactions while executing their function within the protein hubs (Dunker et al. 2005). IDPs execute a crucial function in controlling signaling pathways and other cellular processes that include transcription, translation and the cell cycle (Wright and Dyson 1999; Iakoucheva et al. n.d.; Galea et al. 2008). The nature of IDP prevalence in the cellular system is highly balanced to execute an exact signaling cascade at a precise time and in a precise space. However, any alteration/mutation from this precise modulation often leads to various diseases in humans (Gsponer et al. 2008; Babu et al. 2011). In conjunction with the regulatory function, IDPs perform a crucial role in the macromolecular assembly of ribosomes. This also augments the chromatin organization, macromolecular (microfilaments) assembly and disassembly and the transportation of molecules through the nuclear pore. Moreover, IDPs help in the functional execution of proteins and RNA chaperones as flexible "entropic linkers" (Wright and Dyson 2015; Frey, Richter, and Görlich 2006; Guharoy et al. 2013).

A recent investigation pointed out that most proteins carry a series of low complexity or prion-like sequences that can augment phase separation to promote the formation of membraneless organelles in nucleoplasm/cytoplasm (Kato et al. 2012). Hence, they also contribute to compartmentalization in a controlled manner (Weber and Brangwynne 2012). In addition, they carry the site for posttranslational modifications, the effective use of conserved sequence motifs to modulate interaction and the capability of interacting with partner molecules with specificity and affinity in nature (Wright and Dyson 1999). All of these abilities provide them kinetic advantages while their interaction allows for a rapid and voluntary dissociation to halt signals after the execution of a function (Oldfield et al. 2005). One common instance is that an amino acid sequence is generally utilized to turn on a signaling pathway, and the same sequence can be used in different contexts to turn off the cellular response against particular signals (Wright and Dyson 2015; Mondal et al. 2018).

1.2 IDENTIFICATION AND CHARACTERISTICS OF DISORDER IN PROTEINS

Various web-based servers have emerged to identify the protein disorder that depends upon the amino acid sequences of the protein. A data log has been developed that holds consensus disorder predictions for most of the proteins encoded by the human genome (Oates et al. 2013). The likelihood of the sequence that will be disordered rather than the folded, three-dimensional form is identified by using bioinformatic tools such as DISPROT (Sickmeier et al. 2007), IUPRED (Dosztányi et al. 2005a, 2005b), ANCHOR (Mészáros, Simon, and Dosztányi 2009), PONDR (Obradovic et al. 2003), PrDOS (Ishida and Kinoshita 2007) and ESpritz (Walsh et al. 2012). These programs analyze the local sequence that bears comparatively higher levels of hydrophilic amino acids, which denotes a sequence that has a higher disorder possibility. A Hidden Markov model has been applied in several instances to identify the disordered and structured domains (Fukuchi et al. 2009). Much progress has been accomplished in the development of analytical tools for the characterization of the structural ensembles of disordered proteins. In many instances, modelling has been performed by using nuclear magnetic resonance (NMR) and small angle X-ray scattering (SAXS) data. The dynamic interplay of IDPs and their objects is well illustrated by studying the crystals of a bound target molecule (Wright and Dyson 2015). Presently, single molecule fluorescence energy transfer (smFRET) and simulation studies show a crucial role in the characterization of protein disorder (Nasir et al. 2019; Das and Pappu 2013). In the last decade, sequence-based disorders have been explored mostly in the field of water soluble proteins (WSPs), and very little information is available regarding membrane proteins (Appadurai, Uversky, and Srivastava 2019; van der Lee et al. 2014). The characterization of IDPRs by using different biophysical techniques is increasing for WSPs, and this becomes more difficult for transmembrane proteins (TMPs) (Appadurai, Uversky, and Srivastava 2019). Challenges in the characterization of TMPs arise due to their lipid surroundings. These serve as a suitable platform for the motifs bearing IDPRs in the TMPs. Until now, few such TMP-IDPRs have been explored (Shukla et al. 2013). IDPs are devoid of any particular conformation and transient structure of secondary and tertiary forms (Dyson and Wright 2005a; Dolui et al. 2018). It was observed that IDPs are composed of polar and charged residues, and they lack hydrophobic amino acids (Obradovic et al. 2003; He et al. 2009).

1.3 NATURE OF IDPs

Most IDPs have been found to be polyampholytes in nature, that is, the sequence includes both positively and negatively charged amino acids. The portion and the amount of charged amino acids are responsible for discriminating between strong and weak polyampholytes. Figure 1.1 reports the atomistic simulations that establish that weak polyampholytes prefer globular conformation, whereas structural inclinations of strong polyampholytes are resolved by the joint efforts of the portion of charged amino acids and the linear sequence dispersal of reverse-charged residues. Moreover, residues that are accountable for destabilizing the secondary structure, such as glycine and proline, are frequent in IDPs, and usually, the sequence is low and complex, and repetition is usual in nature. The folding nature of proteins is generally guided by hydrophobicity with the liberation of water; this phenomenon, in turn, elaborates the dynamics of protein ensembles that harbor hydrophilic residues. This sequence signature enables bioinformatic tools to identify IDPs based upon the heuristic disorder score (Dosztányi et al. 2005a; Jones and Cozzetto 2015; Xue, Dunbrack, et al. 2010). All of these properties indicate that proteins comprise both folded and disordered regions. In Figure 1.2, this phenomenon is well illustrated. Protein disorder normally affects the selection methods for experimental studies, and it is therefore vital to determine the disordered region with accuracy; moreover, special attention is needed on the TMP during its expression and solubilization in the model system (J.A. Lyons et al. 2016). The sequences of IDPRs are generally poorly conserved because the selection pressure conferred upon them is lighter in nature (Brown et al. 2002, 2011). Alternative splicing and phosphorylation are very much recurrent in the disordered region of a protein (Buljan et al. 2013; Iakoucheva et al. 2004).

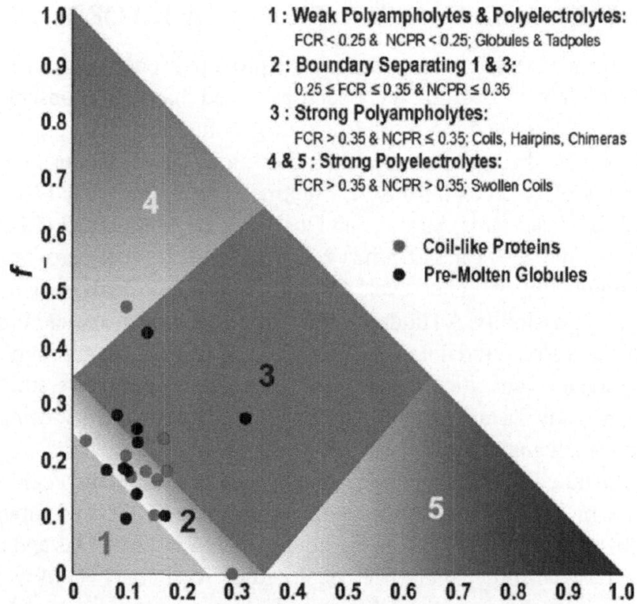

FIGURE 1.1 States of IDPs based on their charged distribution. Region 1 relates to either weak polyampholytes that prefer globule or tadpole-like structures. Region 3 denotes strong polyampholytes that particularly form non-globular structures, such as coils, hairpins, or hybrids. A boundary depicting 2 that separates 1 and 3 shows a continuum of possibilities between the two types of conformations. Sequences related to regions 4 and 5 are strong polyelectrolytes with a fraction of charged residues (FCR) >0.35 and a net charge per residue (NCPR) >0.3 having coil-like conformations. The figure comprises an annotation by properties of sequences that have been depicted as being "coils" or "pre-molten-globules" based on their hydrodynamic radii (Uversky 2002).

Source: This plot was reproduced from Das et al. (R.K. Das and Pappu 2013)

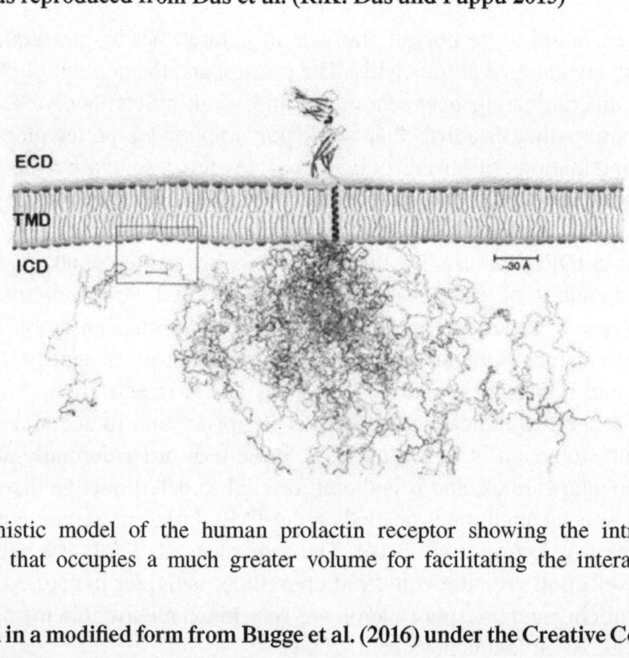

FIGURE 1.2 Atomistic model of the human prolactin receptor showing the intrinsically disordered intracellular domain that occupies a much greater volume for facilitating the interaction in the cellular microenvironment.

Source: Reproduced in a modified form from Bugge et al. (2016) under the Creative Commons Attribution 4.0 International License

A recent investigation in the database of 1,059 proteins report presence of 5,173 phosphorylation sites. Interestingly, more than 90% of these sites are detected within the disordered sequence (Tyanova et al. 2013). The structural plasticity of IDPs significantly enhances the accessibility of the enzymes to the modification sites. However, there is a lacuna regarding the experimental descriptions of the events of phosphorylation in the protein as a local or global event. Local and global modifications can induce changes in the structural propensity during cellular interaction with other molecules (Bah et al. 2015; Grosely et al. 2013; Malaney et al. 2013).

1.3.1 Biophysical Characters

Numerous biophysical studies have been performed on the structural disorder of IDPs by using various experimental approaches. Natively unstructured regions rapidly shift from extended to globular conformation and are regarded as dynamic ensembles. The essential factors that allow this target-specific interaction of IDPs/IDRs are solubility, phase transition and biomineralization (van der Lee et al. 2014).

1.3.2 Solubility

The nature of the solubility of proteins depends upon their favourable tendency to interact with water. Hydrophobic surface residues serve as interaction positions for other molecules in globular proteins. In contrast, IDPs or unstructured proteins generally do not fold to form globular conformations because they lack the hydrophobic amino acids that are abundant in globular proteins and that act as a driver behind protein folding (Uversky, Gillespie, and Fink 2000; P. Romero et al. 2001). The enrichment of disordered proteins with charged hydrophilic amino acids makes them soluble in an aqueous medium. IDPs are generally found to be tolerant to aggregation and precipitation induced by high temperature. The resistance against this thermal denaturation is possible because IDPs lack extensive secondary and tertiary conformations. Heat stability has been observed in intrinsically unstructured proteins; for example, cyclin-dependent kinase (CDK) inhibitor p21 remains structurally disordered and soluble within the temperature range of 5–90^{0}C (Kriwacki et al. 1996). In silico analysis reveals that thermostable proteins are mostly disordered in their nature. From the work of Galea and coworkers, it is well established that disordered proteins are more thermostable and soluble than the part that is presently folded within them (Galea et al. 2009). When considering the evolutionary aspect, it could be concluded that the low abundance of disordered proteins that are involved in regulation and other cellular crosstalk becomes less dependent on the solubility. This solubility facilitates extracellular IDPs in sequestering inorganic molecules in the extracellular environment (van der Lee et al. 2014). Proteins that have a higher degree of disorder have a higher level of expression in the system of cell free protein synthesis (Kurotani et al. 2009). Dunker et al. showed that the fusion of different types of unstructured polypeptide badges bearing negatively charged amino acids, which are called "entropic bristles", can boost the solubility of many proteins (Santner et al. 2012). However, IDRs within multidomain proteins that also contain folded domains might exert some effects on the solubility of proteins.

1.3.3 Phase Transition

The role of IDRs/IDPs in phase transition has unearthed other key insights of their biophysical characters. Li et al. noticed that the interaction between recombinant proteins that harbor numerous copies of the SH3 domain and IDRs along with different occasions of proline-rich SH3 binding motifs generate sharp liquid-liquid demixing (Figure 1.3), which forms micrometer-sized liquid protein-based droplets (Li et al. 2012).

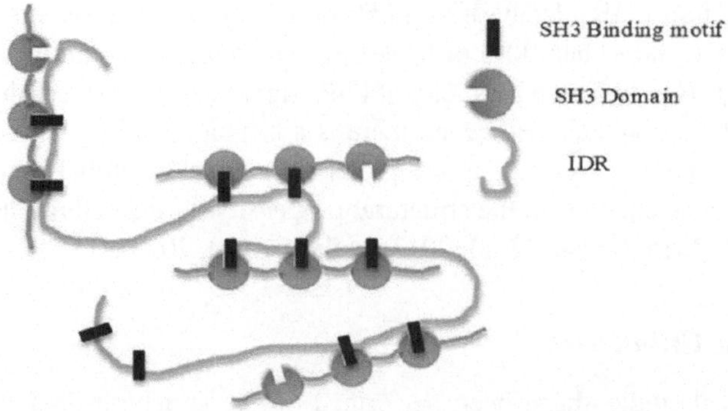

FIGURE 1.3 Involvement of IDR in the process of phase transition. The interplay between proteins that bear multiple copies of a distinct domain (the SH3 domain in the figure) and IDRs with multiple instances of their binding motif (proline-rich SH3 motif) at a proper concentration results in distinct liquid-liquid demixing phase separations (van der Lee et al. 2014).

A similar phenomenon happens with RNA-binding proteins that are abundant in IDRs with a low sequence complexity. Interestingly, hydrogels that are made up of low complexity IDRs of a single purified member of the granules can interact with different members of the IDRs and create heterogeneous macromolecular assembly structures. Most of the unstructured regions that can form this kind of functional aggregates remain under prudent regulation to control their abundance within the cell (Gsponer and Babu 2012). IDP-mediated phase transition also occurs in the case of extracellular proteins, such as mucin family proteins. They rely very much on the disordered portion for the creation of a gel-like network made of mucus (Silverman et al. 2003). The controlled transition of Mucin-2 from order to disorder is necessary while forming colon mucus aggregates (Verdugo 2012; Fink 2005).

1.3.4 Biomineralization

The majority of living organisms can induce crystallization and depositions of minerals in the extracellular matrix for their various physiological processes. The predominant form of the mineral in most instances is carbonate (George and Veis 2008). However, the extent of the mineralization varies according to the tissue and organ. Mineral crystals of bone and dentine form in the extracellular matrix. The nucleation of crystals depends upon the interactions of macromolecules with other secretive macromolecules that place themselves exactly on the structural framework. Bone and teeth are formed of collagen and different molecules of proteins in concurrence with calcium phosphate (Barrère, van Blitterswijk, and de Groot 2006). Proteins that actively participate in the mineralization of the tissue are found to have a highly disordered nature (Wojtas, Dobryszycki, and Ożyhar 2012). IDPs play a decisive role in the homeostasis of minerals during the deposition within the extracellular matrix. Unordered phosphoproteins expropriate calcium phosphate and create stable complexes where the side chains of the protein that was phosphorylated are placed on the surface of the cluster formed of calcium and phosphate (Holt 2013). The disordered form of the proteins allows adjusting their shapes rapidly according to the surroundings and solubilizing the cluster of calcium phosphate. For example, casein belongs to the disordered secretory calcium-binding phosphoprotein gene family (SCCP) in which tooth, milk and bone proteins are found (Kalmar et al. 2012; Kawasaki, Lafont, and Sire 2011).

1.4 EVOLUTION OF IDPs

The pathways behind the evolution of DNA and protein sequences are interrelated. Identification of the pathways can decode the reason behind the differences among homologous sequences (Brown,

Johnson, and Daughdrill 2010). The frequency of amino acid substitution in proteins provides the main aspect of their evolution. The process of protein evolution is not a simple one-step additive process but rather a complex divergent process. Unstructured regions of proteins evolve faster than globular proteins. The amino acid substitution frequency for the secondary structure of globular proteins varies according to its structures, that is, there are different frequencies for alpha helixes, betasheets, random coils and turns (Goldman, Thorne, and Jones 1998). However, this frequency also varies according to the positions in the sequence of amino acids and whether they are positioned on the surface or on the interior region of the folded globular proteins.

IDRs/IDPs that have similar functions and characteristics might have undergone similar evolutionary forces. Generally, IDPs are enriched in polar residues and are fewer in non-polar residues. However, some of them are enriched with low complexity regions. Disordered proteins evolved in a different lineage that accumulated a high amount of glycine (G) and asparagine (N). The faster rate of evolution in disordered proteins resulted in changes in amino acid substitution, deletions or insertions and an expansion of repeats (Huntley and Golding 2000).

1.4.1 Conservation of the Amino Acid Sequence

Amino acid sequences of IDRs evolve rapidly, but their character remains conserved for the functional sequences. IDPs/IDRs differ in sequence conservation according to their functions. There are three main biologically distinct classes (Chen et al. 2006) of IDRs as follows:

i. The flexible disordered class describes the region of conserved amino acid sequences with a high evolution rate;
ii. Constrained disordered species have regions of a highly conserved sequence; and
iii. Unconserved disordered species have regions where the character of being disordered is not conserved.

Numerous studies have identified the sequence conservation of IDRs and compared the compositions of amino acids between humans and seven other eukaryotes, namely, dogs, rats, chimpanzees, mice, flies, worms and yeast (van der Lee et al. 2014; Moesa et al. 2012).

1.5 RACE AND SPECIES SPECIFICITY

Interestingly, higher organisms have a high number of disordered proteins within their proteome (Ward et al. 2004). On an average, 2% of archeal, 4% of bacterial and 33% of eukaryotic proteins are found to harbor disordered regions of 30 amino acids. However, in humans, 31% of the proteome is enriched with more than 35% of the disordered portion (Gsponer et al. 2008). Human IDPs are distributed relatively in a uniform manner ranging from 38% to 50% within the chromosomes (Pentony and Jones 2010). Systematic analysis of the IDPs showed that 53 archeal species that harbor disorder are highly species-specific (Xue, Williams, Oldfield, Dunker, et al. 2010). One possible reason behind their specificity is that different environmental conditions might have exerted different selection pressures over the abundance of disorder in different clades of species.

Notably, conformational disorder is very predominant in viruses compared to prokaryotes (Tokuriki et al. 2009). These disordered regions serve as a buffer against the detrimental effects of mutations caused by low fidelity virus polymerases (Tokuriki et al. 2009). The flexibility of IDRs augments the interaction of viral proteins with other proteins of host immune systems (Xue, Williams, Oldfield, Goh, et al. 2010). For example, RIG-I like receptors and MDA5 allow various types of viral double stranded RNA (dsRNA) (Goubau, Deddouche, and Reis e Sousa 2013). Activated RIG-I and MDA5 that augment the generation of prion-like aggregates of the domain caspase activation and recruitment (CARD) belong to mitochondrial antiviral signaling (MAVS) types (Hou et al. 2011). MAVS carry a distinctly disordered central portion that bears many phosphorylation places and connects with different proteins such as TRAF2 and TRAF6 via their corresponding consensus-interacting motifs (Xu et al. 2005). These signaling cascades turn on transcription factors IRF3/7 and NF-κB, which instigate the release of proinflammatory cytokines such as IFN-α/β and other proteins that have antiviral activity.

Identification of retrovirus capsids by the restriction factor TRIM5α is performed by the disordered region in the SPRY domain, which carry many positively picked residues responsible for the antiviral activity (Sawyer et al. 2005). The SPRY domain subsists as an assembly of disordered structures that directs specificity along with the affinity of the binding between TRIM5α and viral capsids (Biris et al. 2012; Kovalskyy and Ivanov 2014). This pathway of evolutionary flexibility of the disordered regions serves as a pivotal player for the host immune system to prudently combat the switching pathogen during which they maintain their functionality. However, there is a variation in the prevalence of the disordered region in different species belonging to different kingdoms. Interestingly, most of them conserve IDRs for different functions, such as prokaryotes use disorder for long-lasting interactions during the formation of complexes, and eukaryotes use it for transient interaction during the propagation of their signal (Chen et al. 2006).

1.6 MEGA ROLES OF IDPs

The functions of IDP domains are perpetually unfolding. Their main functions include the regulation of translation and transcription, the phosphorylation of proteins, signal transduction and the control of the self-assembly of large multiprotein complexes such as ribosome (Wright and Dyson 2015). Several intrinsic disordered proteins undergo transition to form a more compact or ordered structure upon interacting with their targets. This phenomenon is regarded as coupled folding and binding.

1.6.1 Multiple Interaction Motifs

The dynamics of inter- and intracellular signaling are performed by a network of multiple interacting proteins. IDPs play a pivotal role in modulating these interactions. These IDPs/IDRs carry multiple conserved sequence motifs that interact with different proteins and nucleic acids (Dyson and Wright 2002). These conserved sequence motifs are amphipathic in nature. Various bioinformatics analyses suggest that our human proteome may harbor more than 100,000 short linear interacting motifs situated within the IDPS/IDRs (Tompa et al. 2014). The most peculiar feature of this recognition element is that they can adopt different conformations depending upon their targets. This phenomenon is well observed in the nuclear co-activator binding domain (NCBD) of CBP proteins where it folds into two separate conformations when interacting with the activator domain of p160 nuclear receptor co-activators or the interferon regulatory factor3 (IRF3) (Qin et al. 2005).

A recent investigation suggests that the preformed helical structure that is modulated by the IDP sequence is a crucial factor in performing different biological functions. This is well established in the population of p27 cyclin-dependent kinase inhibitors (CDKs), where their linker region regulates the cell cycle through preformed helical conformation (Otieno and Kriwacki 2012). The preformed helical structure in the amino terminal domain of p53 is the determinant of the binding affinity of MDM2 (Borcherds et al. 2014). Some of the disordered sequences appear to function as tensile linkers between the globular and disordered interaction domains, while some IDPs prefer to remain in disordered conformation after interacting with their partner and create a *fuzzy complex* (Tompa and Fuxreiter 2008).

1.6.2 Hubs of the Signaling Network

IDPs/IDRs are multifunctional proteins that carry multiple interaction motifs for mediating the binding of diverse targets and serve as central hubs in a signaling cascade (Kim et al. 2008). The functional interactions of IDPs depend upon the distribution of binding free energy. The interaction between IDPs that carry multiple interaction motifs and their target is heterogeneous in nature. The frequency of binding will be higher in local hotspots that generate higher free energy. The existence of a weak dynamic interaction has advantages in cellular signaling. These structural oscillations help in posttranslational modifications by transiently exposing the interaction motifs. This is well observed in p120 catenin, which modulates the stability of cell-cell adhesion by interacting the disordered cytoplasmic tail of cadherin via static and dynamic interfaces (Ishiyama et al. 2010). In the absence of high affinity binding motifs, IDPs/IDRs can firmly interact with their target through the *synergistic* action of multiple weak interacting regions. Ordered phosphorylation in a sequential manner can function as a logical "AND" gate in signaling networks (Ferrell and Ha 2014). Such cascade is well observed in the degradation of the establishment of cohesion1 (Eco1) protein in yeast (N.A. Lyons et al. 2013). Alternation in the net charge and charge distribution can highly influence the structural propensity of IDP through the electrostatic effects, for example, the time-delayed regulation response in the *circadian clock* of *Neurospora sp*. (Querfurth et al. 2011). Regulatory proteins often form higher-order assemblies-signalosomes, which intensify the signal and facilitate threshold signaling responses by providing spatiotemporal control (Wu 2013).

1.6.3 Formation of Isoforms

Different types of isoforms are created by proteins in two separate ways. The first one is through gene variants, and the second is through the splicing of mRNA. The latter one, specifically, alternative splicing, is explicitly linked to the disordered regions of proteins (P.R. Romero et al. 2006). However, in the case of gene variants, this is very difficult to verify. One fact complicating the situation is whether the protein is a variant or is an entire gene. However, it is well evidenced with investigation that differences in the gene variant primarily occur in the disordered region of the protein (Kjaergaard and Kragelund 2017). One common example is the human sodium proton exchanger family. For a folded protein to remain effective, a portion of the fold needs to be conserved, which, in turn, leads to the exertion of selective pressure in different regions, for example, a binding/interacting site. Lacking this preservation of the fold often leads to nihilo (Davey, Cyert, and Moses 2015), and a suitably functional specialization may crop up by altering the disordered sites with the addition or removal of binging sites. The known nine isoforms of the human sodium proton exchanger (NHE1–9) exhibit such a specialization property (Hendus-Altenburger, Kragelund, and Pedersen 2014). It has been observed that the membrane-embedded portions of isoforms are highly conserved, but they differ in the distal portion in terms of the length and sequence of the disordered regions. These isoforms are located in different tissues. NHE1 is located in the plasma membrane (Wakabayashi et al. 1994). In NHEs, the disordered regions serve as a major role in ion transport, which is executed in the transmembrane domain where they are also vital for controlling pH and growth factor signaling (Wakabayashi et al. 1994; Wakabayashi, Shigekawa, and Pouyssegur 1997). Earlier studies report that electroneutral sodium-bicarbonate co-transporter NBCn1 (SLC4A7) involved in cancer and has a composite gene organization of 18 major NBCn1 variants (Gorbatenko et al. 2014). Variation is formed by skipping two exons in the N and C terminals, along with the inclusion of two alternative start sequences. Splicing occurring in the interaction sites of the disordered sites is a repetitive event that is found to be replicated across different membrane tails including the N-methyl-D-aspartate (NMDA) receptor subunit 1. Deletion of PDZ binding motifs through alternative splicing has been well depicted for the G protein-activated inward rectifier potassium channel 2 (Lunn et al. 2007). In contrast, the removal of the sorting signal has been observed in neural cell adhesion molecule L1 (Kamiguchi and Lemmon 1998). This has also been observed in the dopamine receptor D2(D2R), and splicing of the gene Drd2 creates two isoforms, namely, a long (D2L) and a short (D2S) isoform. They differ in their third intracellular loop and are coupled with different G-proteins to instigate various signaling cascades in which D2L perturbs the AKT/PKB pathway and D2S initiates the ERK1/2 cascade (Guiramand et al. 1995; Radl et al. 2013).

1.7 STRUCTURAL TRIO ADAPTED BY PROTEINS

IDPs/IDRs exhibit a broad range of conformational subtypes. Proteins are thought to execute their function within a structural perpetuity that ranges from fully structured to entirely unstructured (Dyson and Wright 2005b). The continuum covers a compactly folded domain that shows neither disorder nor only locally unstructured regions in the tail. The functional region of any protein may persist in any form of the following three dynamic states – ordered conformation, molten globule forms and the conformation of random coils (Dunker et al. 2001). The main point of this trio conformation is that the specific function of a protein depends upon any one of these conformations or a transition conformation between any two of these conformations. In this context, not only the conformation of ordered states but also any conformation of the earlier (Figure 1.4) conformation can serve as a native structure for the protein. This trio configuration can be compared to the solid (ordered), liquid (molten globule) and gaseous (random coil) states of matter. Here, the interconversions between the three states is analogous to the phase transition. The interconversion between the conformations of ordered states and random coil or between the ordered states and molten globule simply reveal quantifiable cooperativity (Ptitsyn 1987).

Gases and liquids generally exhibit distinct phase transition, whereas the transition between random coil and molten globules do not exhibit the same phenomenon (Ptitsyn and Uversky 1994). However, some artificial helical bundles display molten globule-like non rigid packing side chains and cooperative unfolding (Handel, Williams, and DeGrado 1993). The artificial bundles are rich only in one type of hydrophobic core, whereas the natural bundle is rich in polar and nonpolar cores (Betz and DeGrado 1996). In ordered structures, protein pliability alludes to the excursions of atoms to a higher magnitude from their equilibrium states, whereas in unstructured proteins, an alternation in flexibility denotes the speed of the interconversions (Figure 1.5) among various members of the structural ensembles (Dunker et al. 2001).

The transiently localized secondary structure that is frequently found to be helical is rich in amphipathic sequences. These sites are found to play a central role in the binding process. Low and high amino acid charges normally lead to disordered globule and swollen coil formation. Unstructured sites in native unbound conditions persist as ensembles of rapidly interchangeable dynamic conformations (Jahn and Radford 2005). However, the modifications during the posttranslational and binding states alter the relative free energy distribution of each structure (Boehr, Nussinov, and Wright 2009; Ma and Nussinov 2009). The individual conformation within the ensemble

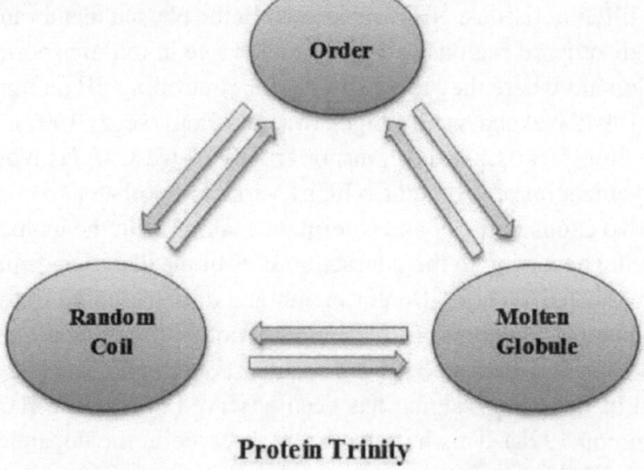

FIGURE 1.4 Protein trinity indicates that native protein structures include any of the previously mentioned structures, and the function can arise from either the mentioned conformation or transition conformation (Dunker et al. 2001).

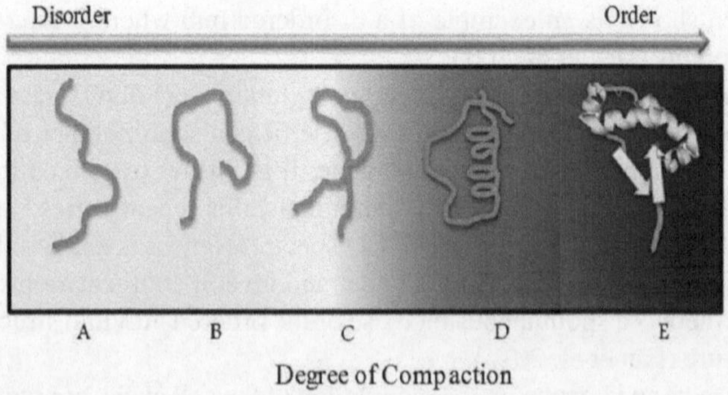

FIGURE 1.5 Schematic representation depicting the model structure of proteins. The shading represents the changes of conformational states from highly disordered (lighter) to ordered states (darker). A dynamically disordered state is represented in (A) as extended conformation, then it changes to (B) a transiently secondary structure, followed by (C) a compact globule; ultimately, through (D) a molten globule conformation, it is stably folded to (E) a compact globule structure. The protein can configure itself in any of these conformations during its course of interaction.

could change depending upon the surroundings. IDPs/IDRs possess great structural heterogeneity; proteins are sometimes also regarded as modular assemblies of foldons (independently foldable regions), inducible foldons (regions that attain structure upon binding to its partner molecule), semi foldons (partially folded regions) and non foldons (regions that never fold) (Uversky 2013). The role of these multiple conformations of disordered regions can be well understood with the example of FG nucleoporins. The central part of the nuclear pore complex (NPC) is formed by porins. Here, the IDRs with multiple phenylalanine-glycine (FG) motifs form a large portion of the NPC gates. These FG regions can adopt different structures while executing their tasks (Mohamed et al. 2017). This structural variation between tertiary and quaternary conformations of multiprotein complexes is regarded as a supertertiary structure (Tompa 2012). This type of structural conformation is well illustrated in the case of the Src-family tyrosine kinase Hck (Yang et al. 2010).

1.8 ROLES OF IDPs IN DISEASE BIOLOGY

The main roles of disordered proteins include a broad spectrum of biological tasks such as cellular differentiation, transcription, DNA condensation, cell cycle, mRNA splicing, wnt signaling, neurogenesis and growth regulation. In addition, several disease-related genes that encode IDPs are found to play a key role in various diseases. This includes disordered proteins, mutations that can augment some diseases, such as α-synuclein, a protein engaged in Parkinson's disease, p53 tumor-suppressor proteins occurring in a variety of cancers, huntingtin that is linked to Huntington's disease, BRCA1 that is associated with breast/ovarian cancer and EWS-FLI1, which is found with Ewing's sarcom, a family of tumors (Midic et al. 2009).

α-Synuclein is a type of intrinsically disordered protein that is connected to various synucleopathies, and most of the neurodegenerative diseases are associated with the deposition of aggregated α-Synuclein in neuronal cells (Goedert 1997). Clinical symptoms of most of the synucleopathies are a progressive loss of motor, cognitive, behavioral and autonomic functions. Alzheimer's disease is also a neurodegenerative disease caused by the deposition of Aβ peptide in the neuronal cells. Some direct evidence indicates the major role of α-Synuclein in the neurodegenerative process and Lewy body dementia. Familial Parkinson's disease and dementia are induced by a missense mutation in the synuclein gene. Synuclein gets atypically phosphorylated, ubiquitinated and nitrated in the formation of pathology kindred inclusion bodies. All of this evidence clearly indicates that IDP α-Synuclein is the key player behind various neurodegenerative diseases (Midic et al. 2009). Furthermore, α-Synuclein has been shown as an example of a disordered hub where it interacts with at least 50 ligands and other proteins (Dev et al. 2003).

Huntingtin is a protein that bears an expanded polyglutamine (PolyQ) tract near its N-terminus, which is the main reason behind Huntington's disease. It is also a member of the neural degenerative diseases linked to high CAG repeat in the gene. It is further translated into the expansion of the PolyQ tract of related proteins. In healthy individuals, this repeat varies between 16–37, but in affected individuals, it increases. From Far UV CD spectroscopy, it is evident that secondary structural conformations of the PolyQ repeats tend to be random coil conformations. With the expansion of the PolyQ repeat, there is a spontaneous increase of the ordered amyloid structure formation from the disordered structure (Fan et al. 2014).

The transcription factor p53 occurs in the modulation of the cell cycle, apoptosis and other cellular functions. If the function of protein p53 is altered by a mutation, then any other cells often undergo the process of cancerous transformation. It is now quiet evident that most cancers are caused by the altered expression of p53. Generally, missense mutations within the DNA binding core domain result in abnormal expression and function (Hollstein et al. 1991). p53 has three structural domains: an N terminal translational activation domain, a central DNA binding domain and a C terminal tetramerization domain (Midic et al. 2009). Structural study indicates that the DNA binding domain of protein p53 is intrinsically structured, whereas the terminal domains are unstructured or disordered. Most of the interactions (~ 70%) of p53 and its interacting partners are moderated by their disordered regions (Oldfield et al. 2008). For example, only 5–10% of ovarian cancers are hereditary in nature, and 30–50% of these cancers are caused by a mutation in the autosomal dominant genes BRCA1 and BRCA2 (Szabo and King 1995). The most widely investigated BRCA1 isoform comprises 1,863 amino acids and is enriched with an extremely disordered central portion that is flanked by ordered domains at the two ends. Structural characterization of BRCA1 has revealed that the fully disordered central region bears recognition regions for both DNA and other binding partners such as p53, retinoblastoma, c-Myc and JunB. Accordingly, the disordered region in the protein plays a crucial role in all of the previously mentioned diseases while modulating the functions of the protein. These disordered regions are key sites for interaction with other binding partners.

1.9 FUTURE PERSPECTIVES

The structural information regarding the disorder in proteins deeply infiltrates molecular biology. The most interesting development is the advancement of the biophysical studies of IDPs within cells. Tagging of IDPs helps in probing their dynamic interactions within the cellular system. Much information is needed regarding the abundance, concentration and subcellular localization of IDPs to better understand their behavioral and functional roles. To decipher the mechanistic details of this complex system, it is necessary to take a more holistic approach while analyzing the structural ensembles. IDPs play a crucial role in cellular crosstalk and are now emerging as a new therapeutic target for drug molecules because they are associated with several diseases (Cheng et al. 2006). IDPs/IDRs often interact with target molecules with modest affinity predominantly through hydrophobic interactions, which makes them promising as a novel therapeutic target. It is now feasible to design drug molecules for IDPs rather than for their globular target, and recent favorable results have been reported regarding this (Hammoudeh et al. 2009). Viruses frequently mimic the IDR motif to hijack cellular machinery and subvert the signaling network (Hagai et al. 2014). Despite the promise of IDPs as a therapeutic target, critical issues remain regarding the development of target molecules by choosing the IDPs or their binding pocket.

1.9.1 Acknowledgements

The author is thankful to GGDC, Mangalkote for providing the research facility and to WBDSTBT (STBT-11012(12)/1/2022-ST SEC) for providing grant support.

REFERENCES

Appadurai, Rajeswari, Vladimir N. Uversky, and Anand Srivastava. 2019. "The Structural and Functional Diversity of Intrinsically Disordered Regions in Transmembrane Proteins." *The Journal of Membrane Biology* 252 (4): 273–92. https://doi.org/10.1007/s00232-019-00069-2.

Babu, M. Madan, Robin van der Lee, Natalia Sanchez de Groot, and Jörg Gsponer. 2011. "Intrinsically Disordered Proteins: Regulation and Disease." *Current Opinion in Structural Biology* 21 (3): 432–40. https://doi.org/10.1016/j.sbi.2011.03.011.

Bah, Alaji, Robert M. Vernon, Zeba Siddiqui, Mickaël Krzeminski, Ranjith Muhandiram, Charlie Zhao, Nahum Sonenberg, et al. 2015. "Folding of an Intrinsically Disordered Protein by Phosphorylation as a Regulatory Switch." *Nature* 519 (7541): 106–9. https://doi.org/10.1038/nature13999.

Barrère, Florence, Clemens A. van Blitterswijk, and Klaas de Groot. 2006. "Bone Regeneration: Molecular and Cellular Interactions with Calcium Phosphate Ceramics." *International Journal of Nanomedicine* 1 (3): 317–32.

Betz, S. F., and W. F. DeGrado. 1996. "Controlling Topology and Native-Like Behavior of de Novo-Designed Peptides: Design and Characterization of Antiparallel Four-Stranded Coiled Coils." *Biochemistry* 35 (21): 6955–62. https://doi.org/10.1021/bi960095a.

Biris, Nikolaos, Yang Setiawati, Alexander B. Taylor, Andrei Tomashevski, Miao Guo, P. John Hart, Felipe Diaz-Griffero, and Dmitri N. Ivanov. 2012. "Structure of the Rhesus Monkey TRIM5α PRYSPRY Domain, the HIV Capsid Recognition Module." *Proceedings of the National Academy of Sciences of the United States of America* 109 (33): 13278–83. https://doi.org/10.1073/pnas.1203536109.

Boehr, David D., Ruth Nussinov, and Peter E. Wright. 2009. "The Role of Dynamic Conformational Ensembles in Biomolecular Recognition." *Nature Chemical Biology* 5 (11): 789–96. https://doi.org/10.1038/nchembio.232.

Borcherds, Wade, François-Xavier Theillet, Andrea Katzer, Ana Finzel, Katie M. Mishall, Anne T. Powell, Hongwei Wu, et al. 2014. "Disorder and Residual Helicity Alter P53-Mdm2 Binding Affinity and Signaling in Cells." *Nature Chemical Biology* 10 (12): 1000–1002. https://doi.org/10.1038/nchembio.1668.

Brown, Celeste J., Audra K. Johnson, and Gary W. Daughdrill. 2010. "Comparing Models of Evolution for Ordered and Disordered Proteins." *Molecular Biology and Evolution* 27 (3): 609–21. https://doi.org/10.1093/molbev/msp277.

Brown, Celeste J., Audra K. Johnson, A. Keith Dunker, and Gary W. Daughdrill. 2011. "Evolution and Disorder." *Current Opinion in Structural Biology* 21 (3): 441–46. https://doi.org/10.1016/j.sbi.2011.02.005.

Brown, Celeste J., Sachiko Takayama, Andrew M. Campen, Pam Vise, Thomas W. Marshall, Christopher J. Oldfield, Christopher J. Williams, and A. Keith Dunker. 2002. "Evolutionary Rate Heterogeneity in Proteins with Long Disordered Regions." *Journal of Molecular Evolution* 55 (1): 104–10. https://doi.org/10.1007/s00239-001-2309-6.

Bugge, Katrine, Elena Papaleo, Gitte W. Haxholm, Jonathan T. S. Hopper, Carol V. Robinson, Johan G. Olsen, Kresten Lindorff-Larsen, and Birthe B. Kragelund. 2016. "A Combined Computational and Structural Model of the Full-Length Human Prolactin Receptor." *Nature Communications* 7 (1): 11578. https://doi.org/10.1038/ncomms11578.

Buljan, Marija, Guilhem Chalancon, A. Keith Dunker, Alex Bateman, S. Balaji, Monika Fuxreiter, and M. Madan Babu. 2013. "Alternative Splicing of Intrinsically Disordered Regions and Rewiring of Protein Interactions." *Current Opinion in Structural Biology* (New Constructs and Expressions of Proteins/ Sequences and Topology), 23 (3): 443–50. https://doi.org/10.1016/j.sbi.2013.03.006.

Chen, Jessica Walton, Pedro Romero, Vladimir N. Uversky, and A. Keith Dunker. 2006. "Conservation of Intrinsic Disorder in Protein Domains and Families: II. Functions of Conserved Disorder." *Journal of Proteome Research* 5 (4): 888–98. https://doi.org/10.1021/pr060049p.

Cheng, Yugong, Tanguy LeGall, Christopher J. Oldfield, James P. Mueller, Ya-Yue J. Van, Pedro Romero, Marc S. Cortese, et al. 2006. "Rational Drug Design via Intrinsically Disordered Protein." *Trends in Biotechnology* 24 (10): 435–42. https://doi.org/10.1016/j.tibtech.2006.07.005.

Das, Rahul K., and Rohit V. Pappu. 2013. "Conformations of Intrinsically Disordered Proteins Are Influenced by Linear Sequence Distributions of Oppositely Charged Residues." *Proceedings of the National Academy of Sciences* 110 (33): 13392–97. https://doi.org/10.1073/pnas.1304749110.

Das, Swagata, Uttam Pal, Supriya Das, Khyati Bagga, Anupam Roy, Arpita Mrigwani, and Nakul C. Maiti. 2014. "Sequence Complexity of Amyloidogenic Regions in Intrinsically Disordered Human Proteins." *PLoS ONE* 9 (3): e89781. https://doi.org/10.1371/journal.pone.0089781.

Davey, Norman E., Martha S. Cyert, and Alan M. Moses. 2015. "Short Linear Motifs – Ex Nihilo Evolution of Protein Regulation." *Cell Communication and Signaling* 13 (1): 43. https://doi.org/10.1186/s12964-015-0120-z.

DeForte, Shelly, and Vladimir N. Uversky. 2016. "Intrinsically Disordered Proteins in PubMed: What Can the Tip of the Iceberg Tell Us About What Lies Below?" *RSC Advances* 6 (14): 11513–21. https://doi.org/10.1039/C5RA24866C.

———. 2017. "Not an Exception to the Rule: The Functional Significance of Intrinsically Disordered Protein Regions in Enzymes." *Molecular BioSystems* 13 (3): 463–69. https://doi.org/10.1039/C6MB00741D.

Dev, Kumlesh K., Katja Hofele, Samuel Barbieri, Vladimir L. Buchman, and Herman van der Putten. 2003. "Part II: Alpha-Synuclein and Its Molecular Pathophysiological Role in Neurodegenerative Disease." *Neuropharmacology* 45 (1): 14–44. https://doi.org/10.1016/s0028-3908(03)00140-0.

Dolui, Sandip, Anupam Roy, Uttam Pal, Achintya Saha, and Nakul C. Maiti. 2018. "Structural Insight of Amyloidogenic Intermediates of Human Insulin." *ACS Omega* 3 (2): 2452–62. https://doi.org/10.1021/acsomega.7b01776.

Dosztányi, Zsuzsanna, Veronika Csizmók, Péter Tompa, and István Simon. 2005a. "The Pairwise Energy Content Estimated from Amino Acid Composition Discriminates between Folded and Intrinsically Unstructured Proteins." *Journal of Molecular Biology* 347 (4): 827–39. https://doi.org/10.1016/j.jmb.2005.01.071.

Dosztányi, Zsuzsanna, Veronika Csizmok, Peter Tompa, and István Simon. 2005b. "IUPred: Web Server for the Prediction of Intrinsically Unstructured Regions of Proteins Based on Estimated Energy Content." *Bioinformatics* 21 (16): 3433–34. https://doi.org/10.1093/bioinformatics/bti541.

Dunker, A. Keith, Celeste J. Brown, J. David Lawson, Lilia M. Iakoucheva, and Zoran Obradović. 2002. "Intrinsic Disorder and Protein Function." *Biochemistry* 41 (21): 6573–82. https://doi.org/10.1021/bi012159+.

Dunker, A. Keith, Marc S. Cortese, Pedro Romero, Lilia M. Iakoucheva, and Vladimir N. Uversky. 2005. "Flexible Nets." *The FEBS Journal* 272 (20): 5129–48. https://doi.org/10.1111/j.1742-4658.2005.04948.x.

Dunker, A. Keith, J. David Lawson, Celeste J. Brown, Ryan M. Williams, Pedro Romero, Jeong S. Oh, Christopher J. Oldfield, et al. 2001. "Intrinsically Disordered Protein." *Journal of Molecular Graphics and Modelling* 19 (1): 26–59. https://doi.org/10.1016/S1093-3263(00)00138-8.

Dyson, H. Jane, and Peter E. Wright. 2002. "Coupling of Folding and Binding for Unstructured Proteins." *Current Opinion in Structural Biology* 12 (1): 54–60.

———. 2005a. "Intrinsically Unstructured Proteins and Their Functions." *Nature Reviews Molecular Cell Biology* 6 (3): 197–208. https://doi.org/10.1038/nrm1589.

———. 2005b. "Intrinsically Unstructured Proteins and Their Functions." *Nature Reviews Molecular Cell Biology* 6 (3): 197–208. https://doi.org/10.1038/nrm1589.

Edsall, J. T., and William Thomas Astbury. 1953. "Some Comments on Proteins and Protein Structure." *Proceedings of the Royal Society of London. Series B – Biological Sciences* 141 (902): 97–103. https://doi.org/10.1098/rspb.1953.0026.

Fan, Hueng-Chuen, Li-Ing Ho, Ching-Shiang Chi, Shyi-Jou Chen, Giia-Sheun Peng, Tzu-Min Chan, Shinn-Zong Lin, and Horng-Jyh Harn. 2014. "Polyglutamine (PolyQ) Diseases: Genetics to Treatments." *Cell Transplantation* 23 (4–5): 441–58. https://doi.org/10.3727/096368914X678454.

Ferrell, James E., and Sang Hoon Ha. 2014. "Ultrasensitivity Part II: Multisite Phosphorylation, Stoichiometric Inhibitors, and Positive Feedback." *Trends in Biochemical Sciences* 39 (11): 556–69. https://doi.org/10.1016/j.tibs.2014.09.003.

Fink, Anthony L. 2005. "Natively Unfolded Proteins." *Current Opinion in Structural Biology* (Folding and Binding/Protein-Nucleic Acid Interactions), 15 (1): 35–41. https://doi.org/10.1016/j.sbi.2005.01.002.

Frey, Steffen, Ralf P. Richter, and Dirk Görlich. 2006. "FG-Rich Repeats of Nuclear Pore Proteins Form a Three-Dimensional Meshwork with Hydrogel-Like Properties." *Science* 314 (5800): 815–17. https://doi.org/10.1126/science.1132516.

Fukuchi, Satoshi, Keiichi Homma, Yoshiaki Minezaki, Takashi Gojobori, and Ken Nishikawa. 2009. "Development of an Accurate Classification System of Proteins into Structured and Unstructured Regions That Uncovers Novel Structural Domains: Its Application to Human Transcription Factors." *BMC Structural Biology* 9 (1): 26. https://doi.org/10.1186/1472-6807-9-26.

Galea, Charles A., Anthony A. High, John C. Obenauer, Ashutosh Mishra, Cheon-Gil Park, Marco Punta, Avner Schlessinger, et al. 2009. "Large-Scale Analysis of Thermostable, Mammalian Proteins Provides Insights into the Intrinsically Disordered Proteome." *Journal of Proteome Research* 8 (1): 211–26. https://doi.org/10.1021/pr800308v.

Galea, Charles A., Yuefeng Wang, Sivashankar G. Sivakolundu, and Richard W. Kriwacki. 2008. "Regulation of Cell Division by Intrinsically Unstructured Proteins: Intrinsic Flexibility, Modularity, and Signaling Conduits." *Biochemistry* 47 (29): 7598–609. https://doi.org/10.1021/bi8006803.

George, Anne, and Arthur Veis. 2008. "Phosphorylated Proteins and Control Over Apatite Nucleation, Crystal Growth, and Inhibition." *Chemical Reviews* 108 (11): 4670–93. https://doi.org/10.1021/cr0782729.

Goedert, Michel. 1997. "Familial Parkinson's Disease: The Awakening of α-Synuclein." *Nature* 388 (6639): 232. https://doi.org/10.1038/40767.

Goldman, N., J. L. Thorne, and D. T. Jones. 1998. "Assessing the Impact of Secondary Structure and Solvent Accessibility on Protein Evolution." *Genetics* 149 (1): 445–58.

Gorbatenko, Andrej, Christina Olesen, Ebbe Boedtkjer, and Stine Pedersen. 2014. "Regulation and Roles of Bicarbonate Transport in Cancer." *Frontiers in Physiology* 5. www.frontiersin.org/article/10.3389/fphys.2014.00130.

Goubau, Delphine, Safia Deddouche, and Caetano Reis e Sousa. 2013. "Cytosolic Sensing of Viruses." *Immunity* 38 (5): 855–69. https://doi.org/10.1016/j.immuni.2013.05.007.

Grosely, Rosslyn, Jennifer L. Kopanic, Sarah Nabors, Fabien Kieken, Gaëlle Spagnol, Mona Al-Mugotir, Sydney Zach, and Paul L. Sorgen. 2013. "Effects of Phosphorylation on the Structure and Backbone Dynamics of the Intrinsically Disordered Connexin43 C-Terminal Domain." *Journal of Biological Chemistry* 288 (34): 24857–70. https://doi.org/10.1074/jbc.M113.454389.

Gsponer, Jörg, Matthias E. Futschik, Sarah A. Teichmann, and M. Madan Babu. 2008. "Tight Regulation of Unstructured Proteins: From Transcript Synthesis to Protein Degradation." *Science* 322 (5906): 1365–68. https://doi.org/10.1126/science.1163581.

Gsponer, Jörg, and M. Madan Babu. 2012. "Cellular Strategies for Regulating Functional and Nonfunctional Protein Aggregation." *Cell Reports* 2 (5): 1425–37. https://doi.org/10.1016/j.celrep.2012.09.036.

Guharoy, Mainak, Beata Szabo, Sara Contreras Martos, Simone Kosol, and Peter Tompa. 2013. "Intrinsic Structural Disorder in Cytoskeletal Proteins." *Cytoskeleton* 70 (10): 550–71. https://doi.org/10.1002/cm.21118.

Guiramand, Janique, Jean-Pierre Montmayeur, Jocelyn Ceraline, Madhav Bhatia, and Emiliana Borrelli. 1995. "Alternative Splicing of the Dopamine D2 Receptor Directs Specificity of Coupling to G-Proteins (∗)." *Journal of Biological Chemistry* 270 (13): 7354–58. https://doi.org/10.1074/jbc.270.13.7354.

Hagai, Tzachi, Ariel Azia, M. Madan Babu, and Raul Andino. 2014. "Use of Host-like Peptide Motifs in Viral Proteins Is a Prevalent Strategy in Host-Virus Interactions." *Cell Reports* 7 (5): 1729–39. https://doi.org/10.1016/j.celrep.2014.04.052.

Hammoudeh, Dalia I., Ariele Viacava Follis, Edward V. Prochownik, and Steven J. Metallo. 2009. "Multiple Independent Binding Sites for Small-Molecule Inhibitors on the Oncoprotein c-Myc." *Journal of the American Chemical Society* 131 (21): 7390–401. https://doi.org/10.1021/ja900616b.

Handel, T. M., S. A. Williams, and W. F. DeGrado. 1993. "Metal Ion-Dependent Modulation of the Dynamics of a Designed Protein." *Science* 261 (5123): 879–85. https://doi.org/10.1126/science.8346440.

He, Bo, Kejun Wang, Yunlong Liu, Bin Xue, Vladimir N. Uversky, and A. Keith Dunker. 2009. "Predicting Intrinsic Disorder in Proteins: An Overview." *Cell Research* 19 (8): 929–49. https://doi.org/10.1038/cr.2009.87.

Hendus-Altenburger, Ruth, Birthe B. Kragelund, and Stine Falsig Pedersen. 2014. "Chapter Two – Structural Dynamics and Regulation of the Mammalian SLC9A Family of Na+/H+ Exchangers." In *Current Topics in Membranes*, edited by Mark O. Bevensee, 73:69–148. Exchangers. Academic Press. https://doi.org/10.1016/B978-0-12-800223-0.00002-5.

Hollstein, M., D. Sidransky, B. Vogelstein, and C. C. Harris. 1991. "P53 Mutations in Human Cancers." *Science* 253 (5015): 49–53. https://doi.org/10.1126/science.1905840.

Holt, Carl. 2013. "Unfolded Phosphopolypeptides Enable Soft and Hard Tissues to Coexist in the Same Organism with Relative Ease." *Current Opinion in Structural Biology* (New Constructs and Expressions of Proteins/Sequences and Topology), 23 (3): 420–25. https://doi.org/10.1016/j.sbi.2013.02.010.

Hou, Fajian, Lijun Sun, Hui Zheng, Brian Skaug, Qiu-Xing Jiang, and Zhijian J. Chen. 2011. "MAVS Forms Functional Prion-Like Aggregates to Activate and Propagate Antiviral Innate Immune Response." *Cell* 146 (3): 448–61. https://doi.org/10.1016/j.cell.2011.06.041.

Huntley, Melanie, and G. Brian Golding. 2000. "Evolution of Simple Sequence in Proteins." *Journal of Molecular Evolution* 51 (2): 131–40. https://doi.org/10.1007/s002390010073.

Iakoucheva, Lilia M., Celeste J. Brown, J. David Lawson, Zoran Obradović, and A. Keith Dunker. n.d. "Intrinsic Disorder in Cell-Signaling and Cancer-Associated Proteins." *Journal of Molecular Biology* 323: 573–84.

Iakoucheva, Lilia M., Predrag Radivojac, Celeste J. Brown, Timothy R. O'Connor, Jason G. Sikes, Zoran Obradovic, and A. Keith Dunker. 2004. "The Importance of Intrinsic Disorder for Protein Phosphorylation." *Nucleic Acids Research* 32 (3): 1037–49. https://doi.org/10.1093/nar/gkh253.

Ishida, Takashi, and Kengo Kinoshita. 2007. "PrDOS: Prediction of Disordered Protein Regions from Amino Acid Sequence." *Nucleic Acids Research* 35 (suppl_2): W460–64. https://doi.org/10.1093/nar/gkm363.

Ishiyama, Noboru, Seung-Hye Lee, Shuang Liu, Guang-Yao Li, Matthew J. Smith, Louis F. Reichardt, and Mitsuhiko Ikura. 2010. "Dynamic and Static Interactions between P120 Catenin and E-Cadherin Regulate the Stability of Cell-Cell Adhesion." *Cell* 141 (1): 117–28. https://doi.org/10.1016/j.cell.2010.01.017.

Jahn, Thomas R., and Sheena E. Radford. 2005. "The Yin and Yang of Protein Folding." *The FEBS Journal* 272 (23): 5962–70. https://doi.org/10.1111/j.1742-4658.2005.05021.x.

Jones, David T., and Domenico Cozzetto. 2015. "DISOPRED3: Precise Disordered Region Predictions with Annotated Protein-Binding Activity." *Bioinformatics* 31 (6): 857–63. https://doi.org/10.1093/bioinformatics/btu744.

Kalmar, Lajos, Daniel Homola, Gabor Varga, and Peter Tompa. 2012. "Structural Disorder in Proteins Brings Order to Crystal Growth in Biomineralization." *Bone* 51 (3): 528–34. https://doi.org/10.1016/j.bone.2012.05.009.

Kamiguchi, Hiroyuki, and Vance Lemmon. 1998. "A Neuronal Form of the Cell Adhesion Molecule L1 Contains a Tyrosine-Based Signal Required for Sorting to the Axonal Growth Cone." *Journal of Neuroscience* 18 (10): 3749–56. https://doi.org/10.1523/JNEUROSCI.18-10-03749.1998.

Kato, Masato, Tina W. Han, Shanhai Xie, Kevin Shi, Xinlin Du, Leeju C. Wu, Hamid Mirzaei, et al. 2012. "Cell-Free Formation of RNA Granules: Low Complexity Sequence Domains Form Dynamic Fibers within Hydrogels." *Cell* 149 (4): 753–67. https://doi.org/10.1016/j.cell.2012.04.017.

Kawasaki, Kazuhiko, Anne-Gaelle Lafont, and Jean-Yves Sire. 2011. "The Evolution of Milk Casein Genes from Tooth Genes Before the Origin of Mammals." *Molecular Biology and Evolution* 28 (7): 2053–61. https://doi.org/10.1093/molbev/msr020.

Keith Dunker, A., Celeste J. Brown, and Zoran Obradovic. 2002. "Identification and Functions of Usefully Disordered Proteins." In *Advances in Protein Chemistry*, 62:25–49. Unfolded Proteins. Academic Press. https://doi.org/10.1016/S0065-3233(02)62004-2.

Kim, Philip M., Andrea Sboner, Yu Xia, and Mark Gerstein. 2008. "The Role of Disorder in Interaction Networks: A Structural Analysis." *Molecular Systems Biology* 4 (March): 179. https://doi.org/10.1038/msb.2008.16.

Kjaergaard, Magnus, and Birthe B. Kragelund. 2017. "Functions of Intrinsic Disorder in Transmembrane Proteins." *Cellular and Molecular Life Sciences* 74 (17): 3205–24. https://doi.org/10.1007/s00018-017-2562-5.

Kovalskyy, Dmytro B., and Dmitri N. Ivanov. 2014. "Recognition of the HIV Capsid by the TRIM5α Restriction Factor Is Mediated by a Subset of Pre-Existing Conformations of the TRIM5α SPRY Domain." *Biochemistry* 53 (9): 1466–76. https://doi.org/10.1021/bi4014962.

Kriwacki, R. W., L. Hengst, L. Tennant, S. I. Reed, and P. E. Wright. 1996. "Structural Studies of P21Waf1/Cip1/Sdi1 in the Free and Cdk2-Bound State: Conformational Disorder Mediates Binding Diversity." *Proceedings of the National Academy of Sciences* 93 (21): 11504–9. https://doi.org/10.1073/pnas.93.21.11504.

Kurotani, Atsushi, Tetsuo Takagi, Mitsutoshi Toyama, Mikako Shirouzu, Shigeyuki Yokoyama, Yasuo Fukami, and Alexander A. Tokmakov. 2009. "Comprehensive Bioinformatics Analysis of Cell-Free Protein Synthesis: Identification of Multiple Protein Properties That Correlate with Successful Expression." *The FASEB Journal* 24 (4): 1095–1104. https://doi.org/10.1096/fj.09-139527.

Lee, Robin van der, Marija Buljan, Benjamin Lang, Robert J. Weatheritt, Gary W. Daughdrill, A. Keith Dunker, Monika Fuxreiter, et al. 2014. "Classification of Intrinsically Disordered Regions and Proteins." *Chemical Reviews* 114 (13): 6589–6631. https://doi.org/10.1021/cr400525m.

Li, Pilong, Sudeep Banjade, Hui-Chun Cheng, Soyeon Kim, Baoyu Chen, Liang Guo, Marc Llaguno, et al. 2012. "Phase Transitions in the Assembly of Multivalent Signalling Proteins." *Nature* 483 (7389): 336–40. https://doi.org/10.1038/nature10879.

Lunn, Marie-Louise, Rounak Nassirpour, Christine Arrabit, Joshua Tan, Ian Mcleod, Carlos M. Arias, Paul E. Sawchenko, et al. 2007. "A Unique Sorting Nexin Regulates Trafficking of Potassium Channels via a PDZ Domain Interaction." *Nature Neuroscience* 10 (10): 1249–59. https://doi.org/10.1038/nn1953.

Lyons, Joseph A., Azadeh Shahsavar, Peter Aasted Paulsen, Bjørn Panyella Pedersen, and Poul Nissen. 2016. "Expression Strategies for Structural Studies of Eukaryotic Membrane Proteins." *Current Opinion in Structural Biology* 38 (June): 137–44. https://doi.org/10.1016/j.sbi.2016.06.011.

Lyons, Nicholas A., Bryan R. Fonslow, Jolene K. Diedrich, John R. Yates, and David O. Morgan. 2013. "Sequential Primed Kinases Create a Damage-Responsive Phosphodegron on Eco1." *Nature Structural & Molecular Biology* 20 (2): 194–201. https://doi.org/10.1038/nsmb.2478.

Ma, Buyong, and Ruth Nussinov. 2009. "Regulating Highly Dynamic Unstructured Proteins and Their Coding MRNAs." *Genome Biology* 10 (1): 204. https://doi.org/10.1186/gb-2009-10-1-204.

Malaney, Prerna, Ravi Ramesh Pathak, Bin Xue, Vladimir N. Uversky, and Vrushank Davé. 2013. "Intrinsic Disorder in PTEN and Its Interactome Confers Structural Plasticity and Functional Versatility." *Scientific Reports* 3 (1): 2035. https://doi.org/10.1038/srep02035.

Mészáros, Bálint, István Simon, and Zsuzsanna Dosztányi. 2009. "Prediction of Protein Binding Regions in Disordered Proteins." *PLoS Computational Biology* 5 (5): e1000376. https://doi.org/10.1371/journal.pcbi.1000376.

Midic, Uros, Christopher J. Oldfield, A. Keith Dunker, Zoran Obradovic, and Vladimir N. Uversky. 2009. "Protein Disorder in the Human Diseasome: Unfoldomics of Human Genetic Diseases." *BMC Genomics* 10 (1): S12. https://doi.org/10.1186/1471-2164-10-S1-S12.

Moesa, Harry Amri, Shunichi Wakabayashi, Kenta Nakai, and Ashwini Patil. 2012. "Chemical Composition Is Maintained in Poorly Conserved Intrinsically Disordered Regions and Suggests a Means for Their Classification." *Molecular BioSystems* 8 (12): 3262–73. https://doi.org/10.1039/C2MB25202C.

Mohamed, Mahmoud Shaaban, Akiko Kobayashi, Azuma Taoka, Takahiro Watanabe-Nakayama, Yosuke Kikuchi, Masaharu Hazawa, Toshinari Minamoto, et al. 2017. "High-Speed Atomic Force Microscopy Reveals Loss of Nuclear Pore Resilience as a Dying Code in Colorectal Cancer Cells." *ACS Nano* 11 (6): 5567–78. https://doi.org/10.1021/acsnano.7b00906.

Mondal, Animesh, Uttam Pal, Anupam Roy, and Nakul C. Maiti. 2018. "Structural Intricacy of Disordered Regions in Transcription Factors Imparting Colon Cancer." *Journal of Proteins & Proteomics* 9 (3).

Nasir, Irem, Paulo L. Onuchic, Sergio R. Labra, and Ashok A. Deniz. 2019. "Single-Molecule Fluorescence Studies of Intrinsically Disordered Proteins and Liquid Phase Separation." *Biochimica et Biophysica Acta (BBA) – Proteins and Proteomics* (Intrinsically Disordered Proteins: Amyloid Formation and Phase Separation), 1867 (10): 980–87. https://doi.org/10.1016/j.bbapap.2019.04.007.

Oates, Matt E., Pedro Romero, Takashi Ishida, Mohamed Ghalwash, Marcin J. Mizianty, Bin Xue, Zsuzsanna Dosztányi, et al. 2013. "D2P2: Database of Disordered Protein Predictions." *Nucleic Acids Research* 41 (D1): D508–16. https://doi.org/10.1093/nar/gks1226.

Obradovic, Zoran, Kang Peng, Slobodan Vucetic, Predrag Radivojac, Celeste J. Brown, and A. Keith Dunker. 2003. "Predicting Intrinsic Disorder from Amino Acid Sequence." *Proteins: Structure, Function, and Bioinformatics* 53 (S6): 566–72. https://doi.org/10.1002/prot.10532.

Oldfield, Christopher J., Yugong Cheng, Marc S. Cortese, Pedro Romero, Vladimir N. Uversky, and A. Keith Dunker. 2005. "Coupled Folding and Binding with α-Helix-Forming Molecular Recognition Elements." *Biochemistry* 44 (37): 12454–70. https://doi.org/10.1021/bi050736e.

Oldfield, Christopher J., Jingwei Meng, Jack Y. Yang, Mary Qu Yang, Vladimir N. Uversky, and A. Keith Dunker. 2008. "Flexible Nets: Disorder and Induced Fit in the Associations of P53 and 14-3-3 with Their Partners." *BMC Genomics* 9 (1): S1. https://doi.org/10.1186/1471-2164-9-S1-S1.

Otieno, Steve, and Richard Kriwacki. 2012. "Probing the Role of Nascent Helicity in P27 Function as a Cell Cycle Regulator." *PLoS One* 7 (10): e47177. https://doi.org/10.1371/journal.pone.0047177.

Pal, Uttam, Mritunjoy Maity, Nitin Khot, Swagata Das, Supriya Das, Sandip Dolui, and Nakul C. Maiti. 2016. "Statistical Insight into the Binding Regions in Disordered Human Proteome." *Journal of Proteins & Proteomics* 7 (1). www.proteomicssociety.in/jpp-journal-master/www.jpp.org.in/index.php/jpp/article/view/172.html.

Pal, Uttam, Anupam Roy, Supriya Das, Swagata Das, Mangaldeep Kundu, Khyati Bagga, and Nakul Chandra Maiti. 2016. "Origin of Protein Aggregation: Identification of Some Characteristic Traits in Structured and Intrinsically Disordered Proteins." *Journal of Proteins & Proteomics* 7 (4). https://proteomicssociety.in/jpp-journal-master/www.jpp.org.in/index.php/jpp/article/view/206.html.

Pentony, Melissa M., and David T. Jones. 2010. "Modularity of Intrinsic Disorder in the Human Proteome." *Proteins: Structure, Function, and Bioinformatics* 78 (1): 212–21. https://doi.org/10.1002/prot.22504.

Piovesan, Damiano, Marco Necci, Nahuel Escobedo, Alexander Miguel Monzon, András Hatos, Ivan Mičetić, Federica Quaglia, et al. 2021. "MobiDB: Intrinsically Disordered Proteins in 2021." *Nucleic Acids Research* 49 (D1): D361–67. https://doi.org/10.1093/nar/gkaa1058.

Ptitsyn, O. B., and V. N. Uversky. 1994. "The Molten Globule Is a Third Thermodynamical State of Protein Molecules." *FEBS Letters* 341 (1): 15–18. https://doi.org/10.1016/0014-5793(94)80231-9.

Ptitsyn, Oleg B. 1987. "Protein Folding: Hypotheses and Experiments." *Journal of Protein Chemistry* 6 (4): 273–93. https://doi.org/10.1007/BF00248050.

Qin, Bin Y., Cheng Liu, Hema Srinath, Suvana S. Lam, John J. Correia, Rik Derynck, and Kai Lin. 2005. "Crystal Structure of IRF-3 in Complex with CBP." *Structure* 13 (9): 1269–77. https://doi.org/10.1016/j.str.2005.06.011.

Querfurth, Christina, Axel C. R. Diernfellner, Elan Gin, Erik Malzahn, Thomas Höfer, and Michael Brunner. 2011. "Circadian Conformational Change of the Neurospora Clock Protein Frequency Triggered by Clustered Hyperphosphorylation of a Basic Domain." *Molecular Cell* 43 (5): 713–22. https://doi.org/10.1016/j.molcel.2011.06.033.

Radl, Daniela, Claudia De Mei, Eric Chen, Hyuna Lee, and Emiliana Borrelli. 2013. "Each Individual Isoform of the Dopamine D2 Receptor Protects from Lactotroph Hyperplasia." *Molecular Endocrinology* 27 (6): 953–65. https://doi.org/10.1210/me.2013-1008.

Romero, Pedro, Zoran Obradovic, Xiaohong Li, Ethan C. Garner, Celeste J. Brown, and A. Keith Dunker. 2001. "Sequence Complexity of Disordered Protein." *Proteins: Structure, Function, and Bioinformatics* 42 (1): 38–48. https://doi.org/10.1002/1097-0134(20010101)42:1<38::AID-PROT50>3.0.CO;2-3.

Romero, Pedro R., Saima Zaidi, Ya Yin Fang, Vladimir N. Uversky, Predrag Radivojac, Christopher J. Oldfield, Marc S. Cortese, et al. 2006. "Alternative Splicing in Concert with Protein Intrinsic Disorder Enables Increased Functional Diversity in Multicellular Organisms." *Proceedings of the National Academy of Sciences* 103 (22): 8390–95. https://doi.org/10.1073/pnas.0507916103.

Santner, Aaron A., Carrie H. Croy, Farha H. Vasanwala, Vladimir N. Uversky, Ya-Yue J. Van, and A. Keith Dunker. 2012. "Sweeping Away Protein Aggregation with Entropic Bristles: Intrinsically Disordered Protein Fusions Enhance Soluble Expression." *Biochemistry* 51 (37): 7250–62. https://doi.org/10.1021/bi300653m.

Sawyer, Sara L., Lily I. Wu, Michael Emerman, and Harmit S. Malik. 2005. "Positive Selection of Primate TRIM5alpha Identifies a Critical Species-Specific Retroviral Restriction Domain." *Proceedings of the National Academy of Sciences of the United States of America* 102 (8): 2832–37. https://doi.org/10.1073/pnas.0409853102.

Shukla, Arun K., Aashish Manglik, Andrew C. Kruse, Kunhong Xiao, Rosana I. Reis, Wei-Chou Tseng, Dean P. Staus, et al. 2013. "Structure of Active β-Arrestin-1 Bound to a G-Protein-Coupled Receptor Phosphopeptide." *Nature* 497 (7447): 137–41. https://doi.org/10.1038/nature12120.

Sickmeier, Megan, Justin A. Hamilton, Tanguy LeGall, Vladimir Vacic, Marc S. Cortese, Agnes Tantos, Beata Szabo, et al. 2007. "DisProt: The Database of Disordered Proteins." *Nucleic Acids Research* 35 (suppl_1): D786–93. https://doi.org/10.1093/nar/gkl893.

Silverman, Howard Scott, Mark Sutton-Smith, Kimberley McDermott, Paul Heal, Shih-Hsing Leir, Howard R. Morris, Michael A. Hollingsworth, et al. 2003. "The Contribution of Tandem Repeat Number to the O-Glycosylation of Mucins." *Glycobiology* 13 (4): 265–77. https://doi.org/10.1093/glycob/cwg028.

Szabo, C. I., and M. C. King. 1995. "Inherited Breast and Ovarian Cancer." *Human Molecular Genetics* 4 (suppl_1): 1811–17. https://doi.org/10.1093/hmg/4.suppl_1.1811.

Tokuriki, Nobuhiko, Christopher J. Oldfield, Vladimir N. Uversky, Igor N. Berezovsky, and Dan S. Tawfik. 2009. "Do Viral Proteins Possess Unique Biophysical Features?" *Trends in Biochemical Sciences* 34 (2): 53–59. https://doi.org/10.1016/j.tibs.2008.10.009.

Tompa, Peter. 2012. "On the Supertertiary Structure of Proteins." *Nature Chemical Biology* 8 (7): 597–600. https://doi.org/10.1038/nchembio.1009.

Tompa, Peter, Norman E. Davey, Toby J. Gibson, and M. Madan Babu. 2014. "A Million Peptide Motifs for the Molecular Biologist." *Molecular Cell* 55 (2): 161–69. https://doi.org/10.1016/j.molcel.2014.05.032.

Tompa, Peter, and Monika Fuxreiter. 2008. "Fuzzy Complexes: Polymorphism and Structural Disorder in Protein-Protein Interactions." *Trends in Biochemical Sciences* 33 (1): 2–8. https://doi.org/10.1016/j.tibs.2007.10.003.

Tyanova, Stefka, Jürgen Cox, Jesper Olsen, Matthias Mann, and Dmitrij Frishman. 2013. "Phosphorylation Variation During the Cell Cycle Scales with Structural Propensities of Proteins." *PLoS Computational Biology* 9 (1): e1002842. https://doi.org/10.1371/journal.pcbi.1002842.

Uversky, Vladimir N. 2002. "What Does It Mean to Be Natively Unfolded?" *European Journal of Biochemistry* 269 (1): 2–12. https://doi.org/10.1046/j.0014-2956.2001.02649.x.

———. 2013. "Unusual Biophysics of Intrinsically Disordered Proteins." *Biochimica Et Biophysica Acta* 1834 (5): 932–51. https://doi.org/10.1016/j.bbapap.2012.12.008.

———. 2016. "Dancing Protein Clouds: The Strange Biology and Chaotic Physics of Intrinsically Disordered Proteins *." *Journal of Biological Chemistry* 291 (13): 6681–88. https://doi.org/10.1074/jbc.R115.685859.

———. 2019. "Intrinsically Disordered Proteins and Their 'Mysterious' (Meta)Physics." *Frontiers in Physics* 7. https://doi.org/10.3389/fphy.2019.00010.

Uversky, Vladimir N., Joel R. Gillespie, and Anthony L. Fink. 2000. "Why Are 'Natively Unfolded' Proteins Unstructured under Physiologic Conditions?" *Proteins: Structure, Function, and Bioinformatics* 41 (3): 415–27. https://doi.org/10.1002/1097-0134(20001115)41:3<415::AID-PROT130>3.0.CO;2-7.

Verdugo, Pedro. 2012. "Supramolecular Dynamics of Mucus." *Cold Spring Harbor Perspectives in Medicine* 2 (11): a009597. https://doi.org/10.1101/cshperspect.a009597.

Wakabayashi, S., B. Bertrand, T. Ikeda, J. Pouysségur, and M. Shigekawa. 1994. "Mutation of Calmodulin-Binding Site Renders the Na+/H+ Exchanger (NHE1) Highly H(+)-Sensitive and Ca2+ Regulation-Defective." *The Journal of Biological Chemistry* 269 (18): 13710–15.

Wakabayashi, S., M. Shigekawa, and J. Pouyssegur. 1997. "Molecular Physiology of Vertebrate Na+/H+ Exchangers." *Physiological Reviews* 77 (1): 51–74. https://doi.org/10.1152/physrev.1997.77.1.51.

Walsh, Ian, Alberto J. M. Martin, Tomàs Di Domenico, and Silvio C. E. Tosatto. 2012. "ESpritz: Accurate and Fast Prediction of Protein Disorder." *Bioinformatics* 28 (4): 503–9. https://doi.org/10.1093/bioinformatics/btr682.

Ward, J. J., J. S. Sodhi, L. J. McGuffin, B. F. Buxton, and D. T. Jones. 2004. "Prediction and Functional Analysis of Native Disorder in Proteins from the Three Kingdoms of Life." *Journal of Molecular Biology* 337 (3): 635–45. https://doi.org/10.1016/j.jmb.2004.02.002.

Weber, Stephanie C., and Clifford P. Brangwynne. 2012. "Getting RNA and Protein in Phase." *Cell* 149 (6): 1188–91. https://doi.org/10.1016/j.cell.2012.05.022.

Wojtas, Magdalena, Piotr Dobryszycki, and Andrzej Ożyhar. 2012. "Intrinsically Disordered Proteins in Biomineralization." *Advanced Topics in Biomineralization*, February. https://doi.org/10.5772/31121.

Wright, Peter E., and H. Jane Dyson. 1999. "Intrinsically Unstructured Proteins: Re-Assessing the Protein Structure-Function Paradigm." *Journal of Molecular Biology* 293 (2): 321–31. https://doi.org/10.1006/jmbi.1999.3110.

Wright, Peter E., and H. Jane Dyson. 2015. "Intrinsically Disordered Proteins in Cellular Signaling and Regulation." *Nature Reviews: Molecular Cell Biology* 16 (1): 18–29. https://doi.org/10.1038/nrm3920.

Wu, Hao. 2013. "Higher-Order Assemblies in a New Paradigm of Signal Transduction." *Cell* 153 (2): 287–92. https://doi.org/10.1016/j.cell.2013.03.013.

Xu, Liang-Guo, Yan-Yi Wang, Ke-Jun Han, Lian-Yun Li, Zhonghe Zhai, and Hong-Bing Shu. 2005. "VISA Is an Adapter Protein Required for Virus-Triggered IFN-Beta Signaling." *Molecular Cell* 19 (6): 727–40. https://doi.org/10.1016/j.molcel.2005.08.014.

Xue, Bin, Roland L. Dunbrack, Robert W. Williams, A. Keith Dunker, and Vladimir N. Uversky. 2010. "PONDR-FIT: A Meta-Predictor of Intrinsically Disordered Amino Acids." *Biochimica et Biophysica Acta (BBA) – Proteins and Proteomics* 1804 (4): 996–1010. https://doi.org/10.1016/j.bbapap.2010.01.011.

Xue, Bin, A. Keith Dunker, and Vladimir N. Uversky. 2012. "Orderly Order in Protein Intrinsic Disorder Distribution: Disorder in 3500 Proteomes from Viruses and the Three Domains of Life." *Journal of Biomolecular Structure & Dynamics* 30 (2): 137–49. https://doi.org/10.1080/07391102.2012.675145.

Xue, Bin, Robert W. Williams, Christopher J. Oldfield, A. Keith Dunker, and Vladimir N. Uversky. 2010. "Archaic Chaos: Intrinsically Disordered Proteins in Archaea." *BMC Systems Biology* 4 (1): S1. https://doi.org/10.1186/1752-0509-4-S1-S1.

Xue, Bin, Robert W. Williams, Christopher J. Oldfield, Gerard Kian-Meng Goh, A. K. Dunker, and Vladimir N. Uversky. 2010. "Viral Disorder or Disordered Viruses: Do Viral Proteins Possess Unique Features?" *Protein and Peptide Letters* 17 (8): 932–51. https://doi.org/10.2174/092986610791498984.

Yang, Sichun, Lydia Blachowicz, Lee Makowski, and Benoît Roux. 2010. "Multidomain Assembled States of Hck Tyrosine Kinase in Solution." *Proceedings of the National Academy of Sciences* 107 (36): 15757–62. https://doi.org/10.1073/pnas.1004569107.

2 Network Pharmacology as a Potential Tool to Unlock the Treasure Trove of Pharmaceuticals in Natural Products

Pritam Biswas[1,*] *and Uttam Pal*[2,**]

[1]Department of Microbiology, St. Xavier's College, 30 Mother Teresa Sarani, Kolkata 700016, India

[2]Structural Biology & Bioinformatics Division, CSIR-Indian Institute of Chemical Biology, 4 Raja S.C. Mullick Road, Kolkata 700032, India

*Correspondence to: biswaspritam1376@gmail.com

**Correspondence to: uttam2707@gmail.com

2.1 INTRODUCTION

Drug discovery has followed the model of the "one drug, one target" and focused on designing drug molecules selectively for the target (Sams-Dodd 2005). With the development of large-scale functional genomics, studies have also revolutionised the process of drug discovery and developed more target-based designing of drug candidates, a process called reverse pharmacology (Zambrowicz and Sands 2004; Chandran et al. 2017). This method relies on the principle that the alteration of specific protein functions will lead to therapeutic effects. Drug candidate molecules are screened to study their effects on the target protein function. Once a candidate molecule has been selected, it is further modified to augment the therapeutic effect through enhanced target molecule binding by the drug candidate molecule (Hacker, Messer, and Bachmann 2009). This approach of one drug, one target has been the staple for drug discovery for decades. However, this has led to lowering both drug efficiency and safety in drug candidates (Kola and Landis 2004).

Advancements in computational biology changed the drug discovery landscape at the turn of the century. The advent of omics data sciences was used for the development and design of novel drug candidates (Buriani et al. 2012). Development of molecular docking and other computational tools changed the strategy of drug discovery for predicting the binding behavior of candidate molecules to the target molecule (Clark and Pickett 2000). The inclusion of ADMET tools would further boost the screening and prediction of novel drug candidates (Guan et al. 2019; van de Waterbeemd and Gifford 2003). However, such an approach would suffer from several drawbacks such as drug-associated complications and drug resistance.

Natural products in the form of traditional medicines are used all over the world and contain a plethora of bioactive compounds, which makes them a treasure trove for drug discovery and rational drug design. Drugs based on natural products act as multi-target therapeutics rather than synthetic drugs based on a single target, as they are designed to be target-specific (Runzhi Zhang et al. 2019). Diseases generally are a result of multiple dysfunctional proteins, which makes natural product-based drug formulation, or rather the bioactive compounds obtained from them, a perfect alternative to the

 DOI: 10.1201/9781003302131-3

synthetic singular target-based therapy. Hence, understanding the mechanism of multi-target therapy by natural products could lead to the holistic amelioration of a disease. Therefore, the paradigm shift approach in drug discovery from one drug, one target to more multi-target therapies is the need of the hour. This has prompted a renewed interest in the research of natural products for their pharmaco-potential and to "rediscover" traditional medicines by integrating age-old medicinal systems into our present-day technologies for robust scientific rationalization (Patwardhan and Chandran 2015). The field of systems biology has been at the forefront of this paradigm shift in drug discovery. Advances in the field of systems biology led to the development of network pharmacology. Network pharmacology can effectively predict and determine a "compound-protein/gene-disease" network and reveal the regulatory principles of bioactive molecules in a high-throughput manner (Runzhi Zhang et al. 2019). Accordingly, this makes network pharmacology a highly powerful method of choice for the analysis of natural products as potential therapeutics.

2.1.1 What Is Network Pharmacology?

Systems biology, which arose due to the paradigm shift in drug discovery, focuses on the complex interactions involved in biological systems in a holistic manner (Kitano 2002). Network pharmacology, a method based on network theory and systems biology, emphasises substituting the singular target concept of drug discovery with searching for novel drug molecules that act on multiple biological networks (Hopkins 2008; Roth, Sheffler, and Kroeze 2004). Network pharmacology can catalogue the different molecular interactions that a drug candidate can undergo in a biological system, which makes network pharmacology an important tool in modern drug discovery and in leading candidates (Zhang, Wang, and Li 2013). The increased understanding of the role of network pharmacology challenges the dominant assumption of single-target drug discovery (Wermuth 2004; Keith, Borisy, and Stockwell 2005). Network pharmacology has the potential to combine the knowledge of natural product formulation in the form of traditional medicines to rationally design the next generation of drug molecules (Ellingson, Smith, and Baudry 2014; Cho, Kim, and Przytycka 2012). The holistic perspective of network pharmacology analysis not only opens new avenues of drug research but also reduces the chances of drug resistance and drug-related complications.

2.1.2 The Concepts of Systems Biology and Network Pharmacology

The availability of biomedical data in the postgenomic era, which coincided with the rapid development of computational tools, has led to an analysis of the available biomedical data for a fresh perspective on rational drug design and drug discovery (Friboulet and Thomas 2005; Metz and Hajduk 2010). The main aim was to index all drug candidates and the different molecular interactions that they undertake within the biological system. Such analysis and indexing also led to renewed interest in understanding the implications that such interactions have on complex living systems. This resulted in the development of network biology. Network biology depends on computational tools and literature mining to gain insights into disease relationships with phenotype and genotype. This novel method of literature mining combined with microarray analysis gave rise to the construction of biological networks. Network biology utilizing literature mining and microarray analysis (LMMA) can help construct complex biological networks not only for large-scale data analysis but also for understanding various diseases regarding gene interactions (Li, Wu, and Zhang 2006). With the indexing of biomedical data, a more holistic picture of the disease cycle in biological systems has started to emerge. Analysis of complex biological networks has provided a fresh perspective on drug action and opened new avenues to design drug candidates. Advancement in systems biology has brought about an understanding that complex diseases are a result of multiple dysfunctional proteins within the biological system, which cannot be effectively treated using the conventional singular target therapeutics. A concurrent modulation of multiple targets is required for effective treatment. This caused the development of polypharmacology, a radical shift in drug design where drug candidates were now screened for their ability to affect multiple targets rather than one specific target (Yıldırım et al. 2007). Development in polypharmacology can also result in increased drug efficacy and the lowering of drug-associated complications.

The amalgamation of network biology and polypharmacology made possible the development of network pharmacology. Network pharmacology has the potential to lead the paradigm shift in drug discovery research. Hopkins, in 2007, devised the fundamental concept of network pharmacology where many effective drugs in therapeutic areas act on multiple rather than on single targets (Hopkins 2007). Network pharmacology can integrate multidisciplinary concepts ranging from biochemicals and bioinformatics to systems biology to construct molecular networks that can predict the off-target efficiency of drug candidates at a higher efficiency, which significantly improves the chances of multi-target therapeutic candidate discovery (Hopkins 2008). Multi-channel regulation of signalling pathways, enhanced drug efficacy, reduced drug complications, reduced costs of drug discovery and increased chances of success of clinical trials are some of the advantages of network pharmacology (Wu et al. 2008). Since diseases are manifestations of dysfunctional pathways, network pharmacology aims at determining where in the network the drug candidate acts, either by activating or inhibiting the disease phenotype. Such an approach could lead to therapies with a lower risk of drug resistance and drug-associated complications due to the drug molecule acting at the systems level of the disease network through lethal interactions in a synergistic way (Y. Chen et al. 2008). Strategies of drug discovery should involve a determination of the regulated pathology networks and a reduction of the higher attrition rates in the disease network (Y. Chen et al. 2008). According to Hopkins (Hopkins 2007), there are three strategies for designing a multi-target therapeutic molecule. First, multiple individual drug molecules are introduced as a multi-drug cocktail. However, complications arising from drug-drug interaction is a major drawback of this strategy. Second, multicomponent drug formulation is developed. Changes in the pharmacokinetics, bioavailability and metabolism of the formulation, however, are major concerns regarding such an approach. Third, a single molecule having multiple target sites at the disease network is developed, which overcomes the drawbacks of the first and second strategies. To determine the relationship between genotype and phenotype, network pharmacology can integrate human protein-protein interactions, disease phenotype similarities and known gene-phenotype associations. Such an approach is guided by the fact that phenotypically analogous diseases are related to one another by functionally similar genes (Chandran et al. 2017). Network pharmacology has also been used to create disease-drug networks. Later, genotype information was integrated into this network. Such networks provide insights into the relationship among multiple genes with different diseases. Analysis of such networks can provide new drug targets for the development of novel candidates.

2.2 DATABASES AND TOOLS FOR NETWORK PHARMACOLOGY RESEARCH

A variety of databases and tools have been developed through the years of developing network pharmacology as a novel approach to drug discovery. The most commonly used databases for network pharmacology include the Traditional Chinese Medicine (TCM) databases such as HIT (Ye et al. 2011), TCMSP (Ru et al. 2014), TCM-Mesh (Run-zhi Zhang et al. 2017) and TCM-ID (Ye et al. 2011), the chemical and drug information databases that include Drugbank (Wishart 2006), STITCH (Kuhn et al. 2014) and ChEMBL (Gaulton et al. 2012), the target interaction databases such as STRING (Szklarczyk et al. 2015), HPRD (Peri et al. 2003), MINT (Chatr-aryamontri et al. 2007), IntAct (Kerrien et al. 2012) and REACTOME (Matthews et al. 2009) and others including the gene-disease association databases such as OMIM (Hamosh et al. 2005) and GAD (Becker et al. 2004). Through data analysis from such databases combined with clinical trial data, an integrated network of drug candidate-gene/protein-disease networks can be mapped, which can provide insights into the drug targets and their potential effects on the disease (Runzhi Zhang et al. 2019). Several algorithms and tools have also been developed for data mining from these databases. One

TABLE 2.1
Databases and Software for Network Pharmacology Research

Name	Description	Reference
Cytoscape	Visualization Software for Molecular Interaction Networks	(Shannon et al. 2003)
HPRD	Database of Human Proteins Relevant to Health and Disease	(Peri et al. 2003)
OMIM	Database on Human Genes and Genetic Phenotypes	(Hamosh et al. 2005)
MINT	Database of Protein-Protein Interaction from Literature Mining	(Chatr-aryamontri et al. 2007)
PubChem	Public Database for the Bioactivity of Small Molecules	(Wang et al. 2009)
Reactome	Open Source Pathway Database	(Matthews et al. 2009)
Random Walk	Algorithm for the Prediction of Potential Drug-Target Interaction	(Brinda, Vishveshwara, and Vishveshwara 2010)
HIT	Database of Protein Targets for FDA-Approved Drugs and Precursor Compounds	(Ye et al. 2011)
TCM-Database@Taiwan	Small Molecule Database for TCM screening	(C. Y.-C. Chen 2011)
ChEMBL	Bioactive Compound Database Containing Binding and ADMET Information	(Gaulton et al. 2012)
AfroDb	A Select, Highly Potent and Diverse Natural Product Library on African Medicinal Plants	(Ntie-Kang et al. 2013)
STITCH	Database of Known Chemical and Protein Interactions	(Kuhn et al. 2014)
TCMSP	Database for TCM Exploring the Drug-Disease-Target Relationship	(Ru et al. 2014)
STRING	Database of Known Protein-Protein Interaction	(Szklarczyk et al. 2015)
TCM-Mesh	Database for TCM prescription analysis	(Run-zhi Zhang et al. 2017)
Pajek	Software for Network Analysis	(De Nooy, Mrvar, and Batagelj 2018)

example is algorithms such as Random Walk (Xing Chen, Liu, and Yan 2012), which is used to construct drug target-disease networks. The algorithm constructs a random node (comprising a drug, target or disease) and calculates the similarity to an adjacent node to generate the drug target-disease network. Other algorithms such as PRINCE (Vanunu et al. 2010) can predict gene and protein complex associations with disease. Several tools have been developed to uncover such disease-causing genes. Correlating protein Interaction network and PHEnotype network to pRedict disease genes (CIPHER) is a network tool that deciphers disease genes (Wu et al. 2008). This tool can also predict the human disease landscape to study genotype-phenotype relationships by creating a molecular interaction network. Such an interaction network will help in uncovering disease genes and in predicting novel susceptible candidates. However, to understand the closeness of two nodes in a heterogeneous network, a modified Hitting-Time-based CIPHER (CIPHER-HIT) was also developed (Yao et al. 2011). Tools such as Cytoscape (Shannon et al. 2003) and Pajek (De Nooy, Mrvar, and Batagelj 2018) have also been created for the visualization of molecular interaction networks, biological pathways and network analysis. Several databases and software are described in Table 2.1.

2.3 NATURAL PRODUCTS: A TREASURE TROVE OF POTENTIAL THERAPEUTICS?

Natural products in the form of traditional medicine (TCM and Ayurvedic medicine) are some of the most ancient systems of medicine. Natural products have played a pivotal role in drug discovery, particularly for cancer, cardiovascular diseases, multiple sclerosis and infectious diseases (Atanasov et al. 2015; Harvey, Edrada-Ebel, and Quinn 2015; Newman and Cragg 2016; Pal 2016; Waltenberger et al. 2016; Tintore, Vidal-Jordana, and Sastre-Garriga 2019).

Limonoids (azadiradione, azadirachtin, etc.) and flavonoids found in *Azadirachta indica* (Alam et al. 2012; Yadav et al. 2017), curcumin from *Curcuma longa* (Ghosh et al. 2017; Rani et al. 2020), alkaloid sanguinarine from *Sanguinaria canadensis* (Chakraborty et al. 2018), thymol from *Thymus* vulgaris (Bose et al. 2019), sinapic acid from *Brassica nigra* (Sengupta et al. 2019), cinnamic acid from *Cinnamomum sp.* (Mondal et al. 2021), dietary flavonoids riboflavin (Sau, Sanyal, et al. 2018; Sau, Sen, et al. 2018), rutin (Sengupta et al. 2019, 2022) and other phytochemicals such as verbascoside (Biswas et al. 2021; Talapatra et al. 2016; Kashyap et al. 2021) and melatonin, which can be obtained from several plant and animal sources (Rudra et al. 2013), are some of the very well-known natural products that have been shown by different researchers worldwide in independent studies to target several different proteins and potentially be effective against multiple diseases.

Natural products are different from conventional systems of therapeutics. They offer enormous chemical diversity and complexity. Currently, there are several hundred natural product-based drug leads in clinical trials (Harvey 2008). Many bioactive compounds from natural products could potentially serve as leading candidates for rational drug design. Formulations based on natural products modulate multiple targets rather than a highly specific single target site. In this way, they influence and alter the disease pathways and shift the biological network from a diseased state to a healthy one (Gu et al. 2013).

Natural products are characterised by a higher molecular mass, with a large number of sp3 carbon atoms and a significantly higher number of oxygen atoms. However, natural products have lower numbers of nitrogen and halogen atoms. They also contain high H-bond donors and acceptors, high hydrophilicity due to a low water-octanol coefficient (cLogP) and have higher molecular rigidity than conventional drug molecules (Feher and Schmidt 2003; Barnes, Kumar, and Davis 2016; J.W.-H. Li and Vederas 2009; Clardy and Walsh 2004). Such properties are advantageous as potential therapeutics because of the improved structure and function due to natural selection and evolution (Atanasov et al. 2015). Therefore, these properties contribute to the enhanced function of such molecules on an infectious disease and their inherent compatibility with the immune system and the relevant biological pathway. Such compatibility also provides insights into the efficacy, toxicity and safety of natural products as potential therapeutics (Atanasov et al. 2021). The natural products pool of bioactive compounds is far more diverse than the synthetic drug molecule library (Lachance et al. 2012). Despite such advantages, large-scale drug discovery programmes pertaining to natural products have yet to be conducted (Atanasov et al. 2021). Drug candidates based on natural products are typically incompatible with conventional target-based assays (Henrich and Beutler 2013). The identification and isolation of bioactive drug compounds and filtering of known compounds can be quite challenging (Atanasov et al. 2021). Intellectual property rights (IPRs) for unmodified compounds from natural sources may not always be patentable (Harrison 2014). Accessing the yield of bioactive compounds from natural products/sources can also be challenging (Atanasov et al. 2021). The synthesis of structural analogues to access the structure-activity of natural product-based compounds are also a considerable challenge. The molecular mechanisms of action of drug candidates from natural products are accessed through phenotypic assays, which can be time-consuming (Corson and Crews 2007). Advances in the screening assays and the development of assays to ascertain the molecular mechanisms of action of bioactive compounds have led to significant research on natural products (Moffat et al. 2017). Technological advancements and the development of network pharmacology have been at the heart of natural product research.

2.4 STRATEGIES OF BIOACTIVE COMPOUND EXPLORATION

Natural products are a valuable pool of bioactive compounds with varied pharmaceutical applications (Rayan, Raiyn, and Falah 2017). Historically, natural product-based drugs such as quinine, pilocarpine, morphine, artemisinin, etc. have been widely used by the pharmaceutical industry (Newman and Cragg 2016; Pal 2016; L. Zhang et al. 2020). Moreover, almost 50 percent of the drug molecules approved by the FDA are derived or modified from compounds obtained from natural products (Newman and Cragg 2020); the systematic search of the natural products landscape for bioactive compounds with commercial and industrial potential (bioprospecting) still remains the greatest challenge (Skirycz et al. 2016; Roumpeka et al. 2017; Cubillos et al. 2019). The conventional methods of discovering bioactive compounds from natural products typically consist of extraction, fractionation, isolation/purification and screening assays of the isolated bioactive compounds (L. Zhang et al. 2020). Structural optimization involves chemical modification of the isolated bioactive compounds for enhanced pharmacokinetics and pharmacodynamic properties to amplify biological activities (Huffman and Shenvi 2019). The bioprospecting strategies involving network pharmacology are efficient, economical and require less labour and time than the traditional approaches of drug discovery from natural products (J.W.-H. Li and Vederas 2009; Wingert and Camacho 2018; Trujillo-Correa et al. 2019).

2.4.1 Research Approaches and Processes in Network Pharmacology Research

Network pharmacology can influence the drug development landscape through two approaches. First, it can establish a network model for predicting drug targets based on available research data from public databases. The mechanism of action for drug candidates can be ascertained by using the network forecasting method (G. Zhang et al. 2013). Such an approach has successfully determined a new mechanism of action for Ephedra decoction and for the effect of Sennoside C and Rheidin A and C on type II diabetes (Yan et al. 2011; Gu et al. 2011). The second approach is the utilization of high-throughput screening (HTS) and bioinformatics to construct drug target disease networks as a prediction model. Such an approach can predict drug targets in a biological network. Using these approaches, salvianolic acid B was found to be a feasible treatment for cardiovascular disease (Ma et al. 2011). TCM research on Fufang Danshen formulation found nine different bioactive compounds that can effectively modulate 42 genes associated with cardiovascular function (X. Li et al. 2011). Network pharmacology utilises the approaches mentioned previously to optimise the efficiency and safety of a bioactive compound. Network pharmacology employs two different processes for drug discovery from natural products as described in the following sections.

2.4.1.1 Collection and Validation of Data

Network pharmacology research begins with the building of a biological network from the available experimental data. The next step is the validation of the predicted biological network (G. Zhang et al. 2013). Proteomics, metabolomics, genomics and HTS/high-content screening (HCS) technologies can be utilised to quantify the validated data (Runzhi Zhang et al. 2019). HTS/HCS particularly can detect and identify molecular pathways through the screening of data samples and drug candidate molecules (Edwards et al. 2004; Miscevic, Rotstein, and Wen, 2012). Such technologies can collect biological network data from available experiments and validate the network model. The advent of molecular interaction validation technology has been used to validate and reveal drug action mechanisms and drug networks by exploring the relationship between a drug candidate and target macromolecule (G. Zhang et al. 2013).

Drug-repositioning studies, involving host-host, host-pathogen, and drug-target interactions, are a time-consuming effort that delays the development of life-saving therapies in the face of a pandemic such as the SARS-CoV-2. Integrating critical data resources as soon as the new data sources emerge is pivotal in a drug discovery effort. To address this, recently, Zahoránszky-Kőhalmi et al. described a workflow for the semi-automated integration of rapidly emerging data sets that can be adopted in a broad network pharmacology research setting (Figure 2.1).

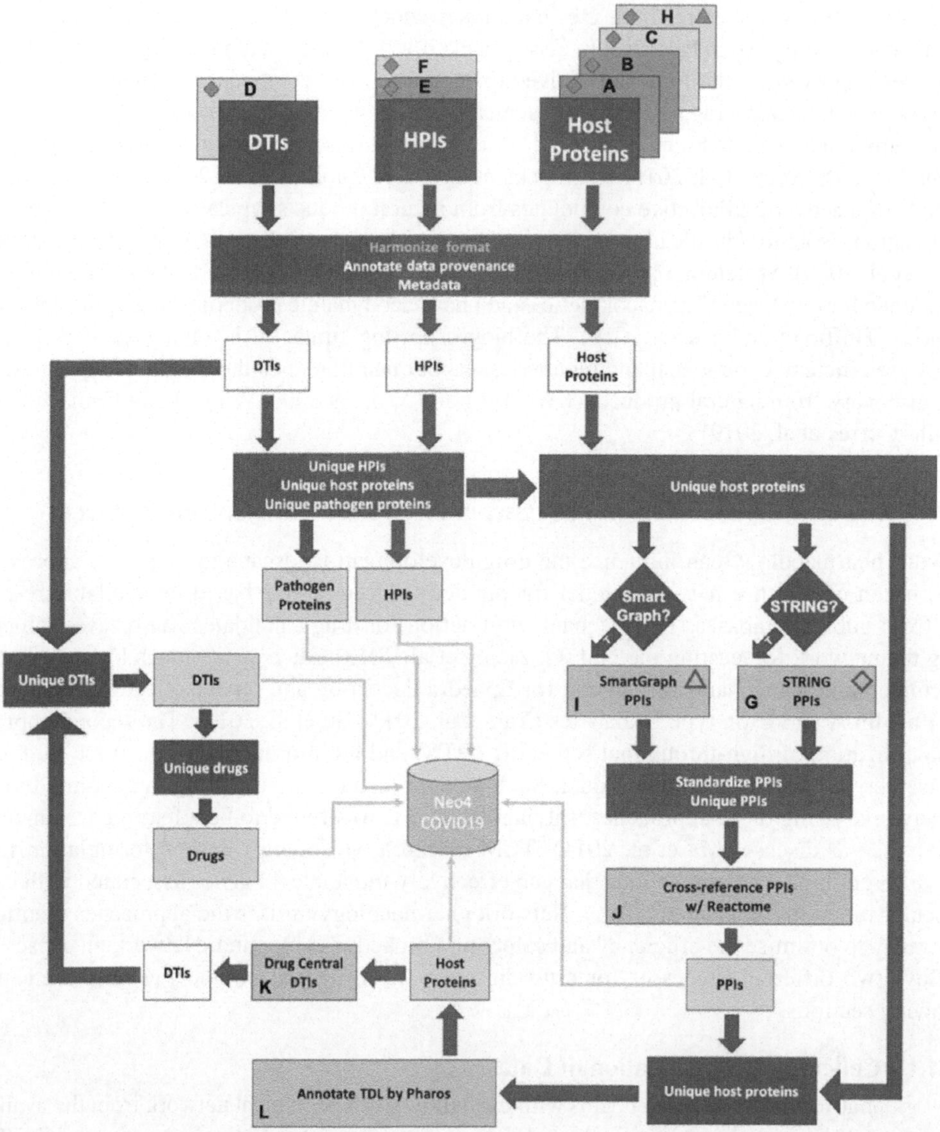

FIGURE 2.1 Schema of a workflow of network pharmacology-driven COVID-19 research. The most important steps of data processing are highlighted. Inputs are labeled as follows: PPIs, interactions between host proteins; HPIs, interactions between host and pathogen (here, SARS-CoV-2) proteins; DTIs, drug-target interactions; TDLs, target development levels; and T, True. The solid triangle indicates the proteins to be used as starting nodes in a SmartGraph analysis, whereas the solid diamond indicates the host proteins to be used for integrating relevant PPIs from the STRING DB. The destination of human proteins routed towards a specific resource (SmartGraph as starting nodes or STRING for PPI expansion) is marked by the open triangle and diamond symbols, respectively.

Source: Reprinted from Zahoránszky-Kőhalmi et al. 2022 with permission © 2022 American Chemical Society.

2.4.1.2 Network Analysis and Visualization

The built network is analysed in three different ways to extract relevant information. The first method involves the calculation of the optimal topological structure of the network and statistical analysis of the extracted network data (Cagnolo, Salvo, and Valladares 2011). The second method involves a comparison of randomly generated networks to assess the reliability of the existing network through the induction of acceptable modulation (Brinda, Vishveshwara, and Vishveshwara 2010). The hierarchical clustering of the generated network is performed in the last method (Hou et al. 2012). Here, the network is pre-digested by applying an algorithm, and information is extracted.

As mentioned earlier, visualization tools are applied for network visualization. The visualization tools convert the extracted information into a visual network (Wu, Wang, and Fan 2011). The visualization process consists of two steps: first, the visualization tools enrich the network attributes through the addition of network nodes, and second, the visualization tools describe the architectural integrity of the network and generate the final network (G. Zhang et al. 2013).

2.5 NETWORK PHARMACOLOGY IN NATURAL PRODUCT RESEARCH

The network pharmacology approach has opened a new avenue in the study of natural product-based research. Network pharmacology not only utilises and recovers traditional knowledge but also provides a scientific rationale in solving the issues regarding natural product research (Zhang, Wang, and Li 2013). This section explains both the applications of network pharmacology in natural product research and some of the current research on network pharmacology to develop a scientific rationale of natural product-based therapeutics.

TCM and Ayurvedic medicine are ancient forms of traditional medicine that employ natural products. Natural products as therapeutics are effective in a holistic way for disease control (J. Xu and Yang 2009; Cheung 2011). However, insights into the pharmacological mechanisms governing the therapeutic potential are still improperly explored. With increasing research on molecular interactions and the development of biological networks, researchers have adopted a network pharmacological approach towards drug discovery from natural products.

2.5.1 CONSTRUCTION OF A DATABASE ON NATURAL PRODUCTS AND TARGETS

For network pharmacology study, building a natural product database is absolutely critical (X Chen et al. 2006). Information regarding tissue distribution, disease targets and pathways are gathered depending upon the type of study conducted. The universal natural products database (UNPD) is one of the foremost information databases regarding bioactive compounds (Gu et al. 2013). Databases such as the TCM-ID database include not only ingredients of TCM formulations but also the 3D structure of different herbal ingredients. The TCM-ID database is used to demonstrate the effect mechanism of different herbal ingredients (Ye et al. 2011). The herbal medicine targets' database can provide information regarding protein targets for bioactive compounds. The database also offers a platform for the analysis of compound formula and protein sequence similarity. SuperNatural (Banerjee et al. 2015) and Dr. Duke's phytochemical and ethnobotanical database ("Dr. Duke's Phytochemical and Ethnobotanical Databases at NAL" 2022) are some other databases that provide information regarding bioactive compounds. Binding databases such as ChEMBL (Gaulton et al. 2012) retrieve information regarding drug targets by searching for bioactive compounds in the database and the corresponding protein target for the bioactive compound. The query structure is compared with other structurally similar compounds. Such binding databases are connected to other databases such as UniProt (Bairoch et al. 2005). Other databases such as Therapeutics Target Database (TTD) (F. Zhu et al. 2012) and the Disease and Gene Annotation (DGA) database (Peng et al. 2013) give information regarding the protein target of disease, along with the corresponding bioactive compound for the targeted protein; furthermore, information regarding human genes in different disease networks is also provided. The integration of literature mining and knowledge base databases helps in gathering information for network pharmacology studies (S. Li et al. 2014). Several of these databases were mentioned in the previous section. Such databases are important in the screening of bioactive compounds for novel drug candidates.

Relevant databases can be used for screening natural products and provide a theoretical foundation for detecting novel drug molecules from natural products. Network target, a methodology proposed by S. Li (2011) has revealed interactions between bioactive compounds and complex syndrome systems. HTS and connectivity map databases were used by Yu et al. (2012) to forecast the new therapeutic effect of the Fuzheng Huayu capsule. Similarly, Zhu and Yao (2011) found new targets for Xiaochaihu decoction by using chemical structure screening, molecular target determination and enrichment analysis. The study found 21 different compounds involved in potential metabolic and inflammation processes. Analysis of Bushenhuoxue formula found six compounds that are potentially effective against 62 different targets for the amelioration of chronic kidney disease (Shi et al. 2014). Studies involving Dragon's Blood (DB) pills found 48 different components effective in the treatment of colitis (H. Xu et al. 2014).

2.5.2 Network Construction and Visualization

The representation of different molecular interactions is a network. The interaction between different entities is called a node. The nodes in network pharmacology represent bioactive compounds, disease targets, tissues and tissue types, disease, disease pathways and their types. The various nodes in a network are connected through lines called edges (Chandran et al. 2017). The edges signify the relationship among the different nodes. There are two approaches towards building a network. The first is the bottom up approach. This method is based on established biological knowledge. The second is the top down approach, which is based on statistical analysis of available data (Chandran et al. 2017; Adhikari et al. 2020). Although there are several ways of generating biological networks, the most generalised method is the de novo network assembly from available experimental data or computational analysis (screening of genes/proteins). Biological networks present a reliable idea regarding biological pathways, genes, proteins and diseases (Xu, Qu, and Pelkonen 2012).

As mentioned previously (see section 2.2), tools such as Cytoscape (Shannon et al. 2003) and Pajek (De Nooy, Mrvar, and Batagelj 2018) are used for visualization and analysis of molecular interaction networks. Apart from visualization, such tools can also integrate data and analysis and profile biological networks and molecular interaction. Other than Cytoscape and Pajek, tools such Visual Network Pharmacology (VNP) can be used to analyse the relationship among drugs-target-diseases (Hu et al. 2014). Tools such as Gephi (Bastian, Heymann, and Jacomy, 2009) and cell illustrator (Nagasaki et al. 2003) can also be used for generating and analysing biological networks.

2.6 SYSTEMS TOXICOLOGY

Network toxicity utilises network analysis to understand and forecast the toxicity and side effects of drugs. It can estimate the toxic reaction of natural product-based compounds in complex biological systems, when integrated with the behaviour of known toxins at the systems level would aid in the evaluation of drug toxicity (G. Zhang et al. 2013; Xu, Qu, and Pelkonen 2012). Studies by Fan et al. (2011) utilised network pharmacology to describe the toxicological properties of potential drug candidates in a complex system. The nephrotoxicity of aristolochic acid was identified using systems toxicology (Zhou et al. 2006). Several databases such as Comparative Toxicogenomics Database, National Toxicology Program and Toxicology Data Network are widely used for screening potential drug candidates from natural products (Wolfgang and Johnson 2002; Davis et al. 2011). In addition to the development of systems toxicology databases, toxicity forecasting software such as DEREK, HazardExpert, Prediction System of Carcinogenic Toxicity and other tools is available for systems toxicology analysis (Lagunin et al. 2009).

2.7 APPLICATIONS AND LIMITATIONS OF NETWORK PHARMACOLOGY

Network pharmacology has been at the forefront of drug discovery because of its cost efficiency. Network pharmacology has its applications in the field of natural products (traditional medicine) research, pharmacology and drug research.

In the field of natural product research, network pharmacology is used for understanding the molecular mechanism of traditional formulations, thereby developing a scientific evidence-based rationale for their use. Network pharmacology is used for analysis of multiple bioactive compounds and gaining insights into their synergistic action. System toxicology is employed for the safety evaluation of natural product formulations. In pharmacology, network pharmacology is used to determine novel lead molecules from natural products by understanding their mode of action and forecasting possible side effects. Network pharmacology is also utilised for predicting drug-drug interactions and repurposing old drug molecules. Meanwhile, in drug research, network pharmacology has significantly improved cost efficiencies and generated a paradigm shift in drug research from a singular target to multi-target therapeutics. Analysis of biological networks have aided significantly the discovery of disease-causing genes and the diagnosis of disease biomarkers.

Despite all the advantages of network pharmacology in drug discovery and the rediscovery of traditional medicine/natural product-based therapy, there are limitations of using network pharmacology. Some of them are listed as follows.

a. Network pharmacology depends upon literature and component mining from different databases. Inconsistencies in the information among databases especially for natural product-based therapeutics may occur. Natural product formulations that undergo preparation such as boiling or acid/alkali reactions may cause the bioactive molecules to change chemically. Utilization of high throughput techniques such as ultra-performance liquid chromatography-electrospray ionization-tandem mass spectroscopy (UPLC-ESIMS/MS) can circumvent this problem (Chandran et al. 2017).
b. The use of delivery vehicles during the administration of natural product formulation (Pramanik et al. 2019) can cause a change in the absorption, distribution, metabolism, excretion and toxic effects (ADMET) parameters. A carefully designed experimental setup is required to address this issue (Chandran et al. 2017).
c. Natural product formulations work by the synergy of multiple bioactive compounds modulating multiple targets simultaneously. In silico studies such as network pharmacology are unable to give a clear picture regarding the exact nature of such interactions (Chandran et al. 2017).

2.8 CONCLUSION

Natural product-based formulations are considered the best alternative to contemporary medicine. However, a lack of understanding of the molecular mechanisms governing their mode of action due to the presence of multidrug components and the synergism among them serves as the largest stumbling block to their development. This lack of a scientific evidence-based understanding of their mode of action limits their utility in modern medicine. In silico tools such as network pharmacology are used extensively in drug discovery to address issues such as drug efficiency, side effects, drug-drug interaction and mode of action. Network pharmacology utilises cross-disciplinary fields to decipher the details of the drug mode of action for multi-drug component natural product formulations. In this chapter, we have highlighted the successful approaches in unlocking the therapeutic potential of natural products. In addition, rationalizing natural product-based therapeutic network pharmacology can also be used in repurposing old drug molecules, new drug molecule discovery and drug toxicity studies. Network pharmacology provides insight into the pharmacological action of natural products based on the constituent bioactive molecules. However, the modern approach of network pharmacology depends on studies already conducted and literature mining. Since studies on natural products are still underway and more data are being generated, this approach is inadequate. Despite all of its inadequacies, network pharmacology is still the most favourable approach available for understanding and rationalizing natural product-based formulations for therapeutic purposes. With the rapid development of computational technology, several network pharmacology tools have been developed and are already available. These tools are becoming more accurate in terms of molecular interactions, protein targets and pharmacological action prediction. Drug discovery armed with network pharmacology is slowly moving towards more precision and a personalised approach to treatment. It is essential that network pharmacology development be translated into experimental verification to unlock the treasure trove of pharmaceuticals that are natural products.

REFERENCES

Adhikari, Aniruddha, Soumendra Darbar, Monojit Das, Susmita Mondal, Siddhartha Sankar Bhattacharya, Debasish Pal, and Samir Kumar Pal. 2020. "Rationalization of a Traditional Liver Medicine Using Systems Biology Approach and Its Evaluation in Preclinical Trial." *Computational Biology and Chemistry* 84 (February): 107196. doi:10.1016/j.compbiolchem.2019.107196.

Alam, Athar, Saikat Haldar, Hirekodathakallu V. Thulasiram, Rahul Kumar, Manish Goyal, Mohd Shameel Iqbal, Chinmay Pal, et al. 2012. "Novel Anti-Inflammatory Activity of Epoxyazadiradione against Macrophage Migration Inhibitory Factor: Inhibition of Tautomerase and Proinflammatory Activities of Macrophage Migration Inhibitory Factor." *The Journal of Biological Chemistry* 287 (29): 24844–61. doi:10.1074/jbc.M112.341321.

Atanasov, Atanas G., Birgit Waltenberger, Eva-Maria Pferschy-Wenzig, Thomas Linder, Christoph Wawrosch, Pavel Uhrin, Veronika Temml, et al. 2015. "Discovery and Resupply of Pharmacologically Active Plant-Derived Natural Products: A Review." *Biotechnology Advances* 33 (8): 1582–614. doi:10.1016/j.biotechadv.2015.08.001.

Atanasov, Atanas G., Sergey B. Zotchev, Verena M. Dirsch, and Claudiu T. Supuran. 2021. "Natural Products in Drug Discovery: Advances and Opportunities." *Nature Reviews Drug Discovery* 20 (3): 200–16. doi:10.1038/s41573-020-00114-z.

Bairoch, Amos, Rolf Apweiler, Cathy H. Wu, Winona C. Barker, Brigitte Boeckmann, Serenella Ferro, El Isabeth Gasteiger, et al. 2005. "The Universal Protein Resource (UniProt)." *Nucleic Acids Research* 33 (suppl_1): D154–59. doi:10.1093/nar/gki070.

Banerjee, Priyanka, Jevgeni Erehman, Björn-Oliver Gohlke, Thomas Wilhelm, Robert Preissner, and Mathias Dunkel. 2015. "Super Natural II—A Database of Natural Products." *Nucleic Acids Research* 43 (D1): D935–39. doi:10.1093/nar/gku886.

Barnes, Emma C., Rohitesh Kumar, and Rohan A. Davis. 2016. "The Use of Isolated Natural Products as Scaffolds for the Generation of Chemically Diverse Screening Libraries for Drug Discovery." *Natural Product Reports* 33 (3): 372–81. doi:10.1039/C5NP00121H.

Bastian, Mathieu, Sebastien Heymann, and Mathieu Jacomy. 2009. "Gephi: An Open Source Software for Exploring and Manipulating Networks," *Proceedings of the International AAAI Conference on Web and Social Media* 3(1): 361–62. Retrieved from https://ojs.aaai.org/index.php/ICWSM/article/view/13937.

Becker, Kevin G., Kathleen C. Barnes, Tiffani J. Bright, and S. Alex Wang. 2004. "The Genetic Association Database." *Nature Genetics* 36 (5): 431–32. doi:10.1038/ng0504-431.

Biswas, Pritam, Aniruddha Adhikari, Susmita Mondal, Monojit Das, Siddhartha Sankar Bhattacharya, Debasish Pal, Sudeshna Shyam Choudhury, and Samir Kumar Pal. 2021. "Synthesis and Spectroscopic Characterization of a Zinc Oxide-Polyphenol Nanohybrid from Natural Resources for Enhanced Antioxidant Activity with Less Cytotoxicity." *Materials Today: Proceedings, International Conference on Nanoelectronics, Nanophotonics, Nanomaterials, Nanobioscience & Nanotechnology* 43 (January): 3481–86. doi:10.1016/j.matpr.2020.09.567.

Bose, Adity, Priti Sengupta, Uttam Pal, Sanjib Senapati, Mohd. Ahsan, Santanu Roy, Upasana Das, and Kamalika Sen. 2019. "Encapsulation of Thymol in Cyclodextrin Nano-Cavities: A Multi Spectroscopic and Theoretical Study." *Spectrochimica Acta Part A: Molecular and Biomolecular Spectroscopy* 208 (February): 339–48. doi:10.1016/j.saa.2018.10.011.

Brinda, K. V., Saraswathi Vishveshwara, and Smitha Vishveshwara. 2010. "Random Network Behaviour of Protein Structures." In *Molecular BioSystems* 6 (2): 391–98. doi:10.1039/B903019K.

Buriani, Alessandro, Maria L. Garcia-Bermejo, Enrica Bosisio, Qihe Xu, Huige Li, Xuebin Dong, Monique S. J. Simmonds, et al. 2012. "Omic Techniques in Systems Biology Approaches to Traditional Chinese Medicine Research: Present and Future." *Journal of Ethnopharmacology* 140 (3): 535–44. doi:10.1016/j.jep.2012.01.055.

Cagnolo, Luciano, Adriana Salvo, and Graciela Valladares. 2011. "Network Topology: Patterns and Mechanisms in Plant-Herbivore and Host-Parasitoid Food Webs." *Journal of Animal Ecology* 80 (2): 342–51. doi:10.1111/j.1365-2656.2010.01778.x.

Chakraborty, Sudeshna Das, Abhishek Sau, Arnab Maity, Uttam Pal, Chaitrali Sengupta, Maireyee Bhattacharya, Gourab Bhattacharjee, et al. 2018. "Low Magnetic Field Induced Surface Enhanced Transient Spin-Trajectory Modulation of a Prototype Anticancer Drug Sanguinarine on a Single Domain Superparamagnetic Nanosurface." *The Journal of Physical Chemistry C* 122 (36): 20619–31. doi:10.1021/acs.jpcc.8b06129.

Chandran, Uma, Neelay Mehendale, Saniya Patil, Rathnam Chaguturu, and Bhushan Patwardhan. 2017. "Network Pharmacology." In *Innovative Approaches in Drug Discovery*. Elsevier, 127–64. doi:10.1016/B978-0-12-801814-9.00005-2.

Chatr-Aryamontri, Andrew, Arnaud Ceol, Luisa Montecchi Palazzi, Giuliano Nardelli, Maria Victoria Schneider, Luisa Castagnoli, and Gianni Cesareni. 2007. "MINT: The Molecular INTeraction Database." *Nucleic Acids Research* 35 (suppl_1): D572–74. doi:10.1093/nar/gkl950.

Chen, Calvin Yu-Chian. 2011. "TCM Database@Taiwan: The World's Largest Traditional Chinese Medicine Database for Drug Screening in Silico." *PLoS One* 6 (1). Public Library of Science: e15939. doi:10.1371/journal.pone.0015939.

Chen, X., H. Zhou, Y. B. Liu, J. F. Wang, H. Li, C. Y. Ung, L. Y. Han, et al. 2006. "Database of Traditional Chinese Medicine and Its Application to Studies of Mechanism and to Prescription Validation." *British Journal of Pharmacology* 149 (8): 1092–103. doi:10.1038/sj.bjp.0706945.

Chen, Xing, Ming-Xi Liu, and Gui-Ying Yan. 2012. "Drug–Target Interaction Prediction by Random Walk on the Heterogeneous Network." *Molecular BioSystems* 8 (7): 1970–78. doi:10.1039/C2MB00002D.

Chen, Yanqing, Jun Zhu, Pek Yee Lum, Xia Yang, Shirly Pinto, Douglas J. MacNeil, Chunsheng Zhang, et al. 2008. "Variations in DNA Elucidate Molecular Networks That Cause Disease." *Nature* 452 (7186): 429–35. doi:10.1038/nature06757.

Cheung, Felix. 2011. "TCM: Made in China." *Nature* 480 (7378): S82–83. doi:10.1038/480S82a.

Cho, Dong-Yeon, Yoo-Ah Kim, and Teresa M. Przytycka. 2012. "Chapter 5: Network Biology Approach to Complex Diseases." *PLoS Computational Biology* 8 (12): e1002820. doi:10.1371/journal.pcbi.1002820.

Clardy, Jon, and Christopher Walsh. 2004. "Lessons from Natural Molecules." *Nature* 432 (7019): 829–37. doi:10.1038/nature03194.

Clark, David E., and Stephen D. Pickett. 2000. "Computational Methods for the Prediction of 'Drug-Likeness.'" *Drug Discovery Today* 5 (2): 49–58. doi:10.1016/S1359-6446(99)01451-8.

Corson, Timothy W., and Craig M. Crews. 2007. "Molecular Understanding and Modern Application of Traditional Medicines: Triumphs and Trials." *Cell* 130 (5): 769–74. doi:10.1016/j.cell.2007.08.021.

Cubillos, Francisco A., Brian Gibson, Nubia Grijalva-Vallejos, Kristoffer Krogerus, and Jarkko Nikulin. 2019. "Bioprospecting for Brewers: Exploiting Natural Diversity for Naturally Diverse Beers." *Yeast* 36 (6): 383–98. doi:10.1002/yea.3380.

Davis, Allan Peter, Benjamin L. King, Susan Mockus, Cynthia G. Murphy, Cynthia Saraceni-Richards, Michael Rosenstein, Thomas Wiegers, and Carolyn J. Mattingly. 2011. "The Comparative Toxicogenomics Database: Update 2011." *Nucleic Acids Research* 39 (suppl_1): D1067–72. doi:10.1093/nar/gkq813.

De Nooy, Wouter, Andrej Mrvar, and Vladimir Batagelj. 2018. *Exploratory Social Network Analysis with Pajek: Revised and Expanded Edition for Updated Software* (3rd ed.). Cambridge University Press. (Structural Analysis in the Social Sciences). doi:10.1017/9781108565691.

"Dr. Duke's Phytochemical and Ethnobotanical Databases at NAL." Accessed April 24, 2022. https://phytochem.nal.usda.gov/phytochem/search.

Edwards, Bruce S., Tudor Oprea, Eric R. Prossnitz, and Larry A. Sklar. 2004. "Flow Cytometry for High-Throughput, High-Content Screening." *Current Opinion in Chemical Biology* 8 (4): 392–98. doi:10.1016/j.cbpa.2004.06.007.

Ellingson, Sally R., Jeremy C. Smith, and Jerome Baudry. 2014. "Polypharmacology and Supercomputer-Based Docking: Opportunities and Challenges." *Molecular Simulation* 40 (10–11): 848–54. doi:10.1080/08927022.2014.899699.

Fan, Xiaohui, Xiaoping Zhao, Yecheng Jin, Xiuping Shen, and Changxiao Liu. 2011. "[Network Toxicology and Its Application to Traditional Chinese Medicine]." *Zhongguo Zhong Yao Za Zhi = Zhongguo Zhongyao Zazhi = China Journal of Chinese Materia Medica* 36 (21): 2920–22.

Feher, Miklos, and Jonathan M. Schmidt. 2003. "Property Distributions: Differences between Drugs, Natural Products, and Molecules from Combinatorial Chemistry." *Journal of Chemical Information and Computer Sciences* 43 (1): 218–27. doi:10.1021/ci0200467.

Friboulet, Alain, and Daniel Thomas. 2005. "Systems Biology—An Interdisciplinary Approach." *Biosensors and Bioelectronics* 20 (12): 2404–7. doi:10.1016/j.bios.2004.11.014.

Gaulton, Anna, Louisa J. Bellis, A. Patricia Bento, Jon Chambers, Mark Davies, Anne Hersey, Yvonne Light, et al. 2012. "ChEMBL: A Large-Scale Bioactivity Database for Drug Discovery." *Nucleic Acids Research* 40 (D1): D1100–7. doi:10.1093/nar/gkr777.

Ghosh, Supratim, Sumana Mallick, Upasana Das, Ajay Verma, Uttam Pal, Sabyasachi Chatterjee, Abhishek Nandy, et al. 2017. "Curcumin Stably Interacts with DNA Hairpin Through Minor Groove Binding and Demonstrates Enhanced Cytotoxicity in Combination with FdU Nucleotides." *Biochimica et Biophysica Acta (BBA) – General Subjects*, October. https://doi.org/10.1016/j.bbagen.2017.10.018.

Gu, Jiangyong, Yuanshen Gui, Lirong Chen, Gu Yuan, Hui-Zhe Lu, and Xiaojie Xu. 2013. "Use of Natural Products as Chemical Library for Drug Discovery and Network Pharmacology." *PLoS One* 8 (4): e62839. doi:10.1371/journal.pone.0062839.

Gu, Jiangyong, Hu Zhang, Lirong Chen, Shun Xu, Gu Yuan, and Xiaojie Xu. 2011. "Drug–Target Network and Polypharmacology Studies of a Traditional Chinese Medicine for Type II Diabetes Mellitus." *Computational Biology and Chemistry* 35 (5): 293–97. doi:10.1016/j.compbiolchem.2011.07.003.

Guan, Longfei, Hongbin Yang, Yingchun Cai, Lixia Sun, Peiwen Di, Weihua Li, Guixia Liu, and Yun Tang. 2019. "ADMET-Score – A Comprehensive Scoring Function for Evaluation of Chemical Drug-Likeness." *MedChemComm* 10 (1). The Royal Society of Chemistry: 148–57. doi:10.1039/C8MD00472B.

Hacker, Miles, William S. Messer, and Kenneth A. Bachmann. 2009. *Pharmacology: Principles and Practice*. Academic Press.

Hamosh, Ada, Alan F. Scott, Joanna S. Amberger, Carol A. Bocchini, and Victor A. McKusick. 2005. "Online Mendelian Inheritance in Man (OMIM), a Knowledgebase of Human Genes and Genetic Disorders." *Nucleic Acids Research* 33 (suppl_1): D514–17. doi:10.1093/nar/gki033.

Harrison, Charlotte. 2014. "Patenting Natural Products Just Got Harder." *Nature Biotechnology* 32 (5): 403–4. doi:10.1038/nbt0514-403a.

Harvey, Alan L. 2008. "Natural Products in Drug Discovery." *Drug Discovery Today* 13 (19): 894–901. doi:10.1016/j.drudis.2008.07.004.

Harvey, Alan L., RuAngelie Edrada-Ebel, and Ronald J. Quinn. 2015. "The Re-Emergence of Natural Products for Drug Discovery in the Genomics Era." *Nature Reviews Drug Discovery* 14 (2): 111–29. doi:10.1038/nrd4510.

Henrich, Curtis J., and John A. Beutler. 2013. "Matching the Power of High Throughput Screening to the Chemical Diversity of Natural Products." *Natural Product Reports* 30 (10): 1284–98. doi:10.1039/C3NP70052F.

Hopkins, Andrew L. 2007. "Network Pharmacology." *Nature Biotechnology* 25 (10): 1110–11. doi:10.1038/nbt1007-1110.

Hopkins, Andrew L. 2008. "Network Pharmacology: The Next Paradigm in Drug Discovery." *Nature Chemical Biology* 4 (11): 682–90. doi:10.1038/nchembio.118.

Hou, Lin, Lin Wang, Arthur Berg, Minping Qian, Yunping Zhu, Fangting Li, and Minghua Deng. 2012. "Comparison and Evaluation of Network Clustering Algorithms Applied to Genetic Interaction Networks." *Frontiers in Bioscience-Elite* 4 (6): 2150–61. doi:10.2741/532.

Hu, Q.-N., Z. Deng, W. Tu, X. Yang, Z.-B. Meng, Z.-X. Deng, and J. Liu. 2014. "VNP: Interactive Visual Network Pharmacology of Diseases, Targets, and Drugs." *CPT: Pharmacometrics & Systems Pharmacology* 3 (3): 105. doi:10.1038/psp.2014.1.

Huffman, Benjamin J., and Ryan A. Shenvi. 2019. "Natural Products in the 'Marketplace': Interfacing Synthesis and Biology." *Journal of the American Chemical Society* 141 (8): 3332–46. doi:10.1021/jacs.8b11297.

Kashyap, Bhaswati, Sagar Ramrao Barge, Simanta Bharadwaj, Barsha Deka, Seydur Rahman, Aparajita Ghosh, Prasenjit Manna, et al. 2021. "Evaluation of Therapeutic Effect of Premna Herbacea in Diabetic Rat and Isoverbascoside against Insulin Resistance in L6 Muscle Cells Through Bioenergetics and Stimulation of JNK and AKT/MTOR Signaling Cascade." *Phytomedicine* 93 (December): 153761. doi:10.1016/j.phymed.2021.153761.

Keith, Curtis T., Alexis A. Borisy, and Brent R. Stockwell. 2005. "Multicomponent Therapeutics for Networked Systems." *Nature Reviews Drug Discovery* 4 (1): 71–78. doi:10.1038/nrd1609.

Kerrien, Samuel, Bruno Aranda, Lionel Breuza, Alan Bridge, Fiona Broackes-Carter, Carol Chen, Margaret Duesbury, et al. 2012. "The IntAct Molecular Interaction Database in 2012." *Nucleic Acids Research* 40 (D1): D841–46. doi:10.1093/nar/gkr1088.

Kitano, Hiroaki. 2002. "Systems Biology: A Brief Overview." *Science* 295 (5560): 1662–64. doi:10.1126/science.1069492.

Kola, Ismail, and John Landis. 2004. "Can the Pharmaceutical Industry Reduce Attrition Rates?" *Nature Reviews Drug Discovery* 3 (8): 711–16. doi:10.1038/nrd1470.

Kuhn, Michael, Damian Szklarczyk, Sune Pletscher-Frankild, Thomas H. Blicher, Christian von Mering, Lars J. Jensen, and Peer Bork. 2014. "STITCH 4: Integration of Protein–Chemical Interactions with User Data." *Nucleic Acids Research* 42 (D1): D401–7. doi:10.1093/nar/gkt1207.

Lachance, Hugo, Stefan Wetzel, Kamal Kumar, and Herbert Waldmann. 2012. “Charting, Navigating, and Populating Natural Product Chemical Space for Drug Discovery.” *Journal of Medicinal Chemistry* 55 (13): 5989–6001. doi:10.1021/jm300288g.

Lagunin, Alexey, Dmitrii Filimonov, Alexey Zakharov, Wei Xie, Ying Huang, Fucheng Zhu, Tianxiang Shen, et al. 2009. “Computer-Aided Prediction of Rodent Carcinogenicity by PASS and CISOC-PSCT.” *QSAR & Combinatorial Science* 28 (8): 806–10. doi:10.1002/qsar.200860192.

Li, Jesse W.-H., and John C. Vederas. 2009. “Drug Discovery and Natural Products: End of an Era or an Endless Frontier?” *Science* 325 (5937): 161–65. doi:10.1126/science.1168243.

Li, Shao. 2011. “[Network Target: A Starting Point for Traditional Chinese Medicine Network Pharmacology].” *Zhongguo Zhong Yao Za Zhi = Zhongguo Zhongyao Zazhi = China Journal of Chinese Materia Medica* 36 (15): 2017–20.

Li, Shao, Tai-Ping Fan, Wei Jia, Aiping Lu, and Weidong Zhang. 2014. “Network Pharmacology in Traditional Chinese Medicine.” *Evidence-Based Complementary and Alternative Medicine* 2014 (February): e138460. doi:10.1155/2014/138460.

Li, Shao, Lijiang Wu, and Zhongqi Zhang. 2006. “Constructing Biological Networks Through Combined Literature Mining and Microarray Analysis: A LMMA Approach.” *Bioinformatics* 22 (17): 2143–50. doi:10.1093/bioinformatics/btl363.

Li, Xiang, Leihong Wu, Xiaohui Fan, Boli Zhang, Xiumei Gao, Yi Wang, and Yiyu Cheng. 2011. “[Network Pharmacology Study on Major Active Compounds of Fufang Danshen Formula].” *Zhongguo Zhong Yao Za Zhi = Zhongguo Zhongyao Zazhi = China Journal of Chinese Materia Medica* 36 (21): 2911–15.

Ma, Chao, Yan Yao, Qing-Xi Yue, Xin-Wen Zhou, Peng-Yuan Yang, Wan-Ying Wu, Shu-Hong Guan, et al. 2011. “Differential Proteomic Analysis of Platelets Suggested Possible Signal Cascades Network in Platelets Treated with Salvianolic Acid B.” *PLoS One* 6 (2): e14692. doi:10.1371/journal.pone.0014692.

Matthews, Lisa, Gopal Gopinath, Marc Gillespie, Michael Caudy, David Croft, Bernard de Bono, Phani Garapati, et al. 2009. “Reactome Knowledgebase of Human Biological Pathways and Processes.” *Nucleic Acids Research* 37 (suppl_1): D619–22. doi:10.1093/nar/gkn863.

Metz, James T., and Philip J. Hajduk. 2010. “Rational Approaches to Targeted Polypharmacology: Creating and Navigating Protein–Ligand Interaction Networks.” *Current Opinion in Chemical Biology* 14 (4): 498–504. doi:10.1016/j.cbpa.2010.06.166.

Miscevic, Filip, Ori Rotstein, and Xiao-Yan Wen. 2012. “Advances in Zebrafish High Content and High Throughput Technologies.” *Combinatorial Chemistry & High Throughput Screening* 15 (7): 515–21.

Moffat, John G., Fabien Vincent, Jonathan A. Lee, Jörg Eder, and Marco Prunotto. 2017. “Opportunities and Challenges in Phenotypic Drug Discovery: An Industry Perspective.” *Nature Reviews Drug Discovery* 16 (8): 531–43. doi:10.1038/nrd.2017.111.

Mondal, Prasenjit, Priti Sengupta, Uttam Pal, Sutapa Saha, and Adity Bose. 2021. “Biophysical and Theoretical Studies of the Interaction between a Bioactive Compound 3,5-Dimethoxy-4-Hydroxycinnamic Acid with Calf Thymus DNA.” Spectrochimica Acta Part A: Molecular and Biomolecular Spectroscopy 245 (January): 118936. doi:10.1016/j.saa.2020.118936.

Nagasaki, Masao, Atsushi Doi, Hiroshi Matsuno, and Satoru Miyano. 2003. “Genomic Object Net: I. A Platform for Modelling and Simulating Biopathways.” *Applied Bioinformatics* 2 (3): 181–84.

Newman, David J., and Gordon M. Cragg. 2016. “Natural Products as Sources of New Drugs from 1981 to 2014.” *Journal of Natural Products* 79 (3): 629–61. doi:10.1021/acs.jnatprod.5b01055.

———. 2020. “Natural Products as Sources of New Drugs Over the Nearly Four Decades from 01/1981 to 09/2019.” *Journal of Natural Products* 83 (3): 770–803. doi:10.1021/acs.jnatprod.9b01285.

Ntie-Kang, F., D. Zofou, S. B. Babiaka, R. Meudom, M. Scharfe, L. L. Lifongo, et al. 2013 AfroDb: A Select Highly Potent and Diverse Natural Product Library from African Medicinal Plants. *PLoS One* 8(10): e78085. doi:10.1371/journal.pone.0078085.

Pal, Uttam. 2016. “Interaction of Proteins with Small Molecules and Peptides.” Doctoral dissertation, Kolkata: Jadavpur University. ark:/13960/t7jq5vb4k.

Patwardhan, Bhushan, and Uma Chandran. 2015. “Network Ethnopharmacology Approaches for Formulation Discovery.” *Indian Journal of Traditional Knowledge* 14 (4): 574–80.

Peng, Kai, Wei Xu, Jianyong Zheng, Kegui Huang, Huisong Wang, Jiansong Tong, Zhifeng Lin, et al. 2013. “The Disease and Gene Annotations (DGA): An Annotation Resource for Human Disease.” *Nucleic Acids Research* 41 (D1): D553–60. doi:10.1093/nar/gks1244.

Peri, Suraj, J. Daniel Navarro, Ramars Amanchy, Troels Z. Kristiansen, Chandra Kiran Jonnalagadda, Vineeth Surendranath, Vidya Niranjan, et al. 2003. “Development of Human Protein Reference Database as an Initial Platform for Approaching Systems Biology in Humans.” *Genome Research* 13 (10): 2363–71. doi:10.1101/gr.1680803.

Pramanik, Sumit Kumar, Uttam Pal, Preety Choudhary, Harwinder Singh, Russel J. Reiter, Anitha Ethirajan, Snehasikta Swarnakar, et al. 2019. "Stimuli-Responsive Nanocapsules for the Spatiotemporal Release of Melatonin: Protection against Gastric Inflammation." *ACS Applied Bio Materials* 2 (12): 5218–26. doi:10.1021/acsabm.9b00236.

Rani, Swati, Damayanti Bagchi, Uttam Pal, Mamta Kumari, Manisha Sharma, Arpan Bera, Javaid Shabir, et al. 2020. "The Role of Imidazolium-Based Surface-Active Ionic Liquid to Restrain the Excited-State Intramolecular H-Atom Transfer Dynamics of Medicinal Pigment Curcumin: A Theoretical and Experimental Approach." *ACS Omega* 5 (40): 25582–92. doi:10.1021/acsomega.0c02438.

Rayan, Anwar, Jamal Raiyn, and Mizied Falah. 2017. "Nature Is the Best Source of Anticancer Drugs: Indexing Natural Products for Their Anticancer Bioactivity." *PLoS One* 12 (11): e0187925. doi:10.1371/journal.pone.0187925.

Roth, Bryan L., Douglas J. Sheffler, and Wesley K. Kroeze. 2004. "Magic Shotguns versus Magic Bullets: Selectively Non-Selective Drugs for Mood Disorders and Schizophrenia." *Nature Reviews Drug Discovery* 3 (4): 353–59. doi:10.1038/nrd1346.

Roumpeka, Despoina D., R. John Wallace, Frank Escalettes, Ian Fotheringham, and Mick Watson. 2017. "A Review of Bioinformatics Tools for Bio-Prospecting from Metagenomic Sequence Data." *Frontiers in Genetics* 8. www.frontiersin.org/article/10.3389/fgene.2017.00023.

Ru, Jinlong, Peng Li, Jinan Wang, Wei Zhou, Bohui Li, Chao Huang, Pidong Li, et al. 2014. "TCMSP: A Database of Systems Pharmacology for Drug Discovery from Herbal Medicines." *Journal of Cheminformatics* 6 (1): 13. doi:10.1186/1758-2946-6-13.

Rudra, Deep Sankar, Uttam Pal, Nakul Chandra Maiti, Russel J. Reiter, and Snehasikta Swarnakar. 2013. "Melatonin Inhibits Matrix Metalloproteinase-9 Activity by Binding to Its Active Site." *Journal of Pineal Research* 54 (4): 398–405. doi:10.1111/jpi.12034.

Sams-Dodd, Frank. 2005. "Target-Based Drug Discovery: Is Something Wrong?" *Drug Discovery Today* 10 (2): 139–47. doi:10.1016/S1359-6446(04)03316-1.

Sau, Abhishek, Sulagna Sanyal, Kallol Bera, Sabyasachi Sen, Amrit Krishna Mitra, Uttam Pal, Prabal Kumar Chakraborty, et al. 2018. "DNA Damage and Apoptosis Induction in Cancer Cells by Chemically Engineered Thiolated Riboflavin Gold Nanoassembly." *ACS Applied Materials & Interfaces*. https://doi.org/10.1021/acsami.7b18837.

Sau, Abhishek, Sabyasachi Sen, Kallol Bera, Uttam Pal, Biswarup Satpati, Chandrima Das, and Samita Basu. 2018. "Nuclear Uptake of Thiolated Riboflavin Gold Nanoassembly: DNA Damage and Apoptosis Induction in Cancer Cell." *Biophysical Journal* 114 (3): 360a. doi:10.1016/j.bpj.2017.11.2002.

Sengupta, Priti, Uttam Pal, Prasenjit Mondal, and Adity Bose. 2019. "Multi-Spectroscopic and Computational Evaluation on the Binding of Sinapic Acid and Its Cu(II) Complex with Bovine Serum Albumin." *Food Chemistry* 301 (December): 125254. doi:10.1016/j.foodchem.2019.125254.

Sengupta, Priti, Uttam Pal, Pritam Roy, Tapendu Samanta, Nitin Chattopadhyay, Kamalika Sen, and Adity Bose. 2022. "Effect of a Metal Ion in Modulating the Binding Interaction of a Dietary Flavonoid with Bovine Serum Albumin and DNA: A Spectroscopic and Theoretical Approach." *ACS Food Science & Technology* 2 (1): 114–24. doi:10.1021/acsfoodscitech.1c00361.

Shannon, Paul, Andrew Markiel, Owen Ozier, Nitin S. Baliga, Jonathan T. Wang, Daniel Ramage, Nada Amin, et al. 2003. "Cytoscape: A Software Environment for Integrated Models of Biomolecular Interaction Networks." *Genome Research* 13 (11): 2498–504. doi:10.1101/gr.1239303.

Shi, Shao-Hua, Yue-Piao Cai, Xiao-Jun Cai, Xiao-Yong Zheng, Dong-Sheng Cao, Fa-Qing Ye, and Zheng Xiang. 2014. "A Network Pharmacology Approach to Understanding the Mechanisms of Action of Traditional Medicine: Bushenhuoxue Formula for Treatment of Chronic Kidney Disease." *PLoS One* 9 (3): e89123. doi:10.1371/journal.pone.0089123.

Skirycz, Aleksandra, Sylwia Kierszniowska, Michaël Méret, Lothar Willmitzer, and George Tzotzos. 2016. "Medicinal Bioprospecting of the Amazon Rainforest: A Modern Eldorado?" *Trends in Biotechnology* 34 (10): 781–90. doi:10.1016/j.tibtech.2016.03.006.

Szklarczyk, Damian, Andrea Franceschini, Stefan Wyder, Kristoffer Forslund, Davide Heller, Jaime Huerta-Cepas, Milan Simonovic, et al. 2015. "STRING V10: Protein–Protein Interaction Networks, Integrated Over the Tree of Life." *Nucleic Acids Research* 43 (D1): D447–52. doi:10.1093/nar/gku1003.

Talapatra, Soumendra Nath, Partha Talukdar, Uttam Pal, Nakul Chandra Maiti, and Snehasikta Swarnakar. 2016. "Interaction of T-State Haemoglobin and Phytochemicals of Hygrophila Spinosa T. Anders: An Approach by Molecular Docking." *World Journal of Pharmaceutical Research* 5 (8): 1354–69.

Tintore, Mar, Angela Vidal-Jordana, and Jaume Sastre-Garriga. 2019. "Treatment of Multiple Sclerosis – Success from Bench to Bedside." *Nature Reviews Neurology* 15 (1): 53–58. doi:10.1038/s41582-018-0082-z.

Trujillo-Correa, Andrea Isabel, Diana Carolina Quintero-Gil, Fredyc Diaz-Castillo, Winston Quiñones, Sara M. Robledo, and Marlen Martinez-Gutierrez. 2019. "In Vitro and in Silico Anti-Dengue Activity of Compounds Obtained from Psidium Guajava Through Bioprospecting." *BMC Complementary and Alternative Medicine* 19 (1): 298. doi:10.1186/s12906-019-2695-1.

Vanunu, Oron, Oded Magger, Eytan Ruppin, Tomer Shlomi, and Roded Sharan. 2010. "Associating Genes and Protein Complexes with Disease via Network Propagation." Edited by Wyeth W. Wasserman. *PLoS Computational Biology* 6 (1): e1000641. doi:10.1371/journal.pcbi.1000641.

Waltenberger, Birgit, Andrei Mocan, Karel Šmejkal, Elke H. Heiss, and Atanas G. Atanasov. 2016. "Natural Products to Counteract the Epidemic of Cardiovascular and Metabolic Disorders." *Molecules* 21 (6): 807. doi:10.3390/molecules21060807.

Waterbeemd, Han van de, and Eric Gifford. 2003. "ADMET in Silico Modelling: Towards Prediction Paradise?" *Nature Reviews Drug Discovery* 2 (3): 192–204. doi:10.1038/nrd1032.

Wang, Yanli, Jewen Xiao, Tugba O. Suzek, Jian Zhang, Jiyao Wang, and Stephen H. Bryant. 2009. "PubChem: A Public Information System for Analyzing Bioactivities of Small Molecules." *Nucleic Acids Research* 37 (suppl_2): W623–33. doi:10.1093/nar/gkp456.

Wermuth, Camille G. 2004. "Multitargeted Drugs: The End of the 'One-Target-One-Disease' Philosophy?" *Drug Discovery Today* 9 (19): 826–27. doi:10.1016/S1359-6446(04)03213-1.

Wingert, Bentley M., and Carlos J. Camacho. 2018. "Improving Small Molecule Virtual Screening Strategies for the Next Generation of Therapeutics." *Current Opinion in Chemical Biology* 44 (June): 87–92. doi:10.1016/j.cbpa.2018.06.006.

Wishart, D. S. 2006. "DrugBank: A Comprehensive Resource for in Silico Drug Discovery and Exploration." *Nucleic Acids Research* 34 (90001): D668–72. doi:10.1093/nar/gkj067.

Wolfgang, Grushenka H. I., and Dale E. Johnson. 2002. "Web Resources for Drug Toxicity." *Toxicology* 173 (1): 67–74. doi:10.1016/S0300-483X(02)00022-7.

Wu, Leihong, Yi Wang, and Xiaohui Fan. 2011. "[Tools for Network Pharmacology Study: Network Visualization and Network Analysis]." *Zhongguo Zhong Yao Za Zhi = Zhongguo Zhongyao Zazhi = China Journal of Chinese Materia Medica* 36 (21): 2923–25.

Wu, Xuebing, Rui Jiang, Michael Q. Zhang, and Shao Li. 2008. "Network-Based Global Inference of Human Disease Genes." *Molecular Systems Biology* 4 (1): 189. doi:10.1038/msb.2008.27.

Xu, Haiyu, Yanqiong Zhang, Yun Lei, Xiumei Gao, Huaqiang Zhai, Na Lin, Shihuan Tang, et al. 2014. "A Systems Biology-Based Approach to Uncovering the Molecular Mechanisms Underlying the Effects of Dragon's Blood Tablet in Colitis, Involving the Integration of Chemical Analysis, ADME Prediction, and Network Pharmacology." *PLoS One* 9 (7): e101432. doi:10.1371/journal.pone.0101432.

Xu, Judy, and Yue Yang. 2009. "Traditional Chinese Medicine in the Chinese Health Care System." *Health Policy* 90 (2): 133–39. doi:10.1016/j.healthpol.2008.09.003.

Xu, Qihe, Fan Qu, and Olavi Pelkonen. 2012. "Network Pharmacology and Traditional Chinese Medicine." *Alternative Medicine*. doi:10.5772/53868.

Yadav, Dharmendra K., Yogesh P. Bharitkar, Abhijit Hazra, Uttam Pal, Sugreev Verma, Sayantan Jana, Umesh P. Singh, et al. 2017. "Tamarixetin 3-O-β-d-Glucopyranoside from Azadirachta Indica Leaves: Gastroprotective Role Through Inhibition of Matrix Metalloproteinase-9 Activity in Mice." *Journal of Natural Products* 80 (5): 1347–53. doi:10.1021/acs.jnatprod.6b00957.

Yan, Jing, Yun Wang, Si-jun Luo, and Yan-jiang Qiao. 2011. "TCM Grammar Systems: An Approach to Aid the Interpretation of the Molecular Interactions in Chinese Herbal Medicine." *Journal of Ethnopharmacology* 137 (1): 77–84. doi:10.1016/j.jep.2011.04.057.

Yao, Xin, Han Hao, Yanda Li, and Shao Li. 2011. "Modularity-Based Credible Prediction of Disease Genes and Detection of Disease Subtypes on the Phenotype-Gene Heterogeneous Network." *BMC Systems Biology* 5 (1): 79. doi:10.1186/1752-0509-5-79.

Ye, H., L. Ye, H. Kang, D. Zhang, L. Tao, K. Tang, X. Liu, et al. 2011. "HIT: Linking Herbal Active Ingredients to Targets." *Nucleic Acids Research* 39 (Database): D1055–59. doi:10.1093/nar/gkq1165.

Yıldırım, Muhammed A., Kwang-Il Goh, Michael E. Cusick, Albert-László Barabási, and Marc Vidal. 2007. "Drug—Target Network." *Nature Biotechnology* 25 (10): 1119–26. doi:10.1038/nbt1338.

Yu, Shuhao, Zhizhong Guo, Yan Guan, Yi-Yu Lu, Pei Hao, Yixue Li, and Shi-Bing Su. 2012. "Combining ZHENG Theory and High-Throughput Expression Data to Predict New Effects of Chinese Herbal Formulae." *Evidence-Based Complementary and Alternative Medicine* 2012 (May): e986427. doi:10.1155/2012/986427.

Zahoránszky-Kőhalmi, Gergely, Vishal B. Siramshetty, Praveen Kumar, Manideep Gurumurthy, Busola Grillo, Biju Mathew, Dimitrios Metaxatos, et al. 2022. "A Workflow of Integrated Resources to Catalyze Network Pharmacology Driven COVID-19 Research." *Journal of Chemical Information and Modeling* 62 (3): 718–29. doi:10.1021/acs.jcim.1c00431.

Zambrowicz, Brian P., and Arthur T. Sands. 2004. "Modeling Drug Action in the Mouse with Knockouts and RNA Interference." *Drug Discovery Today: Targets* 3 (5): 198–207. doi:10.1016/S1741-8372(04)02454-5.

Zhang, Bo, Xu Wang, and Shao Li. 2013. "An Integrative Platform of TCM Network Pharmacology and Its Application on a Herbal Formula, Qing-Luo-Yin." *Evidence-Based Complementary and Alternative Medicine* 2013 (April): e456747. doi:10.1155/2013/456747.

Zhang, Gui-Biao, Qing-Ya Li, Qi-Long Chen, and Shi-Bing Su. 2013. "Network Pharmacology: A New Approach for Chinese Herbal Medicine Research." *Evidence-Based Complementary and Alternative Medicine* 2013 (May): e621423. doi:10.1155/2013/621423.

Zhang, Li, Junke Song, Linglei Kong, Tianyi Yuan, Wan Li, Wen Zhang, Biyu Hou, et al. 2020. "The Strategies and Techniques of Drug Discovery from Natural Products." *Pharmacology & Therapeutics* 216 (December): 107686. doi:10.1016/j.pharmthera.2020.107686.

Zhang, Run-Zhi, Shao-Jun Yu, Hong Bai, and Kang Ning. 2017. "TCM-Mesh: The Database and Analytical System for Network Pharmacology Analysis for TCM Preparations." *Scientific Reports* 7 (1): 2821. doi:10.1038/s41598-017-03039-7.

Zhang, Runzhi, Xue Zhu, Hong Bai, and Kang Ning. 2019. "Network Pharmacology Databases for Traditional Chinese Medicine: Review and Assessment." *Frontiers in Pharmacology* 10 (February): 123. doi:10.3389/fphar.2019.00123.

Zhou, M. M., Ping Liu, Wei Jia, A. Zhao, M. Qui, Y. Qui, and M. Su. 2006. "Evaluation of Therapeutic Effects of TCM Based on Metabolic Variation." *Modernization of Traditional Chinese Medicine and Materia Materia-World Science and Technology* 8 (6): 113–19.

Zhu, Feng, Zhe Shi, Chu Qin, Lin Tao, Xin Liu, Feng Xu, Li Zhang, et al. 2012. "Therapeutic Target Database Update 2012: A Resource for Facilitating Target-Oriented Drug Discovery." *Nucleic Acids Research* 40 (D1): D1128–36. doi:10.1093/nar/gkr797.

Zhu, W., and L. M. Yao. 2011. "Potential Targets and Molecular Mechanisms of, 'Xiaochaihu Decoction' by Metadrug Software." *Shanghai Journal of Traditional Chinese Medicine* 45 (1): 79–82.

3 Drug Discovery Approaches Targeting Dengue NS2B/NS3 Protease

A Comprehensive Review on Recent Developments

Rajdip Misra[1], *Shubham Kundu*[1], *Anupam Maity*[1], *Uttam Pal*[1,*] *and Nakul C. Maiti*[1,**]
[1]CSIR – Indian Institute of Chemical Biology
*Correspondence to: uttam2707@gmail.com
**Correspondence to: ncmaiti@iicb.res.in

3.1 INTRODUCTION

Dengue, the arthropod-borne viral fever, is a neglected tropical disease. The disease is transmitted by female aedes mosquitoes (*Aedes aegypti and Aedes polynesiensis*) to humans (Cao-Lormeau 2009). The causative agent for this disease is the well-known dengue virus (DENV). The virus belongs to the family 'Flaviviridae' and 'Flavi-virus' genus (Rees et al. 2019; Raafat, Blacksell, and Maude 2019). Four different dengue viral serotypes or strains, namely, DENV1, DENV2, DENV3 and DENV4, were known until 2013, before the discovery of the latest variant of serotype 4, DENVT4 from Thailand (Normile 2013). Infection against a strain gives a lifetime immunity against that particular strain, but it also increases the risk of secondary infection when contracting other strains (Katzelnick et al. 2017). The intenseness of the disease varies from a normal 'dengue fever' to a more serious, life-threatening, severe condition known as dengue hemorrhagic fever (DHF) or dengue shock syndrome (DSS). The virus has a positive sense single-stranded RNA, is 11 kb long in size and translates and synthesizes a linear polyprotein chain (Figure 3.1). The proteins can be classified as structural (S) and non-structural (NS) proteins. Structural proteins help in the viral assembly and entry inside hosts, whereas the function of non-structural proteins is mainly to assist the viral replication and protein synthesis. There are three structural proteins, specifically, capsid (C), membrane (prM) and envelope (E). The seven non-structural proteins are NS1, NS2A, NS2B, NS3, NS4A, NS4B and NS5 (Henchal and Putnak 1990). The orientation of these proteins on the polyprotein chain is NH_2–C-PrM-E-NS1-NS2A-NS2B-NS3-NS4A-NS4B-NS5-COOH (Chambers et al. 1997). The function of NS1 is still not clear. NS2B plays a crucial cofactor role of non-structural protein 3 (NS3), the 'serine protease' of the virus. A 47-amino acid-long hydrophilic stretch of NS2B is necessary for the proper proteolytic activity of NS3 protease. NS4A functions as a host RNAi suppressor, and NS5, the largest multi-protein complex of the virus, functions as viral RNA-dependent RNA polymerase and methyl transferase (Rawlinson et al. 2006).

The disease causes more than 400 million infections (Chambers et al. 1997) and more than 15,000 deaths per year worldwide. Although there is a considerable infection rate and an alarming number of deaths yearly, progress on the prevention of the disease remains cryptic. Currently, it is considered no more than a simple 'tropical disease'. So far, proper treatment against the disease is unavailable. The only vaccine, Dengvaxia (CYT-TDV), was approved in 2015 (Alagarasu 2016). The

DOI: 10.1201/9781003302131-4

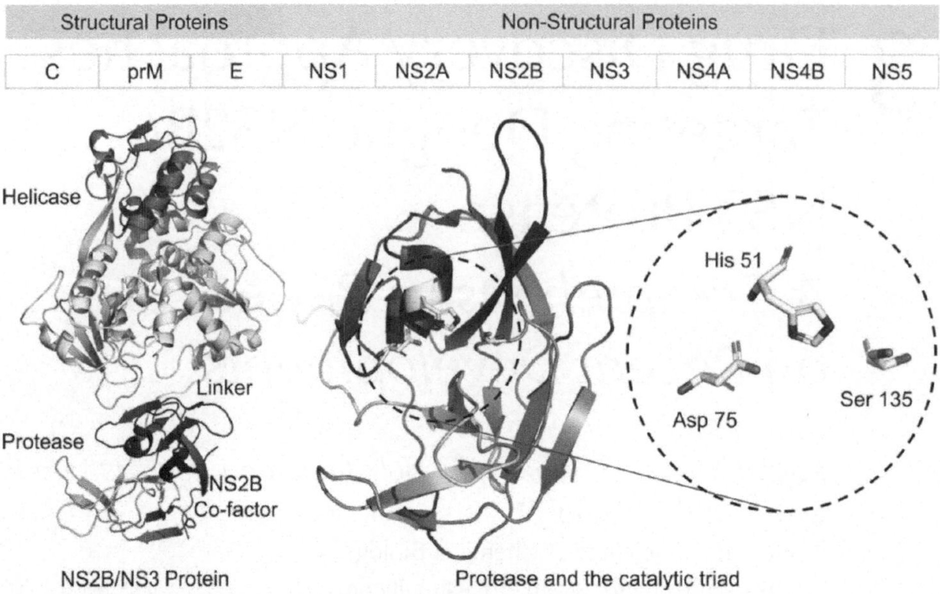

FIGURE 3.1 Schematic diagram of a dengue polyprotein chain and the crystal structure of NS2B-NS3 protease-helicase (PDB: 2VBC). The serine protease domain and the catalytic residues are shown separately.

Source: Reprinted from Uday et al. 2021 under a Creative Commons Attribution 4.0 International (CC BY 4.0) license.

vaccine was found to be safe and effective for those who already have had a natural dengue infection before vaccination. Nevertheless, the vaccine has its own drawbacks. According to the World Health Organization (WHO), the CYT-TDV vaccine increased the risk of developing a severe form of dengue for individuals who did not have a natural dengue infection previously. In these circumstances, small molecule-based drug research can be a better way to treat dengue infections. Not a single molecule has cleared the phase III human trial until now, and many are trapped in between phases I and II. In this chapter, we discuss the recent small molecule-based and advanced approaches attempted to inhibit the viral NS2B/NS3 protease and thus prevent viral transmission. A backdrop on the history of dengue viral disease is included, followed by the genome organization and life cycle of the virus and finally, the small molecule protease inhibitor-based antiviral approaches using both computational and *in vitro* advanced techniques to target the viral serine protease. In the conclusion, the potent inhibitors and future directions in inhibitor discovery are discussed.

3.2 HISTORY

The word 'dengue' originated from a Swahili word 'ki-dinga-pepo' meaning 'cramp-like pain'. The Swahili people believed that it is a curse given by an evil spirit (Christie 1872). A reference from an ancient Chinese medical book around 610 AD mentions this disease with the name 'shuidu', an ancient Chinese word meaning 'water poison' (Gubler 1998). Throughout history, it can be determined that there were many occasions of dengue epidemics. The 1779 Indonesia epidemic was one of them. The epidemic originated from Indonesia and spread to Cairo, Egypt. Another epidemic was observed during a similar time (1779–1780) in Philadelphia (Gubler 1998). There could be more cases of dengue epidemics; however, most of them are hardly reported.

DENV1 and DENV2, the first two identified dengue virus serotypes, were discovered by Dr. R. Kimura and Dr. Hatta in 1943 and 1945, respectively (Konishi and Kuno 2013). Within the next two decades of this revolutionary discovery, two other new strains, DENV3 and DENV4, were identified

(1953) in the Philippines and Thailand, respectively (Messina et al. 2014). The most recent dengue strain, known as the dengue Thailand variant or DNV4T, was recently identified in 2013 (Normile 2013).

3.3 GENOMIC ORGANIZATION AND LIFE CYCLE OF DENV

To develop highly potent vaccines and drugs against dengue virus, understanding its genomic organization/constructions and life cycle is a prerequisite. As previously discussed, dengue virus (DENV) is the causative agent of dengue fever, which belongs to the '*Flavivirus*' genus under the family of '*Flaviviridae*'. The 11kb-long RNA of the virus contains a single open reading frame (ORF). The ORF is surrounded by both 5' and 3' untranslated regions (UTR). The 3' end is non-polyadenylated. The RNA synthesizes a 3,400-amino acid-long polyprotein chain from a single ORF (Potisopon et al. 2014). Once the chain is translated, it is then processed through proteolytic cleavage by signal peptidase and host proteases to release the viral NS2B/NS3 serine protease, which then further cleaves the polyprotein into the functional proteins (Figure 3.1).

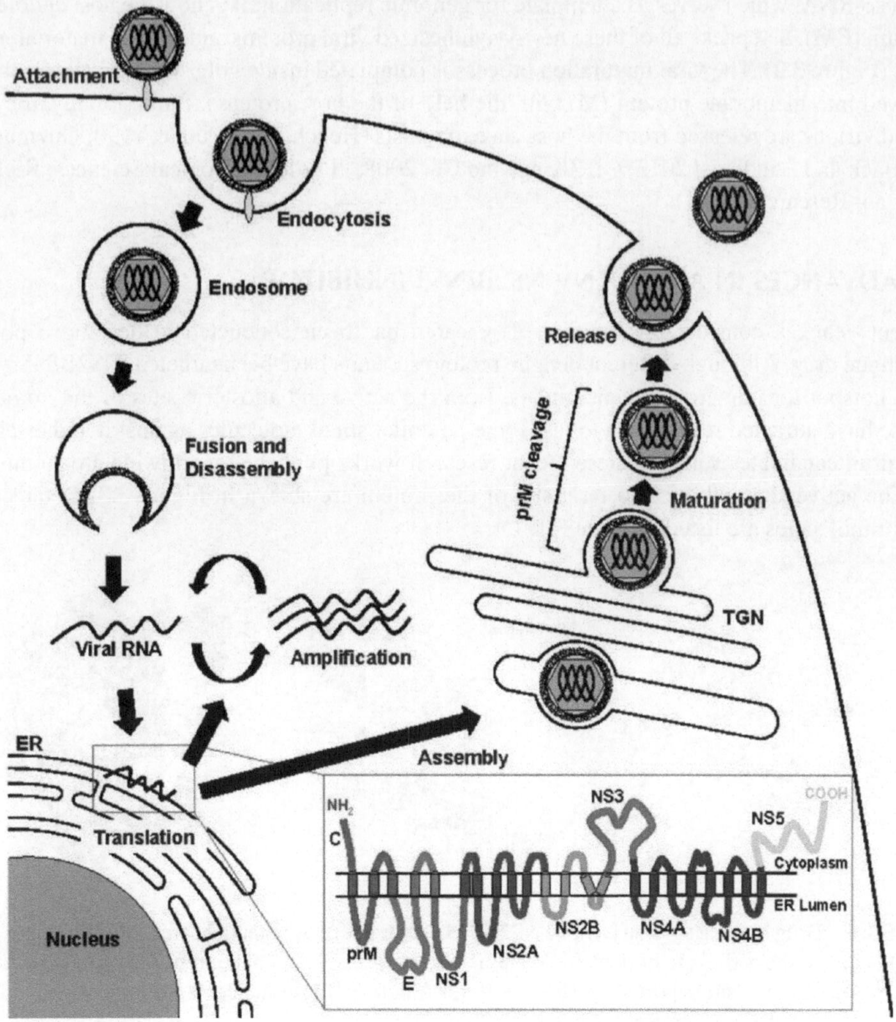

FIGURE 3.2 Life cycle of dengue virus revealing the potential stages for antiviral therapy development.

Source: Reproduced from (Kato and Hishiki 2016) under a Creative Commons Attribution (CC BY) license.

The dengue virus is transmitted with the bite of a female aedes mosquito harboring the virus. Platelets, a type of blood cell present in the host, function as the primary targets of the virus. The life cycle of the virus contains some major events: viral entry, fusion and disassembly of viral particles, genome (RNA) replication, translation of viral RNA, assembly of viral particles, maturation, exocytosis and release of the virus (Figure 3.2). The first stage of the viral entry into the host cell is the attachment of viral envelope proteins with its host cell surface receptor. A proper receptor ligand interaction triggers the process of host receptor-mediated endocytosis. However, instead of a routine degradation of viral proteins inside the endosome, fusion occurs. The low pH inside the endosome activates the conformational change in viral envelope (E) proteins, which, in turn, allow E proteins to be fused with the endosomal membrane and to release the encoded viral genetic material into the host cytosol. Inside the cytosol, the uncoating of the viral core particle takes place, which results in the release of viral genetic material and proteins. The positive-sense-RNA, the genetic material of the virus, translates into about a 3,400-amino acid-long polyprotein with the help of host ribosomal machinery. Simultaneously, the positive-sense-RNA is converted into an intermediate negative-sense-RNA, which serves as a template for genomic replication. It is now the host endoplasmic reticulum (EM), that packs all of these newly synthesized viral proteins and genetic material into the virions (Figure 3.2). The viral maturation process is completed inside golgi vesicles, where the prM is cleaved into membrane protein (M) with the help of the host protease, funin. Finally, the newly matured virions are released from the host via exocytosis (Henchal and Putnak 1990; Guzman et al. 2010; Bäck and Lundkvist 2013; Qi, Zhang, and Chi 2008; "Purdue Biological Sciences: Richard J. Kuhn Lab: Research" 2021).

3.4 ADVANCES IN ANTI-DENV NS2B/NS3 INHIBITORS

In recent years, a considerable amount of research has been conducted to develop a potential anti-dengue drug. Although different dengue protein subunits have been targeted, NS2B/NS3 seems to be a hotspot for numerous recent studies. Both the active and allosteric sites of the viral serine protease have attracted researchers to test some potential small molecules against it and establish a strong drug candidate. Some of these recent research works published worldwide are summarized here. The active site and the allosteric site of the protein are shown in Figure 3.3. Details of the inhibitor molecules are listed in Table 3.1.

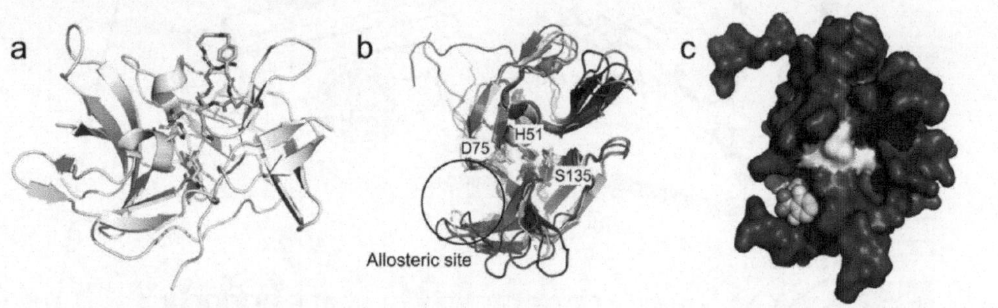

FIGURE 3.3 Drug binding sites on Dengue NS2B-NS3 protease. (a) A potential competitive inhibitor (cyclic peptide) bound to the active site of the DENV NS2B-NS3 proteins' homologue from Zika virus (Braun et al. 2020). (b) An allosteric site reported by Uday et al. (c) A potential lead molecule (dabrafenib) bound to the allosteric site of the DENV3 protease (Uday et al. 2021).

TABLE 3.1
List of Reported Potential Inhibitors of Dengue NS2B/NS3 Protease

Name of compound	Chemical structure	Interacting region on NS2B/NS3	Nature of compound	Reference
Rutin	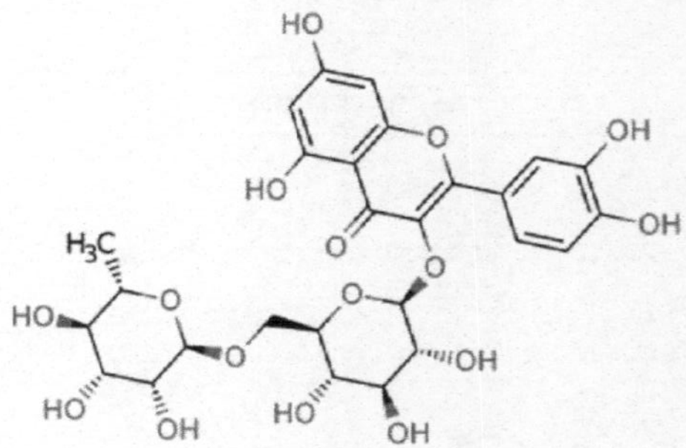	Active site	Plant flavonoid	Dwivedi et al. 2021
Kaempferol-3-o-rutinoside	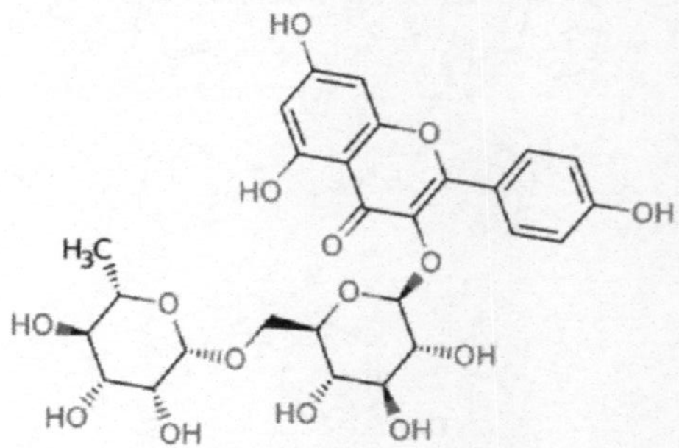	Active site	Plant flavonoid	Dwivedi et al. 2021
α-Mangostin	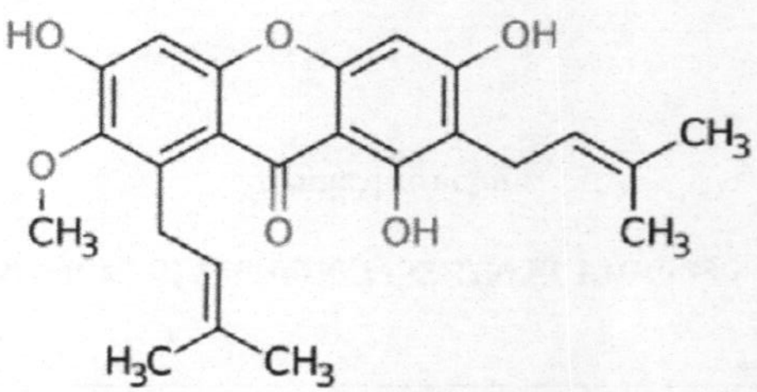	Active site	Phytochemical	Panda et al. 2021

(Continued)

TABLE 3.1 (*Continued*)
List of Reported Potential Inhibitors of Dengue NS2B/NS3 Protease

Name of compound	Chemical structure	Interacting region on NS2B/NS3	Nature of compound	Reference
Tolcapone	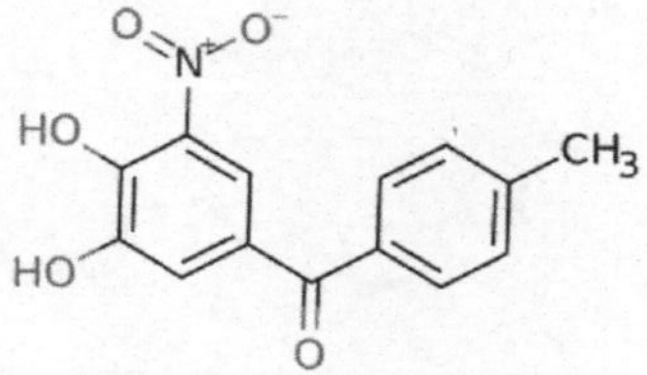	Active site	Repurposed drug	Balasubramanian et al. 2016
4-[(E)-{[2-(4-methoxyphenyl) ethyl]imino}methyl]phenol	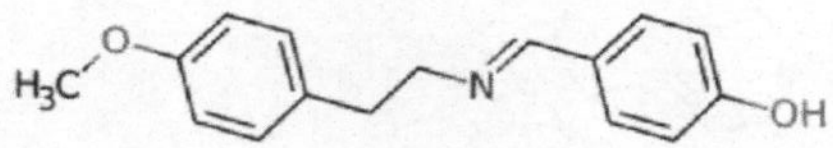	Active site	Synthetic imine compounds	Frimayanti et al. 2017
2-[(E)-{[2-(4-methoxyphenyl) ethyl]imino}methyl] naphthalen-1-ol	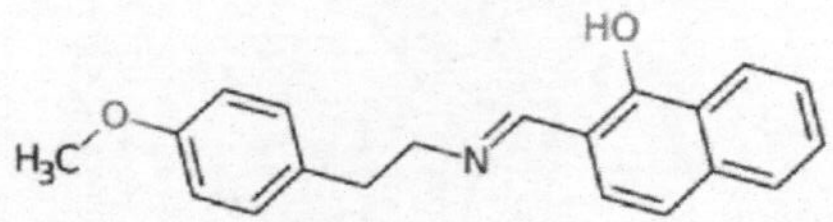	Active site	Synthetic imine compounds	Frimayanti et al. 2017
Circumin	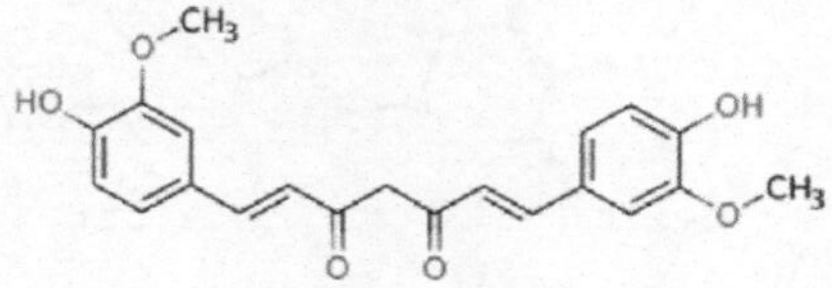	Allosteric site	Phytochemical	Lim et al. 2020
Dabrafenib	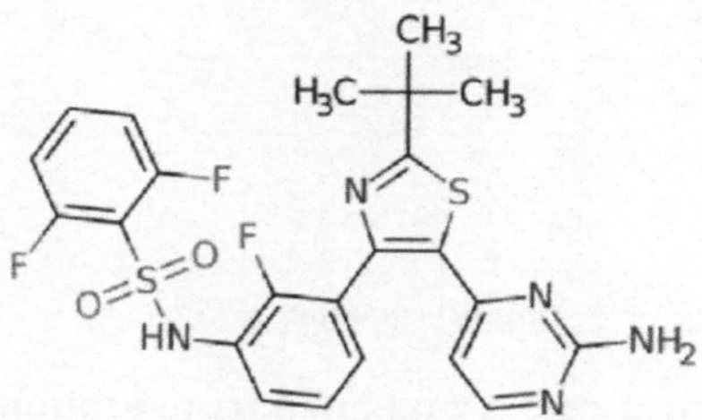	Allosteric site	Repurposed drug	Uday et al. 2021

Idelalisib	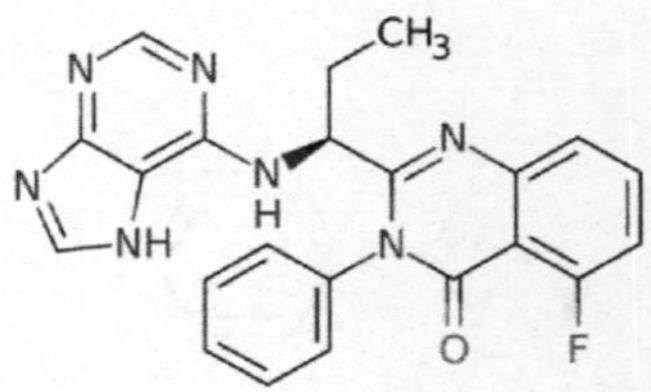	Allosteric site	Repurposed drug	Uday et al. 2021
Nintedanib	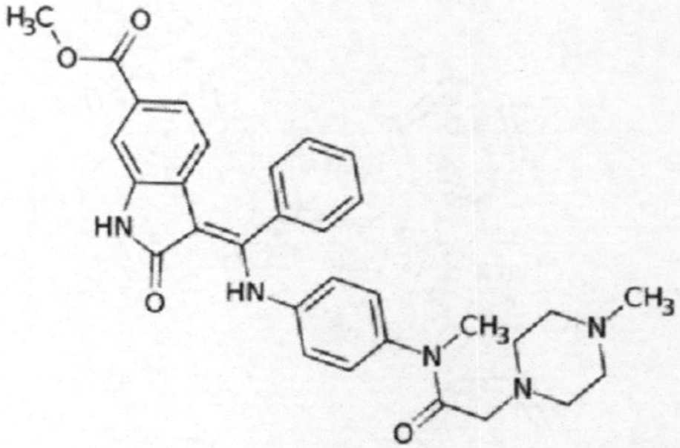	Allosteric site	Repurposed drug	Uday et al. 2021
(2S)-1-(benzenesulfonyl)-N-(5,6-dihydroxy-1,3-benzothiazol-2-yl) pyrrolidine-2-carboxamide	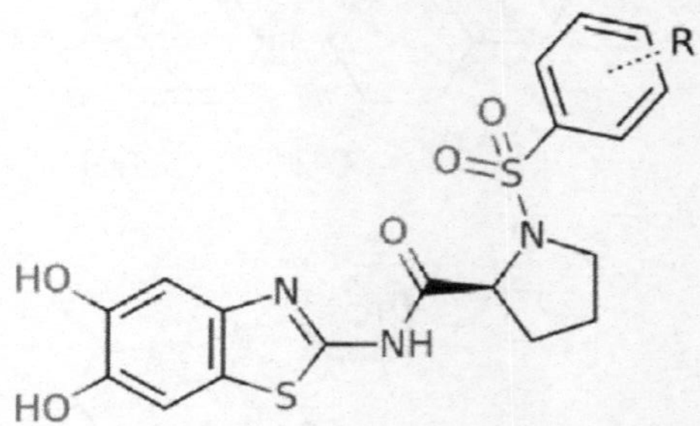	Allosteric site	Synthetic proline derivative	Millies et al. 2019

(Continued)

TABLE 3.1 (*Continued*)
List of Reported Potential Inhibitors of Dengue NS2B/NS3 Protease

Name of compound	Chemical structure	Interacting region on NS2B/NS3	Nature of compound	Reference
Ganodermanotriol	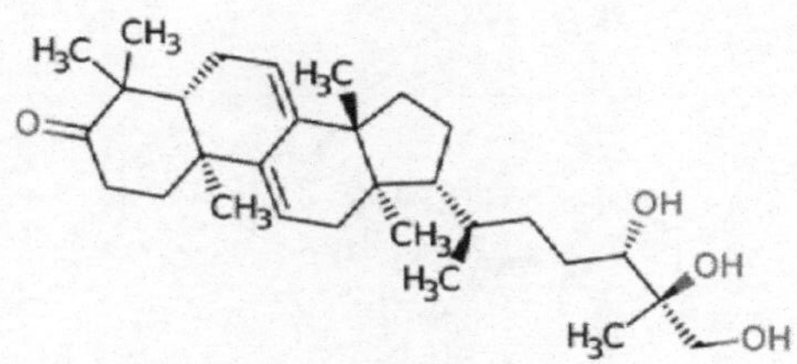	Catalytic triads	Bioactive fungal compound	Bharadwaj et al. 2019
Epigallocatechin	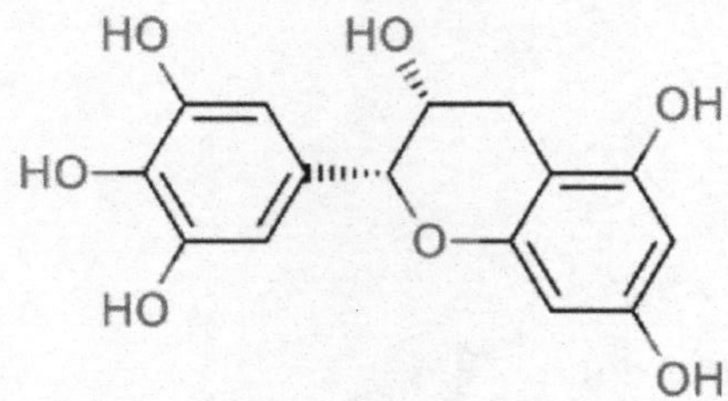	Catalytic triads	Phytochemical	Farooq et al. 2020
Catechin	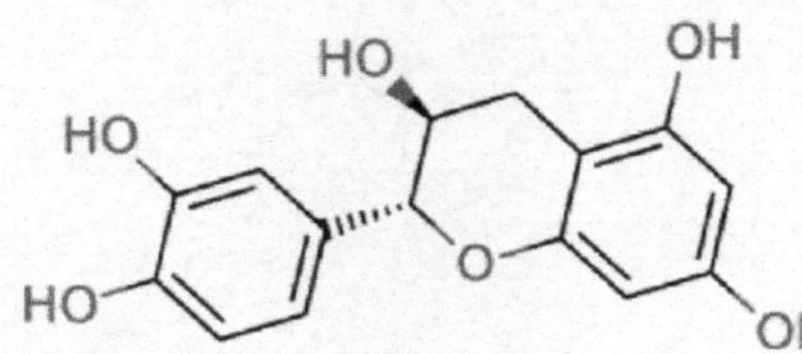	Catalytic triads	Phytochemical	Farooq et al. 2020
Protocatechuric acid	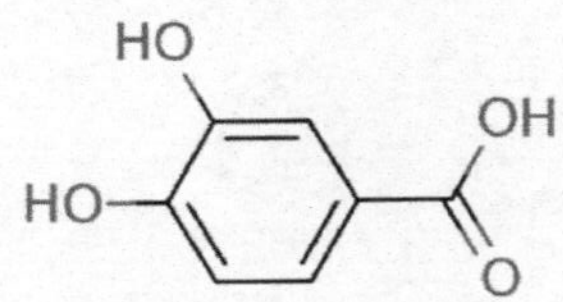	Catalytic triads	Phytochemical	Farooq et al. 2020

Uncinanone B	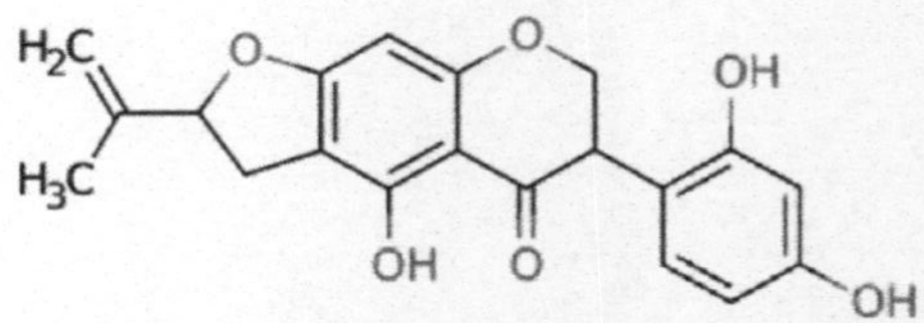	Catalytic triads	Plant flavonoid	Qamar et al. 2017
5-hydroxybowdichione	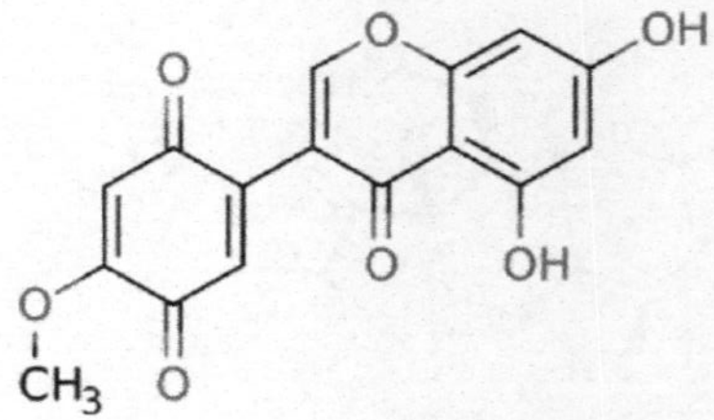	Catalytic triads	Plant flavonoid	Qamar et al. 2017
Prunetin	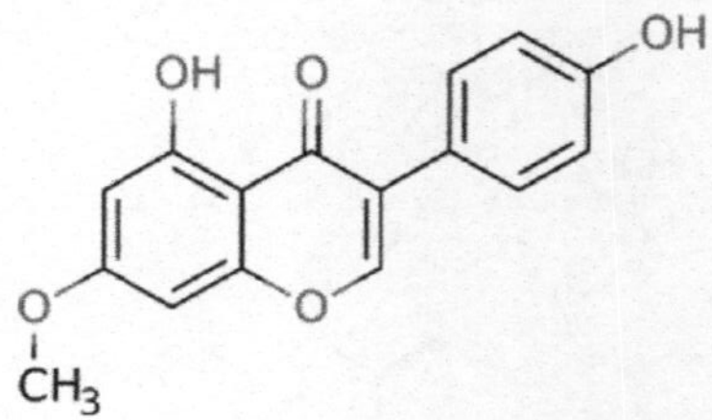	Catalytic triads	Plant flavonoid	Qamar et al. 2017
5,7,3',4'-tetrahydroxyisoflavone	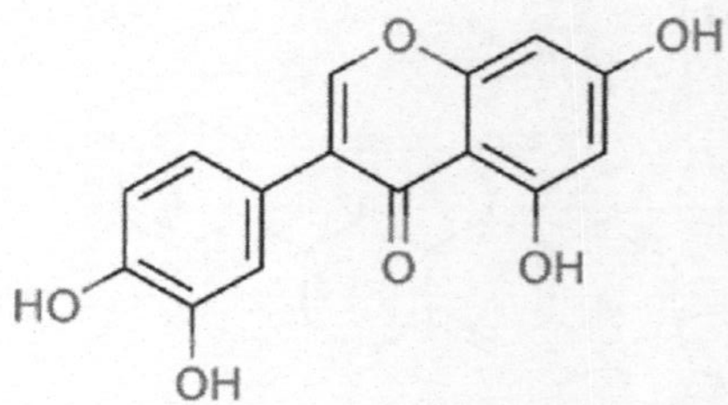	Catalytic triads	Plant flavonoid	Qamar et al. 2017

(Continued)

TABLE 3.1 (*Continued*)
List of Reported Potential Inhibitors of Dengue NS2B/NS3 Protease

Name of compound	Chemical structure	Interacting region on NS2B/NS3	Nature of compound	Reference
Alpinumisoflavone	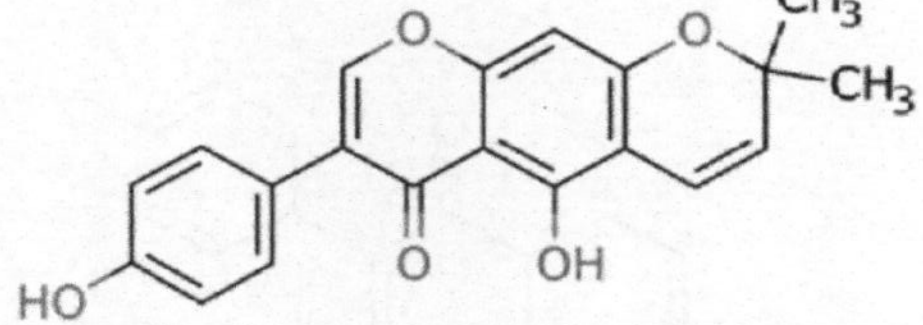	Catalytic triads	Plant flavonoid	Qamar et al. 2017
Glicoisoflavanone	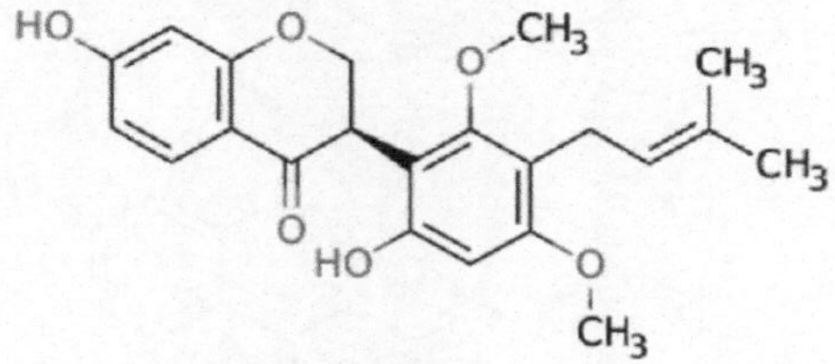	Catalytic triads	Plant flavonoid	Qamar et al. 2017
Nimbin	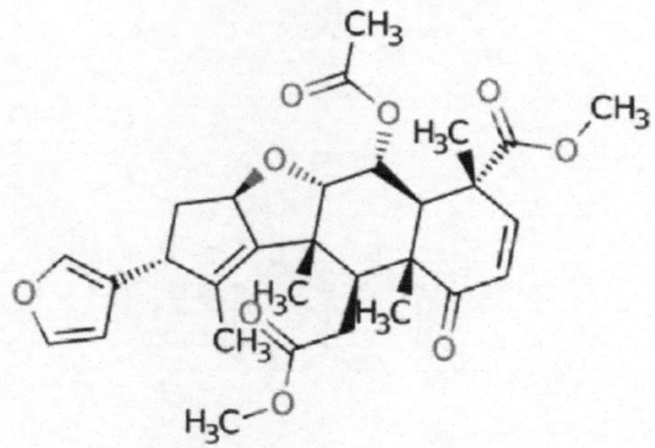	Catalytic triads	Plant triterpenoids	Dwivedi et al. 2016

Desacetylnimbin	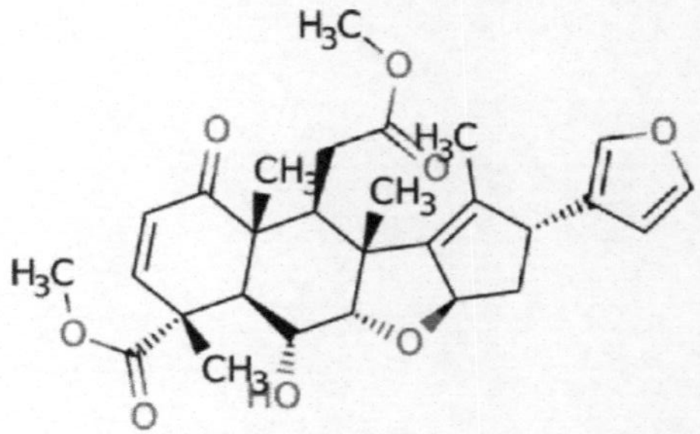	Catalytic triads	Plant triterpenoids	Dwivedi et al. 2016
Desacetylsalannin	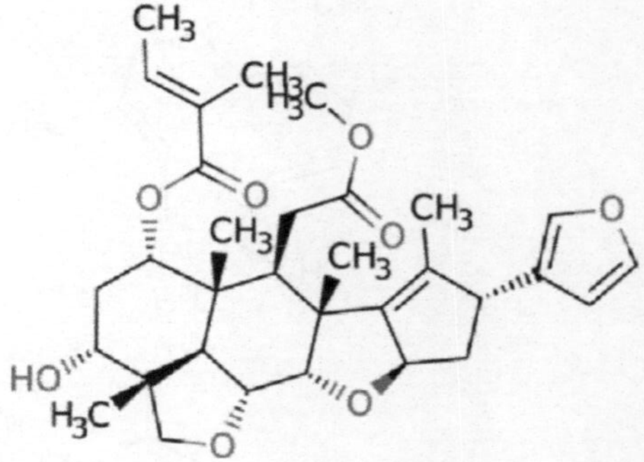	Catalytic triads	Plant triterpenoids	Dwivedi et al. 2016
1-(3-{[2-(1H-1,3-benzodiazol-2-yl)hydrazin-1-yl]methyl}-4-hydroxy-5-[(E)-[2-(3H-indol-2-yl)hydrazin-1-ylidene]methyl]phenyl)butane-1,2-dione	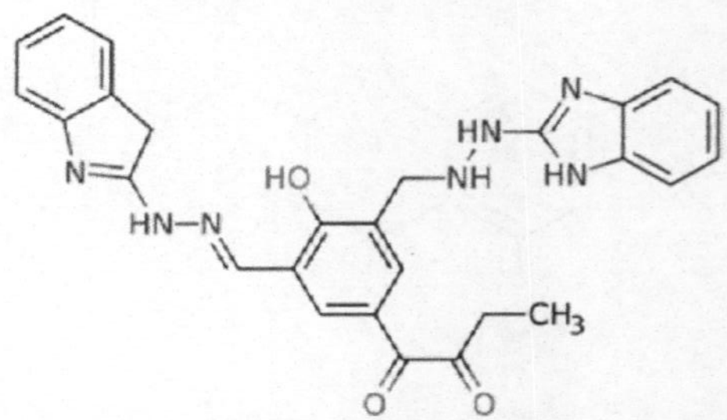	Catalytic triads	Synthetic	Balajee et al. 2016

(Continued)

TABLE 3.1 (*Continued*)
List of Reported Potential Inhibitors of Dengue NS2B/NS3 Protease

Name of compound	Chemical structure	Interacting region on NS2B/NS3	Nature of compound	Reference
Hesperetin	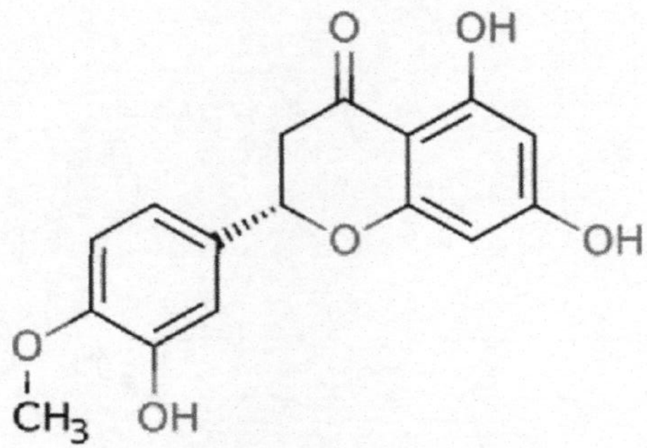	Catalytic triads	Plant flavonoid	Lim et al. 2020
α-caryophyllene	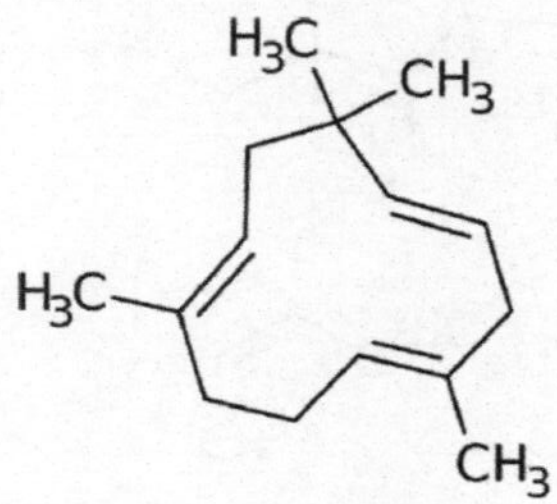	His51 of active site	Phytochemical	Kharisma et al. 2021

δ-Selinene	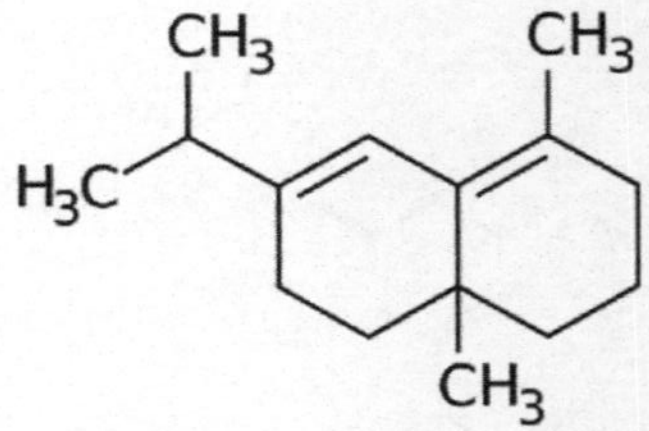	His51 of active site	Phytochemical	Kharisma et al. 2021
4,5-Dinitrochrysazin	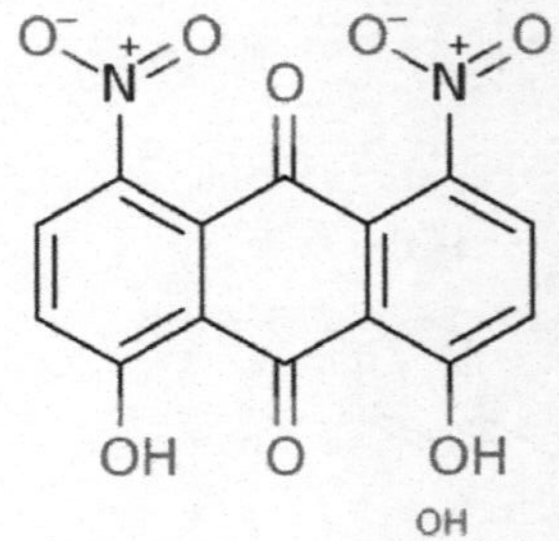	NS3-NS4 junction	Chemical compound	Constant et al. 2018
Quercetin 3-O-(2″,3″-digalloyl)-β-D-galactopyranoside	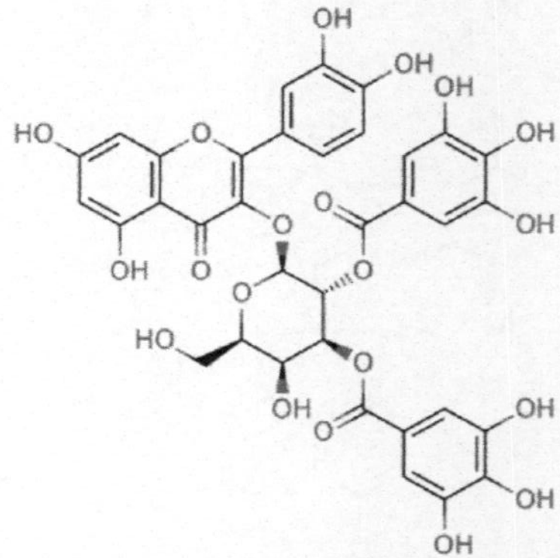	Unspecified	Bioactive flavonoid	Sarwar et al. 2018

(Continued)

TABLE 3.1 (*Continued*)
List of Reported Potential Inhibitors of Dengue NS2B/NS3 Protease

Name of compound	Chemical structure	Interacting region on NS2B/NS3	Nature of compound	Reference
Quercetin 3-O-α-(6"'-caffeoylglucosyl-β-1,2-rhamnoside)	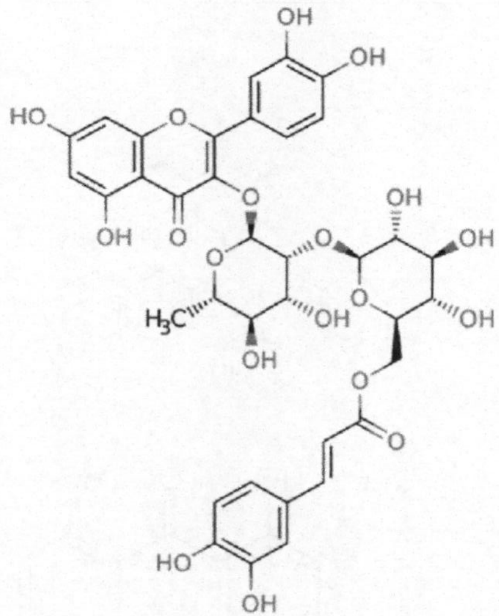	Unspecified	Bioactive flavonoid	Sarwar et al. 2018
Schaftoside	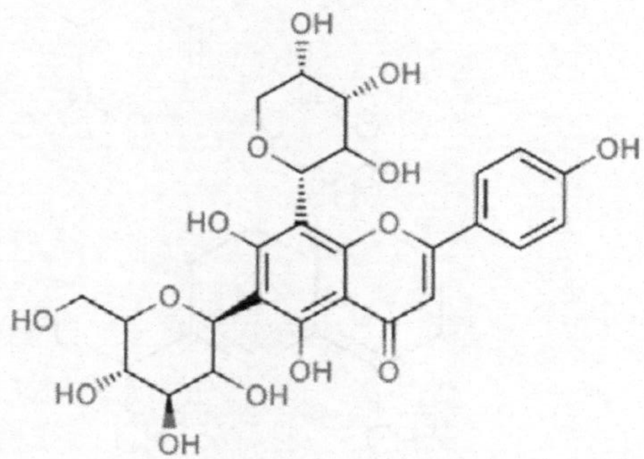	Unspecified	Bioactive Flavonoid	Sarwar et al. 2018

Myricetin	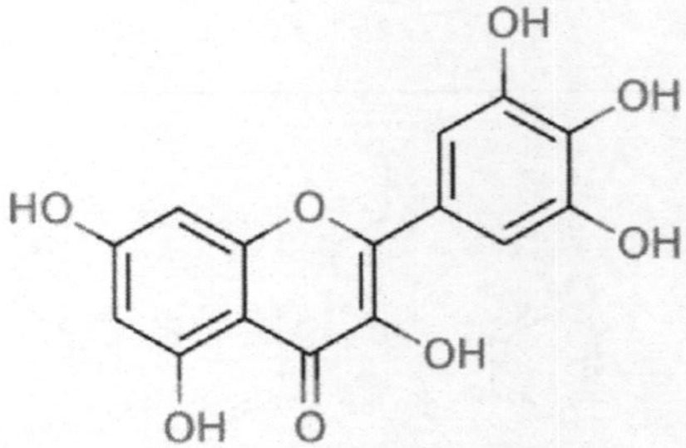	Unspecified	Bioactive flavonoid	Sarwar et al. 2018
5,7-Dimethoxycoumanrins	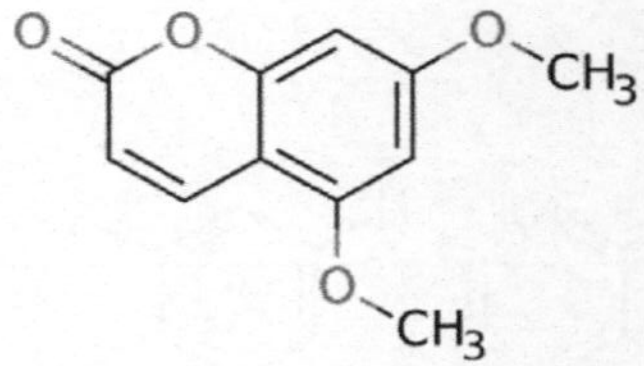	Unspecified	Phytochemical	Rathnayake et al. 2020
p-Coumaric acid	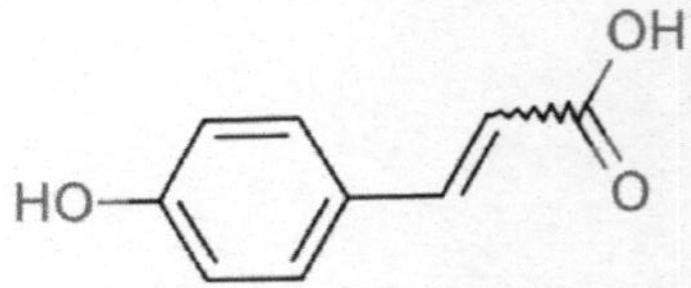	Unspecified	Phytochemical	Rathnayake et al. 2020
Gallic acid	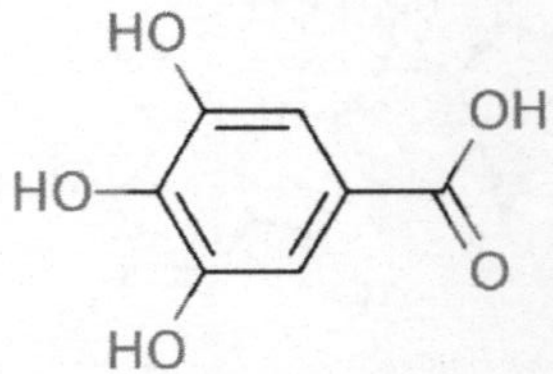	Unspecified	Phytochemical	Nefsu 2020

(Continued)

TABLE 3.1 (*Continued*)
List of Reported Potential Inhibitors of Dengue NS2B/NS3 Protease

Name of compound	Chemical structure	Interacting region on NS2B/NS3	Nature of compound	Reference
Astragaloside III	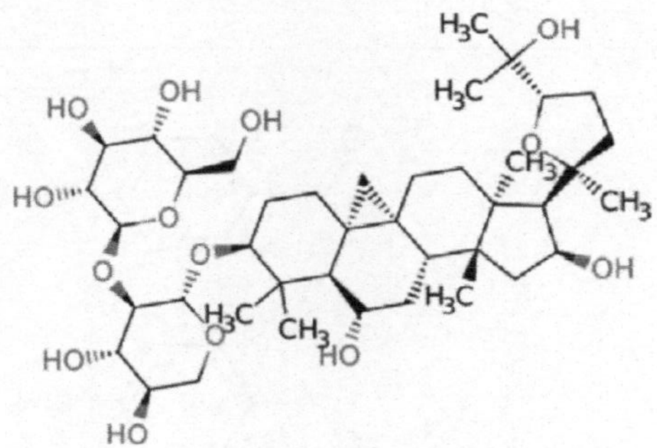	Unspecified	Phytochemical	Indu et al. 2020
N-[5-nitro-6-(piperazin-1-yl) pyridin-2-yl]-1H-indazol-5-amine	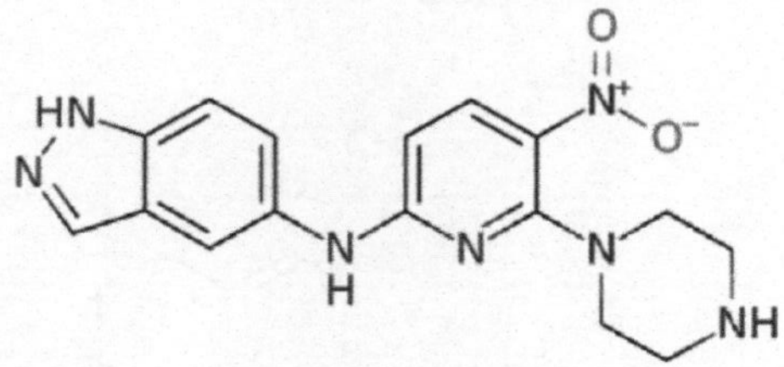	Unspecified	Synthetic	Shin et al. 2021
(3E,5E)-1-(3-chloropropyl)-3,5-bis[(4-methoxyphenyl) methylidene]piperidin-4-one	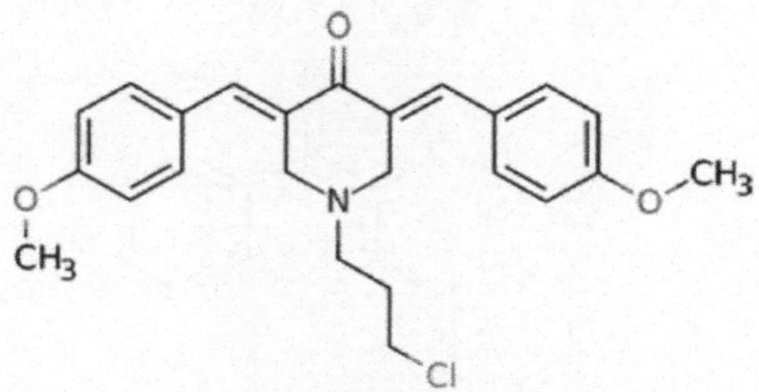	Unspecified	Synthetic	Habibi et al. 2020

3.4.1 Small Molecule NS2B/NS3 Inhibitors

In the fields of pharmacology and molecular biology, small molecules are organic compounds having a molecular weight of less than 900 daltons (Macielag 2012), which have a role in the regulation of different biological processes. They serve many functions starting from cell signaling and enzyme modulation to drug delivery. Polysaccharides, nucleic acids and proteins are relatively larger compounds and are not considered small molecules. However, the monomers dNTPs and rNTPs belong to the category of small molecules. Small molecules are primarily used as a research tool to investigate biological functions. Scientists are trying to develop different potential therapeutics based on small molecular research. In the following portion of the chapter, we summarize the current small molecule-based research on dengue virus NS2B/NS3 protease. The small molecules have been categorized into two broad sections according to their primary sources.

3.4.1.1 Natural Small Molecule Inhibitors of NS2B/NS3

Natural molecules are chemicals that can be obtained from different herbal, plant or organic origins. In ancient times, humans were deeply dependent on natural resources to treat different diseases. As the pharmacological industry grew, the focus shifted to synthetic drug molecules. Figure 3.4 presents some natural sources of small molecules that are found to be effective against DENV protease.

3.4.1.1.1 Plant-Derived NS2B/NS3 Inhibitors

Approximately 75% of the world's inhabitants use medicinal plants as a source of life-saving medicine. According to the International Union for Conservation of Nature (IUCN) Species Survival Commission Medicinal Plant 2007, there are more than 2.5 lakhs of plant species present worldwide, among which at least 80,000 species have medicinal virtue. Based on traditional knowledge, many countries such as India, China, etc. use these medicinal plants as a source of primary treatment for different diseases. Currently, it is known that these medicinal plants are rich in various phytochemicals (Calixto 2000). The range of these phytochemicals is very wide, and they consist of different flavonoids, peptides, limonoids, coumarins, terpenoids, furyl compounds, saponins, etc. They serve as a first line of defense for plants as a scavenging viral entry to plant cells (Doss et al. 2011; Idrees and Ashfaq 2013). Some potent plant-derived NS2B/NS3 inhibitors are discussed in the following.

For example, bioactive compounds present on sweet potato leaves can be effective antiviral agents. A recently published computation study by Kharisma et al. (Kharisma et al. 2021) showed the potency of δ-selinene and α-caryophyllene, two compounds present on sweet potato leaves, to be an ideal anti-dengue molecule (*vide* Table 3.1). The docking study suggested that these molecules interact with His 51, one of the three residues of the enzyme active site catalytic triad. Both molecules showed favorable binding-free energies upon docking in the study and can thus be potential drug leads. However, an *in vitro* experimental study such as an enzyme assay could have strengthened the research.

Dengue serine protease is a prime target in therapeutics. Molecular modeling is a clever approach that gives some accurate results on receptor ligand binding. Dwivedi, Tripathi and Mishra (2016) identified some triterpenoids from the 'Neem' tree (*Azadirachta indica*) to treat dengue fever (Figure 3.4). The neem tree is considered to have innumerable medicinal benefits according to Indian traditional medicine (Ayurveda). Among the selected triterpenoids, nimbin (-5.56 Kcal/mol), desacetylnimbin (-5.24 Kcal/mol) and desacetylsalannin (-3.43 Kcal/mol) showed good binding-free energies against the selected structure of dengue NS2B/NS3 protease (*vide* Table 3.1). All three molecules interacted with the catalytic triad of the enzyme and formed hydrogen bonds and hydrophobic interactions in computer simulations. These interactions seem to disrupt the electron transfer between the catalytic triad residues, resulting in the inhibition of enzyme activity. Continuing in this study, the authors further identified the anti-dengue roles of some other bioactive flavonoids present on neem tree leaves (Dwivedi et al. 2021). Out of the pre-identified 49 flavonoids molecules from neem tree leaves, some molecules such as epicatechin (-7.222 kcal/mol), hyperoside

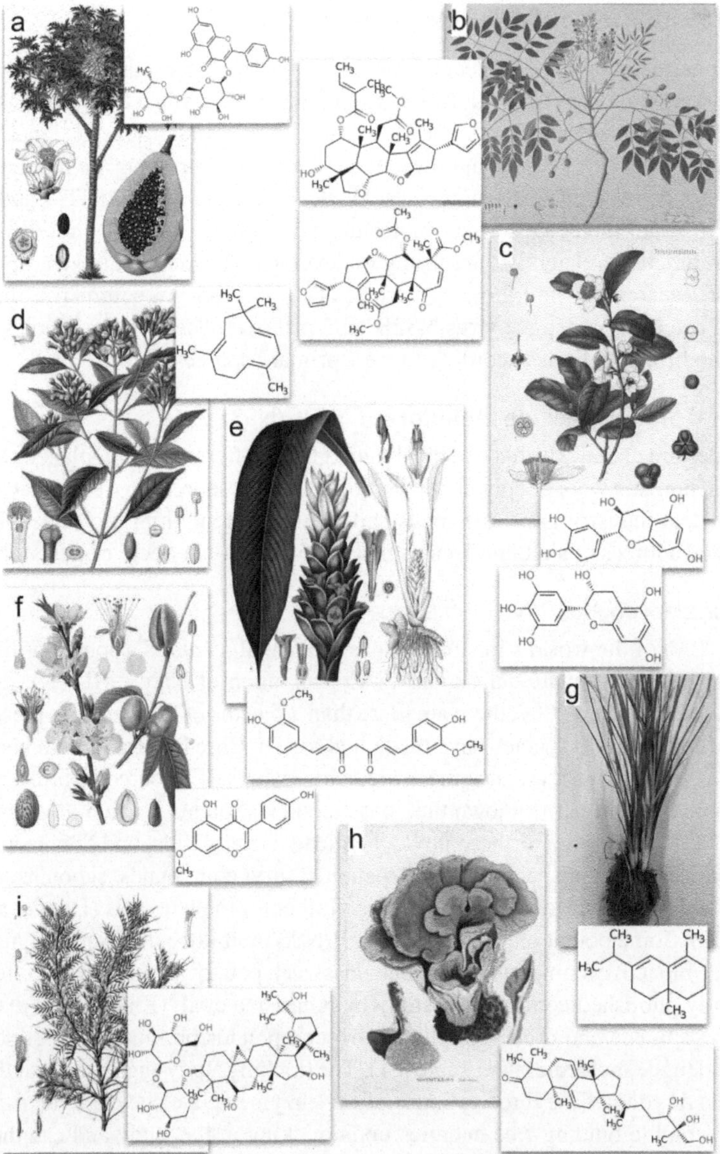

FIGURE 3.4 Some active compounds have reportedly showed significant antiprotease activity against dengue and their natural sources. (a) Papaya (*Carica papaya*) contains kaempferol. Plant illustration from Koehler's Medicinal-Plants (1887) is in the public domain. (b) Neem tree (*Azadirachta indica*) is a major source of nimbin and salanin. Plant illustration by William Jackson Hooker (1867) is in the public domain. (c) Tea leaves (*Camellia sinensis*) contain a large amount of catechin and epigallocatechin. Plant illustration from Koehler's Medicinal-Plants (1887) is in the public domain. (d) Clove (*Syzygium aromaticum*) is a source of the essential oil α-caryophyllene. Plant illustration from Koehler's Medicinal-Plants (1887) is in the public domain. (e) Turmeric (*Curcuma sp.*) is the major source of curcumin. Plant illustration from Koehler's Medicinal-Plants (1887) is in the public domain. (f) Prunus (*Prunus sp.*) is full of prunetin. Plant illustration from Koehler's Medicinal-Plants (1887) is in the public domain. (g) Vetiver grass (*Vetiveria zizanioides*) is another source of essential oils such as δ-Selinene. Plant image is reprinted from (Gebashe et al. 2020) under a Creative Commons Attribution 4.0 (CC BY 4.0) license. (h) Reishi mushroom (*Ganoderma lingzhi*) produces ganodermanontriol. Drawing of the Ganoderma mushroom in Synopsis methodica fungorum (1803) by the mycologist Christiaan Hendrik Persoon is in the public domain. (i) Milkvetch (*Astragalus sp.*) contains astragaloside. Plant illustration from Koehler's Medicinal-Plants (1887) is in the public domain.

(-7.879 kcal/mol), rutin (-9.324 kcal/mol) and Kaempferol-3-o-rutinoside (-9.555 kcal/mol) showed good binding interactions with the NS2B/NS3 protease and could be potential drug leads (*vide* Table 3.1). After further in vitro analysis, Kaempferol-3-o-rutinoside was identified to be a more promising lead than quercetin, a drug known for its antiviral properties.

α-Mangostin (*vide* Table 3.1), a xanthonoid, can inhibit dengue NS2B/NS3 protease and thus help reduce infection according to a recent in vitro study (Panda et al. 2021). α-Mangostin completely inhibited DENV with an 8 μM IC_{50} concentration under a co-treatment condition. Molecular docking of several natural products against NS2B/NS3 protease and NS5 methyltransferase followed by viral plaque assay, qRT-PCR and cell line-based immunofluorescence assays were carried out to determine the inhibitory potential of the screened molecules. α-Mangostin showed promising results as a future anti-dengue drug.

Bioactive compounds present in *Carica papaya* (Figure 3.4) are frequently used in pharmacology. The leaf extract of papaya trees is often used to treat jaundice. A group of researchers recently identified the antiprotease roles of some selected bioactive compounds from *C. papaya*. Farooq et al. performed a molecular docking-based computational screening of 900 selected compounds of *C. papaya* and observed that among them, nine interacted with DENV NS2B/NS3 and NS5 methyltransferase. Three among these selected nine compounds, specifically, epigallocatchin, catechin and protocatechuric acid, bound strongly to the catalytic triad residues of the serine protease (Farooq et al. 2020). A similar effect of *C. papaya* leaves was also reported by Ratnayake et al. (Rathnayake et al. 2020). Their *in silico* analysis further divulges two more potential bioactive anti-dengue compounds from *C. papaya* leaf extracts: 5,7-Dimethoxycoumanrins and p-Coumaric acid (*vide* Table 3.1). These compounds can therefore be useful leads against dengue protease. Nevertheless, experimental studies are required to validate the findings.

Curcumin, a bright yellow-colored chemical molecule present on the *Curcuma longa* species (Figure 3.4), has been shown to be a potent allosteric inhibitor of dengue serine protease. According to recently published research by Lim et al. (L. Lim et al. 2020), curcumin directs the unfolding process of the C-terminally located β-hairpin of the NS2B cofactor. The β-hairpin of NS2B is essential because it helps to keep the NS2B/NS3 serine protease in close (active) configuration. Interacting with the hairpin results in the formation of an open configuration and a consequent inactivation of the enzyme (*vide* Table 3.1).

Momordica charantia, a vegetable commonly known as bitter melon, is an example of a medicinal herb. Previously, the medicinal role of *M. charantia* was published by several research articles. It has been shown to have both anti-diabetic and anti-tumor activities. The fruits of the plant contain several bioactive compounds such as alkaloids, fixed oils, glycosides, triterpenes, steroids, etc. Recent research (Nesfu 2020) highlighted the antiviral activities of *M. charantia* fruits. This study identified the importance of the ethyl acetate extract of bitter melon fruit flesh against dengue NS3 protease. A GC-MS analysis suggested the presence of gallic acid as the main component of the fruit (*vide* Table 3.1).

One more group of inhibitors that can potentially be effective against DENV is a series of astragaloside compounds. These compounds are derived from *Astragalus sp.* (commonly known as milkvetch; *vide* Figure 3.4), an herb traditionally known for its anti-aging, anti-diabetic and anti-heart disease roles in traditional Chinese medicine. To search for the active phytochemicals present in the herb, a group of researchers (Indu et al. 2021) virtually screened all the available compounds against different dengue protein targets (NS2B/NS3 protease, NS5 and envelope). Three compounds showed promising binding energies against different targets. Astragaloside III showed a binding free energy of -8.447 kcal/mol against NS2B/NS3 protease. The other two compounds were effective against the envelope and RNA cap of the virus. Vero cells were used to measure cytotoxicity in viral assays. The results of such assays showed that the compounds were non-toxic and non-mutagenic. Astragaloside III may therefore be considered a leading molecule in the anti-dengue drug discovery pipeline (*vide* Table 3.1).

The relationship of the chemical structure of a bioactive molecule with its biological activity is termed the structure activity relationship (SAR). SAR analysis allows the interpretation of different chemical groups responsible for inducing a well-targeted biological effect in living organisms. These SAR and quantitative SAR (QSAR) studies play a vital role in the search for a new generation of serine protease inhibitors. A molecular docking-based virtual screening study (Sarwar et al. 2018) was run to check the dengue NS2b/NS3 inhibitory potential of some plant-derived flavonoids. Out of 100 potential molecules, the 10 best-docked molecules were chosen. A correlation coefficient (r) of 0.95 in post docking QSAR and SAR studies suggested that there could be a positive relation between these flavonoids structures and selected properties. Based on their results, it is possible that molecules such as quercetin -3-O-(2″,3″-digalloyl)-β-D-galactopyranoside, quercetin -3-O-α-(6‴-caffeoylglucosyl-β-1,2-rhamnoside), schaftoside and myricetin could show promising results as effective dengue protease inhibitors (*vide* Table 3.1).

Plant-derived coumarins are secondary metabolites synthesized via phenylpropanoid pathways. The compound plays an effective role in the iron uptake of plants. In soil with an iron deficit, this compound is secreted by plant root cells and acts as iron mobilizing compounds. Apart from their role in iron regulation in plants, scientists have identified some derivatives of coumarins that may function as both antiviral and antibiotic agents. Yusufzai et al. (2018) has synthesized some conjugated compounds such as SKYf ((Z)-2-((E)-(1-(8-Methoxy-2H-chromen-3-yl)ethylidene)hydrazono)thiazolidin-4-one), SKYd ((Z)-2-((E)-(1-(6-Methoxy-2H-chromen-3-yl)ethylidene)hydrazono)thizolidin-4-one), SKYc ((Z)-2-((E)-(1-(7-Hydroxy-2H-chromen-3-yl)ethylidene)hydrazono)thiazolidin-4-one), etc., which exhibited enhanced inhibitory effects towards DENV NS2B/NS3 protease in computer simulations. The 4-thiazolidinone inhibitors containing the core structure of coumarins were produced via facile synthesis. Favorable binding-free energies against viral serine protease were observed in these compounds; SKYf (-4.014 Kcal/mol) showed the highest inhibition score followed by SKYd (-3.964Kcal/mol) and SKYc (-3.905 Kcal/mol).

Qamar et al. (Tahir Ul Qamar et al. 2017) identified some important flavonoids from medicinal plants. Their *in silico* analysis revealed that six molecules, namely, Uncinanone B, 5-hydroxybowdichione, Prunectin, 5,7,3',4'-tetrahydroxyisoflavone, Alpinumisoflavone and Glicoisoflavanone, strongly bind to the catalytic triads of DENV NS2B/NS3 protease (*vide* Table 3.1). In a summary of their study, they virtually screened 2,500 available plant flavonoids against the catalytic triad of the protease. These six molecules showed strong hydrophobic interaction with the catalytic triad and also passed Lipinski's rule of five (Lipinski et al. 2001). They did not show any adverse score on the computational ADMET profile, which makes them likely drug leads against the virus.

The gramineous, evergreen and perennial herb, *Vetiveria zizanoides*, commonly known as vetiver grass (Figure 3.4), is recognized for its commercial value. The essential oils produced from this grass are routinely used as an ingredient and fragrance in aromatic products and cosmetics (Chou, Shih, and Lin 2016). Vetiver also has some potential medicinal value. In 2016, a computational study by Lavanya and his co-workers (Lavanya, Ramaiah, and Anbarasu 2016) revealed the antiviral properties of this herb. The study identified the compound Ethyl -4(4-methylphenyl)-4-pentenoate as a potent dengue NS2B/NS3 protease inhibitor. The bioactive compound from the plant showed favorable binding energy against all four serotypes of dengue virus.

3.4.1.1.2 Fungi-Derived NS2B/NS3 Inhibitors

Lacquered bracket fungus (Figure 3.4) is unique among other species of fungus as it produces some bio-active flavonoids that can potentially ameliorate dengue NS2B/NS3 protease. A group of scientists (Lim et al. 2020) detected some bioactive compounds from Lacquered bracket finger extracts (*Ganoderma lucidum*). Six molecules were confirmed by liquid chromatography-mass spectrometry (LC-MS) in their study. A further docking analysis showed that each molecule interacted with the three residues of catalytic triads and the oxyanion hole. The major interactions involved in the process were hydrogen bonding, hydrophobic interactions and van der waal forces. Among the six molecules, hesperetin seems to be a step ahead in being a suitable anti-dengue inhibitor lead.

Recent *in silico* anti-dengue research published by Bharadwaj et al. (2019) described a virtual screening- and molecular dynamics-based study and found some biologically active compounds from *Ganoderma lucidum*, a medically recognized fungus, to be useful against dengue NS2B/NS3 protease. Initially, they identified four potential inhibitors, specifically, Ganodermanontriol (-6.291 kcal/mol), Lucidumol (-5.993 kcal/mol), Ganosporeric acid A (-5.983 kcal/mol) and Ganoderic acid C2 (-5.948 kcal/mol). The molecules interacted with the catalytic triads of the enzyme. After detailed binding assays and molecular dynamics simulation studies against the viral protease, Ganodermanotriol performed exceptionally well and can thus be a potent drug lead (*vide* Table 3.1).

3.4.1.2 Chemically Modified/Synthetic NS2B/NS3 Inhibitors

A recent study by Shin et al. (2021) showed a classic example of a synthetic inhibitor. The authors used virtual screening within a large number of available chemical compounds against dengue NS2B/NS3 protease. 123 compounds showed significant inhibitory activities against viral protease. After a detailed fluorescence-based protease assay, one lead compound was identified, specifically, N-[5-nitro-6-(piperazin-1-yl)pyridin-2-yl]-1H-indazol-5-amine (*vide* Table 3.1), which could be a promising anti-dengue drug.

Phytoconstituents are primarily responsible for protecting plants against various pathogens. These antimicrobial chemicals are non-nutrients in nature. Apart from their importance in plants' immune response, they are also responsible for aroma, color and other organic properties (Alamgir 2018). Ganji et al. (2021) designed some scaffold molecules modified from the phytoconstituents, mainly the analog compounds of chalcone and caffeic acid. *In silico* docking study and protease inhibition assay showed that one caffeic acid derivative (3-(2-chlorophenyl) prop-2-enoic acid) was able to bind the catalytic triad of dengue NS2B/NS3 and subsequently inhibited its function. The role of curcumin in anti-dengue therapeutics has already been discussed earlier. However, a major drawback is its poor stability. The β-diketone group, which is largely present in curcumin, is vulnerable to physiological pH. Habibi et al. (2020) therefore replaced the β-diketone group with mono kitone and observed that the stability of the synthetic compound increased dramatically. 1-(3-Chloropropyl)-3,5-Bis((*E*)-4-Methoxybenzylidene) Piperidin-4-One, a mono kitone circumin derivative (vide Table 3.1), was synthesized from piperidone via two steps, which showed lower binding energy in molecular docking (-6.40 Kcal/mol) against the serine protease of the virus and is thus promising as an anti-dengue agent.

4-benzyloxyphenylglycine derivatives can inhibit DENV NS2B/NS3 at micro molar concentrations. A combination of different cell-based assays, molecular modeling, cytotoxicity tests and viral titer assays confirmed that these newly synthesized small molecule DENV inhibitors were very much selective against the viral trypsin-like serine protease and relatively less toxic to cell lines (Kühl et al. 2020). A group of scientists led by Millies et al. (2019) synthesized some proline-based small molecules and evaluated their role as anti-dengue agents. When docked against the NS3 protease, they found that these molecules targeted the allosteric site of the enzyme. In the conclusion of their study, they identified some novel inhibitor scaffolds that can effectively bind and increase allosteric inhibition both in *in vitro* and in cellular models.

Combined approaches seem to be a useful tool. The host furin plays an important role along with the viral protease in the maturation of dengue virus. Kouretova et al. (2017) developed a combined anti-NS2B/NS3 anti-furin inhibition system. When the protease was targeted with selected potential antiviral molecules, viral titers were reduced up to 10,000 folds.

Frimayanti et al. (2017) demonstrated a docking study against DENV2 NS2B/NS3 protease and observed the inhibitory effect of some imine derivatives. The authors synthesized different imine derivatives from aldehyde compounds. They tested the effectiveness of 10 different imine derivatives against dengue NS2B/NS3, and 2-[(E)-{[2-(4-methoxyphenyl)ethyl]imino}methyl]naphthalen-1-ol and 4-[(E)-{[2-(4-methoxyphenyl)ethyl]imino}methyl]phenol were found to be the most productive inhibitors according to their research (vide Table 3.1). Balajee et al. (2016) demonstrated *in silico* screening to identify some valuable NS2B/NS3 inhibitors. The identified compound 19 (1-(3-{[2-(1H-1,3-benzodiazol-2-yl)hydrazin-1-yl]methyl}-4-hydroxy-5-[(E)-[2-(3H-indol-2-yl) hydrazin-1-ylidene]methyl]phenyl)butane-1,2-dione), a promising anti-dengue agent, interacted with the catalytic triad and some other associated residues of the enzyme (vide Table 3.1).

3.4.2 RNA-Based NS2B/NS3 Inhibitors

Significant examples of DNA editing by the CRISPR-Cas9 method were already known. However, it is extremely difficult to edit RNA with this system. A modification of this present system may lead to a new window into antiviral research. Li et al. used a modified CRISPR-Cas13a-based reprogrammable RNA-induced silencing system. In their study, they targeted the DENV-NS3 region along with other targetable regions of viral RNA by using more than 10 types of CRISPR RNAs (crRNA). They observed the highest percentage of viral inhibition when the crRNAs degraded the dengue NS3 part of the RNA (Li et al. 2020).

3.4.3 Trans-Dominant NS2B/NS3 Inhibitors

The DENV NS2B/NS3 serine protease is responsible for at least seven different positional cleavages on the polyprotein chain. Through an experiment, Constant et al. (2018) showed that among these seven cleavages, at least three are intramolecular. In between these three intramolecular cleavage sites, one cleavage junction is flanked by the viral NS3 domain. They observed that when an inhibitor, namely, 4,5-Dinitrochrysazin, bound to this cleavage junction and prevented the intra-molecular cleavage of NS3-NS4A, the function of NS2B/NS3 was also inhibited. Consequently, the viral polyprotein chain remained uncleaved. This unprocessed viral precursor is toxic to the virus itself and thus has an ability to inhibit entire viral growth. This phenomenon is known as trans-domination. This is a unique approach that may help to stop the spread of multi-drug resistance to dengue viruses (Constant et al. 2018).

3.4.4 Peptide NS2B/NS3 Inhibitors

Antiviral peptides (AVPs) are defense peptides of the host that are part of the inborn or innate immune response. AVPs can be identified among all levels of higher life forms. Though there are considerable structural differences between the eukaryotic and prokaryotic cells derived from AVPs, their function remains more or less the same. AVPs act as a broad spectrum antibiotic, as they are both antifungal and antiviral agents, and can therefore be considered a novel therapeutic weapon against microorganisms. Currently, AVPs have gained attention predominantly in therapeutics due to their better safety, easy bioavailability and lower cost. Bz-Nle-Lys-Arg-Arg-NHSO2CF3 is one such AVP that targets dengue NS3 protease (Yin et al. 2006).

3.4.4.1 Linear Peptide Inhibitors of NS2B/NS3

Linear peptides, also called marginins, are mostly charged amphipathic molecules that show effectiveness against microbes and viruses. Ji et al. (2019) revealed the efficacy of a defense peptide against DENV. First, in their study, the authors collected venom from a spider, *Alopecosa nagpag*. After detailed analysis, they identified a defensin peptide, Av-LCTX-An1a (Antiviral-Lycotoxin-An1a), that showed *in vitro* inhibitory activities against DENV NS2B/NS3 protease. Later, the result was further verified by using a fluorescence-based protease inhibition assay.

3.4.4.2 Cyclic Peptide Inhibitors of NS2B/NS3

Cyclic peptides can also be an alternative to treat dengue fever. Takagi and his coworkers synthesized a series of different cyclic peptides through solid phase synthesis. When these peptides were introduced experimentally against DENV, the peptides were found to be promising in inhibiting the enzymatic activities of the viral NS2B/NS3 with an IC_{50} value of 0.95 μM. These synthesized cyclic peptides mimicked the structure of the interacting interface of the substrates of the enzyme.

The proper positioning of aromatic rings and key amino acid residue of the substrate plays a pivotal role behind their antiviral properties. Among the synthesized peptides, cyclic [-(D)Pro-Lys-Arg-Lys-Ser-homoPhe-Ser-(D)Phe-] showed the most promising results based on their research (Takagi et al. 2017).

Another series of cyclic inhibitors was designed by Braun et al. (2020). The inhibitors, which are basically NS2B/NS3 substrate analogs, contain reversible D-Lys at the P1 position. Even though their study could not demonstrate the *in vivo* efficiency of these inhibitors, the research provides direction to other researchers for discovering a cyclic peptide-based pan-flavivirus serine protease inhibitor in the near future (Braun et al. 2020).

3.4.5 Repurposed Drugs as DENV NS2B/NS3 Inhibitors

Drug repurposing can fast-track the drug development process by bringing the drugs from bench to bedside and bypassing large-scale safety trials as the drugs are already available on the market for some different purpose. In a repurposing study by Uday et al. (2021), a number of potential drugs were identified through virtual screening. A few drugs generally used as anti-cancer medicines were found to be potential inhibitors of the viral protease. The three main identified leads, namely, dabrafenib, idelalisib and nintedanib, showed promising results against the NS3 protease of all 5 serotypes of dengue viruses. An *in silico* molecular docking and molecular dynamics study revealed that upon allosteric binding of these drugs, significant structural modulation occurs in the NS3 protease, which may result in allosteric inhibition of the protease activity. Among the three molecules, dabrafenib (-9.2 to -8.1 kcal/mol) showed the lowest binding energies against all 5 serotypes and can thus be considered for further experimental studies.

In the late 1990s, the FDA approved catechol-o-methyltransferase inhibitor tolcapone for the treatment of Parkinson's disease. Later, the catechol-derived drug was banned for its horrible side effects on the liver. The hepatotoxicity caused more than three deaths. A group of scientists reinvestigated the efficacy of tolcapone, along with some other small molecules, and found that these molecules can effectively block DENV2 replication. The computational studies further suggested that the binding of tolcapone and two other molecules bound to NS2B/NS3 protease disturb its stability (Balasubramanian et al. 2016). However, safety remains an issue and defeats the repurposing objective in this case.

3.5 CONCLUSION

The severity of the dengue virus disease is increasing as new strains emerge. As vaccine development progresses, mutant strains often escape the acquired immunity, which necessitates alternate post-exposure drug-based therapeutics. Although some lead molecules are under clinical trials, no potent antivirals for dengue are available on the market. In this chapter, the recent research progress towards the development of a suitable drug candidate against dengue NS2B/NS3 protease is highlighted. Significant lead molecules that target dengue viral serine protease and inhibit its action either *in silico*, *in vitro* or *in vivo* setup are discussed. The receptor-ligand interactions, modes of action and experimental procedures described here could serve as a basic reference and database for future works. The *in silico* studies mentioned here need further experimental evidence; thus, this chapter provides a stepping stone to researchers worldwide to embark on this endeavor. The articles reviewed here are expected to bring useful tips to researchers and continue the conversation on this neglected tropical disease.

REFERENCES

Alagarasu, Kalichamy. 2016. "Introducing Dengue Vaccine: Implications for Diagnosis in Dengue Vaccinated Subjects." *Science Direct*. www.sciencedirect.com/science/article/pii/S0264410X16302353?via%3Dihub.

Alamgir, A. N. M. 2018. "Phytoconstituents—Active and Inert Constituents, Metabolic Pathways, Chemistry and Application of Phytoconstituents, Primary Metabolic Products, and Bioactive Compounds of Primary Metabolic Origin." In *Therapeutic Use of Medicinal Plants and Their Extracts: Volume 2: Phytochemistry and Bioactive Compounds*, edited by A. N. M. Alamgir, 25–164. Progress in Drug Research. Cham: Springer International Publishing. https://doi.org/10.1007/978-3-319-92387-1_2.

Bäck, Anne Tuiskunen, and Åke Lundkvist. 2013. "Dengue Viruses – An Overview." *Infection Ecology & Epidemiology* 3 (1): 19839. https://doi.org/10.3402/iee.v3i0.19839.

Balajee, R., V. Srinivasadesikan, M. Sakthivadivel, and P. Gunasekaran. 2016. "In Silico Screening, Alanine Mutation, and DFT Approaches for Identification of NS2B/NS3 Protease Inhibitors." *Biochemistry Research International* 2016 (February): e7264080. https://doi.org/10.1155/2016/7264080.

Balasubramanian, Anuradha, Mark Manzano, Tadahisa Teramoto, Rajendra Pilankatta, and Radhakrishnan Padmanabhan. 2016. "High-Throughput Screening for the Identification of Small-Molecule Inhibitors of the Flaviviral Protease." *Antiviral Research* 134 (October): 6–16. https://doi.org/10.1016/j.antiviral.2016.08.014.

Bharadwaj, Shiv, Kyung Eun Lee, Vivek Dhar Dwivedi, Umesh Yadava, Aleksha Panwar, Stuart J. Lucas, Amit Pandey, et al. 2019. "Discovery of Ganoderma Lucidum Triterpenoids as Potential Inhibitors against Dengue Virus NS2B-NS3 Protease." *Scientific Reports* 9 (1): 19059. https://doi.org/10.1038/s41598-019-55723-5.

Braun, Niklas J., Jun P. Quek, Simon Huber, Jenny Kouretova, Dorothee Rogge, Heike Lang-Henkel, Ezekiel Z. K. Cheong, et al. 2020. "Structure-Based Macrocyclization of Substrate Analogue NS2B-NS3 Protease Inhibitors of Zika, West Nile and Dengue Viruses." *Chemmedchem* 15 (15): 1439–52. https://doi.org/10.1002/cmdc.202000237.

Calixto, J. B. 2000. "Efficacy, Safety, Quality Control, Marketing and Regulatory Guidelines for Herbal Medicines (Phytotherapeutic Agents)." *Brazilian Journal of Medical and Biological Research = Revista Brasileira De Pesquisas Medicas E Biologicas* 33 (2): 179–89. https://doi.org/10.1590/s0100-879x2000000200004.

Cao-Lormeau, Van-Mai. 2009. "Dengue Viruses Binding Proteins from Aedes Aegypti and Aedes Polynesiensis Salivary Glands." *Virology Journal* 6 (1): 35. https://doi.org/10.1186/1743-422X-6-35.

Chambers, T. J., T. F. Tsai, Y. Pervikov, and T. P. Monath. 1997. "Vaccine Development against Dengue and Japanese Encephalitis: Report of a World Health Organization Meeting." *Vaccine* 15 (14): 1494–1502. https://doi.org/10.1016/s0264-410x(97)00195-3.

Chou, Su-Tze, Ying Shih, and Chih-Chien Lin. 2016. "Chapter 96 – Vetiver Grass (Vetiveria Zizanioides) Oils." In *Essential Oils in Food Preservation, Flavor and Safety*, edited by Victor R. Preedy, 843–48. San Diego, CA: Academic Press. https://doi.org/10.1016/B978-0-12-416641-7.00096-1.

Christie, James. 1872. "Remarks on 'Kidinga Pepo': A Peculiar Form of Exanthematous Disease." *British Medical Journal* 1 (596): 577–79. https://doi.org/10.1136/bmj.1.596.577.

Constant, David A., Roberto Mateo, Claude M. Nagamine, and Karla Kirkegaard. 2018. "Targeting Intramolecular Proteinase NS2B/3 Cleavages for Trans-Dominant Inhibition of Dengue Virus." *Proceedings of the National Academy of Sciences* 115 (40): 10136–41. https://doi.org/10.1073/pnas.1805195115.

Doss, A., V. Parivuguna, M. Vijayasanthi, and Sruthi Surendran. 2011. "Antibacterial Evaluation and Phytochemical Analysis of Medicago Sativa L. against Some Microbial Pathogens." *Indian Journal of Science and Technology* 4 (5): 550–52. https://doi.org/10.17485/ijst/2011/v4i5.23.

Dwivedi, Vivek, Indra Tripathi, and Sarad Mishra. 2016. "In Silico Evaluation of Inhibitory Potential of Triterpenoids from Azadirachta Indica against Therapeutic Target of Dengue Virus, NS2B-NS3 Protease." *Journal of Vector Borne Diseases* 53 (June): 156–61.

Dwivedi, Vivek Dhar, Shiv Bharadwaj, Sumbul Afroz, Nooruddin Khan, Mairaj Ahmed Ansari, Umesh Yadava, Ramesh Chandra Tripathi, et al. 2021. "Anti-Dengue Infectivity Evaluation of Bioflavonoid from Azadirachta Indica by Dengue Virus Serine Protease Inhibition." *Journal of Biomolecular Structure and Dynamics* 39 (4): 1417–30. https://doi.org/10.1080/07391102.2020.1734485.

Farooq, Umar, Bushra Munir, Sadaf Naeem, Muhammad Yameen, Shahzad Iqbal, Aftab Ahmad, Muhammad Mustaan, et al. 2020. "Exploration of Carica Papaya Bioactive Compounds as Potential Inhibitors of Dengue NS2B, NS3 and NS5 Protease." *Pakistan Journal of Pharmaceutical Sciences* 33 (January): 355–60. https://doi.org/10.36721/PJPS.2020.33.1.SUP.355-360.1.

Frimayanti, Neni, Benni Iskandar, Marzieh Yaeghoobi, Heh Han, Sharifuddin Zain, Rohana Yusof, and Noor Rahman. 2017. "Docking, Synthesis and Bioassay Studies of Imine Derivatives as Potential Inhibitors for Dengue NS2B/NS3 Serine Protease." *Asian Pacific Journal of Tropical Disease* 7 (November): 792–96. https://doi.org/10.12980/apjtd.7.2017D7-177.

Ganji, Lata R., Lekha Gandhi, Venkataramana Musturi, and Meena A. Kanyalkar. 2021. "Design, Synthesis, and Evaluation of Different Scaffold Derivatives against NS2B-NS3 Protease of Dengue Virus." *Medicinal Chemistry Research* 30 (1): 285–301. https://doi.org/10.1007/s00044-020-02660-y.

Gebashe, Fikisiwe, Adeyemi O. Aremu, Jiri Gruz, Jeffrey F. Finnie, and Johannes Van Staden. 2020. "Phytochemical Profiles and Antioxidant Activity of Grasses Used in South African Traditional Medicine." *Plants* 9 (3): 371. https://doi.org/10.3390/plants9030371.

Gubler, D. J. 1998. "Dengue and Dengue Hemorrhagic Fever." *Clinical Microbiology Reviews* 11 (3): 480–96. https://doi.org/10.1128/CMR.11.3.480.

Guzman, Maria G., Scott B. Halstead, Harvey Artsob, Philippe Buchy, Jeremy Farrar, Duane J. Gubler, Elizabeth Hunsperger, et al. 2010. "Dengue: A Continuing Global Threat." *Nature Reviews Microbiology* 8 (12): S7–16. https://doi.org/10.1038/nrmicro2460.

Habibi, Romi, Noval Herfindo, Rudi Hendra, Hilwan Y. Teruna, and Adel Zamri. 2020. "Synthesis and Molecular Docking Study of 1-(3-Chloropropyl)-3,5-Bis((E)-4-Methoxybenzylidene)Piperidin-4-One as Dengue Virus Type 2 (DEN2) NS2B/NS3 Protease Inhibitor Candidate." *Habibi Pharmacology and Clinical Pharmacy Research.* http://jurnal.unpad.ac.id/pcpr/article/view/25624/0.

Henchal, E. A., and J. R. Putnak 1990. "The Dengue Viruses." *Clinical Microbiology Reviews.* https://cmr.asm.org/content/3/4/376.

Idrees, Sobia, and Usman A. Ashfaq. 2013. "RNAi: Antiviral Therapy against Dengue Virus." *Asian Pacific Journal of Tropical Biomedicine* 3 (3): 232–36. https://doi.org/10.1016/S2221-1691(13)60057-X.

Indu, Purushothaman, Narasingam Arunagirinathan, Marimuthu Ragavan Rameshkumar, Kodhandan Sangeetha, Angamuthu Divyadarshini, and Swaminathan Rajarajan. 2021. "Antiviral Activity of Astragaloside II, Astragaloside III and Astragaloside IV Compounds against Dengue Virus: Computational Docking and in Vitro Studies." *Microbial Pathogenesis* 152 (March): 104563. https://doi.org/10.1016/j.micpath.2020.104563.

Ji, Mengyao, Tengyu Zhu, Meichen Xing, Ning Luan, James Mwangi, Xiuwen Yan, Guoxiang Mo, et al. 2019. "An Antiviral Peptide from Alopecosa Nagpag Spider Targets NS2B–NS3 Protease of Flaviviruses." *Toxins* 11 (10): 584. https://doi.org/10.3390/toxins11100584.

Kato, Fumihiro, and Takayuki Hishiki. 2016. "Dengue Virus Reporter Replicon Is a Valuable Tool for Antiviral Drug Discovery and Analysis of Virus Replication Mechanisms." *Viruses* 8 (5): 122. https://doi.org/10.3390/v8050122.

Katzelnick, Leah C., Lionel Gresh, M. Elizabeth Halloran, Juan Carlos Mercado, Guillermina Kuan, Aubree Gordon, Angel Balmaseda, and Eva Harris. 2017. "Antibody-Dependent Enhancement of Severe Dengue Disease in Humans." *Science (New York, N.Y.)* 358 (6365): 929–32. https://doi.org/10.1126/science.aan6836.

Kharisma, Viol, Arif Muhammad, Rasyadan Probojati, Muhammad Tamam, and Yulanda Antonius. 2021. "Revealing Potency of Bioactive Compounds as Inhibitor of Dengue Virus (DENV) NS2B/NS3 Protease from Sweet Potato (Ipomoea Batatas L.) Leaves." *Indian Journal of Forensic Medicine and Toxicology* 15 (January): 1627. https://doi.org/10.37506/ijfmt.v15i1.13644.

Konishi, Eiji, and Goro Kuno. 2013. "In Memoriam: Susumu Hotta (1918–2011)." *Emerging Infectious Diseases* 19 (May). https://doi.org/10.3201/eid1905.im0986.

Kouretova, Jenny, M. Zouhir Hammamy, Anton Epp, Kornelia Hardes, Stephanie Kallis, Linlin Zhang, Rolf Hilgenfeld, et al. 2017. "Effects of NS2B-NS3 Protease and Furin Inhibition on West Nile and Dengue Virus Replication." *Journal of Enzyme Inhibition and Medicinal Chemistry* 32 (1): 712–21. https://doi.org/10.1080/14756366.2017.1306521.

Kühl, Nikos, Dominik Graf, Josephine Bock, Mira A. M. Behnam, Mila-Mareen Leuthold, and Christian D. Klein. 2020. "A New Class of Dengue and West Nile Virus Protease Inhibitors with Submicromolar Activity in Reporter Gene DENV-2 Protease and Viral Replication Assays." *Journal of Medicinal Chemistry* 63 (15): 8179–97. https://doi.org/10.1021/acs.jmedchem.0c00413.

Lavanya, P., Sudha Ramaiah, and Anand Anbarasu. 2016. "Ethyl 4-(4-Methylphenyl)-4-Pentenoate from Vetiveria Zizanioides Inhibits Dengue NS2B–NS3 Protease and Prevents Viral Assembly: A Computational Molecular Dynamics and Docking Study." *Cell Biochemistry and Biophysics* 74 (3): 337–51. https://doi.org/10.1007/s12013-016-0741-x.

Li, Hao, Shan Wang, Xue Dong, Qiao Li, Min Li, Junfeng Li, Yan Guo, et al. 2020. "CRISPR-Cas13a Cleavage of Dengue Virus NS3 Gene Efficiently Inhibits Viral Replication." *Molecular Therapy – Nucleic Acids* 19 (March): 1460–69. https://doi.org/10.1016/j.omtn.2020.01.028.

Lim, Liangzhong, Mei Dang, Amrita Roy, Jian Kang, and Jianxing Song. 2020. "Curcumin Allosterically Inhibits the Dengue NS2B-NS3 Protease by Disrupting Its Active Conformation." *ACS Omega* 5 (40): 25677–86. https://doi.org/10.1021/acsomega.0c00039.

Lim, Wui Zhuan, Poh Guat Cheng, Ammar Yasir Abdulrahman, and Teow Chong Teoh. 2020. "The Identification of Active Compounds in Ganoderma Lucidum Var. Antler Extract Inhibiting Dengue Virus Serine Protease and Its Computational Studies." *Journal of Biomolecular Structure and Dynamics* 38 (14): 4273–88. https://doi.org/10.1080/07391102.2019.1678523.

Lipinski, C. A., F. Lombardo, B. W. Dominy, and P. J. Feeney. 2001. "Experimental and Computational Approaches to Estimate Solubility and Permeability in Drug Discovery and Development Settings." *Advanced Drug Delivery Reviews* 46 (1–3): 3–26. https://doi.org/10.1016/s0169-409x(00)00129-0.

Macielag, Mark J. 2012. "Chemical Properties of Antimicrobials and Their Uniqueness." In *Antibiotic Discovery and Development*, edited by Thomas J. Dougherty and Michael J. Pucci, 793–820. Boston, MA: Springer US. https://doi.org/10.1007/978-1-4614-1400-1_24.

Messina, Jane P., Oliver J. Brady, Thomas W. Scott, Chenting Zou, David M. Pigott, Kirsten A. Duda, Samir Bhatt, et al. 2014. "Global Spread of Dengue Virus Types: Mapping the 70 Year History." *Trends in Microbiology* 22 (3): 138–46. https://doi.org/10.1016/j.tim.2013.12.011.

Millies, Benedikt, Franziska von Hammerstein, Andrea Gellert, Stefan Hammerschmidt, Fabian Barthels, Ulrike Göppel, Melissa Immerheiser, et al. 2019. "Proline-Based Allosteric Inhibitors of Zika and Dengue Virus NS2B/NS3 Proteases." *Journal of Medicinal Chemistry* 62 (24): 11359–82. https://doi.org/10.1021/acs.jmedchem.9b01697.

Nesfu, Nadirah Zawani Mohd. 2020. "The Studies on Momordica Charantia Fruits and the Synthesis of Pyrazolyl Substituted Benzylidene Indanone Derivatives as Dengue Virus Type-2 NS2B/NS3 Protease Inhibitor." PhD thesis, Université de Lorraine; Universiti Sains Malaysia (Malaisie). https://hal.univ-lorraine.fr/tel-02969953.

Normile, D. 2013. "Tropical Medicine. Surprising New Dengue Virus Throws a Spanner in Disease Control Efforts." *Science* 342 (6157): 415. https://doi.org/10.1126/science.342.6157.415.

Panda, Kingshuk, Kalichamy Alagarasu, Poonam Patil, Megha Agrawal, Ashwini More, Naveen V. Kumar, Prathama S. Mainkar, et al. 2021. "In Vitro Antiviral Activity of α-Mangostin against Dengue Virus Serotype-2 (DENV-2)." *Molecules* 26 (10): 3016. https://doi.org/10.3390/molecules26103016.

Potisopon, Supanee, Stéphane Priet, Barbara Selisko, and Bruno Canard. 2014. "Comparison of Dengue Virus and HCV: From Impact on Global Health to Their RNA-Dependent RNA Polymerases." *Future Virology* 9 (January): 53–67. https://doi.org/10.2217/fvl.13.121.

"Purdue Biological Sciences: Richard J. Kuhn Lab: Research." n.d. Accessed June 1, 2021. www.bio.purdue.edu/lab/kuhn/research.html.

Qi, Rui-Feng, Ling Zhang, and Cheng-Wu Chi. 2008. "Biological Characteristics of Dengue Virus and Potential Targets for Drug Design." *Acta Biochimica et Biophysica Sinica*. https://academic.oup.com/abbs/article/40/2/91/640.

Raafat, Nader, Stuart D. Blacksell, and Richard J. Maude. 2019. "A Review of Dengue Diagnostics and Implications for Surveillance and Control." *Transactions of The Royal Society of Tropical Medicine and Hygiene* 113 (11): 653–60. https://doi.org/10.1093/trstmh/trz068.

Rathnayake, Samith, Ayesh Madushanka, Asha Wijegunawardana, Harthika Mylvaganam, Ajith Rathnayake, Eranga Geethanjana Perera, Ishara Jayamanna, et al. 2020. "In Silico Study of 5,7-Dimethoxycoumarin and p-Coumaric Acid in Carica Papaya Leaves as Dengue Virus Type 2 Protease Inhibitors †" 83 (December): 1–13. https://doi.org/10.3390/IECBM2020-08820.

Rawlinson, Stephen M., Melinda J. Pryor, Peter J. Wright, and David A. Jans. 2006. "Dengue Virus RNA Polymerase NS5: A Potential Therapeutic Target?" *Current Drug Targets* 7 (12): 1623–38. https://doi.org/10.2174/138945006779025383.

Rees, Chris A., Peter J. Hotez, Michael C. Monuteaux, Michelle Niescierenko, and Florence T. Bourgeois. 2019. "Neglected Tropical Diseases in Children: An Assessment of Gaps in Research Prioritization." *PLoS Neglected Tropical Diseases* 13 (1): e0007111. https://doi.org/10.1371/journal.pntd.0007111.

Sarwar, Muhammad Waseem, Adeel Riaz, Syed Muhammad Raihan Dilshad, Ahmed Al-Qahtani, Muhammad Shah Nawaz-Ul-Rehman, and Muhammad Mubin. 2018. "Structure Activity Relationship (SAR) and Quantitative Structure Activity Relationship (QSAR) Studies Showed Plant Flavonoids as Potential Inhibitors of Dengue NS2B-NS3 Protease." *BMC Structural Biology* 18 (1): 6. https://doi.org/10.1186/s12900-018-0084-5.

Shin, Hye Jin, Mi-Hwa Kim, Joo-Youn Lee, Insu Hwang, Gun Young Yoon, Hae Soo Kim, Young-Chan Kwon, et al. 2021. "Structure-Based Virtual Screening: Identification of a Novel NS2B-NS3 Protease Inhibitor with Potent Antiviral Activity against Zika and Dengue Viruses." *Microorganisms* 9 (3): 545. https://doi.org/10.3390/microorganisms9030545.

Tahir Ul Qamar, Muhammad, Usman Ashfaq, Kishver Tusleem, Arooj Mumtaz, Quratulain Tariq, Alina Goheer, and Bilal Ahmed. 2017. "In-Silico Identification and Evaluation of Plant Flavonoids as Dengue NS2B/NS3 Protease Inhibitors Using Molecular Docking and Simulation Approach." *Pakistan Journal of Pharmaceutical Sciences* 30 (November): 2119–37.

Takagi, Youhei, Kouhei Matsui, Haruaki Nobori, Haruka Maeda, Akihiko Sato, Takeshi Kurosu, Yasuko Orba, et al. 2017. "Discovery of Novel Cyclic Peptide Inhibitors of Dengue Virus NS2B-NS3 Protease with Antiviral Activity." *Bioorganic & Medicinal Chemistry Letters* 27 (15): 3586–90. https://doi.org/10.1016/j.bmcl.2017.05.027.

Uday, R., Rajdip Misra, Annaram Harika, Sandip Dolui, Achintya Saha, Uttam Pal, V. Ravichandiran, and Nakul Maiti. 2021. "Dabrafenib, Idelalisib and Nintedanib Act as Significant Allosteric Modulator for Dengue NS3 Protease." *PLoS One* 16 (September): e0257206. https://doi.org/10.1371/journal.pone.0257206.

Yin, Zheng, Sejal J. Patel, Wei-Ling Wang, Gang Wang, Wai-Ling Chan, K. R. Ranga Rao, Jenefer Alam, et al. 2006. "Peptide Inhibitors of Dengue Virus NS3 Protease. Part 1: Warhead." *Bioorganic & Medicinal Chemistry Letters* 16 (1): 36–39. https://doi.org/10.1016/j.bmcl.2005.09.062.

Yusufzai, Samina Khan, Hasnah Osman, Mohammad Shaheen Khan, Basma M. Abd Razik, Mohammed Oday Ezzat, Suriyati Mohamad, Othman Sulaiman, et al. 2018. "4-Thiazolidinone Coumarin Derivatives as Two-Component NS2B/NS3 DENV Flavivirus Serine Protease Inhibitors: Synthesis, Molecular Docking, Biological Evaluation and Structure–Activity Relationship Studies." *Chemistry Central Journal* 12 (1): 69. https://doi.org/10.1186/s13065-018-0435-0.

4 Modern Computational Tools for Vaccine Engineering

Debabrata Ghosh Dastidar[1,#], *Kamalesh Roy*[2,#], *Disha Bhattacharya*[1], *and Amlan Das*[3,*]

[1]Guru Nanak Institute of Pharmaceutical Science and Technology, 157/F Nilgunj Road, Panihati, Kolkata – 700114, West Bengal, India

[2]Department of Genetics, Institute of Genetic Engineering, No. 30, Thakurhat Road, Badu, Kolkata – 700128, West Bengal, India

[3]Department of Biochemistry, School of Biosciences, Royal Global University, Guwahati, India

*Correspondence to: amlan.res1980@gmail.com

#These author contributed equally to this work as the first author.

4.1 INTRODUCTION

4.1.1 Vaccines

Immunization is the key to primary health care. It saves millions of lives every year. Vaccination is a type of active immunization process to build up the natural defense mechanism of the body against a specific pathogen (Pollard, n.d.). It prevents 3.5 to 5 million deaths per year from diseases such as diphtheria, tetanus, pertussis, influenza, and measles. Vaccines are antigenic preparations that upon administration per protocol, mimic natural infection without causing harm to the recipients. They contain adjuvants such as MF59, AS01, and AS04 (Apostólico et al., 2016) that improve the immunogenic response. Thus, vaccines confer active protection against infection and/or disease upon subsequent exposure to the same pathogen. Vaccines are the major weapon in preventing and controlling the outbreak of infectious diseases. The COVID-19 pandemic is the major health crisis of this century. The vaccines against SARS-CoV-2 have saved the world from massive disasters (Hufsky et al., 2020).

4.1.2 History

The British physician Edward Jenner is regarded as the Father of Vaccination (Riedel, 2005). His experiments in the 18th century gave birth to the concept of modern vaccination. Jenner knew that dairy maids have natural protection against smallpox, as they must have suffered from cowpox. Based on his observations and knowledge, he designed an experiment. In May 1796, he inoculated the pus from a fresh cowpox lesion of a young dairy maid, Sarah Nelmes, and administered it to Jems Phipps, the 8-year-old son of his gardener. The boy developed mild lesions and recovered from cowpox. In July 1796, the boy was inoculated with pus from fresh smallpox lesions. He did not develop any lesions. Dr Jenner published these findings in 3 parts in 1798 (Smith, 2011). As the Latin word 'vacca' means cow and 'cowpox' was called vaccinia, he coined the term 'vaccination'. In 1980, the World Health Assembly declared the world free of naturally occurring smallpox (Plotkin, 2014).

4.1.3 Vaccine Types

Gradually, scientists understood that infections are caused by microbes and that the Jenner trick worked because the causative agent of cowpox was virulent in a cow but naturally attenuated in

 DOI: 10.1201/9781003302131-5

humans. It was Louis Pasteur who first established the basic rules of vaccinology. He proposed to use the inactivated pathogen as the vaccine. His findings were followed by pioneer scientists like Jonas Salk who developed vaccines against pathogens such as the poliovirus. However, due to the poor antigenicity of the killed pathogen, researchers focused on using live but attenuated pathogens (pathogens with reduced virulence) as the vaccine. Albert Sabin attenuated poliovirus by serial passage in vitro and formulated it as the vaccine (Kallerup & Foged, 2005). Hilleman developed vaccines against measles, mumps, and rubella by attenuating the viruses. Live cells from these vaccines replicate sufficiently to produce a strong immune response without causing the manifestation of the disease. Therefore, live attenuated vaccines are administered in multiple doses to induce relatively short-lived immunity (Dai et al., 2001). Toxoids are also vaccines (e.g., for tetanus and diphtheria). They are protein toxins purified from the pathogen and inactivated with formaldehyde (Vetter et al., 2017).

Generally, the formulation and development of a vaccine take more than a decade, but the COVID-19 outbreak has demanded urgency for novel vaccine technologies that support rapid development, production, and upscaling. The recent advances in systems biology, genomics and bioinformatics, immunology, machine learning, and artificial intelligence have developed new platforms for vaccines that have enhanced antigenicity and adjuvants with improved immunogenicity.

New platforms include viral vector vaccines and nucleic acid-based vaccines. They are constructed by using the genetic sequence of the pathogen (Ura et al., 2014). Viral vector-based vaccines consist of recombinant viruses packed with the genetically engineered genome to express the target antigen within the host. The virus may or may not replicate within the host, but mimicked infection leads to strong humoral and cellular immune responses. However, this type of vaccine does not require any adjuvant, and the host should not have any pre-existing immunity against the virus vector. Nucleic acid-based vaccines consist of either DNA or RNA encoding the target antigen (Leitner et al., 2000). The nucleic acids are delivered directly into the host cell where they are expressed in the target antigen. Although these vaccines are versatile and quickly produced in the case of an emerging pathogen, the availability of suitable formulations (for the enhancement of stability) and delivery devices is of prime importance (Jiskoot et al., 2019). For example, the mRNA-1273 vaccine against SARS-CoV-2 was tested just 2 months after the publication of its whole genome sequence. This is possible because of reverse vaccinology technology (Figure 4.1).

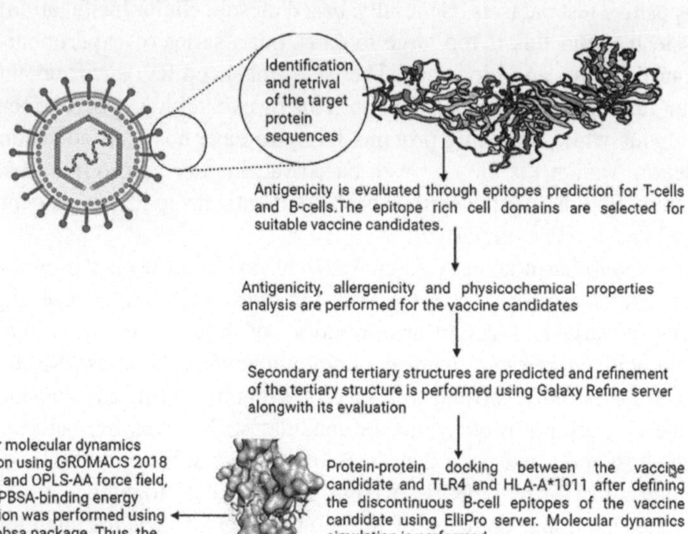

FIGURE 4.1 Different steps of reverse vaccinology in the development of COVID-19 vaccines. HLA-Human leukocyte antigens, TLR-Toll-like receptors, GROMACS-Groningen Machine for Chemical Simulation, MMPBSA-Molecular mechanics Poisson-Boltzmann surface area.

4.2 REVERSE VACCINOLOGY FOR IN SILICO IDENTIFICATION OF POTENTIAL VACCINE CANDIDATES (PVCs)

Subunit vaccines, viral vector vaccines, and nucleic acid vaccines are modern vaccines. The first step of their development requires the selection of one or more protective antigens, which may be a gene or the corresponding protein. The selection is based on the functioning, immunogenicity, and abundance of the antigens. The abundance of an antigen under natural conditions of growth and the laboratory culture medium may differ. An antigen that has poor expression under laboratory conditions may be highly abundant under natural conditions during colonization and infections. Hence, sequencing of the whole genome is an intelligent approach to obtaining a proper vaccine candidate. This is known as reverse vaccinology (RV). It emerged in the year 2000. RV has dramatically reduced the time required to identify protective antigens from 5–15 years to 1–2 years. Different steps of in silico vaccine design are shown in Figure 4.2.

The well-known high-throughput genomic sequencing technologies are Illumina sequencing technology (Thakur et al., 2018), ion semiconductor sequencing, Pyrosequencing (Ronaghi & Ronaghi, 2001), single-molecule real-time (SMRT) sequencing (Ardui et al., 2018), nanopore DNA sequencing (Wang et al., 2021), and massively parallel signature sequencing (Reinartz et al., 2002). There is a growing database of the complete genome sequences of microorganisms, and multiple genome sequences are available for particular species of bacteria. The comparison of genomes from pathogenic and non-pathogenic strains provides information regarding the genes responsible for pathogenesis. The corresponding antigenic protein may be a real candidate for vaccine development. Generally, at first, the open reading frame (ORF) of target genome sequences is found. This is followed by screening the proteins based on criteria such as the association with the membrane surface, transmembrane domains, homology to known surface proteins, outer membrane anchoring motif, leader peptides, and lipoprotein signatures (Dalsass et al., 2019).

PSORT.org provides a family of programs for the analysis and prediction of the subcellular localization of proteins. PSORTb 3.0 covers all prokaryotes and has been reported to predict subcellular localization with more than 95% precision and with a recall of more than 90% for both Gram-negative and -positive test datasets. Generally, based on subcellular localization, ~30% of the whole proteome is selected. But this is too large to carry out a series of experiments including cloning, expressions, purifications, and bioassays. The New Enhanced RV Environment (NERVE) (Vivona et al., 2006) can further narrow the selected proteins by excluding multipass membrane proteins and human homologous while including proteins likely to have host cell adhesion properties. Vaxign (https://violinet.org/vaxign/) is the first web-based vaccine design program that provides the additional functioning of predicting the epitope binding of antigens to major histocompatibility complex (MHC) classes I and II (He et al., 2010).

VaxiJen (www.ddg-pharmfac.net/vaxijen/VaxiJen/VaxiJen.html) is a program based on the principle of physicochemical properties rather than on conventional sequence alignment. It is based on the auto-cross-covariance (ACC) transformations of protein sequences into uniform vectors of principal amino acid properties (www.ncbi.nlm.nih.gov/pmc/articles/PMC1780059/). The most common feature of these software and web servers is that they primarily consider the adhesion likeliness for the identification of protein vaccine candidates. However, non-adhesion proteins can also interact with the host and cause pathogenesis (Doytchinova & Flower, 2007).

Jenner-Predict outperforms NERVE, Vaxign, and VaxiJen. Together with adhesion proteins, it also considers the functional domains of different classes of proteins such as invasion, flagellin, porin, virulence, and proteins that bind to penicillin, fibronectin, and transferrin. It predicts the non-cytosolic proteins with the above domains as antigens (Jaiswal et al., 2013).

Eukaryotic pathogens cause diseases such as malaria, leishmaniasis, mycoses, etc. Vacceed is a high-throughput in silico vaccine candidate discovery pipeline for eukaryotic pathogens. It offers supporting evidence on whether a protein is secreted or membrane-associated.

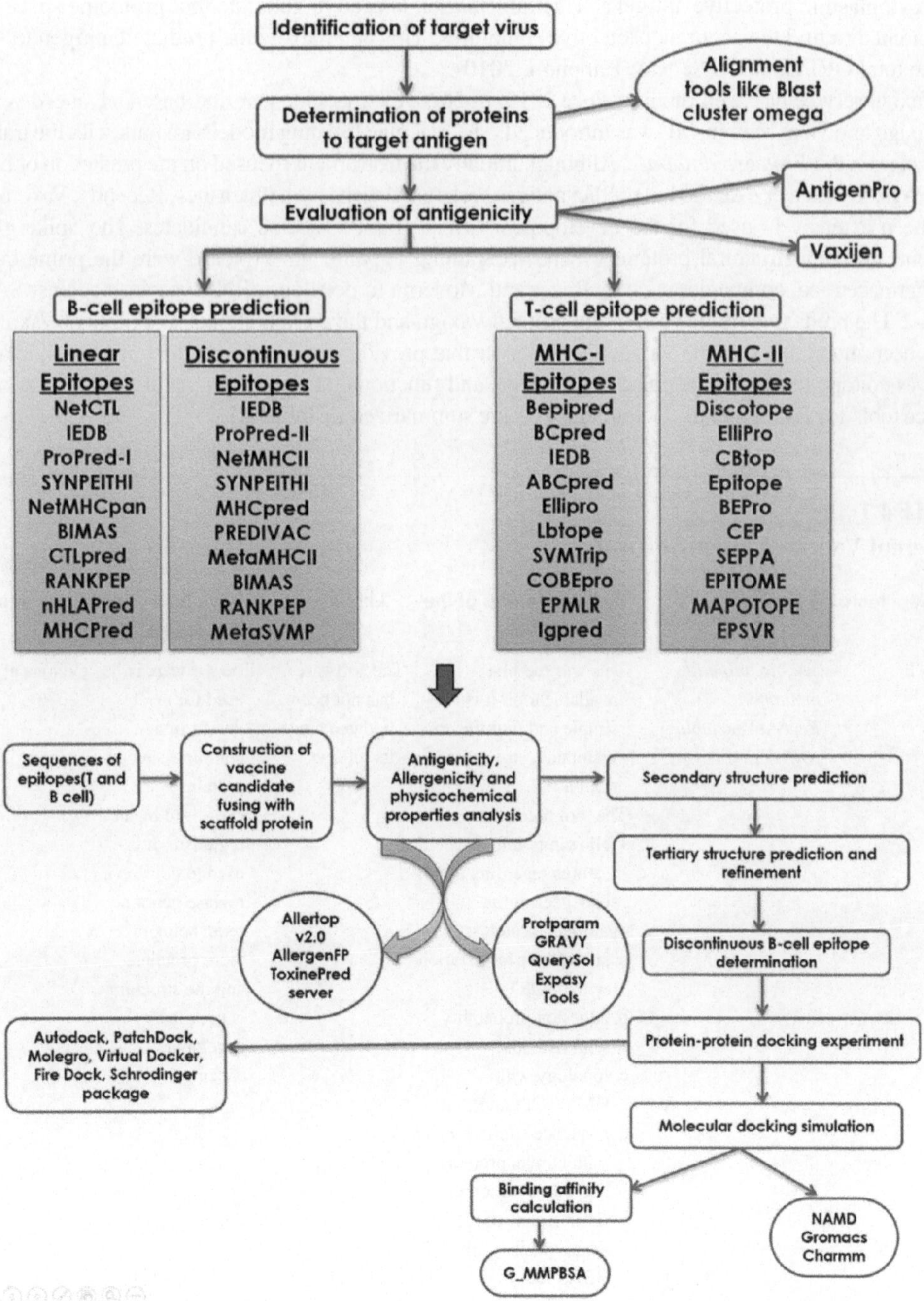

FIGURE 4.2 Schematic representation of the different steps of in silico vaccine design.

The objective of RV is to single out an antigen accurately for vaccine development. Such immunogenic antigens have structural/functional features known as protective signatures that distinguish them from immunologically inactive proteins. 'Proteome analysis' is a newer technology that is based on the biological role of proteins. It uses the Pfam database and the SMART prediction tool. This approach is helpful for the identification of not only surface-exposed and secreted antigens but also cytoplasmic protective antigens. This information is used to generate the 'proteome space' that is searched to find the accurate protective signatures. This can narrow the predicted antigen to ~3 % of the total ORF content (Sette & Rappuoli, 2010).

Vaxign service has been running since 2010. In 2020, the machine learning-based advanced version of Vaxign known as Vaxign-ML was introduced. The machine learning models are built with the training dataset from the *Protegen database*. Although initially the instrument focused on the prediction of bacterial PAgs, it was later extended to predict antigens related to viruses and parasites. Recently, Vaxign-ML has been extensively used for the development of COVID-19 vaccine candidates. The Spike glycoproteins and non-structural protein 3 (nsp3)-containing Papain-Like Protease were the prime targets of pharmaceutical companies such as Pfizer and Moderna to develop mRNA vaccines against SARS-CoV-2. The predictive framework of the original Vaxign and the computational workflow of Vaxin-ML have been integrated into the Vaxign2 web server that provides facilities with post-prediction analysis such as epitope prediction, population coverage, and functional analysis (Ong et al., 2021). The open-source tools for RV to identify potential PVCs are summarized in Table 4.1.

TABLE 4.1
Different Vaccine Design Software

Software tools	Homepage link	Salient features of the tools	Limitations	Usage in vaccine design	Reference
NERVE	*www.bio.unipd.it/molbinfo/ReverseVaccinology_NERVE.html*	This was the first available tool. It is simple and intuitive to maintain a structured database. The tool predicts the following biological features regarding the sequences of the pathogen proteins: a. subcellular localization with pSORT; b. adhesion probability with SPAAN; c. topology with HMMTOP; and d. sequence similarity with human proteins and conservation in other strains of the same species with BLASTp.	The software has not been updated since its release.	The software is used for the design of a complete set of proteins expressed by an organism. It is used to develop a reverse genomic sequencing of these proteins, and the structure is used to develop a vaccine.	(Vivona et al., 2006)
VaxiJen	*www.ddg-pharmfac.net/vaxijen/VaxiJen/VaxiJen.html*	This machine learning type of tool has a graphical user interface. It helps in classifying the selected proteins on the basis of their physicochemical properties without any biological or functional data.	The training datasets are unalterable, that is, they are limited to 100.	Predicts bacterial, viral, and tumor antigens that help to design antibodies. It further helps to design a vaccine.	(Doytchinova & Flower, 2007)

TABLE 4.1 (*Continued*)
Different Vaccine Design Software

Software tools	Homepage link	Salient features of the tools	Limitations	Usage in vaccine design	Reference
Vaxign	www.violinet.org/vaxign/, *www.violinet.org/*	This web-based vaccine design software is a part of VIOLIN, which is a web-based vaccine database and analysis resource. It predicts target candidates from wet lab confirmations. Users can input the data as *'Query'* and explore the results after the completion of the computation. Moreover, the *Dynamic Vaxign Analysis* module allows users to dynamically execute and visualize the results.	It is limited to exploring all of the protein elements of a pathogen.	When the genome sequences of all four Uropathogenic *Escherichia coli* (UPEC) strains (CFT073, 536, UT189, and F11) were analyzed, Vaxign successfully predicted 22 outer membrane proteins as vaccine targets.	(Smith, 2011) (He et al., 2010)
Jenner-Predict	*http://117.211.115.67/vaccine/home.html*	Predicts PVCs from proteome/protein sequences.		Predicted non-adhesin candidates STK, NanA, and PsaA in *S. pneumoniae.*	(Jaiswal et al., 2013; Dalsass et al., 2019)
VacSol	*https://sourceforge.net/projects/vacsol/*	It has a user-friendly interface that has four different modules for predicting homology, subcellular location, transmembrane helices, and B-cell and T-cell epitopes.		The proteins predicted by using this pipeline against the *H. pylori* strain may serve as promising PVCs against gastric pathogens.	(Rizwan et al., 2017)

4.3 EPITOPE MODELING BY IMMUNE-INFORMATICS

B- and T-cells comprising the acquired immune system are known to bind to the specific part of antigen molecules, which are called epitopes (Chaplin, 2010). Although both B- and T-cells use antigen-specific molecules on their surface to recognize the epitopes, they perform this action in different ways. The B-cell antigen receptors recognize free antigens, whereas T-cell antigen receptors recognize only the portion of the antigen that is presented on the surface of antigen-presenting cells (APCs) with MHC molecules. A wide range of tools available for in silico vaccine discovery (Raoufi et al., 2020) is shown in Table 4.2.

TABLE 4.2
The Tools and Databases for the Prediction of Bacterial and Viral Antigen Epitopes

Prediction of T-cell epitopes		Prediction of B-cell epitopes	
Tools for MHC-I binding	**Tools for MHC-II binding**	**Tools for the prediction of linear epitopes**	**Tools for the prediction of discontinuous epitopes**
NetCTL	IEDB	Bepipred	Discotope
IEDB	Propred-II	BCpred	ElliPro
ProPred-I	NetMHCII	IEDB	CBTop
SYFPEITHI	SYFPEITHI	ABCpred	Epitope
NetMHCpan	MHCpred	Ellipro	BEPro
BIMAS	PREDIVAC	LBtopee	CEP
CTLpred	MetaMHCII	SVMTrip	SEPPA
RANKPEP	MetaSVMP	COBEpro	EPITOME
nHLAPred	RANKPEP	EPMLR	MAPOTOPE
MHCpred	BIMAS	Igpred	EPSVR

4.3.1 Details of a Few Popular Software

4.3.1.1 IEDB

The Immune Epitope Database (IEDB) (*www.iedb.org/*) is a manually curated database of experimentally characterized immune epitopes. This is an important resource for epitope prediction for vaccine design. Along with the openly accessible database, it has several tools that are used with a remarkable success rate.

4.3.1.2 MHCPred

MHCPred is the tool used for the prediction of the binding affinity of class I and class II MHC molecules. It also predicts the capability of epitopes to be processed with TAP. The tool works on the additive method where the input sequence can be a minimum of 1,000 residues. It uses the partial least squares method (PLS) to generate allele-specific Quantitative Structure-Activity Relationship (QSAR) models.

4.3.1.3 NetCTL

The tool integrates the prediction of MHC class I binding, proteasomal cleavage, and TAP transport efficiency. MHC class I binding and proteasomal cleavage are performed by using Artificial Neural Network (ANN). TAP transport efficiency is predicted with the weight matrix method as described by Peters et al. (2003).

4.3.1.4 SYFPEITHI

SYFPEITHI is based on datasets of published reports. The database comprises more than 7,000 peptide sequences that are subjected to bind class I and class II MHC molecules. It consists of parameters such as peptide sequences, anchor positions, MHC specificity, source proteins, source organisms, and publication references.

4.3.1.5 RANKPEP

Sequence similarity is the key factor for the prediction of peptides that bind to MHC-I and MHC-II molecules. The basic idea is based on the fact that the peptides must share sequence similarities to bind to a common MHC molecule. However, the binding motifs of a protein are so complex that they cannot be interpreted by only the sequence residues. This compositional complexity can be overcome by using position-specific scoring matrices (PSSM) to find suitable binding motifs. PSSM specifies a score for each amino acid or nucleotide at a particular position. The sequence length is represented by the number of columns in the matrix, which is then transformed into a probability matrix by dividing it by the total number of sequences. The descriptor of the binding motif is based on the sequence or structural similarity that gives rise to functional similarity. RANKPEP is a search algorithm that ranks all possible peptides from an input protein using the PSSM coefficients.

4.3.2 Prediction of B-Cell Epitopes

The prediction of discontinuous B-cell epitopes is a major challenge in vaccine design (Van Regenmortel, 2009). Previously, the B-cell epitope prediction methods mostly relied on protein sequence similarity and were less effective (Haste Andersen et al., 2006).

Most of the B-cell epitopes were discontinuous, that is, they consist of segments that are distantly separated in the pathogen protein sequence. But it is rare to find various discontinuous epitopes in the same antigen. As the field of the structural biology of antigen-antibody interactions is rapidly growing and the in silico tools are increasing, this allows for a broader and more precise analysis of discontinuous epitopes. Together with viral, bacteria, fungal, and parasitic epitopes, even tumorigenic antigen epitopes can be predicted with the bioinformatics approach (Kardani et al., 2020).

Among the many methods available, the Hopp and Wood method is one of the important methods that is based on the alignment score to predict antigenicity (Hopp & Woods, 1981). Welling's method assigns the antigenicity value of each amino acid for the relative occurrence of amino acid in the epitope (Welling et al., 1985). The Kolaskar and Tongaokar method is based on the physiological properties of amino acid residue and their frequency of occurrences (Kolaskar & Tongaonkar, 1990).

4.3.2.1 Important Linear B-Cell Epitope Prediction Tools

4.3.2.1.1 BepiPred-2.0

BepiPred-2.0 (*https://services.healthtech.dtu.dk/service.php?BepiPred-2.0*) is a web server for predicting B-cell epitopes from antigen sequences. This method is somewhat an outperformer over other available tools. BepiPred-2.0 uses a random forest algorithm training set based on epitopes annotated from antibody-antigen protein structures (Jespersen et al., 2017).

4.3.2.1.2 ABCpred

ABCpred is a web server used to predict linear B-cell epitope regions in an antigen sequence by using an artificial neural network (ANN). In addition to selecting synthetic vaccine candidates by locating epitope regions, this server is also used in disease diagnosis and allergy research. ANN is the key method for this server to work on B-cell linear epitope prediction. ABCpred has been trained on B-cell epitopes obtained from the Bcipep database (BCIPEP).

4.3.2.1.3 LBtopee

LBtopee (*http://crdd.osdd.net/raghava/LBtopee/*) uses a larger dataset of experimentally validated B-cell and non-B-cell epitopes to develop predictive models for linear B-cell epitopes. Thus, it has overcome the limitations of tools such as ABCpred that uses a very small dataset library and random peptides.

4.3.2.1.4 ElliPro

ElliPro (*http://tools.iedb.org/ellipro/*) is used to predict linear and discontinuous antibody epitopes based on the 3D structure of an antigen. The discontinuous epitopes are defined based on the protrusion index (PI) values that quantify the extent to which a residue protrudes from the surface of a protein. Large scores are associated with greater solvent accessibility of the predicted protein. Higher PI values are clustered based on the distance R (in Å between the residue's centers of mass). The larger R is associated with the larger discontinuous epitopes being predicted (Ponomarenko et al., 2008).

4.3.2.2 Discontinuous B-Cell Epitope Prediction Tools

Structure-based tools predict the B-cell epitope more efficiently than the sequence-based approach. DiscoTope (*http://tools.iedb.org/discotope/*) is one of the most efficient tools to predict the discontinuous B-cell epitopes from the 3D structures of proteins (Kringelum et al., 2012). The method is based on amino acid statistics, spatial information, and surface accessibility in a compiled dataset of discontinuous epitopes determined by X-ray crystallography of antibody/antigen protein complexes (Haste Andersen et al., 2006). The specificity and sensitivity parameters are used for prediction. The propensity score and the DiscoTope score are used to evaluate the conformation (Haste Andersen et al., 2006).

4.3.3 Some of the Most Promising Approaches for Epitope Determination

4.3.3.1 Molecular Docking

Molecular docking is a key method in both structural molecular biology and computer-assisted drug design. It is a computational method for predicting how the interactomes interact with one another in different conformations and at different binding sites. The predicted complex may be protein-protein, protein-carbohydrate, protein-lipid, protein-small molecules, protein-DNA, etc. The result is analyzed based on scores calculated using free energy or binding energy (Grifoni et al., 2013). Therefore, the binding site and the efficacy of the binding of an antibody molecule to a specific epitope can be predicted using docking studies.

The most popular and accurate tools are Schrodinger and Discovery Studio, which are commercially available. There are also a number of open-source programs, for example, Autodock vina and Autodock 4.

4.3.3.2 Molecular Dynamics (MD) Simulation

MD simulation plays an important role in the in silico approach to addressing a problem related to structural biology. With the progress in computational facilities and algorithms, the use of MD simulation has become the inevitable approach to vaccine design. All successfully determined epitopes and their interactions with the receptor molecules must be simulated to find the stability of the complex. The atomic interactions between the antigen and antibody molecules are simulated in an environment that mimics the biological condition over a specific period through the classical equations of motion. There are several open-source software packages and commercially available platforms for MD studies. iMODS (*https://imods.iqfr.csic.es/*) is a very effective web server for MD simulation studies (Lopéz-Blanco et al., 2011). Vibrational analysis, motion animations, morphing trajectories, and Monte Carlo simulations can be easily carried out at different scales of resolution using this toolkit. GROMACS (Bekker et al., 1993), CHARMM (Brooks et al., 2009), AMBER, and NAMD are the popular software packages for MD simulation studies with high accuracy.

4.4 ANTIGENICITY PREDICTION

A pathogen may produce a considerable number of proteins, but all will not have the property to activate the immune response in our body. Therefore, it is very necessary that vaccine candidates have antigenic properties. Prediction of the antigenicity of protein sequences is a significant feature of vaccine development. The most popularly used tools to measure the antigenicity of vaccine candidates are ANTIGENpro and VaxiJen v2.0. These tools use different approaches to predict antigenicity. Microarray data are used to calculate protein antigenicity in ANTIGENpro (*http://scratch.Proteomics.ics.uci.edu/*), while cross-covariance (ACC) transformation of protein sequences into uniform vectors is used in the VaxiJen v2.0 server (*www.ddg-pharmfac.net/vaxijen/VaxiJen/VaxiJen.html*).

4.5 ALLERGENICITY

Allergenicity refers to the ability of an antigen to induce an abnormal immune response (Zhang & Tao, 2015). It is the manifestation of idiosyncratic reactions that are caused by an immunologic mechanism. The substance that causes allergies is called allergens. Allergenicity identification is a vital factor in the development of vaccines. There are popular computational tools to predict the potential allergenicity of a vaccine candidate. The most efficient and effective tools are AllerTOP v.2.0 and AllergenFP. AllerTOP v2.0 is an online server (*www.ddg-pharmfac.net/AllerTOP*) that uses ACC transformation of protein sequences into uniform equal-length vectors and a k nearest neighbors (kNN) algorithm to predict allergenicity with an accuracy as high as 85.3% (Dimitrov et al., 2014).

AllergenFP (*http://ddg-pharmfac.net/AllergenFP/*) is an alignment-free, descriptor-based fingerprint method for the detection of allergens and non-allergens. This is a four-step procedure where amino acids are first characterized according to their physicochemical properties, for example, hydrophobicity, relative abundance, and secondary structure-forming propensities; then in the second step, the string of these defined proteins is transformed into vectors; in the third step, the vectors are transferred into binary fingerprints and measured according to the Tanimoto coefficient (Shahid et al., 2020).

4.6 TOXICITY

Toxicity is an important selection criterion of a candidate vaccine. Toxicities arise mainly from various factors involved in inflammatory events that are an intrinsic part of the administered antigen and/or adjuvant. The toxicity of a designed peptide can be predicted with a server-based program known as 'ToxinPred' (*http://crdd.osdd.net/raghava/toxinpred/*) (Gupta et al., 2013). This module analyzes toxicity by generating all possible overlapping peptides and their single mutant analogs of submitted peptides (Gupta et al., 2013). The non-toxic allergens are selected as vaccine candidates.

4.7 PREDICTION OF THE PHYSICOCHEMICAL PROPERTIES OF ANTIGENIC PROTEIN

Proteins are characterized by a number of physicochemical properties, for example, theoretical amino acid composition, isoelectric point (pI), instability index, aliphatic index, in vitro and in vivo half-lives, molecular weight (MW), atomic composition, extinction coefficient, and grand average of hydropathicity (GRAVY). The functionality of a vaccine candidate is determined through these physicochemical properties (Wiederschain, 2006). Expasy Protparam (*https://web.expasy.org/protparam/*) is the tool of choice to predict all of these properties from the amino acid sequence input. The solubility of the candidate antigen can be predicted with the Protein–Sol server (*http://protein-sol.manchester.ac.uk*). Solubility is predicted as the scaled solubility value (QuerySol). This value is compared with the population average for the experimental dataset (PopAvrSol).

4.8 CONCLUSION

The lightning speed of technological advancement in protein engineering, molecular biology, genetics, and vaccinology has started a new era in vaccine development. Computation along with the combination of structural biology, synthetic biology, recombinant technology, and RV have made it easy to understand the mechanistic details of the immune response.

RV is the key to the rapid development of vaccines. RV needs a clear knowledge of immunology along with a combination of computational biology and computational chemistry. With the help of RV, prediction of suitable candidate vaccines has become less time-consuming and more efficient (Raoufi et al., 2020). The accurate prediction of epitopes using high-performance immune-analytics is the key factor for in silico vaccine design. The main challenge is to design a B-cell discontinuous epitope. However, presently, the combination of traditional algorithms with the newer Hidden Markov model (HMM) or machine learning approach has made it possible to predict vaccine candidates accurately.

The number of antigenic protein structures in the Protein Data Bank (PDB) is steadily rising. Along with X-ray crystallography and nuclear magnetic resonance (NMR), there are cryo-EM structures of antigen-antibody complexes that provide a wealth of information for the development of computational tools for epitope-based mapping and design.

The freely available computational tools and databases are very helpful. They are as efficient as the commercially available software. A variety of individual tools for different steps can be utilized for vaccine design. Along with epitope prediction tools such as NetCTL, MHCPred, and DiscoTope, the most popular tools for RV are NERVE, Vaxign, and VaxiJen. The advancement of these tools and algorithms and the enrichment of databases will help us fight new and emerging diseases.

REFERENCES

Apostólico, J. D. S., Alves, V., Lunardelli, S., Coirada, F. C., Boscardin, S. B., & Rosa, D. S. (2016). Adjuvants : Classification, modus operandi, and licensing. *Journal of Immunology Research*, *2016*, 1459394. https://doi.org/10.1155/2016/1459394

Arantes, P. R., Saha, A., & Palermo, G. (2020). Fighting covid-19 using molecular dynamics simulations. *ACS Central Science*, *6*(10). https://doi.org/10.1021/acscentsci.0c01236

Ardui, S., Ameur, A., Vermeesch, J. R., & Hestand, M. S. (2018). Single molecule real-time (SMRT) sequencing comes of age: Applications and utilities for medical diagnostics. *Nucleic Acids Research*, *46*(5), 1–10. https://doi.org/10.1093/nar/gky066

Barlow, D. J., Edwards, M. S., & Thornton, J. M. (1986). Continuous and discontinuous protein antigenic determinants. *Nature*, *322*(6081). https://doi.org/10.1038/322747a0

Bekker, H., Berendsen, H., Dijkstra, E., Achterop, S., Van Drunen, R., Van der Spoel, D., Sijbers, A., Keegstra, H., Reitsma, B., & Renardus, M. (1993). Gromacs: A parallel computer for molecular dynamics simulations. *Physics Computing*, *92*(January).

Brooks, B. R., Brooks, C. L., Mackerell, A. D., Nilsson, L., Petrella, R. J., Roux, B., Won, Y., Archontis, G., Bartels, C., Boresch, S., Caflisch, A., Caves, L., Cui, Q., Dinner, A. R., Feig, M., Fischer, S., Gao, J., Hodoscek, M., Im, W., . . . Karplus, M. (2009). CHARMM: The biomolecular simulation program. *Journal of Computational Chemistry*, *30*(10). https://doi.org/10.1002/jcc.21287

Chaplin, D. D. (2010). Overview of the immune response. *Journal of Allergy and Clinical Immunology*, *125*(2 Suppl. 2). https://doi.org/10.1016/j.jaci.2009.12.980

Dai, X., Xiong, Y., Li, N., & Jian, C. (2001). Vaccine types. In *Vaccines-the history and future*. IntechOpen.

Dalsass, M., Brozzi, A., Medini, D., & Rappuoli, R. (2019). Comparison of open-source reverse vaccinology programs for bacterial vaccine antigen discovery. *Frontiers in Immunology*, *10*(February), 1–12. https://doi.org/10.3389/fimmu.2019.00113

Dimitrov, I., Bangov, I., Flower, D. R., & Doytchinova, I. (2014). AllerTOP v.2 – A server for in silico prediction of allergens. *Journal of Molecular Modeling*, *20*(6). https://doi.org/10.1007/s00894-014-2278-5

Dormitzer, P. R., Ulmer, J. B., & Rappuoli, R. (2008). Structure-based antigen design: A strategy for next generation vaccines. *Trends in Biotechnology*, *26*(12). https://doi.org/10.1016/j.tibtech.2008.08.002

Doytchinova, I. A., & Flower, D. R. (2007). VaxiJen: A server for prediction of protective antigens, tumour antigens and subunit vaccines. *BMC Bioinformatics*, *8*. https://doi.org/10.1186/1471-2105-8-4

Grifoni, A., Montesano, C., Patronov, A., Colizzi, V., & Amicosante, M. (2013). Immunoinformatic docking approach for the analysis of KIR3DL1/HLA-B interaction. *BioMed Research International*, *2013*. https://doi.org/10.1155/2013/283805

Gupta, S., Kapoor, P., Chaudhary, K., Gautam, A., Kumar, R., & Raghava, G. P. S. (2013). In silico approach for predicting toxicity of peptides and proteins. *PLoS One*, *8*(9). https://doi.org/10.1371/journal.pone.0073957

Haste Andersen, P., Nielsen, M., & Lund, O. (2006). Prediction of residues in discontinuous B-cell epitopes using protein 3D structures. *Protein Science*, *15*(11). https://doi.org/10.1110/ps.062405906

He, Y., Xiang, Z., & Mobley, H. L. T. (2010). Vaxign: The first web-based vaccine design program for reverse vaccinology and applications for vaccine development. *Journal of Biomedicine and Biotechnology*. https://doi.org/10.1155/2010/297505

Hopp, T. P., & Woods, K. R. (1981). Prediction of protein antigenic determinants from amino acid sequences. *Proceedings of the National Academy of Sciences of the United States of America*, *78*(61). https://doi.org/10.1073/pnas.78.6.3824

Hufsky, F., Lamkiewicz, K., Almeida, A., Aouacheria, A., Arighi, C., Bateman, A., Baumbach, J., Beerenwinkel, N., Brandt, C., Cacciabue, M., Chuguransky, S., Drechsel, O., Finn, R. D., Fritz, A., Fuchs, S., Hattab, G., Hauschild, A., Heider, D., Hoffmann, M., . . . Marz, M. (2020). Computational strategies to combat COVID-19: Useful tools to accelerate SARS-CoV-2 and coronavirus research. *Briefings in Bioinformatics, 22*(2), 1–22. https://doi.org/10.1093/bib/bbaa232

Jaiswal, V., Chanumolu, S. K., Gupta, A., Chauhan, R. S., & Rout, C. (2013). Jenner-predict server: Prediction of protein vaccine candidates (PVCs) in bacteria based on host-pathogen interactions. *BMC Bioinformatics*, *14*(1). https://doi.org/10.1186/1471-2105-14-211

Jameson, B. A., & Wolf, H. (1988). The antigenic index: A novel algorithm for predicting antigenic determinants. *Computer Applications in the Biosciences: CABIOS*, *4*(1). https://doi.org/10.1093/bioinformatics/4.1.181

Jespersen, M. C., Peters, B., Nielsen, M., & Marcatili, P. (2017). BepiPred-2.0: Improving sequence-based B-cell epitope prediction using conformational epitopes. *Nucleic Acids Research*, *45*(W1). https://doi.org/10.1093/nar/gkx346

Jiskoot, W., Kersten, G. F. A., Mastrobattista, E., & Slütter, B. (2019). Vaccines. In: Crommelin, D., Sindelar, R., & Meibohm, B. (eds.), *Pharmaceutical biotechnology*. Cham: Springer. https://doi.org/10.1007/978-3-030-00710-2_14

Kallerup, R. S., & Foged, C. (2005). Classification of vaccines. In Foged, C., Rades, T., Perrie, Y., & Hook, S. (eds.), *Subunit vaccine delivery*. Advances in Delivery Science and Technology. New York, NY: Springer. https://doi.org/10.1007/978-1-4939-1417-3

Kardani, K., Bolhassani, A., & Namvar, A. (2020). An overview of *in silico* vaccine design against different pathogens and cancer. *Expert Review of Vaccines*, *19*(8), 699–726. https://doi.org/10.1080/14760584.2020.1794832

Kolaskar, A. S., & Tongaonkar, P. C. (1990). A semi-empirical method for prediction of antigenic determinants on protein antigens. *FEBS Letters*, *276*(1–2), 172–174.

Kringelum, J. V., Lundegaard, C., Lund, O., & Nielsen, M. (2012). Reliable B cell epitope predictions: Impacts of method development and improved benchmarking. *PLoS Computational Biology*, *8*(12). https://doi.org/10.1371/journal.pcbi.1002829

Leitner, W. W., Ying, H., & Restifo, N. P. (2000). DNA and RNA-based vaccines: Principles, progress and prospects. *Vaccine*, *18*(9–10), 765–777. https://doi.org/10.1016/s0264-410x(99)00271-6.

Lopéz-Blanco, J. R., Garzón, J. I., & Chacón, P. (2011). iMod: Multipurpose normal mode analysis in internal coordinates. *Bioinformatics*, *27*(20). https://doi.org/10.1093/bioinformatics/btr497

Ong, E., Cooke, M. F., Huffman, A., Xiang, Z., Wong, M. U., Wang, H., Seetharaman, M., Valdez, N., & He, Y. (2021). Vaxign2 : The second generation of the first Web-based vaccine design program using reverse vaccinology and machine learning. *Nucleic Acids Research, 49*(W1), 1–8. https://doi.org/10.1093/nar/gkab279

Peters, B., Bulik, S., Tampe, R., Van Endert, P. M., & Holzhütter, H. G. (2003). Identifying MHC class I epitopes by predicting the TAP transport efficiency of epitope precursors. *Journal of Immunology*, *171*(4), 1741–1749. https://doi.org/10.4049/jimmunol.171.4.1741

Plotkin, S. (2014). History of vaccination. *PNAS*, *111*(34), 12283–12287. https://doi.org/10.1073/pnas.1400472111

Pollard, A. J. (n.d.). A guide to vaccinology: From basic principles to new developments. *Nature Reviews Immunology*. https://doi.org/10.1038/s41577-020-00479-7

Ponomarenko, J., Bui, H. H., Li, W., Fusseder, N., Bourne, P. E., Sette, A., & Peters, B. (2008). ElliPro: A new structure-based tool for the prediction of antibody epitopes. *BMC Bioinformatics*, *9*. https://doi.org/10.1186/1471-2105-9-514

Raoufi, E., Hemmati, M., Eftekhari, S., Khaksaran, K., Mahmodi, Z., Farajollahi, M. M., & Mohsenzadegan, M. (2020). Epitope prediction by novel immunoinformatics approach: A State-of-the-art review. *International Journal of Peptide Research and Therapeutics, 26*(2). https://doi.org/10.1007/s10989-019-09918-z

Reinartz, J., Bruyns, E., Lin, J., Burcham, T., Brenner, S., Bowen, B., Kramer, M., & Woychik, R. (2002). Technique review massively parallel signature sequencing (MPSS) as a tool for in-depth quantitative gene expression profling in all organisms. *Briefings in Functional Genomics and Proteomics*, *1*(1), 95–104.

Riedel, S. (2005). Edward Jenner and the history of smallpox and vaccination. *Proceedings Baylor University. Medical Center*, *18*(1), 21–25. https://doi.org/10.1080/08998280.2005.11928028

Rizwan, M., Naz, A., Ahmad, J., Naz, K., Obaid, A., Parveen, T., Ahsan, M., & Ali, A. (2017). VacSol: A high throughput in silico pipeline to predict potential therapeutic targets in prokaryotic pathogens using subtractive reverse vaccinology. *BMC Bioinformatics*, *18*(1), 1–7. https://doi.org/10.1186/s12859-017-1540-0

Ronaghi, M., & Ronaghi, M. (2001). Pyrosequencing sheds light on DNA sequencing pyrosequencing sheds light on DNA sequencing. 3–11. https://doi.org/10.1101/gr.150601

Sette, A., & Rappuoli, R. (2010). Review reverse vaccinology: Developing vaccines in the era of genomics. *Immunity*, *33*(4), 530–541. https://doi.org/10.1016/j.immuni.2010.09.017

Shahid, F., Ashfaq, U. A., Javaid, A., & Khalid, H. (2020). Immunoinformatics guided rational design of a next generation multi epitope based peptide (MEBP) vaccine by exploring Zika virus proteome. *Infection, Genetics and Evolution*, *80*. https://doi.org/10.1016/j.meegid.2020.104199

Smith, K. A. (2011). Edward Jenner and the small pox vaccine. *Frontiers in Immunology*, *2*, 1–6. https://doi.org/10.3389/fimmu.2011.00021

Thakur, N., Shirkot, P., Pandey, H., Thakur, K., Kumar, V., & Rachappanavar. (2018). Next generation sequencing – Techniques and its applications. *Journal of Pharmacognosy and Phytochemistry*, *7*(6), 1316–1320.

Ura, T., Okuda, K., & Shimada, M. (2014). *Developments in Viral Vector-Based Vaccines*, 624–641. https://doi.org/10.3390/vaccines2030624

Van Regenmortel, M. H. V. (2009). What is a b-cell epitope? *Methods in Molecular Biology*, *524*. https://doi.org/10.1007/978-1-59745-450-6_1

Vetter, V., Denizer, G., Friedland, L. R., & Krishnan, J. (2017). Annals of Medicine understanding modern-day vaccines: What you need to know. *Annals of Medicine*, *50*(2), 1–11. https://doi.org/10.1080/07853890.2017.1407035

Vivona, S., Bernante, F., & Filippini, F. (2006). NERVE: New enhanced reverse vaccinology environment. *BMC Biotechnology*, *6*, 1–8. https://doi.org/10.1186/1472-6750-6-35

Wang, Y., Zhao, Y., Bollas, A., Wang, Y., & Au, K. F. (2021). Nanopore sequencing technology, bioinformatics and applications. *Nature Biotechnology*, *39*(11), 1348–1365. https://doi.org/10.1038/s41587-021-01108-x

Welling, G. W., Weijer, W. J., van der Zee, R., & Welling-Wester, S. (1985). Prediction of sequential antigenic regions in proteins. *FEBS Letters*, *188*, 215–218.

Wiederschain, G. Y. (2006). The proteomics protocols handbook. *Biochemistry*, *71*(6). https://doi.org/10.1134/s0006297906060150

Zhang, J., & Tao, A. (2015, November 3). Antigenicity, immunogenicity, allergenicity. *Allergy Bioinformatics*, *8*, 175–186. https://doi.org/10.1007/978-94-017-7444-4_11

5 Precision Medicine – Scope and Significance

Susmita Mukherjee[1] *and Sonali Paul*[1,*]
[1]Department of Biotechnology, University of Engineering and Management, Kolkata, susmita.mukherjee@uem.edu.in
*Correspondence to: sonali.paul@uem.edu.in

5.1 INTRODUCTION

Over the past decade, it has become increasingly clear that no two patients' cancers are exactly the same, and they therefore may have variable responses to treatments such as chemotherapy and radiation (Burney and Lakhtakia, 2017). The concept of "one size fits all" is undergoing a significant change with a more individualistic approach of treatment based on the study of key patient genetic and omic data (transcriptomics, metabolomics, and proteomics). For example, a mutation in the anaplastic lymphoma kinase (ALK) is the causal factor behind the incidence of about 5% of non-small-cell lung cancers (Williams, 2015). These patients are treated with medicines such as crizotinib and ceritinib, which are ALK blockers. Similarly, the use of the poly ADP ribose polymerase inhibitor olaparib holds great promise in the treatment of breast cancer gene 1 mutation-driven ovarian cancer (Ledermann et al., 2014).

5.2 DIFFERENT ASPECTS OF PRECISION MEDICINE

5.2.1 Economics

Government spending on healthcare has increased manifold, but it has not appreciably increased the life expectancy of people. Many supporters of precision medicine are of the opinion that government spending has the potential to reduce healthcare costs and increase the life span and quality of life of patients.

5.2.1.1 Reduction in Treatment Costs

In conventional patient care, clinicians treat patients with first-line therapy based on the clinical symptoms (Stoekle et al., 2018). Generally during this time, the patient statistics or disease-specific genetic markers are not considered. If the first approach fails, then the physician generally chooses a second approach or uses a combined approach to treatment. This approach neither identifies a patient's probability of the rate of success achieved through treatment nor the types or severity of the side effects.

In the current healthcare scenario, precision medicine methodologies are utilized only after other traditional approaches such as radiation and chemotherapy fail. Unfortunately, by this time, the patients are already exhausted and weakened (Stoekle et al., 2018). Moreover, time is required to understand the positivity of first-line treatments. This can be dangerous for advanced-stage patients as they very seldom have time for experimentation.

Several informatics approaches can be used to pinpoint the optimum therapy for a patient. Although the identification of specific mutations and expression of biomarkers in different diseases is important, the work, time, and cost involved are enormous. Studies have shown that patients who underwent genetic testing and were treated based on the customized precision medicine approach had a better prognosis and increased life span. This was apart from the costs saved in the treatment process compared to traditional methods (Marrer and Dieterle, 2008).

DOI: 10.1201/9781003302131-6

Seemingly, the precision medicine perspective holds great promise for both the pharmaceutical industry and patients. However, to make this system successful, there must be better coordination among all stakeholders – drug developers, regulators, clinicians, and patients. Only then will drug development, the approval process, and the marketing that is prescribed by clinicians be smoother and more efficient. Furthermore, patient counseling needs to be conducted to make patients aware of the benefits of this approach of treatment, which will go a long way in ensuring the final success of this strategy.

5.2.1.2 Reduction in Costs of Clinical Trials

Precision medicine has the potential to reduce the costs of clinical trials by identifying a population that is more likely to respond well to the proposed therapy (Cho et al., 2003).

Studies have shown that targeted therapies have higher success rates. Biomarker-assisted clinical trials have been found to have an average 62% success rate, which is almost six-fold greater than the 11% average positivity rate for any conventional drug going into a Phase I clinical trial. These results suggest that success rates in clinical trials and drug development strategies are possible through the use of biomarkers and precision medicine (Falconi et al., 2014).

5.2.2 Ethical Considerations

Alongside the economic considerations of precision medicine are the ethical concerns. With the process of harnessing large amounts of medical data comes the responsibility of protecting and distributing it correctly. The greatest ethical concern is protecting patient privacy, which means that both researchers and pharmaceutical companies have to keep the data that they receive from clinicians private. In this respect, one mandatory aspect is obtaining patient consent. Patients will be the greatest beneficiaries of the success of the precision medicine approach; hence, the best strategy will be their informed participation in the research and application processes (Stoekle et al., 2018).

5.3 FUTURE OUTLOOK

Overall, the promise of precision medicine is exciting and inspiring, and it has the potential to transform the way in which cancer is effectively treated. At present, patients undergoing traditional medications such as radiotherapy and chemotherapy face a lot of adverse side effects. This makes coping with the treatment procedure difficult, apart from the valuable time and resources that they spend. The most unfortunate part of this is that the majority of these efforts are ultimately ineffective. These hardships can be lessened and the benefits of personalized treatments can be imparted to patients if gene-based testing is performed before starting the treatment. This shift toward a more tailored treatment experience would require better awareness and implementation plans (Vogenberg et al., 2010). While proceeding towards the implementation of precision medicine in everyday clinical practice, we must understand that the positive outcome of this process may not be reflected in individual patients but in patient subpopulations (Bayer and Galea, 2015).

5.4 LUNG CANCER

Every year, over 200,000 new cases of lung cancer are diagnosed, and unfortunately, a majority of these patients succumb to this disease (American Cancer Society, 2015). However, the incidence of lung cancer is decreasing, primarily due to increased awareness of the harmfulness of tobacco use. Lung cancers can be categorized into the following subdivisions: lung adenocarcinoma, squamous cell lung carcinoma, large cell carcinoma, and small cell lung carcinoma (SCLC). These cancer subtypes are not only histologically different but also molecularly distinct. This leads to different prognoses and therapeutic possibilities. Even in the recent past, the selection of treatment approaches was largely based on the subgrouping of lung cancer into either non-small cell lung cancers (NSCLC) or SCLC. Currently, molecular profiling is performed to understand which treatment is the best fit for an individual patient based on the correlation of the patient's individual data with the subtype of the cancer to which the patient belongs.

5.5 APPLICATION OF PRECISION MEDICINE IN THE TREATMENT OF LUNG CANCER

5.5.1 Tumor Profiling: Oncogene-Driven Lung Adenocarcinomas

The first molecular alterations leading to lung adenocarcinoma were the epidermal growth factor receptor *(EGFR)* mutations and anaplastic lymphoma kinase (*ALK)* rearrangements. These were responsible for conferring positive response to targeted therapies, namely, tyrosine kinase inhibitors (TKIs) (Soda et al., 2007; Rikova et al., 2007; Lynch et al., 2004; Paez et al, 2004; Pao et al., 2004). The success of these therapeutic strategies led to the inclusion of genetic testing into routine medical practice. Clinical data have revealed that the treatment of advanced-stage lung cancer patients having *EGFR* and *ALK* rearrangements with appropriate TKIs leads to better treatment efficacies, enhanced life span, and delayed relapse rates compared to chemotherapy-based treatment (Mok et al., 2009; Shaw et al., 2013). To the contrary, it has been shown that patients with non-EGFR mutant lung cancers very seldom respond to EGFR TKIs and are instead treated with chemotherapy.

One of the most surprising revelations of recent times has been repetitive gene fusions in NSCLC other than ALK-rearrangements. Gene fusions involving the tyrosine kinases ROS proto-oncogene 1(*ROS1*) and Rearranged during transfection proto-oncogene (*RET*) are found in 1–2% of lung adenocarcinomas (Kohno et al., 2012; Takeuchi et al., 2012; Bergethon et al., 2012). Tumors having *ROS1* fusions were recently found to be sensitive to the TKI crizotinib (18), and RET fusion-positive lung cancers were responsive to TKIs such as cabozantinib (19). Fusions involving the neurotrophic receptor tyrosine kinase 1 (NTRK1) were also recently reported in lung adenocarcinoma (Vaishnavi et al., 2013). Preliminary studies have shown that reduction in tumor cell growth can be achieved by Tropomyosin receptor kinase A (TRKA) inhibitors, which can inhibit phosphorylation of the NTRK1 fusion. Neuregulin 1 *(CD74-NRG1)* fusions have been found to be the causal factor in invasive mucinous lung adenocarcinomas and could potentially be treated with drugs by blocking erythroblastic leukemia viral oncogene (ERBB) receptors and their signaling molecules; this demonstrates that a majority of lung adenocarcinomas harbor alterations in the oncogenic drivers (Fernandez-Cuesta et al., 2014).

In addition to genetic mutations in members of receptor tyrosine kinase (RTK)-induced signaling cascades, mutations in chromatin-modifying genes have been found in lung adenocarcinoma, lung squamous cell carcinoma, and SCLC (Cancer Genome Atlas Research Network, 2014; Rudin et al., 2012; Peifer et al., 2012; Hammerman et al., 2012). Gene alterations in the KEAP1-CUL3-NFE2L2 oxidative stress response pathway are found in lung squamous cell carcinoma and lung adenocarcinomas. Activation of this pathway regulates redox balance, promotes cell proliferation and survival, and becomes a contributing factor for resistance to chemotherapy- and radiotherapy-mediated treatments (Jaramillo and Zhang, 2013).

Amplification of SRY-related HMG-box genes *(SOX2)* is frequently observed in SCLC and has been shown to be an important causal factor (Bass et al., 2009).

5.5.2 The Future Scope of Molecular Profiling in the Treatment of Lung Cancer

The integration of genomic data and data from precision medicine-regulated clinical trials will shape the future of molecular profiling in the treatment of lung cancer. This will include next-generation sequencing and mutational analyses of a panel of oncogenes, in addition to ascertaining the type of alterations and rearrangements. One upcoming domain is of developing drugs based on biomarkers that target the immune system. These therapies, particularly immune checkpoint inhibitors, show remarkably good results in lung cancer patients (Anagnostou and Brahmer, 2015). The immunotherapeutic drug nivolumab was currently approved for second-line treatment of lung squamous cell carcinoma (Rizvi et al., 2015). Since only a certain fraction of the tumors respond to immune checkpoint inhibitors, research is ongoing to identify the biomarkers of the response and resistance to various drugs. Expression of the ligand for the immune checkpoint molecule PD1 and PDL1 on tumor cells is a marker predictive of response to immunotherapies (Herbst et al., 2014). Future studies to determine which biomarker will give the highest response to immunotherapies will shape the future of molecular profiling in lung cancer.

5.6 OVERCOMING DRUG RESISTANCE

Resistance to chemotherapeutic drugs is the main causal factor behind the advancement of lung cancer. Studies on understanding the process of acquired resistance involve sequencing studies of tumors and repeat biopsies of tumors at progression. More extensive studies of tumor changes after treatment with chemotherapy, targeted therapy, or immunotherapy need to be conducted to understand the mechanism of overcoming acquired resistance. This will decide the future of interventional studies and medication strategies for patients in the coming years.

5.7 CONCLUSION

The future of oncotreatment lies in the awareness and adoption of precision medicine. The principal reason behind this approach is the findings that the patients who were responsive to this customized approach of precision medicine-based therapeutic had better survival, delayed relapse, and a better quality of life than those who were not responsive to this approach.

REFERENCES

American Cancer Society. 2015. *Cancer facts & figures*. Atlanta, GA: American Cancer Society.

Anagnostou VK, Brahmer JR. 2015. Cancer immunotherapy: A future paradigm shift in the treatment of non-small cell lung cancer. *Clin Cancer Res*. 21:976–984.

Bass AJ, Watanabe H, Mermel CH, Yu S, Perner S, Verhaak RG. 2009. SOX2 is an amplified lineage-survival oncogene in lung and esophageal squamous cell carcinomas. *Nat Genet*. 41:1238–1242.

Bayer R, Galea S. 2015. Public health in the precision-medicine era. *N Engl J Med*. 373:499–501.

Bergethon K, Shaw AT, Ou SH, Katayama R, Lovly CM, McDonald NT. 2012. ROS1 rearrangements define a unique molecular class of lung cancers. *J Clin Oncol*. 30:863–870.

Burney IA, Lakhtakia R. 2017. Precision medicine: Where have we reached and where are we headed? *Sultan Qaboos Univ Med J*. 17:e255–e258.

Cancer Genome Atlas Research Network. 2014. Comprehensive molecular profiling of lung adenocarcinoma. *Nature*. 511:543–550.

Cho HS, Mason K, Ramyar KX, Stanley AM, Gabelli SB, Denney DW, Leahy DJ. 2003. Structure of the extracellular region of HER2 alone and in complex with the Herceptin Fab. *Nature*. 421:756–760.

Falconi A, Lopes G, Parker JL. 2014. Biomarkers and receptor targeted therapies reduce clinical trial risk in non-small-cell lung cancer. *J Thorac Oncol*. 9:163–169.

Fernandez-Cuesta L, Plenker D, Osada H, Sun R, Menon R, Leenders F. 2014. CD74-NRG1 fusions in lung adenocarcinoma. *Cancer Discov*. 4:415–422.

Hammerman PS, Hayes DN, Wilkerson MD, Schultz N, Bose R, Chu A. 2012. Comprehensive genomic characterization of squamous cell lung cancers. *Nature*. 489:519–525.

Herbst RS, Soria JC, Kowanetz M, Fine GD, Hamid O, Gordon MS. 2014. Predictive correlates of response to the anti-PD-L1 antibody MPDL3280A in cancer patients. *Nature*. 515:563–567.

Jaramillo MC, Zhang DD. 2013. The emerging role of the Nrf2-Keapl signaling pathway in cancer. *Genes Devel*. 27:2179–2191.

Kohno T, Ichikawa H, Totoki Y, Yasuda K, Hiramoto M, Nammo T. 2012. KIF5B-RET fusions in lung adenocarcinoma. *Nat Med*. 18:375–377.

Ledermann J, Harter P, Gourley C, Friedlander M, Vergote I, Rustin G, Scott CL, Meier W, Shapira-Frommer R, Safra T, Matei D, Fielding A, Spencer S. 2014. Olaparib maintenance therapy in patients with platinum-sensitive relapsed serous ovarian cancer: A preplanned retrospective analysis of outcomes by BRCA status in a randomised phase 2 trial. *Lancet Oncol*. 15:852–861.

Lynch TJ, Bell DW, Sordella R, Gurubhagavatula S, Okimoto RA, Brannigan BW, Harris PL, Haserlat SM, Supko JG, Haluska FG, Louis DN, Christiani DC. 2004. Activating mutations in the epidermal growth factor receptor underlying responsiveness of non-small-cell lung cancer to gefitinib. *New Engl J Med*. 350:2129–2139.

Marrer E, Dieterle F. 2008. Biomarkers in oncology drug development: Rescuers or troublemakers? *Expert Opin Drug Metab Toxicol*. 4:1391–1402.

Mok TS, Wu YL, Thongprasert S, Yang CH, Chu DT, Saijo N. 2009. Gefitinib or carboplatin-paclitaxel in pulmonary adenocarcinoma. *N Engl J Med*. 361:947–957.

Paez JG, Janne PA, Lee JC, Tracy S, Greulich H, Gabriel S, Herman P, Kaye FJ, Lindeman N, Meyerson M. 2004. EGFR mutations in lung cancer: Correlation with clinical response to gefitinib therapy. *Science*. 304:1497–500.

Pao W, Miller V, Zakowski M, Doherty J, Politi K, Sarkaria I, Singh B, Heelan R, Rusch V, Fulton L, Mardis E, Kupfer D, Wilson R, Kris M, Varmus H. 2004. EGF receptor gene mutations are common in lung cancers from "never smokers" and are associated with sensitivity of tumors to gefitinib and erlotinib. *Proc Natl Acad Sci U S A*. 101:13306–13311.

Peifer M, Fernandez-Cuesta L, Sos ML, George J, Seidel D, Kasper LH. 2012. Integrative genome analyses identify key somatic driver mutations of small-cell lung cancer. *Nat Genet*. 44:1104–1110.

Rikova K, Guo A, Zeng Q, Possemato A, Yu J, Haack H, Nardone J, Lee K, Reeves C, Li Y, Hu Y, Tan Z, Stokes M, Sullivan L, Mitchell J, Wetzel R, Macneill J, Ren JM, Yuan J, Bakalarski CE, Villen J, Kornhauser JM, Smith B, Li D, Zhou X, Gygi SP, Gu TL, Polakiewicz RD, Rush J, Comb MJ. 2007. Global survey of phosphotyrosine signaling identifies oncogenic kinases in lung cancer. *Cell*. 131:1190–1203.

Rizvi NA, Mazieres J, Planchard D, Stinchcombe TE, Dy GK, Antonia SJ. 2015. Activity and safety of nivolumab, an anti-PD-1 immune checkpoint inhibitor, for patients with advanced, refractory squamous non-small-cell lung cancer: A phase 2, single-arm trial. *Lancet Oncol*. 16:257–265.

Rudin CM, Durinck S, Stawiski EW, Poirier JT, Modrusan Z, Shames DS. 2012. Comprehensive genomic analysis identifies SOX2 as a frequently amplified gene in small-cell lung cancer. *Nat Genet*. 44:1111–1116.

Shaw AT, Kim DW, Nakagawa K, Seto T, Crino L, Ahn MJ. 2013. Crizotinib versus chemotherapy in advanced ALK-positive lung cancer. *N Engl J Med*. 368:2385–2394.

Soda M, Choi YL, Enomoto M, Takada S, Yamashita Y, Ishikawa S, Fujiwara S, Watanabe H, Kurashina K, Bando M, Ohno S, Ishikawa Y, Aburatani H, Niki T, Sohara Y, Sugiyama Y, Mano H. 2007. Identification of the transforming EML4-ALK fusion gene in non-small-cell lung cancer. *Nature*. 448:561–566.

Stoekle HC, Mamzer Bruneel MF, Frouart CH, Le Tourneau C, Laurent-Puig P, Vogt G, Herve C. 2018. Molecular tumor boards: Ethical issues in the new era of data medicine. *Sci Eng Ethics*. 24:307–322.

Takeuchi K, Soda M, Togashi Y, Suzuki R, Sakata S, Hatano S. 2012. RET, ROS1 and ALK fusions in lung cancer. *Nat Med*. 18:378–381.

Vaishnavi A, Capelletti M, Le AT, Kako S, Butaney M, Ercan D. 2013. Oncogenic and drug-sensitive NTRK1 rearrangements in lung cancer. *Nat Med*. 19:1469–1472.

Vogenberg FR, Barash CI, Pursel M. 2010. Personalized medicine part 2: Ethical, legal, and regulatory issues. *P T*. 35:624–642.

Williams SC. 2015. News feature: Capturing cancer's complexity. *Proc Natl Acad Sci USA*. 112:4509–4511.

6 Use of CRISPR-Cas Technology in Medical Biotechnology

Krishnendu Sarkar[1,*]
[1]Krishnagar Government College, Krishnagar, India
*Correspondence to krishnendu4776@gmail.com

6.1 INTRODUCTION

CRISPR/Cas9 gene editing technology developed by Emmanuelle Charpentier and Jennifer Doudna won the 2020 Nobel Prize in Chemistry. CRISPR is an acronym for clustered regularly interspaced short palindromic repeats. In the genomes of most prokaryotes, these short segments of DNA repeats can be found. CRISPR and Cas proteins (CRISPR-associated proteins) serve as the bacterial defence mechanism against invading bacteriophages (Koonin and Makarova 2019). This prokaryotic defense mechanism is redesigned into a powerful genome-editing tool, and because of its specificity and ease of use, it has been adopted by researchers around the world (Jiang and Doudna 2017). The specificity of this technique allows researchers to study any genomic sequence and thus understand the functioning of disease-causing genes (Charpentier and Doudna 2013). Research suggests that the CRISPR-Cas system is also used for tasks such as collective behaviour and virulence regulations (Louwen et al. 2014; Westra, Buckling, and Fineran 2014).

Y. Ishino discovered CRISPR in 1987 during his work on the genome of *Escheretia coli* (Ishino, Krupovic, and Forterre 2018). The CRISPR-Cas system consists of a series of repetitive segments (CRISPR) separated by short non-repeating segments (spacer segments). Cas genes are present adjacent to the CRISPR locus (Jansen et al. 2002). Cas genes are preceded by a trans-activating CRISPR RNA (tracrRNA) gene. Thus, the CRISPR arrays consist of tracrRNA genes, Cas genes, and CRISPR repeats and spacer sequences (Figure 6.1) (Lino et al. 2018).

The similarity of the spacer sequence to bacteriophage genomes suggests that the CRISPR-Cas system functions as an adaptive immune system in prokaryotes (Bolotin et al. 2005; Mojica et al. 2005; Pourcel, Salvignol, and Vergnaud 2005). Upon attack by a phage virus, spacer sequences from invading pathogens are incorporated into the CRISPR array. This is the basis of the adaptive immunity (Barrangou et al. 2007). On subsequent infection, the CRISPR genome is transcribed with the spacer segments that, together with the endonuclease Cas proteins, cleave the invading pathogen. CRISPR and Cas proteins work together to protect prokaryotic cells from phage and plasmid invasion (Makarova et al. 2006).

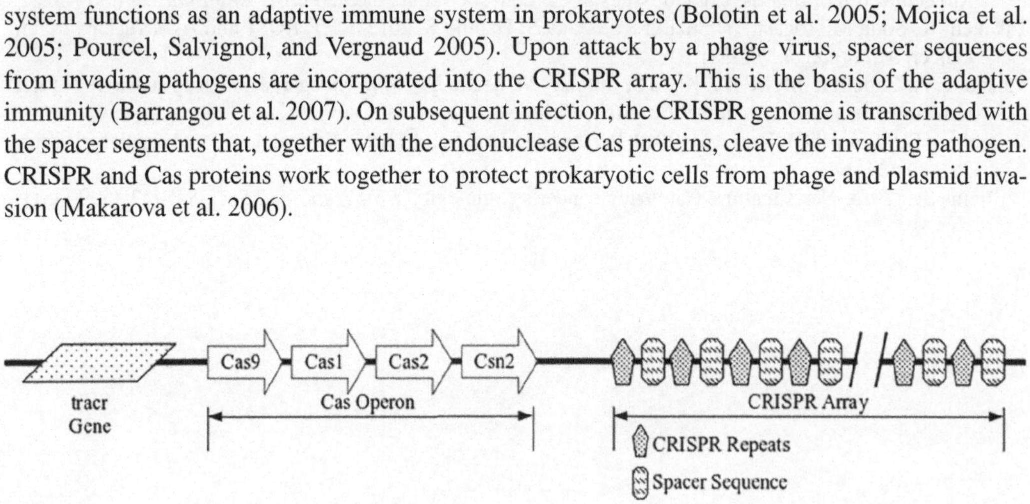

FIGURE 6.1 Schematic illustration of the CRISPR array.

DOI: 10.1201/9781003302131-7

6.2 MECHANISM OF CRISPR

CRISPR systems are divided into two different classes, Class 1 and Class 2 (Wiedenheft et al. 2011; Wright, Nuñez, and Doudna 2016). These are further divided into 6 different types (Type I – Type VI). Class 1 includes types I, III, and IV, and Class 2 consists of types II, V, and VI (Makarova et al. 2015). Each of the different types uses unique Cas proteins. Types I and III require an additional Cas3 helicase to unwind the target DNA (Agrotis and Ketteler 2015), while type II requires only Cas9 nuclease. Therefore, the type II CRISPR/Cas9 system is well designed for genome editing. The type II system is found in the genome of *Streptococcus pyogenes*.

The basis of bacterial adaptive immunity against invading viruses is the integration of short genome segments of pathogens, so-called 'protospacers', into the prokaryotic CRISPR sequences during infection. This provides a 'memory' for all subsequent infections. This immunity involves three distinct mechanistic steps: adaptation, expression, and interference (Barrangou 2015; Makarova et al. 2015). During the adaptation phase, the invading virus and plasmid DNA fragments are incorporated into the CRISPR array. Next is the expression stage where the CRISPR array is transcribed into the pre-crRNA (precursor of CRISPR RNA) (Koonin and Makarova 2019). This pre-crRNA contains the repeats and all the spacer segments from previous infections. The tracrRNA is transcribed separately and binds to the crRNA to act as a scaffold (Agrotis and Ketteler 2015). RNase III cleaves long pre-crRNA into smaller crRNAs (Deltcheva et al. 2011). These crRNA and Cas proteins specifically target invading viruses and cleave them during the interference step.

The crRNA/Cas9 complex then searches the intracellular DNA for a protospacer-adjacent motif (PAM) (Szczelkun et al. 2014). PAM is a short sequence of 2–6 bp that flank the 'protospacer' of the foreign genome. This PAM sequence helps the crRNA/Cas9 complex to distinguish between the host and the foreign genome. PAM is absent from the host CRISPR loci and is therefore not cleaved by this complex (Mojica et al. 2009; Anders et al. 2014; Jinek et al. 2014; Nishimasu et al. 2015).

Different types of Cas proteins vary in their functions within the cell. They can be grouped mainly into four different functions: spacer incorporation, crRNA processing, nuclease activity, and other functions (Makarova et al. 2015). As mentioned earlier, the CRISPR interference mechanisms differ among the CRISPR-Cas systems. However, the way that spacers are incorporated is similar in all CRISPR-Cas systems (Shmakov et al. 2015; Yosef, Goren, and Qimron 2012). In all CRISPR/Cas systems, Cas1 and Cas2 are conserved (Nuñez et al. 2014; Xiao et al. 2017). These two proteins help with incorporating the spacer into the CRISPR array (Xiao et al. 2017; Yosef, Goren, and Qimron 2012). The incorporation of the protospacer into many CRISPR-Cas systems requires recognition of a short PAM sequence in the target DNA (Makarova et al. 2015).

6.3 CRISPR AS A GENOME EDITING TOOL

Genome editing plays an important role in the treatment of disease. Researchers can understand the function of genes in disease development and progression. Gene therapy is the correction of the mutated/defective/missing gene usually for therapeutic purposes (Gonçalves and Paiva 2017). Gene therapy typically involves the generation of double stranded breaks (DSBs), corrections (insertions or deletions) in the region of interest, and repair of the DSBs. In eukaryotes, the breaks are repaired by either non-homologous end joining (NHEJ) or homology-directed repair (HDR). NHEJ does not require a homologous DNA template and usually leads to deletions and mutations of the repaired strands (Lieber 2010). HDR requires a homologous template, so when using this technique, the DSB repair is accurate (Takata et al. 1998).

Zinc finger nuclease (ZFNs) (Urnov et al. 2010) and transcription activator-like nucleases (TALENs) (Joung and Sander 2013) are a few of the gene editing tools that have improved this research field. These tools are expensive and time-consuming. However, the easiness of the type II CRISPR/Cas9 makes it a perfect instrument for genome editing. In 2012, CRISPR/Cas9 was first used to introduce DSBs into a target DNA. A single guide RNA (sgRNA) was developed by fusing the tracrRNA and crRNA. This process further simplified the CRISPR/Cas9 system. It is now a much simpler two-component system with a Cas9 protein and a sgRNA. This streamlines the CRISPR/Cas9 system, making it a convenient and easily programmable tool for gene editing (Jinek et al. 2012). The CRISPR technique basically comprises three components: the sgRNA, which acts as the guide to the target complex in the genome; the Cas9 protein, which is a nuclease and cleaves the specific DNA segment; and the target DNA template for the repair of the break.

Wu et al. (2013) were the first to use CRISPR/Cas9 to treat cataracts in mice. Cataracts are a pathological condition in which vision is impaired due to a clouding of the eye lens. They co-injected the zygote with Cas9 mRNA and sgRNA to target the dominant mutation in the Crygc gene. Correction was made by HDR, and the treated mice were found to be fertile and were able to pass their alleles to their progeny (Wu et al. 2013). This technology even enables us to treat inherited diseases. Mutations in the *dystrophin* gene cause muscle weakness. This leads to the development of Duchenne muscular dystrophy (DMD), an X-linked genetic disease. Gene replacement therapy based on CRISPR has been successfully used to treat the symptoms of DMD in adult mice (Long et al. 2014).

Cas9 can be redesigned for various properties such as gene tagging, silencing, modulator for gene expression, etc. (Martin Jinek et al. 2013; Gilbert et al. 2013; Qi et al. 2013; Mali et al. 2013; Polstein and Gersbach 2015), which is useful for studying genome functions at the precise genome location. Modified Cas9, which are catalytically inactive, can bind to the target genomes without cleaving them. This has helped in understanding genetic disease mechanisms (Zhou et al. 2014; Gilbert et al. 2014). Genome editing with Cas9 allows us to target sites within the genome, helping us understand the genes involved in a specific phenotype (Gilbert et al. 2014; Wang et al. 2014).

CRISPR/Cas9 technology is also used for developing transgenic animal models for investigating the pathogenesis, progression, and treatment of any disease (Wang et al. 2013; Holm, Alstrup, and Luo 2016; Yan et al. 2018; Chen et al. 2015; Niu et al. 2014). To generate transgenic animal models, sgRNA and Cas9 are injected into a zygote of the animal to obtain heritable gene modification (Hsu, Lander, and Zhang 2014). Furthermore, Cas9 can also be used for direct modification of the somatic cells, thus removing the need to edit the embryonic cells, which makes this technique accurate and cost-effective. Researchers have also used this genome editing tool for editing viral genomes. Hepatitis B virus causes chronic hepatitis B. The DNA of this virus is resistant to available antiviral treatments. Study results show that both *in vitro* and *in vivo*, CRISPR/Cas9 was able to destroy the hepatitis B virus-expressing templates (Lin et al. 2014). It has also been used to destroy and block the expression of the human immunodeficiency (HIV)-1 provirus (Ebina et al. 2013).

In November 2016, CRISPR-modified cells were first injected into humans to treat patients with advanced lung cancer (Cyranoski 2016). Another trial used CRISPR to treat patients with β-thalassemia and sickle cell anemia. These are serious monogenic disorders that often manifest into fatal conditions. Frangoul et al. (2021) used this technique in the hematopoietic stem and progenitor cells. Preliminary results showed improvement in the conditions of the patients (Frangoul et al. 2021). Liang et al. (2017) employed CRISPR to treat β-thalassemia. They edited the β-globin gene in a non-viable human embryo. Although this study was not completely successful, it demonstrates the potential of the CRISPR/Cas9 system in treating mutations in human embryos and somatic cells (Liang et al. 2017).

6.4 CONCLUSION

CRISPR was present in bacteria for millions of years until it was discovered accidentally, and then, it took almost two decades for humans to use it to our advantage. It is now the go-to tool for medical biotechnologists in gene editing, cancer therapeutics, the treatment of hereditary diseases, and epigenetics studies. Future application of CRISPR may also help the researcher in understanding the noncoding part of the genome. CRISPR research is progressing at a remarkable pace. It is hoped that in the future, more researchers will be able to work with this technology and address more issues.

The ease with which CRISPR allows genetic modification, in terms of technique and cost, can also encourage controversies such as improving specific human traits for non-medical purposes. Therefore, we need clearer guidelines, that is, ethical lines, for the use of this novel genome editing tool.

REFERENCES

Agrotis, Alexander, and Robin Ketteler. 2015. "A New Age in Functional Genomics Using CRISPR/Cas9 in Arrayed Library Screening." *Frontiers in Genetics* 6 (September): 300. doi:10.3389/fgene.2015.00300.

Anders, Carolin, Ole Niewoehner, Alessia Duerst, and Martin Jinek. 2014. "Structural Basis of PAM-Dependent Target DNA Recognition by the Cas9 Endonuclease." *Nature* 513 (7519): 569–73. doi:10.1038/nature13579.

Barrangou, Rodolphe. 2015. "Diversity of CRISPR-Cas Immune Systems and Molecular Machines." *Genome Biology* 16 (1): 247. doi:10.1186/s13059-015-0816-9.

Barrangou, Rodolphe, Christophe Fremaux, Hélène Deveau, Melissa Richards, Patrick Boyaval, Sylvain Moineau, Dennis A. Romero, et al. 2007. "CRISPR Provides Acquired Resistance Against Viruses in Prokaryotes." *Science* 315 (5819): 1709–12. doi:10.1126/science.1138140.

Bolotin, Alexander, Benoit Quinquis, Alexei Sorokin, and S. Dusko Ehrlich. 2005. "Clustered Regularly Interspaced Short Palindrome Repeats (CRISPRs) Have Spacers of Extrachromosomal Origin." *Microbiology* 151 (8): 2551–61. doi:10.1099/mic.0.28048-0.

Charpentier, Emmanuelle, and Jennifer A. Doudna. 2013. "Rewriting a Genome." *Nature* 495 (7439): 50–51. doi:10.1038/495050a.

Chen, Yongchang, Yinghui Zheng, Yu Kang, Weili Yang, Yuyu Niu, Xiangyu Guo, Zhuchi Tu, et al. 2015. "Functional Disruption of the Dystrophin Gene in Rhesus Monkey Using CRISPR/Cas9." *Human Molecular Genetics* 24 (13): 3764–74. doi:10.1093/hmg/ddv120.

Cyranoski, David. 2016. "CRISPR Gene-Editing Tested in a Person for the First Time." *Nature* 539 (7630): 479. doi:10.1038/nature.2016.20988.

Deltcheva, Elitza, Krzysztof Chylinski, Cynthia M. Sharma, Karine Gonzales, Yanjie Chao, Zaid A. Pirzada, Maria R. Eckert, et al. 2011. "CRISPR RNA Maturation by Trans-Encoded Small RNA and Host Factor RNase III." *Nature* 471 (7340): 602–7. doi:10.1038/nature09886.

Ebina, Hirotaka, Naoko Misawa, Yuka Kanemura, and Yoshio Koyanagi. 2013. "Harnessing the CRISPR/Cas9 System to Disrupt Latent HIV-1 Provirus." *Scientific Reports* 3 (1): 2510. doi:10.1038/srep02510.

Frangoul, Haydar, David Altshuler, M. Domenica Cappellini, Yi-Shan Chen, Jennifer Domm, Brenda K. Eustace, Juergen Foell, et al. 2021. "CRISPR-Cas9 Gene Editing for Sickle Cell Disease and β-Thalassemia." *New England Journal of Medicine* 384 (3): 252–60. doi:10.1056/NEJMoa2031054.

Gilbert, Luke A., Max A. Horlbeck, Britt Adamson, Jacqueline E. Villalta, Yuwen Chen, Evan H. Whitehead, Carla Guimaraes, et al. 2014. "Genome-Scale CRISPR-Mediated Control of Gene Repression and Activation." *Cell* 159 (3): 647–61. doi:10.1016/j.cell.2014.09.029.

Gilbert, Luke A., Matthew H. Larson, Leonardo Morsut, Zairan Liu, Gloria A. Brar, Sandra E. Torres, Noam Stern-Ginossar, et al. 2013. "CRISPR-Mediated Modular RNA-Guided Regulation of Transcription in Eukaryotes." *Cell* 154 (2): 442–51. doi:10.1016/j.cell.2013.06.044.

Gonçalves, Giulliana Augusta Rangel, and Raquel de Melo Alves Paiva. 2017. "Gene Therapy: Advances, Challenges and Perspectives." *Einstein* 15 (3): 369–75. doi:10.1590/S1679-45082017RB4024.

Holm, Ida E., Aage Kristian Olsen Alstrup, and Yonglun Luo. 2016. "Genetically Modified Pig Models for Neurodegenerative Disorders." *The Journal of Pathology* 238 (2): 267–87. doi:10.1002/path.4654.

Hsu, Patrick D., Eric S. Lander, and Feng Zhang. 2014. "Development and Applications of CRISPR-Cas9 for Genome Engineering." *Cell* 157 (6): 1262–78. doi:10.1016/j.cell.2014.05.010.

Ishino, Yoshizumi, Mart Krupovic, and Patrick Forterre. 2018. "History of CRISPR-Cas from Encounter with a Mysterious Repeated Sequence to Genome Editing Technology." *Journal of Bacteriology* 200 (7): e00580–17. doi:10.1128/JB.00580-17.

Jansen, Ruud, Jan D. A. van Embden, Wim Gaastra, and Leo M. Schouls. 2002. "Identification of Genes That Are Associated with DNA Repeats in Prokaryotes." *Molecular Microbiology* 43 (6): 1565–75. doi:10.1046/j.1365-2958.2002.02839.x.

Jiang, Fuguo, and Jennifer A. Doudna. 2017. "CRISPR–Cas9 Structures and Mechanisms." *Annual Review of Biophysics* 46 (1): 505–29. doi:10.1146/annurev-biophys-062215-010822.

Jinek, M., K. Chylinski, I. Fonfara, M. Hauer, J. A. Doudna, and E. Charpentier. 2012. "A Programmable Dual-RNA-Guided DNA Endonuclease in Adaptive Bacterial Immunity." *Science* 337 (6096): 816–21. doi:10.1126/science.1225829.

Jinek, Martin, Alexandra East, Aaron Cheng, Steven Lin, Enbo Ma, and Jennifer Doudna. 2013. "RNA-Programmed Genome Editing in Human Cells." *ELife* 2 (January): e00471. doi:10.7554/eLife.00471.

Jinek, Martin, Fuguo Jiang, David W. Taylor, Samuel H. Sternberg, Emine Kaya, Enbo Ma, Carolin Anders, et al. 2014. "Structures of Cas9 Endonucleases Reveal RNA-Mediated Conformational Activation." *Science (New York)* 343 (6176): 1247997. doi:10.1126/science.1247997.

Joung, J. Keith, and Jeffry D. Sander. 2013. "TALENs: A Widely Applicable Technology for Targeted Genome Editing." *Nature Reviews Molecular Cell Biology* 14 (1): 49–55. doi:10.1038/nrm3486.

Koonin, Eugene V., and Kira S. Makarova. 2019. "Origins and Evolution of CRISPR-Cas Systems." *Philosophical Transactions of the Royal Society B: Biological Sciences* 374 (1772): 20180087. doi:10.1098/rstb.2018.0087.

Liang, Puping, Chenhui Ding, Hongwei Sun, Xiaowei Xie, Yanwen Xu, Xiya Zhang, Ying Sun, et al. 2017. "Correction of β-Thalassemia Mutant by Base Editor in Human Embryos." *Protein & Cell* 8 (11): 811–22. doi:10.1007/s13238-017-0475-6.

Lieber, Michael R. 2010. "The Mechanism of Double-Strand DNA Break Repair by the Nonhomologous DNA End Joining Pathway." *Annual Review of Biochemistry* 79: 181–211. doi:10.1146/annurev.biochem.052308.093131.

Lin, Su-Ru, Hung-Chih Yang, Yi-Ting Kuo, Chun-Jen Liu, Ta-Yu Yang, Ku-Chun Sung, You-Yu Lin, et al. 2014. "The CRISPR/Cas9 System Facilitates Clearance of the Intrahepatic HBV Templates in Vivo." *Molecular Therapy – Nucleic Acids* 3 (8): e186. doi:10.1038/mtna.2014.38.

Lino, Christopher A., Jason C. Harper, James P. Carney, and Jerilyn A. Timlin. 2018. "Delivering CRISPR: A Review of the Challenges and Approaches." *Drug Delivery* 25 (1): 1234–57. doi:10.1080/10717544.2018.1474964.

Long, Chengzu, John R. McAnally, John M. Shelton, Alex A. Mireault, Rhonda Bassel-Duby, and Eric N. Olson. 2014. "Prevention of Muscular Dystrophy in Mice by CRISPR/Cas9-Mediated Editing of Germline DNA." *Science (New York)* 345 (6201): 1184–88. doi:10.1126/science.1254445.

Louwen, Rogier, Raymond H. J. Staals, Hubert P. Endtz, Peter van Baarlen, and John van der Oost. 2014. "The Role of CRISPR-Cas Systems in Virulence of Pathogenic Bacteria." *Microbiology and Molecular Biology Reviews* 78 (1): 74–88. doi:10.1128/MMBR.00039-13.

Makarova, Kira S., Nick V. Grishin, Svetlana A. Shabalina, Yuri I. Wolf, and Eugene V. Koonin. 2006. "A Putative RNA-Interference-Based Immune System in Prokaryotes: Computational Analysis of the Predicted Enzymatic Machinery, Functional Analogies with Eukaryotic RNAi, and Hypothetical Mechanisms of Action." *Biology Direct* 1 (March): 7. doi:10.1186/1745-6150-1-7.

Makarova, Kira S., Yuri I. Wolf, Omer S. Alkhnbashi, Fabrizio Costa, Shiraz A. Shah, Sita J. Saunders, Rodolphe Barrangou, et al. 2015. "An Updated Evolutionary Classification of CRISPR–Cas Systems." *Nature Reviews Microbiology* 13 (11): 722–36. doi:10.1038/nrmicro3569.

Mali, Prashant, John Aach, P. Benjamin Stranges, Kevin M. Esvelt, Mark Moosburner, Sriram Kosuri, Luhan Yang, et al. 2013. "CAS9 Transcriptional Activators for Target Specificity Screening and Paired Nickases for Cooperative Genome Engineering." *Nature Biotechnology* 31 (9): 833–38. doi:10.1038/nbt.2675.

Mojica, C. Díez-Villaseñor, J. García-Martínez, and C. Y. R. Almendros. 2009. "Short Motif Sequences Determine the Targets of the Prokaryotic CRISPR Defence System." *Microbiology* 155 (3): 733–40. doi:10.1099/mic.0.023960-0.

Mojica, Cesar Díez-Villaseñor, Jesús García-Martínez, and Elena Soria. 2005. "Intervening Sequences of Regularly Spaced Prokaryotic Repeats Derive from Foreign Genetic Elements." *Journal of Molecular Evolution* 60 (2): 174–82. doi:10.1007/s00239-004-0046-3.

Nishimasu, Hiroshi, Le Cong, Winston X. Yan, F. Ann Ran, Bernd Zetsche, Yinqing Li, Arisa Kurabayashi, et al. 2015. "Crystal Structure of Staphylococcus Aureus Cas9." *Cell* 162 (5): 1113–26. doi:10.1016/j.cell.2015.08.007.

Niu, Yuyu, Bin Shen, Yiqiang Cui, Yongchang Chen, Jianying Wang, Lei Wang, Yu Kang, et al. 2014. "Generation of Gene-Modified Cynomolgus Monkey via Cas9/RNA-Mediated Gene Targeting in One-Cell Embryos." *Cell* 156 (4): 836–43. doi:10.1016/j.cell.2014.01.027.

Nuñez, James K., Philip J. Kranzusch, Jonas Noeske, Addison V. Wright, Christopher W. Davies, and Jennifer A. Doudna. 2014. "Cas1-Cas2 Complex Formation Mediates Spacer Acquisition During CRISPR-Cas Adaptive Immunity." *Nature Structural & Molecular Biology* 21 (6): 528–34. doi:10.1038/nsmb.2820.

Polstein, Lauren R., and Charles A. Gersbach. 2015. "A Light-Inducible CRISPR/Cas9 System for Control of Endogenous Gene Activation." *Nature Chemical Biology* 11 (3): 198–200. doi:10.1038/nchembio.1753.

Pourcel, C., G. Salvignol, and G. Vergnaud. 2005. "CRISPR Elements in Yersinia Pestis Acquire New Repeats by Preferential Uptake of Bacteriophage DNA, and Provide Additional Tools for Evolutionary Studies." *Microbiology (Reading, England)* 151 (Pt 3): 653–63. doi:10.1099/mic.0.27437-0.

Qi, Lei S., Matthew H. Larson, Luke A. Gilbert, Jennifer A. Doudna, Jonathan S. Weissman, Adam P. Arkin, and Wendell A. Lim. 2013. "Repurposing CRISPR as an RNA-Guided Platform for Sequence-Specific Control of Gene Expression." *Cell* 152 (5): 1173–83. doi:10.1016/j.cell.2013.02.022.

Shmakov, Sergey, Omar O. Abudayyeh, Kira S. Makarova, Yuri I. Wolf, Jonathan S. Gootenberg, Ekaterina Semenova, Leonid Minakhin, et al. 2015. "Discovery and Functional Characterization of Diverse Class 2 CRISPR-Cas Systems." *Molecular Cell* 60 (3): 385–97. doi:10.1016/j.molcel.2015.10.008.

Szczelkun, Mark D., Maria S. Tikhomirova, Tomas Sinkunas, Giedrius Gasiunas, Tautvydas Karvelis, Patrizia Pschera, Virginijus Siksnys, and Ralf Seidel. 2014. "Direct Observation of R-Loop Formation by Single RNA-Guided Cas9 and Cascade Effector Complexes." *Proceedings of the National Academy of Sciences* 111 (27): 9798–803. doi:10.1073/pnas.1402597111.

Takata, M., M. S. Sasaki, E. Sonoda, C. Morrison, M. Hashimoto, H. Utsumi, Y. Yamaguchi-Iwai, et al. 1998. "Homologous Recombination and Non-Homologous End-Joining Pathways of DNA Double-Strand Break Repair Have Overlapping Roles in the Maintenance of Chromosomal Integrity in Vertebrate Cells." *The EMBO Journal* 17 (18): 5497–508. doi:10.1093/emboj/17.18.5497.

Urnov, Fyodor D., Edward J. Rebar, Michael C. Holmes, H. Steve Zhang, and Philip D. Gregory. 2010. "Genome Editing with Engineered Zinc Finger Nucleases." *Nature Reviews Genetics* 11 (9): 636–46. doi:10.1038/nrg2842.

Wang, Haoyi, Hui Yang, Chikdu S. Shivalila, Meelad M. Dawlaty, Albert W. Cheng, Feng Zhang, and Rudolf Jaenisch. 2013. "One-Step Generation of Mice Carrying Mutations in Multiple Genes by CRISPR/Cas-Mediated Genome Engineering." *Cell* 153 (4): 910–18. doi:10.1016/j.cell.2013.04.025.

Wang, Tim, Jenny J. Wei, David M. Sabatini, and Eric S. Lander. 2014. "Genetic Screens in Human Cells Using the CRISPR/Cas9 System." *Science (New York)* 343 (6166): 80–84. doi:10.1126/science.1246981.

Westra, Edze R., Angus Buckling, and Peter C. Fineran. 2014. "CRISPR–Cas Systems: Beyond Adaptive Immunity." *Nature Reviews Microbiology* 12 (5): 317–26. doi:10.1038/nrmicro3241.

Wiedenheft, Blake, Gabriel C. Lander, Kaihong Zhou, Matthijs M. Jore, Stan J. J. Brouns, John van der Oost, Jennifer A. Doudna, et al. 2011. "Structures of the RNA-Guided Surveillance Complex from a Bacterial Immune System." *Nature* 477 (7365): 486–89. doi:10.1038/nature10402.

Wright, Addison V., James K. Nuñez, and Jennifer A. Doudna. 2016. "Biology and Applications of CRISPR Systems: Harnessing Nature's Toolbox for Genome Engineering." *Cell* 164 (1–2): 29–44. doi:10.1016/j.cell.2015.12.035.

Wu, Yuxuan, Dan Liang, Yinghua Wang, Meizhu Bai, Wei Tang, Shiming Bao, Zhiqiang Yan, et al. 2013. "Correction of a Genetic Disease in Mouse via Use of CRISPR-Cas9." *Cell Stem Cell* 13 (6). Elsevier: 659–62. doi:10.1016/j.stem.2013.10.016.

Xiao, Yibei, Sherwin Ng, Ki Hyun Nam, and Ailong Ke. 2017. "How Type II CRISPR-Cas Establish Immunity through Cas1-Cas2-Mediated Spacer Integration." *Nature* 550 (7674): 137–41. doi:10.1038/nature24020.

Yan, Sen, Zhuchi Tu, Zhaoming Liu, Nana Fan, Huiming Yang, Su Yang, Weili Yang, et al. 2018. "A Huntingtin Knockin Pig Model Recapitulates Features of Selective Neurodegeneration in Huntington's Disease." *Cell* 173 (4): 989–1002.e13. doi:10.1016/j.cell.2018.03.005.

Yosef, Ido, Moran G. Goren, and Udi Qimron. 2012. "Proteins and DNA Elements Essential for the CRISPR Adaptation Process in Escherichia Coli." *Nucleic Acids Research* 40 (12): 5569–76. doi:10.1093/nar/gks216.

Zhou, Yuexin, Shiyou Zhu, Changzu Cai, Pengfei Yuan, Chunmei Li, Yanyi Huang, and Wensheng Wei. 2014. "High-Throughput Screening of a CRISPR/Cas9 Library for Functional Genomics in Human Cells." *Nature* 509 (7501): 487–91. doi:10.1038/nature13166.

7 CRISPR-Cas9
Molecular Scissors in Medical Biotechnology

Sankha Banerjee[1], Arunima Sengupta[2] and Santanu Chakraborty[1,]*

[1]Department of Life Sciences, Presidency University, Kolkata, India

[2]Department of Life Science and Biotechnology, Jadavpur University, Kolkata, India

*Correspondence to santanu.dbs@presiuniv.ac.in

7.1 INTRODUCTION

Genome editing is a part of genetic engineering where DNA is intentionally introduced and can be removed or manipulated in living cells (Porteus and Matthew, 2016). The name clustered regularly interspaced short palindromic repeats (CRISPR) is linked to a short stretch of partially repeating DNA sequences that are observed in a prokaryotic genome. CRISPR and a protein (Cas9) act as an adaptive immune response that defends itself against viruses or bacteriophages within prokaryotes (Hille et al., 2016). Japanese scientist Ishino along with his associates observed a series of palindromic sequences that were being interfered with by short spacers within Escherichia coli during analysis of an alkaline phosphatase gene. In 1990, Francisco Mojica categorized an nearly similar genetic constitution and coined the term CRISPR, but the proper functioning of these sequences had not been verified (Ishino et al., 2018). In 2007, CRISPR became a major element acting against viral invasions, hence turning into an adaptive response in prokaryotes.

The CRISPR defence mechanism defends bacteria from frequent viral accumulation through three basic stages, namely, adaptation (spacer acquisition), crRNA synthesis (expression) and target interference. During the adaptation phase, viral DNA (spacers) fragments are introduced within a genomic region of bacterial cells; this is a CRISPR array. Being an integral part of bacterial genetic material, these spacers act as a genetic memory of previous invasions by viruses and bacteriophages (Ibrahim et al., 2019). Accordingly, the CRISPR loci are short stretches of repeated sequences present within the plasmid or chromosomal DNA of prokaryotes. Adjacent to CRISPR loci lies the Cas gene that encodes nuclease proteins (Cas proteins) determined to cleave or destroy viral nucleic acid (Rath et al., 2015).

Prior to CRISPR/Cas9 technology, scientists had to rely upon two other restriction enzyme-based techniques for gene editing, specifically, zinc finger nucleases (ZFN) and transcription activator-like effector nucleases (TALENs) (Gaj et al., 2016). ZFN bears a zinc finger DNA-binding domain that efficiently binds to a stretch of DNA. It also has a restriction endonuclease domain that cleaves the DNA segment at specific target sites. TALENs' modus operandi is the same as ZFN bearing two domains, except that their DNA binding domain shows higher precision for its specified target sequence than ZFN. However, both of these techniques are very expensive and sluggish, which is a major challenge for researchers (Gaj et al., 2013). Therefore, the development of a consistent and efficient tool for gene-editing in living cells has been the eventual aim for biomedical researchers. In 2012, Doudna, J and Charpentier, E first introduced CRISPR/Cas9 for editing segments of genome by using the exact template (Jinek et al., 2012). Subsequently, the efficiency and accuracy of CRISPR/Cas9 have made it an effective tool that enables gene editing in all living cells (Adli, 2018). Researchers have used this technology for treating certain diseases that will be discussed later.

 DOI: 10.1201/9781003302131-8

7.2 GENOMIC MODIFICATIONS USING CRISPR-CAS9

CRISPR/Cas, an association of CRISPR RNA (crRNA) that recognizes the DNA sequence and Cas nuclease for cleaving DNA, has been established as an acquired immune response against viral accumulation (Gasiunas et al., 2012). These are observed in almost 90% of archaea genomes and nearly 40% of Bacterial genomes (Grissa et al., 2007). As stated previously, CRISPR loci consist of highly conserved, repetitive DNA sequences interrupted by discrete nonrepetitive sequences called spacers (Figure 7.1A). This invading viral DNA undergoes processing by Cas proteins to form small fragments of invading DNA that gets inserted within a CRISPR locus of a bacterial host genome sequence in the form of spacers. Invasion of phages and viral genetic elements activates the spacers, which are being utilized as transcriptional templates to generate crRNA. This crRNA, furthermore, directs Cas to cleave the viral and phage DNA segments that are being targeted (Figure 7.1B). Several Cas protein families (presently 40) are associated with crRNA biogenesis, integration of spacers and invasive DNA cleavage (Haft et al., 2005).

Based upon the structures and genetic constitution of Cas protein, the CRISPR/Cas system can be categorised into 3 types: I, II and III (Makarova et al., 2011). Type II bears only one Cas protein, Cas9, which covers an HNH nuclease domain along with another nuclease domain (RuvC-like). CRISPR/Cas9-mediated genome editing acts by generating a double-strand break (DSB) followed by DNA repair. In this case, the mature crRNA becomes associated with transactivating crRNA (tracr-RNA) by showing partial complementarity to construct a tracrRNA:crRNA complex, which helps to direct Cas9 to its target site. Therefore, the overall mechanism of CRISPR/Cas9-based site-specific cleavage engages a short protospacer-adjacent motif (PAM) and a crRNA that carries a similar nucleotide sequence with that of the protospacer. Once the binding of crRNA to the target

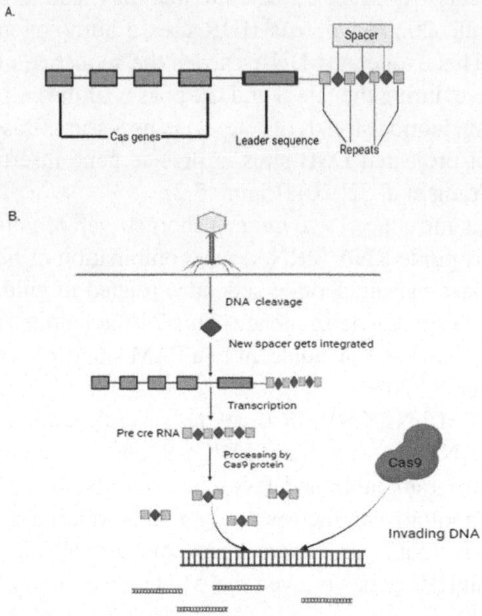

FIGURE 7.1 Overview of CRISPR/Cas9 bacterial immune system. A. Typical structure of CRISPR/Cas9 locus consisting of Direct repeated sequences along with non-repetitive spacer sequences and a Leader sequence flanking by the end of the repeats. B. Invading foreign DNA undergo processing by Cas nuclease that results in small fragments of DNA, which gets incorporated into the CRISPR locus as a spacer. The incorporated spacer is used as a template to form crRNA that guides the Cas nuclease to cleave the invading foreign DNA from bacteriophages and viruses.

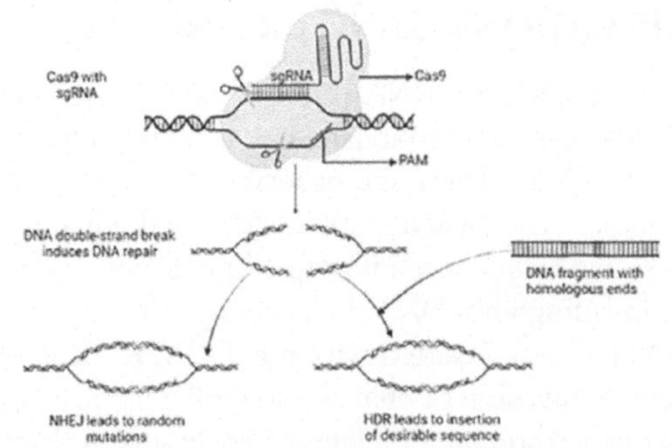

FIGURE 7.2 Cas9 protein that remains inactive initially is guided by sgRNA to induce a double-strand break 3 base pair upstream of PAM. This damage can be repaired by either NHEJ or HDR. NHEJ yields short random insertion deletions (indels) at cleavage sites introducing frame shift mutations or stop codon formations, whereas HDR introduces DNA segments with homology for precise editing.

site is fulfilled, the cleaving within opposite strands takes place by the HNH nuclease domain and RuvC-like nuclease domain of Cas9, which generates a DSB at the target site (Jinek et al., 2012).

The DSB generated by CRISPR/Cas9 will automatically activate the cellular DNA repair mechanism, specifically, non-homologous end joining (NHEJ)-mediated DNA repair with misincorporations and homology directed repair (HDR), without any error (Liu et al., 2019). In NHEJ-mediated DSB repair, the involvement of exogenous Homologous DNA is not necessary, and the process remains active throughout the cell cycle. Though it is an efficient and predominant cell repair system, it has certain disadvantages. This repair mechanism randomly introduces an insertion or deletion (indels) of DNA segments within a cleavage site that may lead to frameshift mutation or stop codon formation (Yang et al., 2020), whereas HDR uses a homologous template DNA to maintain the utmost precision. This exogenous DNA carries the sequence of interest. This mechanism remains predominantly active during the late S and G2 phases of the cell cycle. Here, a large exogenous template bearing the nucleotide stretch of interest is necessary. Based upon the template DNA with sequence homology at projected DSB sites, a precise gene insertion or replacement can be executed (Liu et al., 2019; Yang et al., 2020) (Figure 7.2).

Innovation in this tool has introduced two units: a short stretch of guide RNA and an endonuclease (Jinek et al., 2012). This guide RNA (gRNA) is a combination of both tracrRNA and bacterial crRNA. A 5′-terminal, 20 base pair nucleotide sequence related to guide RNA (homing device) is manufactured to cleave the desired gene sequence. This 24-nucleotide stretch will guide the Cas9-gRNA complex to its exact gene location adjacent to a PAM site (upstream) (Figure 7.2) via RNA–DNA binding (Khadempar et al., 2019).

Compared to ZFNs and TALENs, CRISPR/Cas9 has several advantages. As ZFNs and TALENs use restriction enzymes for DNA cleavage, CRISPR/Cas9 requires a short stretch of gRNA for DNA targeting, which is easily programmable and less expensive. By using Cas9 nuclease and gRNA together, CRISPR/Cas9 technology can incorporate genomic modifications at multiple target locations (Cong et al., 2013). This tool can generate transgenic animals rapidly with gene mutations at multiple sites and distort multiple genes or even the whole gene family for understanding epistatic relationships (Wang et al., 2013).

7.3 GENETIC DISORDERS FOLLOWED BY THERAPEUTIC APPROACHES

There have been 6,000 genetic disorders reported so far. But effective treatment strategies have not been introduced for a majority of diseases (Jackson et al., 2018). From 1998 to August 2019, by using CRISPR-Cas9 technology, about 22 human diseases were cured; thus, it is considered a potent tool in clinical biology (Ma et al., 2019). These major diseases include β-thalassemia, cystic fibrosis, and muscular dystrophy (Pandey et al., 2017; Cai et al., 2016). CRISPR/Cas9 has also been used to cure sickle cell disease (SCD) and is subject to clinical trials (Frangoul et al., 2021).

In SCD, a genetic disorder linked to autosomes is observed in red blood cells, where point mutation in the β-globin chain occurs to form sickle haemoglobin (HbS). At the time of deoxygenation, HbS gets polymerised and generates a severe clinical complication such as haemolytic anaemia (Sharon et al., 2019). The novel approach should be either restoring the function of haemoglobin S or elevating the foetal γ-globin level. The conventional procedure used in clinical trials is to boost foetal haemoglobin. To perform this function, isolating bone marrow cells from patients is conducted followed by using CRISPR-Cas9 to switch off the gene (BCL11A) responsible for hindering the foetal haemoglobin production. These newly constructed cells are reintroduced into the patient's body (Demirci et al., 2019). The responsible gene BCL11A is located on chromosome 2, whose protein product can transform γ-globin into the β-globin chain, thus interfering with the expression of the γ-globin gene (Khosravi et al., 2020). Hence, by using CRISPR-Cas9 technology, the level of γ-globin in foetal haemoglobin gets uplifted by disabling the gene function. Accordingly, the severity of the disease is reduced (Dame et al., 2000).

β-Thalassaemia, another hereditary haemolytic anaemia disease, is considered a very common and deadly disease worldwide. Here, mutation occurs in the b-globin (HBB) gene, causing a decreased haemoglobin (Hb) level leading to severe anaemia (Thompson et al., 2018). Hematopoietic stem cell transplantation (HSCT) is a conventional method for curing this disorder. However, it has its limitations because of the expensive treatment and scarcity of donors (Isgro et al., 2010). Another treatment is blood transfusion that can help patients to live a little longer but is not helpful in curing the disease permanently (Thompson et al., 2018). Because of the outdated procedures to cure such patients, researchers have focused on gene therapy using CRISPR-Cas9 technology. The basic approach is repairing the Beta-globin gene of iPSCs from patients diagnosed with Beta-thalassemia (Xie et al., 2014; Canver et al., 2015). Moreover, knockout of the BCL11A gene can also interfere with foetal haemoglobin functioning. Thus, its production is restored, and β-thalassemia can be cured (Wu et al., 2019; Khosravi et al., 2019).

Cystic fibrosis is another genetic disorder whose cure is currently being investigated by scientists using CRISPR/Cas9 technology. The disorder is based upon a mutation within the CFTR gene that hinders the structural integrity of the protein product (Conese et al., 2018). The cystic fibrosis transmembrane conductance regulator (CFTR) protein remains localised to the epithelial cells of lungs, reproductive tract and pancreas, specifically, at the apical surface that acts as a protein kinase-A-regulated anion channel (Csanády et al., 2019). There is no direct cure for this disorder, but certain symptom-based approaches such as bronchodilators, antibiotics and even mucus-thinning medications are used for basic recovery (Somayaji et al., 2020). Recently, certain approaches involving modifications at a genetic level are also being introduced for such disorders. Therefore, CRISPR/Cas9 has an immense effect upon curing certain disorders with gene editing techniques, although it is in the beginning stages of development (Bergeron and Cantin, 2021). In 2013, researchers isolated and cultured intestinal stem cells from two cystic fibrosis patients. By using CRISPR-Cas9 technology, the mutation within the CFTR locus is corrected, thus restoring the proper function of the protein product (Schwank et al., 2013).

Duchenne muscular dystrophy (DMD), another genetic disorder where mutation occurs in the dystrophin gene associated with muscle weakness, has also been cured successfully using CRISPR/Cas9 technology in patient-induced pluripotent stem cells (Li et al., 2015). Similarly, gene therapy, exon skipping and cell therapy are also being explored for repairing and re-establishing the function of dystrophin in muscles of DMD-affected patients (Fortunato et al., 2021). Therefore, excision/deletion of intragenic DNA followed by removal of the duplicated exon using CRISPR/Cas9 is an innovative approach to correct the DMD gene, thus re-establishing its function as a protein product (Mollanoori et al., 2021).

Moreover, angiogenesis characterised by a loss of vision involves vascular endothelial growth factor receptor 2. This disorder can also be treated genetically using the CRISPR–Cas9 system (Huang et al., 2017). Furthermore, dyskeratosis congenita, an extremely rare and lethal syndrome occurring during childhood characterised by an inability to produce essential blood cells, can be cured using CRISPR-Cas9 (Fok et al., 2017).

Leber Congenital Amaurosis type 10 (LCA10), a retinal dysfunction linked to autosomes, causes a loss of vision at a very young age. Patients diagnosed with this disorder carry a mutation within the CEP290 gene that interrupts the coding region and generates a splice site (Yzer et al., 2012). Ruan et al. conducted a knockdown experiment by removing the intron of the specified gene with the CRISPRCas9 system (Ruan et al., 2017).

Hutchinson-Gilford progeria syndrome (HGPS) is another fatal genetic disorder, where the patient develops the distinctive feature of accelerated aging (Ahmed et al., 2018). It is caused due to a point mutation at exon 11 of the lamin A gene that generates a splice site, thus encoding a malfunctioning lamin A termed progerin (Gonzalo et al., 2017). While standardizing the usage of the CRISPR-Cas9 technique to cure this disorder in mice, researchers administered an Adeno-associated virus (AAV)-mediated delivery of CRISPR-Cas9 components in mice samples with HGPS. This decreased the level of progerin and improved their health condition (Beyret et al., 2019).

The use of CRISPR in removing the HIV genome from every cell is considered an effective method to eliminate the virus. Scientists from McGill University AIDS Centre in Canada applied this technique to cleave the viral DNA within a host cell. The cleaved units were later ligated using cellular repair processes. Accordingly, a genetically modified tissue was generated that can nullify the effect of viral genetic material. Normally, these genetic modifications destroyed the virus, but sometimes, on the contrary, they elevated the viral infection (Wang et al., 2016; Zhu et al., 2015). As this newly repaired stretch of DNA carries sequences that are different from the previous form due to genetic alterations, the CRISPR cutting system was unable to specify the exact sequence and was unable to cleave (Gupta et al., 2015; Ebina et al., 2013; Liao et al., 2015; Hu et al., 2014; Kaminski et al., 2016). Alternatively, genomic manipulations within CCR5 and CXCR4 (coreceptors of HIV) are being initiated with CRISPR-Cas9, and it is thus considered an effective technique for treatment by shielding CD4(+) cells from HIV-1 infection (Liu et al., 2017).

Moreover, advancement in this field has introduced CRISPR/Cas-based single-base modification along with prime editing systems that have the ability of incorporating mutations directly within cellular DNA without using a donor template. As this CRISPR/Cas-base editor or prime editor system does not introduce DSB, the presence of indels within the system gets reduced unlike typical Cas9 activity (Kantor et al., 2020). The base editors can be of two types, namely, cytosine base editor (CBE) and adenine base editor (ABE) (Yang et al., 2021). CBE comprises cytidine deaminase attached to dead or partially inactive Cas9 (dCas9). Which is why it is precise while bases from cytidine (C) to thymidine (T) are replaced (Adlat et al., 2021). However, the CBE base editor has its limitations. It remains restricted to PAM sequences bearing G, T or A bases. ABE, on the contrary, bears adenosine deaminase attached to dCas9 with the purpose of changing the base pairing pattern from purine to purine (A to G) (Kantor et al., 2020). Therefore, repairing point mutations using the single-base editing technique with dCas9 fused to cytidine deaminase or adenosine deaminase is considered an effective method. But the base editors are restricted to transition mutations (purine to purine or pyrimidine to pyrimidine) (Lv et al., 2020). To avoid this deficiency, another CRISPR-Cas9-based tool kit named Prime Editor (PE) has been introduced by researchers (Matsoukas et al., 2020). PE comprises Cas9 nickase, a genetically modified reverse transcriptase along with primer editing guide RNA (pegRNA). The role of pegRNA is to recognize the nucleotide sequence that is later cleaved by using Cas9 nickase, especially at the non-complementary strand just three bases upstream from the PAM site, thus creating a 3′-OH overhang. The 3′ nick is extended by reverse transcriptase by copying the editing sequence of pegRNA. Therefore, PE is not restricted to transitions or transversion mutations but also deals with indels (Scholefield and Harrison, 2021).

7.4 ROLE OF CRISPR-CAS9 IN CANCER BIOLOGY

Recently, the CRISPR system is being applied massively and has proven to be effective in curing tumorigenesis. In investigations that involve the examination of protein domains related to cancer for evaluating drug targets, remodelling the tumorigenesis where multiple genes are involved can be performed with this tool (Maresch et al., 2016; Shi et al., 2015; Weber et al., 2015). Cancer induction takes place through oncogenes or tumour suppressor genes for certain genetic alterations and epigenetic modifications (Khan et al., 2016). Therefore, the CRISPR system has the possibility of treating the disease by incorporating mutations at precise locations, thus switching off the oncogenes and turning on the tumour suppressor genes (Harley, 2008).

Moreover, transgenic mouse models have conventionally incorporated the overexpression of specific genes or even siRNA-based knockdown methods. Although these methods have led to certain innovations in the field of cancer biology, there are limitations. While focusing on cDNA-based expression, there should be a maximised level of gene expression physiologically that portrays an anomaly in various signalling pathways and in cellular mechanisms (DeNicola et al., 2011). Therefore, RNAi is effective for certain experimental applications, but for permanent inactivation, this technique is not effective because of its significant off-target effects. Introducing CRISPR–Cas9 technology to initiate genomic modifications within endogenous loci is a potent technique to overcome these limitations.

There are multiple approaches of delivering CRISPR components. One mechanism involves genetic alterations within endogenous loci by Cas9 conducted by delivering CRISPR components. Researchers have reported successful modifications of endogenous genes in vitro through transfection by using plasmid DNA encoding Cas9 with sgRNAs or Cas9-sgRNA ribonucleoprotein complexes (RNPs) (Mali et al., 2013; Jinek et al., 2013; Cho et al., 2013). In addition, CRISPR components can also be delivered into cells by using retroviruses or lentiviruses (Malina et al., 2013; Shalem et al., 2014). After successful delivery, CRISPR components provide an advantage of a 'hit-and-run' strategy that allows a series of editing within endogenous genes multiple times without a multiple viral genetic element insertion or transfer of CRISPR components. Experimentation upon cell lines with such precisely targeted mutations can be validated using in vitro and in vivo assays to deduce the outcome of point mutations upon cancer induction. A few practical approaches to treat cancer with this technology are stated in the following. Edwin and his colleagues performed loss-of-function genetic screening all over the genome in HCT116 cells (colorectal cancer) that harbour wild-type or mutant KRAS (Yau et al., 2017). Multiple genes are involved in this metabolism including ketohexokinase and NAD kinase, which are considered to be potent therapeutic targets. The growth of colorectal cancer cells can be hindered by inhibition of these enzymes. Moreover, after a basic screening of 492 AML-specific genes, a few have been identified, including BCL2, MEN1, DOT1L, etc., which are potential therapeutic targets, along with a new member, KAT2A, that is being considered a potential focus for AML with regard to therapeutics (Tzelepis et al., 2016). A CRISPR-Cas9 mediated screening of nearly 324 cancer cell lines from 30 types of cancer was conducted to construct a database that selects a potent target in cancer therapeutics (Behan et al., 2019).

Apart from genetic screening, cancer immunotherapy has acquired great success recently. Except for its low responsive output, CRISPR-Cas9-based screening in in vivo models is being considered a potent tool for identifying genes that enable resistance to immunotherapy. Moreover, an experimental approach has been conducted in melanoma-transplanted murine samples to detect the defects in interferon-γ signalling that cause resistance to checkpoint blockade (Manguso et al., 2017). Furthermore, by using CRISPR-Cas9-mediated genome scale screening, certain mechanisms including resistance to the cytotoxic T cell-mediated killing of tumour cells can easily be identified. In fact, the loss of PBAF function by inactivating Brd7, Arid2 and Pbrm1genes leads to enhanced sensitivity of B16F10 cells (mouse melanoma) towards T cell-mediated killing (Pan et al., 2018). Therefore, in vivo screening at a massive scale will be primarily used to study cancer immunotherapy, followed by an effective delivery system for the efficient induction of genetic alterations (Yin et al., 2019; Rogers et al., 2017; Chiou et al., 2015).

7.5 RAPID GENERATION OF MOUSE MODELS

In addition to the applications of CRISPR/Cas9 in manipulating oncogenes and tumour suppressor genes, there is another approach where GEMMs and nGEMMs (non germline) (Heyer et al., 2010) play a crucial role in unravelling the essential features of tumorigenesis and its initiation, retention and progression. Moreover, these models have emerged as a reliable means to test agents with anticancer properties and reveal drug resistance mechanisms (Engelman et al., 2008; Chen et al., 2012). Constructing GEMMs is a costly process in terms of time and money, where ES cell manipulation or pronuclear injection is essential along with precise mouse husbandry to generate animals that bear the alleles of interest. To the contrary, the nGEMMs of cancer is a rather simplified process that bypasses the need to conduct genetic crosses through ES cell retargeting (Heyer et al., 2010). Recently, Jaenisch and colleagues demonstrated that in the ES cells of mice, 8 alleles became disrupted due to the usage of a CRISPR-Cas9 tool in a single step (Wang et al., 2013). In a successive study, they used CRISPR-Cas9 methods to generate mice that carry both conditional Cre–*loxP*-based alleles and reporter alleles. They even introduced the usage of sgRNAs in pairs to generate mice samples that carry small deletions (Yang et al., 2013). Therefore, CRISPR-Cas9 is a simplified way to construct ES cells that harbour gain-of-function and loss-of-function mutation, which has paved the path for developing novel GEMMs and nGEMMs of cancer with precision. Moreover, CRISPR-mediated engineering has helped researchers to construct a large store of ES cell lines that bear several constitutive mutations within tumour-suppressor genes and oncogenes. Later, these newly constructed ES cell lines were utilised in generating the GEMMs and nGEMMs of cancer with multiple and discrete mutant genotypes. Most of the mouse cancer models that are generated have a limited number of mutant alleles or genes, such as G12D or G12V mutations in the *Kras* oncogene (Jackson et al., 2001; Guerra et al., 2003). Thus, the generation of mouse models with multiple alleles that induce tumorigenesis can be initiated using the CRISPR-Cas9 system, which will help in studying therapeutic responses. Shendure laboratory has developed methods for HDR-based editing of targeted genomic regions (Findlay et al., 2014) that will help to analyse 'hotspot' regions with multiple mutations in combinations, thus generating GEMMs and nGEMMs.

7.6 APPLICATIONS OF CRISPR-CAS9 IN EX VIVO GENE THERAPY

The uniqueness of ex vivo gene therapy is based on culturing primary cells derived from patients followed by genome editing and transplanting these cells back into patients. Here, the cells are initially isolated from the peripheral blood of the patient, and after manipulation and expansion, they are reinjected. ZFs is being used as a tool in clinical trials where patients are chosen with certain blood disorders including severe combined immunodeficiency, Wiskott-Aldrich syndrome, Fanconi anaemia and sickle-cell anaemia. Moreover, gene editing techniques can be used in humans diagnosed with HIV for treatment purposes (Tebas et al., 2014). Further studies have indicated that the use of ZFs disrupts the C-C motif chemokine receptor 5 (CCR5), a co-receptor utilised by HIV strains for infecting T cells. The major disadvantage of this approach is that the technique remains functional for a brief time. The elimination of CCR5 from the self-renewing hematopoietic stem cells (HSCs) in humans was demonstrated by Holt et al. (Holt et al., 2010) by using ZFs. Recently, CRISPR/Cas9 gene-editing technology has been introduced for targeting the CCR5 gene within the human CD34+ hematopoietic stem and progenitor stem cells (HSPCs).

The most important aspect of ex vivo gene therapy is based upon the proper selection of the genetically transformed cells that harbour the edited version of the alleles without unwanted off-target effects. Here, the process of selecting precisely edited cells is very important, and only a handful of cells can be infused back into the patient. Thus, the efficiency of CRISPR/Cas9 is more simplified in ex vivo than in in vivo gene therapy. Despite all of the positive aspects, the major drawbacks of an ex vivo system are the introduction of massive genomic variations and cell expansion during cell culture.

7.7 CRISPR-CAS9 IN TRANSLATIONAL MEDICINE

The CRISPR-Cas9 technique being introduced as an editing tool has made industrial, agricultural and most importantly, enormous medical impact. One of the major concerns considering CRISPR/Cas9 technology in clinical usage is the possible off-target effects. Whole genome sequencing and high through-put profiling, both in vitro and in vivo, have provided enough evidence to understand the massive rate of inappropriate off-target binding and off-target editing, which might introduce tumorigenesis and other unforeseen consequences (Cencic et al., 2014; Kuscu et al., 2014; Wu et al., 2014; O'Green et al., 2015, Smith et al., 2014; Veres et al., 2014; Yang et al., 2014; Kim et al., 2015; Tsai et al., 2015; Ran et al., 2015). To lessen the false targets of CRISPR/Cas9 technology, a new approach must be taken. One of these approaches is the involvement of two different Cas9 events concurrently at the same locus for the purpose of cleaving the DNA segment. Furthermore, if one among two of the catalytic residues of Cas9 is being inactivated, then the enzyme gets converted into a nickase that introduces nicks within a single strand of DNA instead of double strands (Ran et al., 2013).

Interestingly, by generating two separate sgRNAs, there is a possibility of DSBs to occur by involving distinct Cas9ns nicking opposite strands while binding to each of them simultaneously (Ran et al., 2013). An alternative approach is the dimerized FokI nuclease-mediated DSB, which is a part of the previously mentioned ZFNs and TALENs, that attaches to the nuclease-deficient dCas9 (Guilinger et al., 2014; Tsai et al., 2014; Wyvekens et al., 2015). Recently, sgRNA-mediated responses have been utilised to limit off-target effects (Fu et al., 2014; Slaymaker et al., 2016; Kleinstiver et al., 2015, 2016). Introducing SpCas9-HF1 and enhanced Nuclease Cas9 (eSpCas9) are the two modified variations of Cas9 to limit such false target effects (Wyvekens et al., 2015; Kleinstiver et al., 2016). A different mechanism to reduce the off-target events and enhance Cas9 specificity depends on reducing Cas9's active time period or its active phase in cells after target locus modification. For example, the usage of Cas9-sgRNA ribonucleotide protein complexes (RNPs) in a tightly titrated amount known to be degraded rapidly can elevate the level of the on-target effects and simultaneously reduce the off-targets while intensely editing the genome in mammalian cells (Lin et al., 2014). Recently, an improvised version of Cas9 was reported named xCas9, which can recognize a variety of PAM sequences such as GAA, GAT and NG (Hu et al., 2018). This particular variant impairs the PAM sequence essentiality at the target site contrary to the limitation faced with the CRISPR/Cas9 system. In 2016, Abudayyeh et al. (2016) categorized another nuclease Cas13a that possesses RNA-guided ribonuclease activity. The mechanism involves the movement of Cas13a to its target site by a single CRISPR RNA to cleave the target RNA (single-strand) present in bacteria and mammalian cells (Abudayyeh et al., 2017). The class 2 CRISPR-Cas can be of 3 types (II, V and VI). Recently, certain VI subtypes have been reported, namely, Cas13a, Cas13b, Cas 13c and Cas13d (Abudayyeh et al., 2016, 2017; Konermann et al., 2018). In addition, Cas13 nuclease possesses non-specific RNAse activity, thus specifying it as a potent tool for clinical diagnosis (Cox et al., 2017; Gootenberg et al., 2017).

7.8 CONCLUSION

Since its discovery as a bacterial immune response system against certain pathogenic invasion, the CRISPR-Cas9 mechanism has been converted to a highly efficient tool for genomic modifications leading to a massive breakthrough in the field of basic biology research that focuses on the clinical side. The usage of this tool in screening genomic variations during tumour induction and in other lethal disorders and its ability to repair such mutations have categorised this technique as indisputable. A better understanding of the mechanisms at an intracellular level involves Cas9 endonuclease that induces DSBs to thus improve the characteristics of target sequences. To initialize this tool, a proper method should be developed for the safe and efficient transfer of CRISPR components. Despite all of its advances in the field of medicine, certain queries remain considering the well-being and efficacy of this technique. The alarming concern regarding this technology is to incorporate modifications in human embryos in line with eugenic goals. Hence, a precise, responsible and humane monitoring in this area is essential by a commanding international organization to proceed on this path in a concerned and operative manner.

REFERENCES

Abudayyeh, Omar O., Jonathan S. Gootenberg, Patrick Essletzbichler, Shuo Han, Julia Joung, Joseph J. Belanto, Vanessa Verdine et al. "RNA targeting with CRISPR–Cas13." *Nature* 550, no. 7675 (2017): 280–284.

Abudayyeh, Omar O., Jonathan S. Gootenberg, Silvana Konermann, Julia Joung, Ian M. Slaymaker, David B. T. Cox, Sergey Shmakov et al. "C2c2 is a single-component programmable RNA-guided RNA-targeting CRISPR effector." *Science* 353, no. 6299 (2016): aaf5573.

Adlat, Salah, Farooq Hayel, Ping Yang, Yang Chen, Zin Mar Oo, May Zun Zaw Myint, Rajiv Kumar Sah et al. "CRISPR-mediated base editing in mice using cytosine deaminase base editor 4." *Electronic Journal of Biotechnology* 52 (2021): 59–66.

Adli, M. "The CRISPR tool kit for genome editing and beyond." *Nature Communications* 9, no. 1911 (2018).

Ahmed, Muhammad Saad, Sana Ikram, Nousheen Bibi, and Asif Mir. "Hutchinson–Gilford progeria syndrome: A premature aging disease." *Molecular Neurobiology* 55, no. 5 (2018): 4417–4427.

Behan, Fiona M., Francesco Iorio, Gabriele Picco, Emanuel Gonçalves, Charlotte M. Beaver, Giorgia Migliardi, Rita Santos et al. "Prioritization of cancer therapeutic targets using CRISPR–Cas9 screens." *Nature* 568, no. 7753 (2019): 511–516.

Bergeron, Christelle, and André M. Cantin. "New therapies to correct the cystic fibrosis basic defect." *International Journal of Molecular Sciences* 22, no. 12 (2021): 6193.

Beyret, Ergin, Hsin-Kai Liao, Mako Yamamoto, Reyna Hernandez-Benitez, Yunpeng Fu, Galina Erikson, Pradeep Reddy et al. "Single-dose CRISPR–Cas9 therapy extends lifespan of mice with Hutchinson–Gilford progeria syndrome." *Nature Medicine* 25, no. 3 (2019): 419–422.

Cai, Liquan, Alfred L. Fisher, Haochu Huang, and Zijian Xie. "CRISPR-mediated genome editing and human diseases." *Genes & Diseases* 3, no. 4 (2016): 244–251.

Canver, Matthew C., Elenoe C. Smith, Falak Sher, Luca Pinello, Neville E. Sanjana, Ophir Shalem, Diane D. Chen et al. "BCL11A enhancer dissection by Cas9-mediated in situ saturating mutagenesis." *Nature* 527, no. 7577 (2015): 192–197.

Cencic, Regina, Hisashi Miura, Abba Malina, Francis Robert, Sylvain Ethier, T. Martin Schmeing, Josee Dostie et al. "Protospacer adjacent motif (PAM)-distal sequences engage CRISPR Cas9 DNA target cleavage." *PLoS One* 9, no. 10 (2014): e109213.

Chen, Zhao, Katherine Cheng, Zandra Walton, Yuchuan Wang, Hiromichi Ebi, Takeshi Shimamura, Yan Liu et al. "A murine lung cancer co-clinical trial identifies genetic modifiers of therapeutic response." *Nature* 483, no. 7391 (2012): 613–617.

Chiou, Shin-Heng, Ian P. Winters, Jing Wang, Santiago Naranjo, Crissy Dudgeon, Fiona B. Tamburini, Jennifer J. Brady et al. "Pancreatic cancer modeling using retrograde viral vector delivery and in vivo CRISPR/Cas9-mediated somatic genome editing." *Genes & Development* 29, no. 14 (2015): 1576–1585.

Cho, Seung Woo, Sojung Kim, Jong Min Kim, and Jin-Soo Kim. "Targeted genome engineering in human cells with the Cas9 RNA-guided endonuclease." *Nature Biotechnology* 31, no. 3 (2013): 230–232.

Conese, Massimo, Elisa Beccia, Stefano Castellani, Sante Di Gioia, Carla Colombo, Antonella Angiolillo, and Annalucia Carbone. "The long and winding road: Stem cells for cystic fibrosis." *Expert Opinion on Biological Therapy* 18, no. 3 (2018): 281–292.

Cong, Le, F. Ann Ran, David Cox, Shuailiang Lin, Robert Barretto, Naomi Habib, Patrick D. Hsu et al. "Multiplex genome engineering using CRISPR/Cas systems." *Science* 339, no. 6121 (2013): 819–823.

Cox, David BT, Jonathan S. Gootenberg, Omar O. Abudayyeh, Brian Franklin, Max J. Kellner, Julia Joung, and Feng Zhang. "RNA editing with CRISPR-Cas13." *Science* 358, no. 6366 (2017): 1019–1027.

Csanády, László, Paola Vergani, and David C. Gadsby. "Structure, gating, and regulation of the CFTR anion channel." *Physiological Reviews* 99, no. 1 (2019): 707–738.

Dame, Christof, and Sandra E. Juul. "The switch from fetal to adult erythropoiesis." *Clinics in Perinatology* 27, no. 3 (2000): 507–526.

Demirci, Selami, Alexis Leonard, Juan J. Haro-Mora, Naoya Uchida, and John F. Tisdale. "CRISPR/Cas9 for sickle cell disease: Applications, future possibilities, and challenges." *Cell Biology and Translational Medicine* 5 (2019): 37–52.

DeNicola, Gina M., Florian A. Karreth, Timothy J. Humpton, Aarthi Gopinathan, Cong Wei, Kristopher Frese, Dipti Mangal et al. "Oncogene-induced Nrf2 transcription promotes ROS detoxification and tumorigenesis." *Nature* 475, no. 7354 (2011): 106–109.

Ebina, Hirotaka, Naoko Misawa, Yuka Kanemura, and Yoshio Koyanagi. "Harnessing the CRISPR/Cas9 system to disrupt latent HIV-1 provirus." *Scientific Reports* 3, no. 1 (2013): 1–7.

Engelman, Jeffrey A., Liang Chen, Xiaohong Tan, Katherine Crosby, Alexander R. Guimaraes, Rabi Upadhyay, Michel Maira et al. "Effective use of PI3K and MEK inhibitors to treat mutant Kras G12D and PIK3CA H1047R murine lung cancers." *Nature Medicine* 14, no. 12 (2008): 1351–1356.

Findlay, Gregory M., Evan A. Boyle, Ronald J. Hause, Jason C. Klein, and Jay Shendure. "Saturation editing of genomic regions by multiplex homology-directed repair." *Nature* 513, no. 7516 (2014): 120–123.

Fok, Wilson Chun, Evandro Luis de Oliveira Niero, Carissa Dege, Kirsten Ann Brenner, Christopher Michael Sturgeon, and Luis Francisco Zirnberger Batista. "p53 mediates failure of human definitive hematopoiesis in dyskeratosis congenita." *Stem Cell Reports* 9, no. 2 (2017): 409–418.

Fortunato, Fernanda, Rachele Rossi, Maria Sofia Falzarano, and Alessandra Ferlini. "Innovative therapeutic approaches for Duchenne muscular dystrophy." *Journal of Clinical Medicine* 10, no. 4 (2021): 820.

Frangoul, Haydar, David Altshuler, M. Domenica Cappellini, Yi-Shan Chen, Jennifer Domm, Brenda K. Eustace, Juergen Foell et al. "CRISPR-Cas9 gene editing for sickle cell disease and β-thalassemia." *New England Journal of Medicine* 384, no. 3 (2021): 252–260.

Fu, Yanfang, Jeffry D. Sander, Deepak Reyon, Vincent M. Cascio, and J. Keith Joung. "Improving CRISPR-Cas nuclease specificity using truncated guide RNAs." *Nature Biotechnology* 32, no. 3 (2014): 279–284.

Gaj, T., and C. A. Gersbach. "Barbas 3rd CF. ZFN, TALEN, and CRISPR/Cas-based methods for genome engineering." *Trends in Biotechnology* 5, no. 15 (2013): 2013.

Gaj, Thomas, Shannon J. Sirk, Sai-lan Shui, and Jia Liu. "Genome-editing technologies: Principles and applications." *Cold Spring Harbor Perspectives in Biology* 8, no. 12 (2016): a023754.

Gasiunas, Giedrius, Rodolphe Barrangou, Philippe Horvath, and Virginijus Siksnys. "Cas9–crRNA ribonucleoprotein complex mediates specific DNA cleavage for adaptive immunity in bacteria." *Proceedings of the National Academy of Sciences* 109, no. 39 (2012): E2579–E2586.

Gonzalo, Susana, Ray Kreienkamp, and Peter Askjaer. "Hutchinson-Gilford Progeria Syndrome: A premature aging disease caused by LMNA gene mutations." *Ageing Research Reviews* 33 (2017): 18–29.

Gootenberg, Jonathan S., Omar O. Abudayyeh, Jeong Wook Lee, Patrick Essletzbichler, Aaron J. Dy, Julia Joung, Vanessa Verdine et al. "Nucleic acid detection with CRISPR-Cas13a/C2c2." *Science* 356, no. 6336 (2017): 438–442.

Grissa, Ibtissem, Gilles Vergnaud, and Christine Pourcel. "The CRISPRdb database and tools to display CRISPRs and to generate dictionaries of spacers and repeats." *BMC Bioinformatics* 8, no. 1 (2007): 1–10.

Guerra, Carmen, Nieves Mijimolle, Alma Dhawahir, Pierre Dubus, Marta Barradas, Manuel Serrano, Victoria Campuzano et al. "Tumor induction by an endogenous K-ras oncogene is highly dependent on cellular context." *Cancer Cell* 4, no. 2 (2003): 111–120.

Guilinger, John P., David B. Thompson, and David R. Liu. "Fusion of catalytically inactive Cas9 to FokI nuclease improves the specificity of genome modification." *Nature Biotechnology* 32, no. 6 (2014): 577–582.

Gupta, Annapurna, and Meenal Agarwal. "Genome editing: Precise and CRISPER." *Genetic Clinics* 8, no. 3 (2015): 10–14.

Haft, Daniel H., Jeremy Selengut, Emmanuel F. Mongodin, and Karen E. Nelson. "A guild of 45 CRISPR-associated (Cas) protein families and multiple CRISPR/Cas subtypes exist in prokaryotic genomes." *PLoS Computational Biology* 1, no. 6 (2005): e60.

Harley, Calvin B. "Telomerase and cancer therapeutics." *Nature Reviews Cancer* 8, no. 3 (2008): 167–179.

Heyer, Joerg, Lawrence N. Kwong, Scott W. Lowe, and Lynda Chin. "Non-germline genetically engineered mouse models for translational cancer research." *Nature Reviews Cancer* 10, no. 7 (2010): 470–480.

Hille, Frank, and Emmanuelle Charpentier. "CRISPR-Cas: Biology, mechanisms and relevance." *Philosophical Transactions of the Royal Society B: Biological Sciences* 371, no. 1707 (2016): 20150496.

Holt, Nathalia, Jianbin Wang, Kenneth Kim, Geoffrey Friedman, Xingchao Wang, Vanessa Taupin, Gay M. Crooks et al. "Human hematopoietic stem/progenitor cells modified by zinc-finger nucleases targeted to CCR5 control HIV-1 in vivo." *Nature Biotechnology* 28, no. 8 (2010): 839–847.

Hu, Johnny H., Shannon M. Miller, Maarten H. Geurts, Weixin Tang, Liwei Chen, Ning Sun, Christina M. Zeina et al. "Evolved Cas9 variants with broad PAM compatibility and high DNA specificity." *Nature* 556, no. 7699 (2018): 57–63.

Hu, Wenhui, Rafal Kaminski, Fan Yang, Yonggang Zhang, Laura Cosentino, Fang Li, Biao Luo et al. "RNA-directed gene editing specifically eradicates latent and prevents new HIV-1 infection." *Proceedings of the National Academy of Sciences* 111, no. 31 (2014): 11461–11466.

Huang, Xionggao, Guohong Zhou, Wenyi Wu, Yajian Duan, Gaoen Ma, Jingyuan Song, Ru Xiao et al. "Genome editing abrogates angiogenesis in vivo." *Nature Communications* 8, no. 1 (2017): 1–8.

Ibrahim, A., Msz Özsöz, G. Tirah, and O. Gideon. "Genome engineering using the CRISPR Cas9 system." *Journal of Biomedical and Pharmaceutical Sciences* 2, no. 2 (2019).

Isgro, Antonella, Javid Gaziev, Pietro Sodani, and Guido Lucarelli. "Progress in hematopoietic stem cell transplantation as allogeneic cellular gene therapy in thalassemia." *Annals of the New York Academy of Sciences* 1202, no. 1 (2010): 149–154.

Ishino, Yoshizumi, Mart Krupovic, and Patrick Forterre. "History of CRISPR-Cas from encounter with a mysterious repeated sequence to genome editing technology." *Journal of Bacteriology* 200, no. 7 (2018): e00580–17.

Jackson, Erica L., Nicholas Willis, Kim Mercer, Roderick T. Bronson, Denise Crowley, Raymond Montoya, Tyler Jacks et al. Tuveson. "Analysis of lung tumor initiation and progression using conditional expression of oncogenic K-ras." *Genes & Development* 15, no. 24 (2001): 3243–3248.

Jackson, Maria, Leah Marks, Gerhard H. W. May, and Joanna B. Wilson. "The genetic basis of disease." *Essays in Biochemistry* 62, no. 5 (2018): 643–723.

Jinek, Martin, Krzysztof Chylinski, Ines Fonfara, Michael Hauer, Jennifer A. Doudna, and Emmanuelle Charpentier. "A programmable dual-RNA–guided DNA endonuclease in adaptive bacterial immunity." *Science* 337, no. 6096 (2012): 816–821.

Jinek, Martin, Alexandra East, Aaron Cheng, Steven Lin, Enbo Ma, and Jennifer Doudna. "RNA-programmed genome editing in human cells." *Elife* 2 (2013): e00471.

Kaminski, Rafal, Yilan Chen, Tracy Fischer, Ellen Tedaldi, Alessandro Napoli, Yonggang Zhang, Jonathan Karn, Wenhui Hu, and Kamel Khalili. "Correction: Corrigendum: Elimination of HIV-1 Genomes from Human T-lymphoid Cells by CRISPR/Cas9 Gene Editing." *Scientific Reports* 6, no. 1 (2016): 1–1.

Kantor, Ariel, Michelle E. McClements, and Robert E. MacLaren. "CRISPR-Cas9 DNA base-editing and prime-editing." *International Journal of Molecular Sciences* 21, no. 17 (2020): 6240.

Khadempar, Saedeh, Shokoufeh Familghadakchi, Roozbeh Akbari Motlagh, Najmeh Farahani, Maryam Dashtiahangar, Hamzeh Rezaei, and Seyed Mohammad Gheibi Hayat. "CRISPR–Cas9 in genome editing: Its function and medical applications." *Journal of Cellular Physiology* 234, no. 5 (2019): 5751–5761.

Khan, Faheem Ahmed, Nuruliarizki Shinta Pandupuspitasari, Huang Chun-Jie, Zhou Ao, Muhammad Jamal, Ali Zohaib, Farhan Ahmed Khan, Muthia Raihana Hakim, and Zhang ShuJun. "CRISPR/Cas9 therapeutics: A cure for cancer and other genetic diseases." *Oncotarget* 7, no. 32 (2016): 52541.

Khosravi, Mohammad Ali, Maryam Abbasalipour, Jean-Paul Concordet, Johannes Vom Berg, Sirous Zeinali, Arash Arashkia, Kayhan Azadmanesh et al. "Targeted deletion of BCL11A gene by CRISPR-Cas9 system for fetal hemoglobin reactivation: A promising approach for gene therapy of beta thalassemia disease." *European Journal of Pharmacology* 854 (2019): 398–405.

Khosravi, Mohammad Ali, Maryam Abbasalipour, Jean-Paul Concordet, Johannes Vom Berg, Sirous Zeinali, Arash Arashkia, Thorsten Buch et al. "Expression analysis data of BCL11A and γ-globin genes in KU812 and KG-1 cell lines after CRISPR/Cas9-mediated BCL11A enhancer deletion." *Data in Brief* 28 (2020): 104974.

Kim, Daesik, Sangsu Bae, Jeongbin Park, Eunji Kim, Seokjoong Kim, Hye Ryeong Yu, Jinha Hwang et al. "Digenome-seq: Genome-wide profiling of CRISPR-Cas9 off-target effects in human cells." *Nature Methods* 12, no. 3 (2015): 237–243.

Kleinstiver, Benjamin P., Vikram Pattanayak, Michelle S. Prew, Shengdar Q. Tsai, Nhu T. Nguyen, Zongli Zheng, and J. Keith Joung. "High-fidelity CRISPR–Cas9 nucleases with no detectable genome-wide off-target effects." *Nature* 529, no. 7587 (2016): 490–495.

Kleinstiver, Benjamin P., Michelle S. Prew, Shengdar Q. Tsai, Ved V. Topkar, Nhu T. Nguyen, Zongli Zheng, Andrew PW Gonzales et al. "Engineered CRISPR-Cas9 nucleases with altered PAM specificities." *Nature* 523, no. 7561 (2015): 481–485.

Konermann, Silvana, Peter Lotfy, Nicholas J. Brideau, Jennifer Oki, Maxim N. Shokhirev, and Patrick D. Hsu. "Transcriptome engineering with RNA-targeting type VI-D CRISPR effectors." *Cell* 173, no. 3 (2018): 665–676.

Kuscu, Cem, Sevki Arslan, Ritambhara Singh, Jeremy Thorpe, and Mazhar Adli. "Genome-wide analysis reveals characteristics of off-target sites bound by the Cas9 endonuclease." *Nature Biotechnology* 32, no. 7 (2014): 677–683.

Li, Hongmei Lisa, Naoko Fujimoto, Noriko Sasakawa, Saya Shirai, Tokiko Ohkame, Tetsushi Sakuma, Michihiro Tanaka et al. "Precise correction of the dystrophin gene in duchenne muscular dystrophy patient induced pluripotent stem cells by TALEN and CRISPR-Cas9." *Stem Cell Reports* 4, no. 1 (2015): 143–154.

Liao, Hsin-Kai, Ying Gu, Arturo Diaz, John Marlett, Yuta Takahashi, Mo Li, Keiichiro Suzuki et al. "Use of the CRISPR/Cas9 system as an intracellular defense against HIV-1 infection in human cells." *Nature Communications* 6, no. 1 (2015): 1–10.

Lin, Steven, Brett T. Staahl, Ravi K. Alla, and Jennifer A. Doudna. "Enhanced homology-directed human genome engineering by controlled timing of CRISPR/Cas9 delivery." *eLife* 3 (2014): e04766.

Liu, Mingjie, Saad Rehman, Xidian Tang, Kui Gu, Qinlei Fan, Dekun Chen, and Wentao Ma. "Methodologies for improving HDR efficiency." *Frontiers in Genetics* 9 (2019): 691.

Liu, Zhepeng, Shuliang Chen, Xu Jin, Qiankun Wang, Kongxiang Yang, Chenlin Li, Qiaoqiao Xiao et al. "Genome editing of the HIV co-receptors CCR5 and CXCR4 by CRISPR-Cas9 protects CD4+ T cells from HIV-1 infection." *Cell & Bioscience* 7, no. 1 (2017): 1–15.

Lv, Xiujuan, Kairui Qiu, Tianxiang Tu, Xiaoxue He, Yuchen Peng, Jinbin Ye, Junhao Fu et al. "Development of a simple and quick method to assess base editing in human cells." *Molecular Therapy-Nucleic Acids* 20 (2020): 580–588.

Ma, Cui-Cui, Zhen-Ling Wang, Ting Xu, Zhi-Yao He, and Yu-Quan Wei. "The approved gene therapy drugs worldwide: From 1998 to 2019." *Biotechnology Advances* 40 (2020): 107502.

Makarova, Kira S., Daniel H. Haft, Rodolphe Barrangou, Stan J. J. Brouns, Emmanuelle Charpentier, Philippe Horvath, Sylvain Moineau et al. "Evolution and classification of the CRISPR–Cas systems." *Nature Reviews Microbiology* 9, no. 6 (2011): 467–477.

Mali, Prashant, Luhan Yang, Kevin M. Esvelt, John Aach, Marc Guell, James E. DiCarlo, Julie E. Norville, and George M. Church. "RNA-guided human genome engineering via Cas9." *Science* 339, no. 6121 (2013): 823–826.

Malina, Abba, John R. Mills, Regina Cencic, Yifei Yan, James Fraser, Laura M. Schippers, Marilène Paquet, Josée Dostie, and Jerry Pelletier. "Repurposing CRISPR/Cas9 for in situ functional assays." *Genes & Development* 27, no. 23 (2013): 2602–2614.

Manguso, Robert T., Hans W. Pope, Margaret D. Zimmer, Flavian D. Brown, Kathleen B. Yates, Brian C. Miller, Natalie B. Collins et al. "In vivo CRISPR screening identifies Ptpn2 as a cancer immunotherapy target." *Nature* 547, no. 7664 (2017): 413–418.

Maresch, R., S. Mueller, C. Veltkamp, R. Öllinger, M. Friedrich, I. Heid, K. Steiger et al. "Multiplexed pancreatic genome engineering and cancer induction by transfection-based CRISPR/Cas9 delivery in mice." *Nature Communications* 2016; 7: 10770."

Matsoukas, Ianis G. "Prime editing: Genome editing for rare genetic diseases without double-strand breaks or donor DNA." *Frontiers in Genetics* 11 (2020): 528.

Mollanoori, H., Y. Rahmati, B. Hassani, Mehr M. Havasi, and S. Teimourian. "Promising therapeutic approaches using CRISPR/Cas9 genome editing technology in the treatment of duchenne muscular dystrophy." *Genes & Diseases* 8 (2021): 146–156.

O'Geen, Henriette, Isabelle M. Henry, Mital S. Bhakta, Joshua F. Meckler, and David J. Segal. "A genome-wide analysis of Cas9 binding specificity using ChIP-seq and targeted sequence capture." *Nucleic Acids Research* 43, no. 6 (2015): 3389–3404.

Pan, Deng, Aya Kobayashi, Peng Jiang, Lucas Ferrari de Andrade, Rong En Tay, Adrienne M. Luoma, Daphne Tsoucas et al. "A major chromatin regulator determines resistance of tumor cells to T cell–mediated killing." *Science* 359, no. 6377 (2018): 770–775.

Pandey, V. K., A. Tripathi, R. Bhushan, A. Ali, P. K. Dubey, and G. Therapy. "Application of CRISPR/Cas9 genome editing in genetic disorders: A systematic review up to date." *Journal of Genetic Syndromes & Gene Therapy* 8, no. 2 (2017): 1–10.

Porteus, Matthew. "Genome editing: A new approach to human therapeutics." *Annual Review of Pharmacology and Toxicology* 56 (2016): 163–190.

Ran, F. A., Le Cong, Winston X. Yan, David A. Scott, Jonathan S. Gootenberg, Andrea J. Kriz, Bernd Zetsche et al. "In vivo genome editing using Staphylococcus aureus Cas9." *Nature* 520, no. 7546 (2015): 186–191.

Ran, F. Ann, Patrick D. Hsu, Chie-Yu Lin, Jonathan S. Gootenberg, Silvana Konermann, Alexandro E. Trevino, David A. Scott et al. "Double nicking by RNA-guided CRISPR Cas9 for enhanced genome editing specificity." *Cell* 154, no. 6 (2013): 1380–1389.

Rath, Devashish, Lina Amlinger, Archana Rath, and Magnus Lundgren. "The CRISPR-Cas immune system: Biology, mechanisms and applications." *Biochimie* 117 (2015): 119–128.

Rogers, Zoë N., Christopher D. McFarland, Ian P. Winters, Santiago Naranjo, Chen-Hua Chuang, Dmitri Petrov, and Monte M. Winslow. "A quantitative and multiplexed approach to uncover the fitness landscape of tumor suppression in vivo." *Nature Methods* 14, no. 7 (2017): 737–742.

Ruan, Guo-Xiang, Elizabeth Barry, Dan Yu, Michael Lukason, Seng H. Cheng, and Abraham Scaria. "CRISPR/Cas9-mediated genome editing as a therapeutic approach for Leber congenital amaurosis 10." *Molecular Therapy* 25, no. 2 (2017): 331–341.

Scholefield, Janine, and Patrick T. Harrison. "Prime editing–an update on the field." *Gene Therapy* 28, no. 7 (2021): 396–401.

Schwank, Gerald, Bon-Kyoung Koo, Valentina Sasselli, Johanna F. Dekkers, Inha Heo, Turan Demircan, Nobuo Sasaki et al. "Functional repair of CFTR by CRISPR/Cas9 in intestinal stem cell organoids of cystic fibrosis patients." *Cell Stem Cell* 13, no. 6 (2013): 653–658.

Shalem, Ophir, Neville E. Sanjana, Ella Hartenian, Xi Shi, David A. Scott, Tarjei S. Mikkelsen, Dirk Heckl et al. "Genome-scale CRISPR-Cas9 knockout screening in human cells." *Science* 343, no. 6166 (2014): 84–87.

Sharon, J. Sha, Gayle K. Deutsch, Lu Tian, Kara Richardson, Maria Coburn, Jennifer L. Gaudioso, Tatiana Marcal et al. "Safety, tolerability, and feasibility of young plasma infusion in the plasma for Alzheimer symptom amelioration study: A randomized clinical trial." *JAMA Neurology* 76, no. 1 (2019): 35–40.

Shi, Junwei, Eric Wang, Joseph P. Milazzo, Zihua Wang, Justin B. Kinney, and Christopher R. Vakoc. "Discovery of cancer drug targets by CRISPR-Cas9 screening of protein domains." *Nature Biotechnology* 33, no. 6 (2015): 661–667.

Slaymaker, Ian M., Linyi Gao, Bernd Zetsche, David A. Scott, Winston X. Yan, and Feng Zhang. "Rationally engineered Cas9 nucleases with improved specificity." *Science* 351, no. 6268 (2016): 84–88.

Smith, Cory, Athurva Gore, Wei Yan, Leire Abalde-Atristain, Zhe Li, Chaoxia He, Ying Wang et al. "Whole-genome sequencing analysis reveals high specificity of CRISPR/Cas9 and TALEN-based genome editing in human iPSCs." *Cell Stem Cell* 15, no. 1 (2014): 12–13.

Somayaji, Ranjani, Dave P. Nichols, and Scott C. Bell. "Cystic fibrosis–Ten promising therapeutic approaches in the current era of care." *Expert Opinion on Investigational Drugs* 29, no. 10 (2020): 1107–1124.

Tebas, Pablo, David Stein, Winson W. Tang, Ian Frank, Shelley Q. Wang, Gary Lee, S. Kaye Spratt et al. "Gene editing of CCR5 in autologous CD4 T cells of persons infected with HIV." *New England Journal of Medicine* 370, no. 10 (2014): 901–910.

Thompson, Alexis A., Mark C. Walters, Janet Kwiatkowski, John E. J. Rasko, Jean-Antoine Ribeil, Suradej Hongeng, Elisa Magrin et al. "Gene therapy in patients with transfusion-dependent β-thalassemia." *New England Journal of Medicine* 378, no. 16 (2018): 1479–1493.

Tsai, Shengdar Q., Nicolas Wyvekens, Cyd Khayter, Jennifer A. Foden, Vishal Thapar, Deepak Reyon, Mathew J. Goodwin et al. "Dimeric CRISPR RNA-guided FokI nucleases for highly specific genome editing." *Nature Biotechnology* 32, no. 6 (2014): 569–576.

Tsai, Shengdar Q., Zongli Zheng, Nhu T. Nguyen, Matthew Liebers, Ved V. Topkar, Vishal Thapar, Nicolas Wyvekens et al. "GUIDE-seq enables genome-wide profiling of off-target cleavage by CRISPR-Cas nucleases." *Nature Biotechnology* 33, no. 2 (2015): 187–197.

Tzelepis, Konstantinos, Hiroko Koike-Yusa, Etienne De Braekeleer, Yilong Li, Emmanouil Metzakopian, Oliver M. Dovey, Annalisa Mupo et al. "A CRISPR dropout screen identifies genetic vulnerabilities and therapeutic targets in acute myeloid leukemia." *Cell Reports* 17, no. 4 (2016): 1193–1205.

Veres, Adrian, Bridget S. Gosis, Qiurong Ding, Ryan Collins, Ashok Ragavendran, Harrison Brand, Serkan Erdin et al. "Low incidence of off-target mutations in individual CRISPR-Cas9 and TALEN targeted human stem cell clones detected by whole-genome sequencing." *Cell Stem Cell* 15, no. 1 (2014): 27–30.

Wang, Haoyi, Hui Yang, Chikdu S. Shivalila, Meelad M. Dawlaty, Albert W. Cheng, Feng Zhang, and Rudolf Jaenisch. "One-step generation of mice carrying mutations in multiple genes by CRISPR/Cas-mediated genome engineering." *Cell* 153, no. 4 (2013): 910–918.

Wang, Zhen, Qinghua Pan, Patrick Gendron, Weijun Zhu, Fei Guo, Shan Cen, Mark A. Wainberg, and Chen Liang. "CRISPR/Cas9-derived mutations both inhibit HIV-1 replication and accelerate viral escape." *Cell Reports* 15, no. 3 (2016): 481–489.

Weber, Julia, Rupert Öllinger, Mathias Friedrich, Ursula Ehmer, Maxim Barenboim, Katja Steiger, Irina Heid et al. "CRISPR/Cas9 somatic multiplex-mutagenesis for high-throughput functional cancer genomics in mice." *Proceedings of the National Academy of Sciences* 112, no. 45 (2015): 13982–13987.

Wu, Xuebing, David A. Scott, Andrea J. Kriz, Anthony C. Chiu, Patrick D. Hsu, Daniel B. Dadon, Albert W. Cheng et al. "Genome-wide binding of the CRISPR endonuclease Cas9 in mammalian cells." *Nature Biotechnology* 32, no. 7 (2014): 670–676.

Wu, Yuxuan, Jing Zeng, Benjamin P. Roscoe, Pengpeng Liu, Qiuming Yao, Cicera R. Lazzarotto, Kendell Clement et al. "Highly efficient therapeutic gene editing of human hematopoietic stem cells." *Nature Medicine* 25, no. 5 (2019): 776–783.

Wyvekens, Nicolas, Ved V. Topkar, Cyd Khayter, J. Keith Joung, and Shengdar Q. Tsai. "Dimeric CRISPR RNA-guided FokI-dCas9 nucleases directed by truncated gRNAs for highly specific genome editing." *Human Gene Therapy* 26, no. 7 (2015): 425–431.

Xie, Fei, Lin Ye, Judy C. Chang, Ashley I. Beyer, Jiaming Wang, Marcus O. Muench, and Yuet Wai Kan. "Seamless gene correction of β-thalassemia mutations in patient-specific iPSCs using CRISPR/Cas9 and piggyBac." *Genome Research* 24, no. 9 (2014): 1526–1533.

Yang, Han, Shuling Ren, Siyuan Yu, Haifeng Pan, Tingdong Li, Shengxiang Ge, Jun Zhang, and Ningshao Xia. "Methods favoring homology-directed repair choice in response to CRISPR/Cas9 induced-double strand breaks." *International Journal of Molecular Sciences* 21, no. 18 (2020): 6461.

Yang, Hui, Haoyi Wang, Chikdu S. Shivalila, Albert W. Cheng, Linyu Shi, and Rudolf Jaenisch. "One-step generation of mice carrying reporter and conditional alleles by CRISPR/Cas-mediated genome engineering." *Cell* 154, no. 6 (2013): 1370–1379.

Yang, Li, Jing Tang, Xuelei Ma, Yuan Lin, Guorong Ma, Minghai Shan, Libin Wang, and Yanhui Yang. "Progression and application of CRISPR-Cas genomic editors." *Methods* 194 (2021): 65–74.

Yang, Luhan, Dennis Grishin, Gang Wang, John Aach, Cheng-Zhong Zhang, Raj Chari, Jason Homsy et al. "Targeted and genome-wide sequencing reveal single nucleotide variations impacting specificity of Cas9 in human stem cells." *Nature Communications* 5, no. 1 (2014): 1–6.

Yau, Edwin H., Indrasena Reddy Kummetha, Gianluigi Lichinchi, Rachel Tang, Yunlin Zhang, and Tariq M. Rana. "Genome-wide CRISPR screen for essential cell growth mediators in mutant KRAS colorectal cancersgenome-wide CRISPR screen of KRAS-mutant tumor xenografts." *Cancer Research* 77, no. 22 (2017): 6330–6339.

Yin, Hao, Wen Xue, and Daniel G. Anderson. "CRISPR–Cas: A tool for cancer research and therapeutics." *Nature Reviews Clinical Oncology* 16, no. 5 (2019): 281–295.

Yzer, Suzanne, Anneke I. den Hollander, Irma Lopez, Jan-Willem R. Pott, Jan Tjeerd HN de Faber, Frans PM Cremers, Robert K. Koenekoop et al. "Ocular and extra-ocular features of patients with Leber congenital amaurosis and mutations in CEP290." *Molecular Vision* 18 (2012): 412.

Zhu, Weijun, Rongyue Lei, Yann Le Duff, Jian Li, Fei Guo, Mark A. Wainberg, and Chen Liang. "The CRISPR/Cas9 system inactivates latent HIV-1 proviral DNA." *Retrovirology* 12, no. 1 (2015): 1–7.

8 Understanding the Efficacy of the Interaction of Herbal Drug-Like Biomolecules with the Virulent Strain of *Mycobacterium Tuberculosis* Proteins Through an In Silico Study

Sonali Paul[1], *Nivedita Chatterjee*[1], *Asmeeta Sircar*[1], *Shimantika Maikap*[1], *Abhilasha Singh*[1] *and Susmita Mukherjee*[1,*]
[1]Department of Biotechnology, University of Engineering & Management, Kolkata
* Coorespondence to: susmita.mukherjee@uem.edu.in

8.1 INTRODUCTION

Tuberculosis (TB) is the leading cause of infectious morbidity and mortality globally. It spreads easily through the inhalation of infected droplets that are exhaled, coughed or sneezed by individuals affected with contagious forms of pulmonary TB. TB infection is caused by *Mycobacterium tuberculosis*, which resides inside the granuloma of affected individuals when the bacteria are present. These cases can be called inactive or latently infected (Méchaï et al., 2020). Recent studies indicate that at a 95% confidence interval (CI), 20–26.4% of the world's population is infected, which amounts to approximately 1.7 billion people. Individuals who were latently infected have an approximately 10% risk of developing active TB through reactivation of the bacteria due to the immuno-compromised system of the host (Houben and Dodd, 2016). Risk factors for the activation of the bacteria can be an immunodeficiency virus, diabetes and malnutrition. Both short- and long-term exposures to ambient air pollution could increase susceptibility to *M. tuberculosis* infection and the development of TB by impairing the immunity of the human respiratory system, thereby resulting in an increased risk of TB (Lönnroth et al., 2010; Narasimhan et al., 2013).

In meta-analyses of different studies, air pollution from passive smoking and biomass fuel has been revealed to cause an increased rate of TB. Evidence has been gathered for a robust association between the risk of getting infected by TB and air pollution, for example, particulate matter such as $PM_{2.5}$, PM_{10}, carbon-monoxide and nitrogen and sulphur oxides (Lin et al., 2007; Sumpter and Chandramohan, 2013; Jassal et al., 2013). A study model has shown that when these pollutants increased in concentration, the number of TB cases increased by 5–6.9% where the optimum pollutants were PM_{10} and NO_2 (Yang et al., 2020).

DOI: 10.1201/9781003302131-9

Out of the strains of TB bacteria, the *Mycobacterium tuberculosis bovis*, a member of the *Mycobacterium tuberculosis* complex, is an integral causative agent of the disease in cattle. In addition, this specific strain can also cause disease in human beings (Van der Werf et al., 2016) through the food chain. The transmission in humans takes place after close contact with the infected animals or if unpasteurized dairy products are consumed (Gupta et al., 2010). The symptoms that this strain causes are very similar to that of *M. tuberculosis*, but in case of *M. bovis*, additional pulmonary symptoms are more likely to be caused. It has been observed that this variant affects people more in developing countries as there is not very adept veterinary control (Bapat et al., 2017). In different research on this strain, three genes were identified as integral components for the cause of virulence. These genes help to sustain major pathways and structural stability in both *M. bovis* and *M. tuberculosis*. The three genes are, *1. aftd* (PDB ID: 6WBY) (Michel et al., 2010; Angala et al., 2014; Jankute et al., 2013), *2. accD2* (PDB ID: 5UHB) (Favrot and Donald, 2014) and *3. adhE1* (PDB ID: 6DVE) (Ayaz and Alpsoy, 2007)

Many drugs already existing on the market have successfully subdued these microbes, but they cannot completely inhibit bacterial activities. Moreover, studies have shown that side effects have also been observed in patients undergoing fixed-dosage therapy for TB (Mejri et al., 2016; Tanavade et al., 2016). These side effects included hepatic cytolysis, cutaneous reactions, hepatotoxicity, hyperuricemia, peripheral neuropathy and even nausea. The average time for these side effects to surface was observed to be 15 to 20 days (Tkhayat et al., 2018; Sana et al., 2016). Plants, namely, *Acalypha indica* and *Allium sativum* (Saha and Ahmed, 2011), have been used consistently in traditional medicines, especially *Allium sativum*, which is commonly known as garlic. Garlic has been used in traditional medicine for the treatment of fevers, colds, blood pressure and blood sugar and helps enhance the immune system; it is effective against parasitic, bacterial, fungal and viral infections (Lyonnet et al., 2017). *Acalypha indica* has been tested previously for its antitubercular activity in studies, where it portrayed a 68% inhibition against H37Rv at 4% v/v concentration (Rienksma, 2019).

In this context, the present in silico drug development overview aimed at understanding the efficacy of the biomolecules (drug-like) of *Acalypha indica* and *Allium sativum* as potential drugs for treatment of the virulent strains of TB through a docking study.

8.2 MATERIALS AND METHODS

To meet the objective of this work through the in silico method, it is important to get not only the proteins of the respective virulent genes but also the FASTA sequence of the amino acids with bioinformatic study. The bioactive molecules of the two herbal plants were searched and tested for their drug-like properties, and these biomolecules were tested against the bacterial proteins for the docking study.

8.2.1 Getting the FASTA Sequence of the Proteins

Information regarding the FASTA sequences of these virulent genes selected from the *Mycobacterium tuberculosis bovis* strain was gathered from NCBI (www.ncbi.nlm.nih.gov/). The protein FASTA sequence was directly searched in NCBI by using the gene names of the variant *bovis*.

8.2.2 Getting the Crystallographic Structure of the Proteins

The crystallographic structures of the proteins were searched in the Protein Data Bank (PDB), which is a centralized database for protein molecules. The 3-dimensional structures of these specific proteins were also searched in the PDB (*www.rcsb.org/*).

8.2.3 Testing the Drug-Like Properties of the Bioactive Compounds from the Medicinal Plants

The drug-likeness was tested by using the simplified molecular-input line-entry system (SMILE) of the two biomolecules selected. For *Acalypha indica*, it is 9, 12-octadecadienoic acid (Z,Z) methyl ester (PubChem CID: 3931), and for *Allium sativum*, it is Alliin (PubChem CID: 121922) according to PubChem (https://pubchem.ncbi.nlm.nih.gov/). This SMILE was then entered into SwissADME (www.swissadme.ch/), and the drug-likeness was observed. We also determined the LD50 or toxicity class from the Pro Tox server (*https://tox-new.charite.de/protox_II/*).

8.2.4 Getting the Structure of the Ligand

The structures of the selected ligands were downloaded in. sdf format from the PDB, which with the help of OpenBabble software, was later converted to. pdbqt format.. pdbqt format is used for the docking interaction study.

8.2.5 Docking Study

The AutoDock suite can be used for the docking analysis of the ligands with the genes selected from the virulent *M. bovis*. The grid parameters for this docking study were 140Å * 180Å * 195 Å. The protein structure was prepared for the docking study by removing all of the water molecules and adding charges and explicit hydrogen molecules. Afterwards, the molecular docking was studied using auto dock tools. Finally, the docking results of these proteins with the ligands were analyzed to understand the efficacy of interaction.

8.3 RESULTS AND DISCUSSION

8.3.1 The Crystallographic Structure of Proteins

From the literature, it has been found that three genes play an integral part in the pathway of the bacterial life maintenance system. Figure 8.1 depicts the PDB structure of these proteins based on the FASTA sequence of the amino acids.

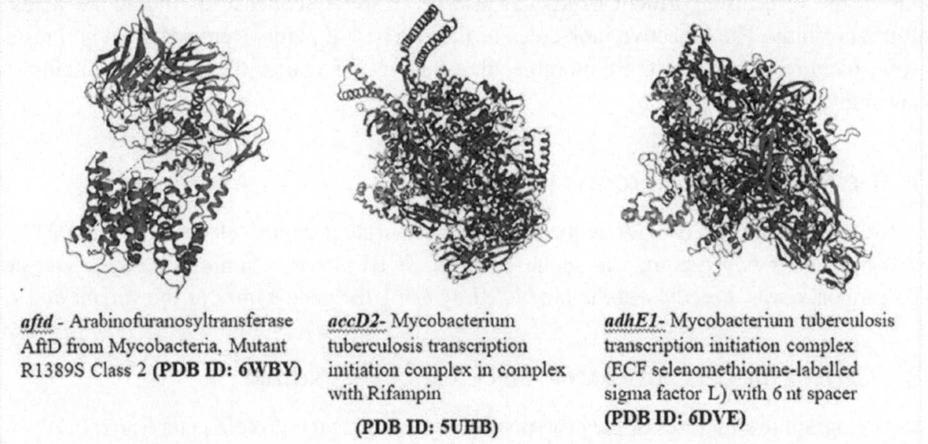

FIGURE 8.1 Crystallographic structures of the proteins selected for this study (PDB).

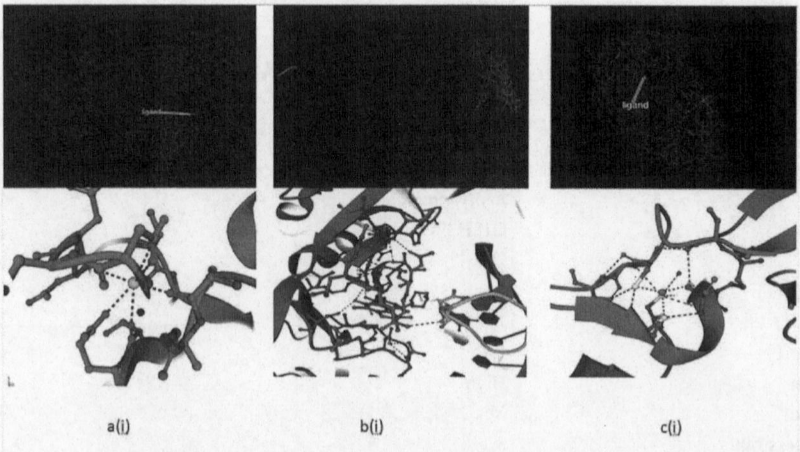

FIGURE 8.2.1 Interaction of Arabinofuranosyltransferase D (AftD) from Mycobacteria, Mutant R1389S Class 2 (**a(i)**), Mycobacterium tuberculosis transcription initiation complex with Rifampin (**b(i)**) and Mycobacterium tuberculosis transcription initiation complex (ECF selenomethionine-labelled sigma factor L) with 6 nt spacer (**c(i)**) with the drug molecule 9, 12-octadecadienoic acid (Z,Z) methyl ester from *Acalypha indica.*

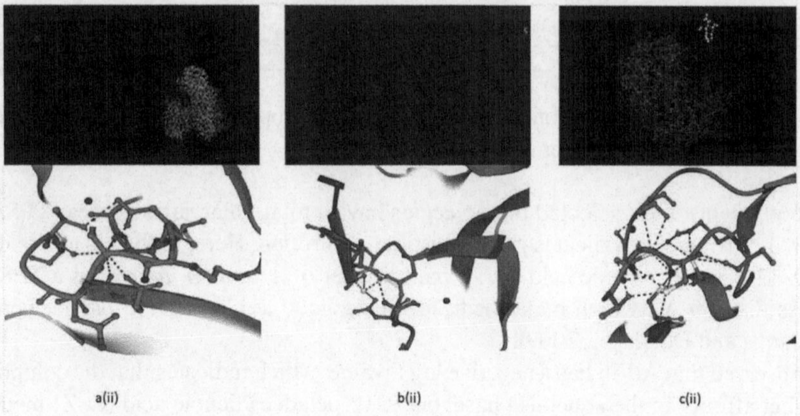

FIGURE 8.2.2 Interaction of Arabinofuranosyltransferase (AftD) from Mycobacteria, Mutant R1389S Class 2 (**a(ii)**), Mycobacterium tuberculosis transcription initiation complex with Rifampin (**b(ii)**) and Mycobacterium tuberculosis transcription initiation complex (ECF selenomethionine-labelled sigma factor L) with 6 nt spacer (**c(ii)**) with the drug molecule Alliin from *Allium sativum.*

8.3.2 The Drug-Likeness Property of the Bioactive Herbal Molecules

The plants selected for this study are *Acalypha indica* and *Allium sativum*, and the bioactive compounds studied were the following:

1. *Acalypha indica:* From this plant, we selected one molecule, a bioactive compound known as 9, 12-octadecadienoic acid (Z, Z) methyl ester (Lyonnet et al., 2017); and
2. *Allium sativum*: From this plant, we chose the bioactive compound Alliin (Rienksma, 2019).

It is important to study the drug-like property of these molecules; Table 8.1 describes the absorption, distribution, metabolism and excretion (ADME) properties of the molecules.

TABLE 8.1
Pharmacological Properties of the Bioactive Molecules (ADME)

Molecule	9, 12-octadecadienoic acid (Z,Z) methyl ester	Alliin
Plant	*Acalypha indica*	*Allium sativum*
Formula	C18H32O2	C6H11NO3S
MW	280.45	177.22
H-bond(acceptor)	2	4
H-bond (donor)	1	2
Consensus log P	5.45	-1.33
GI absorption	High	High
BBB permeant	Yes	No
Synthetic Accessibility	3.1	3.21
TPSA Å2	37.3	99.6
Solubility ESOL	Moderately soluble	soluble
Lipinski violations	1	0
Ghosh violations	1	-
Veber violations	1	-
Lead-likeness violations	2	1
LD50 value (mg/kg)	10,000	8,000
Solubility ESOL	High	High

The ADME results (Table 8.1) on the pharmacokinetic properties of the drug-like biomolecules from selected plants are analyzed as follows:

1. For both plants, their selected biomolecules have a total polar surface area (TPSA) value below 140 Å^2, which indicates good intestinal absorption. Here, in the bioactive molecule, the 9, 12-octadecadienoic acid (Z, Z) methyl ester of *Acalypha indica* has a TPSA value of less than 60 Å^2, which makes it fit for being good at blood-brain barrier penetration (Prasanna and Doerksen, 2009).
2. It is observed that Alliin has a negative log P value, which indicates that the compound has a higher affinity in the aqueous phase, but 9, 12-octadecadienoic acid (Z, Z) methyl ester has a positive value of 5.45, indicating a higher concentration in the lipid phase (Irina and Ruben, 2015).
3. Allin has only one violation in lead-likeness, whereas 9, 12-octadecadienoic acid (Z, Z) methyl ester has a 1 Lipinski rule violation, 1 Ghosh rule violation, 1 Veber rule violation and 2 lead-likeness violations.
4. The LD50 value of Alliin was found to be 8,000, and that of 9, 12-octadecadienoic acid (Z, Z) methyl ester was found to be 10,000, which makes them class 6 in terms of toxicity, and they are therefore safe for ingestion according to the FDA.

8.3.3 Docking Results

Molecular docking studies were used to predict the intermolecular interaction, that is, the efficiency of binding between a ligand and a protein, that can thus account for the ligand molecules' bioactivity (Agwupuye et al., 2022).

The Ki-inhibitory constant (Table 8.2) is the concentration of the drug molecule (inhibitor) that is required to decrease the maximal rate of gene expression (reaction) by half. The best binding energy was found in the interaction between 9, 12-octadecadienoic acid (Z, Z) methyl ester and all three bacterial proteins.

TABLE 8.2
Binding Energy and Inhibition Constant in the Docking Interaction

Drug-like molecule	Protein name	PDB ID	Binding energy (Kcal/mol)	Estimated inhibition constant (Ki)
9, 12-octadecadienoic acid (Z,Z) methyl ester	Arabinofuranosyltransferase (AftD) from Mycobacteria, Mutant R1389S Class 2	6WBY	-5.72	63.95 uM
	Mycobacterium tuberculosis transcription initiation complex with Rifampin	5UHB	-5.06	194.94 uM
	Mycobacterium tuberculosis transcription initiation complex (ECF selenomethionine-labelled sigma factor L) with 6 nt spacer	6DVE	-5.82	53.80 uM
Alliin	Arabinofuranosyltransferase (AftD) from Mycobacteria, Mutant R1389S Class 2	6WBY	-2.21	23.87 uM
	Mycobacterium tuberculosis transcription initiation complex with Rifampin	5UHB	-1.76	50.99 uM
	Mycobacterium tuberculosis transcription initiation complex (ECF selenomethionine-labelled sigma factor L) with 6 nt spacer	6DVE	-1.41	92.67 uM

Molecular docking interaction results (Table 8.2) analysis gave the following findings.

1. Alliin does not bind directly to the proteins, which shows that it is a synergistic compound, as has been noticed in previous studies (Chung et al., 2007).
2. Alliin interacts best with Arabinofuranosyltransferase (AftD) from Mycobacteria, Mutant R1389S Class 2 protein with a Ki of 23.87 uM.
3. We observed good results from 9, 12-octadecadienoicacid (Z, Z) methyl ester for all three of the genes that were taken.
4. The inhibition constant was highest for the protein against gene *accD2* and against drug molecule 9, 12-octadecadienoicacid (Z, Z) methyl ester, which proves to be a more efficient inhibition of gene expression.

8.4 CONCLUSION

This chapter aimed at depicting a drug development strategy from herbal drug-like molecules against the virulent strain of *Mycobacterium tuberculosis bovis* through an *in silico* analysis. The drug-likeness property showed that Alliin does not violate any of the major rules, but the docking result depicted a low inhibition constant, which means that it cannot downregulate gene expression. On the contrary, the docking results showed that the drug-like molecule, 9, 12-octadecadienoic acid (Z,Z) methyl ester has the minimum binding energy with all three proteins, which helps us to predict that this ligand might downregulate the expression of the proteins and can inhibit the pathways regulated by the specific proteins. Thus, it can be predicted that this drug-like molecule might have an inhibitory effect on the virulent genes of *Mycobacterium tuberculosis bovis*.

REFERENCES

Agwupuye, J. A., H. Louis, O. C. Enudi, et al. 2022. Theoretical insight into electronic and molecular properties of halogenated (F, Cl, Br) and hetero-atom (N, O, S) doped cyclooctane. *Mat Chem Phys.*, 275: 125239.

Angala, Shiva Kumar, Belardinelli Juan Manuel, Huc-Claustre Emilie, et al. 2014. The cell envelope glycoconjugates of Mycobacterium tuberculosis. *Crit Rev Biochem Mol Biol.*, 49: 361–399. https://doi.org/10.3109/10409238.2014.925420

Ayaz, Erol, & Hüseyin Can Alpsoy. 2007. Garlic (Allium sativum) and traditional medicine. *Turkiye Parazitol Derg.*, 31(2): 145–149.

Bapat, Prachi R., Dodkey S. Renuka, Shekhawat D. Seema, et al. 2017. Prevalence of zoonotic tuberculosis and associated risk factors in Central Indian populations. *J Epidemiol Glob Health.*, 7(4): 277–283. https://doi.org/10.1016/j.jegh.2017.08.007

Chung, I., S. H. Kwon, S.-T. Shim, et al. 2007. Synergistic antiyeast activity of garlic oil and allyl alcohol derived from alliin in garlic. *J Food Sci.*, 72(9): M437–M440. https://doi.org/10.1111/j.1750-3841.2007.00545.x

Favrot, Lorenza, & Ronning R. Donald. 2014. Targeting the mycobacterial envelope for tuberculosis drug development. *Expert Rev Anti Infect Ther.*, 10(9): 1023–1036. https://doi.org/10.1586/eri.12.91

Gupta, Renu, Thakur Bandana, Singh Pushpendra, et al. 2010. Anti-tuberculosis activity of selected medicinal plants against multi-drug resistant Mycobacterium tuberculosis isolates. *Indian J Med Res.*, 131: 809–813.

Houben, R. M., & P. J. Dodd. 2016. The global burden of latent tuberculosis infection: a re-estimation using mathematical modelling. *PLoS Med.*, 13: e1002152. https://doi.org/10.1371/journal.pmed.1002152

Irina, Kufareva, & Abagyan Ruben. 2015. Methods of protein structure comparison. *Methods Mol Biol.*, 857: 231–257. https://doi.org/10.1007/978-1-61779-588-6_10

Jankute, Monika, Grover Shipra, L. Birch Helen, et al. 2013. Genetics of mycobacterial arabinogalactan and lipoarabinomannan assembly. *MGM2–0013.*, 2(4). https://doi.org/10.1128/microbiolspec

Jassal, M. S., I. Bakman, & B. Jones. 2013. Correlation of ambient pollution levels and heavily-trafficked roadway proximity on the prevalence of smear-positive tuberculosis. *Public Health.*, 127: 268–274. doi:10.1016/j.puhe.2012.12.030

Lin, H. H., M. Ezzati, & M. Murray. 2007. M. Tobacco smoke, indoor air pollution and tuberculosis: a systematic review and meta-analysis. *PLoS Med.*, 4: 173–189. https://doi.org/10.1371/journal.pmed.0040020

Lönnroth, K., K. G. Castro, J. M. Chakaya, et al. 2010. Tuberculosis control and elimination 2010–50: cure, care, and social development. *Lancet.*, 375: 1814–1829. https://doi.org/10.1016/S0140-6736[10]60483-7

Lyonnet, B., L. Diacovich Bazet, G. Gago, et al. 2017. Functional reconstitution of the Mycobacterium tuberculosis long-chain acyl-CoA carboxylase from multiple acyl-CoA subunits. *FEBS J.*, 284(7): 1110–1125. https://doi.org/10.1111/febs.14046

Méchaï, Frédéric, Cordel Hugues, Guglielmetti Lorenzo, et al. 2020. Management of tuberculosis: are the practices homogeneous in high-income countries? *Front. Public Heal.*, 8: 443. https://doi.org/10.3389/fpubh.2020.00443

Mejri, Insaf, Loukil Manel, Khalfallah Ikbelle, et al. 2016. Side effects of tuberculosis treatment with fixed-dose combinations. *Eur Respir J.*, 48: PA2713 https://doi.org/10.1183/13993003.congress-2016.PA2713

Michel, A. L., B. Müller, & P. D. van Helden. 2010. Mycobacterium bovis at the animal–human interface: a problem, or not? *Vet Microbiol.*, 140: 371–381.

Narasimhan, P., J. Wood, C. R. MacIntyre, et al. 2013. Risk factors for tuberculosis. *Pulm Med.*, 2013: 828939. https://doi.org/10.1155/2013/828939

Prasanna, S., & Robert J. Doerksen. 2009. Topological polar surface area: a useful descriptor in 2D-QSAR. *Curr Med Chem.*, 16(1): 21–41. https://doi.org/10.2174/092986709787002817

Rienksma, Rienk A. 2019. Modeling metabolism of mycobacterium tuberculosis, 168 pages. https://doi.org/10.18174/472939

Saha, Rajsekhar, & Azhar Ahmed. 2011. Phytochemical constituents and pharmacological activities of acalyphus Indica Linn: a review. *Int J Pharm Sci Res.*, 2(8): 1900–1904. http://dx.doi.org/10.13040/IJPSR.0975-8232

Sana, Aissa, et al. 2016. Safety and acceptability of a 4-drug fixed-dose combination regimen compared with separate drugs for treatment of pulmonary tuberculosis. *Eur Respir J.*, 48: PA2714. http://dx.doi.org/10.1183/13993003.congress-2016.PA2714

Sumpter, C., & D. Chandramohan. 2013. Systematic review and meta-analysis of the associations between indoor air pollution and tuberculosis. *Trop Med Int Health.*, 18: 101–108. http://dx.doi.org/10.1111/tmi.12013

Tanavade, S. S., D. D. Chougule, & N. S. Naikwade. 2016. Indian scenario on fixed dose combinations. *Res J Pharm Technol.*, 9(5): 587–590. http://dx.doi.org/10.5958/0974-360X.2016.00110.4

Tkhayat, Achraf Ben, Soumaya Ben Saad, Hafaoua Daghfous, & Fatma Tritar. 2018. Fixed-dose combination anti-tuberculosis therapy: is the INH dose adapted for the Tunisian population? *Eur Respir J.*, 52: PA2694. http://dx.doi.org/10.1183/13993003.congress-2018.PA2694

Van der Werf, M. J., C. Kodmon, P. Zucs, et al. 2016. Tuberculosis and HIV coinfection in Europe: looking at one reality from two angles. *AIDS*. 30: 2845–2853. http://dx.doi.org/10.1097/QAD.0000000000001252

Yang, Jiandong, Zhang Mengxi, Chen Yanggui, et al. 2020. A study on the relationship between air pollution and pulmonary tuberculosis based on the general additive model in Wulumuqi, China. *Int J Infect Dis.*, 96: 42–47. https://doi.org/10.1016/j.ijid.2020.03.032

9 The Role of Neuro-Imaging Biomarkers in the Early Detection of Dementia and Alzheimer's Disease

Maitrayee Banerjee Mukherjee[1,*]
[1]Department of Physiology, Krishnagar Government College
*Correspondence to: Maitrayee_77@yahoo.co.in

9.1 INTRODUCTION

Alzheimer's disease (AD) is a neurodegenerative disease that causes a gradual decline in cognitive and behavioral functions including reasoning, memory, comprehension and judgment. Language disorder, a lack of motivation, development of neuropsychiatric symptoms such as apathy, social withdrawal, agitation, psychosis, wandering and the impairment of visuospatial skills are also very common in the middle to late stages. AD is the principal cause of dementia among the aged population worldwide (Tiwari *et al.*, 2019). Assessment of the disease is appropriate for patients with unrevealed persistent impairment of cognition and those with progressive dementia and the early-age onset of similar symptoms (Henry *et al.*, 2010).

The loss of mental processes of perception and cognition and poor memory functions due to the selective accelerated loss of neurons in the forebrain and other brain areas are the principal symptoms of the disorder. The disease is represented by **impairment of episodic memory** (relative preservation of distant memory and amnesia for more recent events), **language deficits**, **apraxia** (inability to carry out skilled motor activities), **agnosia** (failure to recognize objects such as clothing, places and people), **impaired frontal executive function** (inaptitude of organizing, planning and sequencing of work), **posterior cortical atrophy** (the least common presentation of AD, with visual disorientation due to the initial involvement of occipital lobes and occipito parietal regions that result in complex visual symptoms), **anosognosia** (lack of insight to seek medical attention or counseling) and development of some **late non-cognitive features** (myoclonus followed by seizures, reversal of the sleep-wakefulness cycle and impaired swallowing leading to aspiration pneumonia) (Giri *et al.*, 2016, Yiannopoulou and Papageorgiou, 2013). In clinical practice, impairment of two or more symptoms, specifically, memory, calculation, orientation, language or judgment, is considered as criteria for diagnosing dementia as recommended by the *Diagnostic and Statistical Manual of Mental Disorders* (Fourth Edition [*DSM-IV*]). The National Institute of Neurological and Communicative Disorders and Stroke-Alzheimer's Disease and Related Disorders Association (NINCDS-ADRDA) Work Group criteria require the diagnosis of dementia substantiated by clinical examination, deficits in at least 2 cognitive domains, absence of other systemic illness and a progressive decline in memory functions for the diagnosis of "probable AD." Analysis for the global burden of disease (GBD) study 2019 statistically projected and predicted that throughout the world, the number of people with dementia would rise from 57.4 (95% uncertainty interval is 50.4–65.1) million in 2019 to 152.8 (130.8–175.9) million within 2050. The result of this study also proposed that women are diagnosed with dementia more than men globally in 2019 (female-to-male ratio of 1:69 [1:64–1:73]) and predicted this quantitative

DOI: 10.1201/9781003302131-10

relation up to 2050 (female-to-male ratio of 1:67 [1:52–1:85]) (GBD 2019 Dementia Forecasting Collaborators, 2022).

Identification of the proper **non-invasive biomarkers of AD** in the presymptomatic phase is a prerequisite for early diagnosis and proper psychological and pharmaceutical intervention. Several advancements have taken place for developing specific biomarkers for AD and dementia using neuroimaging approaches that provide versatility in terms of targeting distinct **geriatric** and **pathophysiological indicators** such as structural atrophy (e.g., volumetry, shrinkage and cortical thinning), changes of function as evident in fMRI activity, loss of neuronal and synaptic connectivity (e.g., diffusion anisotropy) and pathological accumulations (e.g., amyloid and tau positron emission tomography (PET)). Neuroimaging biomarkers use brain imaging techniques that visually represent the morphological changes (MRI), functional capacity (functional MRI), vasculatures (perfusion MRI) and metabolic imaging (e.g., PET fluorodeoxyglucose (FDG)) of the brain. Through the study of the molecular content of normal and abnormal brain tissue or lesions by using an MRI scanner, for example, MR spectroscopy, proton magnetic resonance spectroscopy gives an idea of the regional deposition of metabolites and neurotransmitters (Elsinga *et al.*, 2006). A number of molecular probes are used to visualize both cellular activity and metabolic or enzymatic activities and to monitor blood flow in specific brain tissues, whereas radiotracer atoms or isotopes are used for imaging. Ongoing clinical trials using specific PET tracers for the diagnosis of AD are the imaging epigenetics, quantification and localization of neuroinflammation and integration of synapses (Márquez and Yassa, 2019).

The neuropathological progression and staging criteria of AD-associated alterations suggest that even if the scattered distribution of beta-amyloid (Aβ) neuritic plaques comprehensively varies, neurofibrillary tangles and neuropil threads show an archetypal dispersal based on what the six-stage classification has done. Stages I-II are the reshaping of the brain structure limited to the transentorhinal region, which gradually progresses to the limbic system region (Stages III-IV) and ultimately, to the isocortical regions of the brain (Stages V-VI) (Serrano-Pozo *et al.*, 2011). For initial intervention and the elimination of other causes of dementia, computed tomography (CT) and then magnetic resonance imaging (MRI) are frequently used (Johnson *et al.*, 2012).

9.2 MRI OF THE BRAIN

MRI can be used in AD to analyze the atrophy (or volumes) and microarchitectural changes of brain tissue responsible for signal alterations on certain sequences, such as white matter hyperintensities on a T2-weighted MRI due to vascular damage. In AD and dementia, numerous MRI sequences that are sensitive to microstructural change (e.g., magnetization transfer or diffusion) reveal striking variations. An MRI of the brain typically shows characteristic atrophy of the mesial temporal lobe structures, including the hippocampi, which subsequently progresses to cerebral atrophy (Vemuri and Jack, 2010). Digital image analysis of medial temporal atrophy and the surrounding soft tissues through MRI is a good marker for AD (Jeong *et al.*, 2020). In the early stages, the imaging may be normal, and in some cases, selective regional atrophy is seen in AD variants such as posterior cortical atrophy (PCA) with occipital lobe atrophy (Tom *et al.*, 1998). Characteristic MRI and pscychometric testing abnormalities are sufficient to make a diagnosis if the clinical history is suggestive. The CSF tau and beta amyloid quantification of patients is helpful in cases of diagnostic difficulty, but the laboratory infrastructure is not yet widely available.

9.3 AMYLOID PET SCANNING

Molecular imaging, particularly PET, is a promising technique for diagnosis and disease staging in presymptomatic and symptomatic AD patients. PET is a minimally invasive procedure and an advanced diagnostic tool in terms of sensitivity as it requires only picomolar concentrations of radiotracers. In this technique, radioisotope-emitting positrons (+ve charge) are injected into the bloodstream of subjects. The mutual annihilation upon collision of a positron and an electron causes the emission of gamma rays, which can be detected in tissues and used to identify the active brain cells. PET is also used to study brain metabolism and drug distribution. The scan visualizes plaques within the brain tissue that are the leading causes for the impairment of neuronal function in AD. Amyloid PET imaging has also emerged as a prospective tool in the assessment of individuals with cognitive impairment. Before amyloid PET, the presence and localization of these plaques could only be ascertained by investigating the patient's brain at autopsy. Amyloid PET tracers cause amyloid plaques to "light up" on a brain PET scan, which enables the perfect identification and tracking of plaques *in vivo*. Several present-day radiotracers have been investigated in experiments for wider use to enable the *in vivo* detection of amyloid depositions in brain tissues (Suppiah *et al.*, 2019). The main indications for investigation of an amyloid scan are to substantiate the diagnosis in indeterminate or atypical cases and to aid in the early detection of AD among subjects with mild cognitive impairment (MCI). Currently, several FDA-approved compounds are detected and bring into diagnostic practice those that possess a high affinity to bind with fibrillary Aβ plaques. The latest benzofuran derivative [18F] FPYBF-2 (5-(5-(2-(2-(2–18F-fluoroethoxy)ethoxy)ethoxy)benzofuran-2-yl)-N-methylpyridin-2-amine(^{18}F-FPYBF-2) is considered to be a relatively useful PET tracer for the imaging and quantitative analysis of Aβ *in vivo*. The diagnostic ability of ^{18}F-FPYBF-2 is feasible and satisfactory as the compound is chemically stable and does not photo degrade (Ben Bouallègue *et al.*, 2017; Higashi *et al.*, 2018).

Another PET imaging agent, ^{11}C-labeled Pittsburgh Compound-B ([^{11}C]PiB), which is also known as 2-(4-N-[^{11}C]methylaminophenyl)-6-hydroxybenzothiazole, is widely used worldwide and has an average 20-minute half-life. This compound has proved its efficacy as it can be retained approximately twice as long in the cortical regions of Alzheimer's patients than controls. Use of this probe is restricted to centers that have an on-site cyclotron facility. Compared to healthy controls, AD patients typically showed a marked retention of [^{11}C]PiB in the prefrontal, parietal and cingulate brain regions that contain a sizeable deposition of fibrillar Aβ plaques. Laboratory evidence suggests an inverse relationship between [^{11}C] PiB cortical retention and cerebral glucose metabolism within the parietal lobes, as determined with [^{18}F]FDG, (Klunk *et al.*, 2004).

[^{18}F]Florbetapir ([^{18}F]FBP), which is also known as ^{18}F-AV-45 or 4-{(E)-2-[6-(2-{2-[2-(18F) Fluoroethoxy]ethoxy}ethoxy)-3-pyridinyl]vinyl}-N-methylaniline, is another ^{18}F-labelled amyloid PET ligand that has shown high accuracy in detecting brain amyloid deposition (Camus *et al.*, 2012). This ligand is promptly absorbed through the blood-brain barrier and is rapidly washed out from grey matter tissue without amyloid. It has a high affinity for the aggregation of Aβ, a distinct separation between the radiotracer amyloid retention and background signal, and a protracted but stable pseudo-equilibrium that enables flexibility in the timing of image acquisition. [^{18}F]FBP received U.S. FDA approval in 2012 as an initial nuclear medicine imaging radiotracer and is effectively used in diagnostic centers for the investigation of dementia and other causes of cognitive decline. It has an advantage over [^{11}C]PiB since it possess a longer half-life (nearly 110 min) that allows its transport from the site of production to regional PET scanner facilities. Florbetaben ([^{18}F]FBB) is another radiotracer that laboratories use in amyloid PET imaging (Filippi *et al.*, 2018).

Amyloid tracers evolved via the chemical moderation of the histological fluorescent dye, thioflavin T. After binding with fibrillar and cerebrovascular amyloid, thioflavin T transmits a fluorescence signal. It crosses the blood-brain barrier in significant amounts, which helps in image acquisition *in vivo*. Imaging of amyloid aggregates of brain tissue was first attempted with carbon-based tracers [^{11}C] such as Pittsburgh Compound B (PiB), which is a radioactive analog of thioflavin T. Currently, long-lasting fluorine-based tracers [^{18}F] such as florbetapir, florbetaben and flutmetamol, which have an extended half-life (~110 minutes) compared to [^{11}C] tracers (20 minutes) (Márquez and Yassa, 2019), are used widely as radioligand for diagnostic purposes (Uzuegbunam *et al.*, 2020).

9.4 TAU PET SCANNING

Tau protein, a microtubule-associated protein derived from a single microtubule-associated protein tau (MAPT) gene (chromosome 17q21), promotes structural stability. Six isoforms of tau protein, namely, τ3L, τ3S, τ3, τ4L, τ4S and τ4, in human brain tissue are established. From these, three of the tau proteins (τ3L, τ3S and τ3) have three (3R) binding domain repeats, and the rest (τ4L, τ4S and τ4) have four (4R) binding domain repeats. Hyperphosphorylation in tau protein causes a disassembly of microtubules in AD, and taupathy ultimately hampers cell signaling through the inactivity of protein kinases and phosphatases. Different tau PET studies have revealed a significant association between the archetype of tau deposition and tissue atrophy. The spatial arrangement of tau impregnation and associated neurodegeneration in cerebral tissues could be assessed by PET tau imaging (Pereira *et al.*, 2020). Based on tissue specificity, a number of tau PET tracers including T807, THK-5117 and PBB3 have been designed for imaging neurofibrillary pathology *in vivo*. Of these, the increased cellular uptake and signal detection of [^{18}F]-1451 (or T-807) was confirmed after the most comprehensive molecular study in prodromal AD patients (Wang and Edison, 2019). An additional compound from these pharmacological probes, F-18 TKH5105, which is an arylquinoline derivative, has shown selective binding to pathologically paired helical filament tau-deposition in living patients with AD. This differentiation helps in the identification of pathological brain tissues from healthy controls. High tissue retention of F-18 TKH5105 in the neurofibrillary tangles of the temporal cortex was confirmed in the neuroimages of patients with AD (Okamura *et al.*, 2013; James *et al.*, 2015).

9.5 FDG-PET-CEREBRAL GLUCOSE METABOLISM

FDG-PET is another category of sensitive imaging techniques using radiopharmaceuticals to detect metabolically active cells in patients suspected to have AD. Initially, 2-Deoxy-2-[^{18}F]fluorodeoxyglucose ([^{18}F]FDG), a glucose analog, was utilized for PET brain imaging. Tissue accumulation of [^{18}F] FDG is based on the enhanced rate of glycolysis, and PET scanning with this radiotracer can

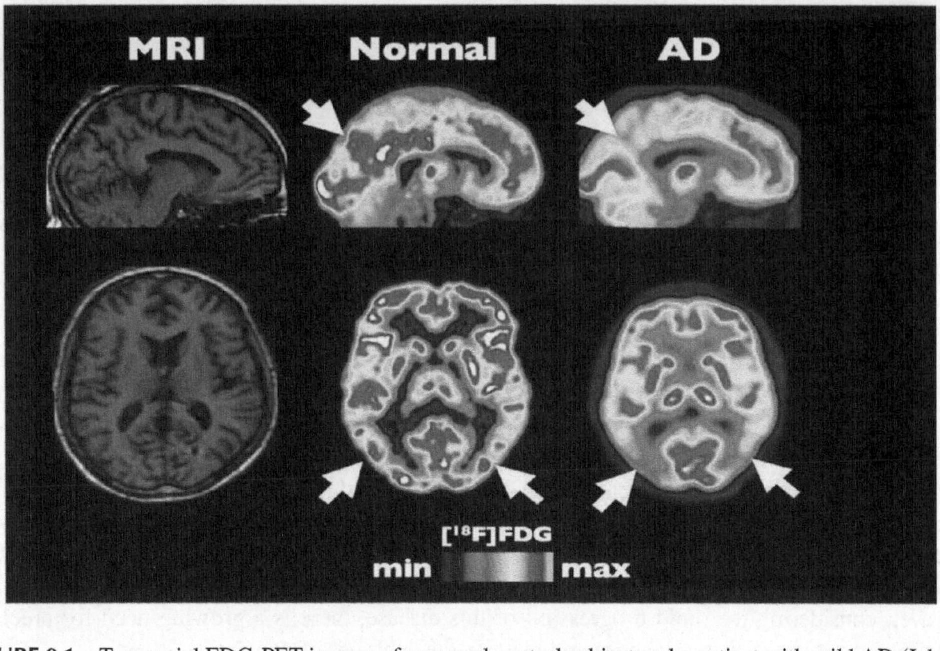

FIGURE 9.1 Transaxial FDG-PET images of a normal control subject and a patient with mild AD (Johnson *et al.*, 2012).

thus be used to determine if the pattern of cerebral hypometabolism is normal or abnormal. Atypical spatial differences of glucose metabolism as evident on brain FDG-PET differentiates AD from other types of dementia such as frontotemporal dementia and Lewy body dementia. After cellular uptake and phosphorylation by hexokinase, FDG is trapped within the neuron and clearly captured during image acquisition. In AD, hypometabolism exists in the prefrontal cortices and sometimes in the frontal lobe. It reliably appears in the temporal and parietal (and possibly frontal) lobes but spares the sensorimotor cortices, visual cortices, basal ganglia, thalamic nuclei and cerebellum. Active brain tissues quickly take up the substance, as they enormously metabolize glucose, but the uptake of substance is remarkably decreased in the temporo-parietal cortical areas of dementia patients. Glucose hypometabolism on brain FDG-PET seems to be a downstream metabolic marker of pathological aggregation and synaptic and neuronal dysfunction. Even though the importance of [^{18}F]FDG has been well recognized for the diagnosis of dementia and Alzheimer's, the accurateness of the diagnosis decreases in cases of aged patients (Berti *et al.*, 2014). Within a relatively younger cohort with a mean age of 64 years, 100% sensitivity of [^{18}F]FDG PET with a specificity of 75% (accuracy 84%) was observed, while a study with older patients reported 20% lower accuracy in late-onset AD (Suppiah *et al.*, 2019).

9.6 FUNCTIONAL MRI

Functional magnetic resonance imaging (fMRI) is a relatively present-day non-invasive imaging technique in the field of biomarkers for AD and provides useful information about the functional integrity of brain networks that support memory and other cognitive domains, including the neural correlates of specific behavioral events such as successful versus failed memory formation. Resting state fMRI is used to image intrinsic functional brain connectivity and is considered a promising biomarker for AD as functional brain changes follow structural brain changes. The small changes in blood flow in affected brain areas such as the hippocampus and related structures in the medial temporal lobe (MTL) that occur with task-based brain activity can be measured by blood-oxygenation-level-dependent (BOLD) contrast fMRI (Sperling, 2007). This determines changes in blood circulation, blood volume and even the proportion of oxyhemoglobin and deoxyhemoglobin in arterioles, which is directly related to neuronal functions. The resting and activation states of fMRI studies could also confirm that compared to healthy controls, AD patients have subsided coordination in the hippocampus, inferior parietal lobes and cingulate cortex (Varghese *et al.*, 2013). Resting-state functional magnetic resonance imaging (rs-fMRI) is largely practiced for the early detection of dementia and Alzheimer's through functional connectivity (FCVoxel-based morphometry (VBM)) and is a very useful tool for finding cellular and synaptic alterations in the preclinical AD stage (Márquez and Yassa, 2019). Different studies suggest that both the imaging of amyloid biomarkers and fMRI are valuable to identify early AD-related neuronal damage (Risacher and Saykin, 2013).

9.7 CONCLUSION

The prospective contribution of neuroimages using radiotracers and pharmacological probes in the early diagnosis and intermittent follow-up of neurodegenerative diseases such as AD are well established. A number of methods that are being used for diagnosis after successful clinical trials provide useful information about the pathological aggregates and neuronal and metabolic dysfunction in terms of diagnosing the disease during the preclinical stage and tracing the routes of neuronal degeneration and synaptic disintegrity. It has been observed that among all the methods, structural imaging biomarkers may be practiced to speculate about the risk in the preclinical stages of AD. However, considering the rapid progression of this disease, there is a growing need for precision neuroimaging approaches to diagnose and classify the subtypes of dementias.

REFERENCES

Ben Bouallègue F, Mariano-Goulart D, Payoux P. 2017. Alzheimer's Disease Neuroimaging Initiative (ADNI). Comparison of CSF markers and semi-quantitative amyloid PET in Alzheimer's disease diagnosis and in cognitive impairment prognosis using the ADNI-2 database. *Alzheimer's Res Ther.* 9(1):32.

Berti V, Mosconi L, Pupi A. 2014. Brain: Normal variations and benign findings in fluorodeoxyglucose-PET/computed tomography imaging. *PET Clin.* 9(2):129–40.

Camus V, Payoux P, Barré L, Desgranges B, Voisin T, Tauber C, La Joie R, Tafani M, Hommet C, Chételat G, Mondon K, de La Sayette V, Cottier JP, Beaufils E, Ribeiro MJ, Gissot V, Vierron E, Vercouillie J, Vellas B, Eustache F, Guilloteau D. 2012. Using PET with 18F-AV-45 (florbetapir) to quantify brain amyloid load in a clinical environment. *Eur J Nucl Med Mol Imaging.* 39(4):621–31.

Elsinga PH, Hatano K, Ishiwata K. 2006. PET tracers for imaging of the dopaminergic system. *Curr Med Chem.* 13(18):2139–53.

Filippi L, Chiaravalloti A, Bagni O, Schillaci O. 2018.^{18}F-labeled radiopharmaceuticals for the molecular neuroimaging of amyloid plaques in Alzheimer's disease. *Am J Nucl Med Mol Imaging.* 8(4):268–81.

GBD 2019 Dementia Forecasting Collaborators. 2022. Estimation of the global prevalence of dementia in 2019 and forecasted prevale nce in 2050: An analysis for the global burden of disease study 2019. *Lancet Public Health.* 7(2):e105–25.

Giri M, Zhang M, Lu Y. 2016. Genes associated with alzheimer's disease: An overview and current status. *Clin Interv Aging.* 11:665–81.

Henry W, Querfurth H, LaFerla F. 2010. Mechanisms of disease Alzheimer's disease. *New Engl J Med.* 362:329–44.

Higashi T, Nishii R, Kagawa S, Kishibe Y, Takahashi M, Okina T, Suzuki N, Hasegawa H, Nagahama Y, Ishizu K, Oishi N, Kimura H, Watanabe H, Ono M, Saji H, Yamauchi H. 2018. ^{18}F-FPYBF-2, a new F-18-labelled amyloid imaging PET tracer: First experience in 61 volunteers and 55 patients with dementia. *Ann Nucl Med.* 32(3):206–16.

James OG, Doraiswamy PM, Borges-Neto S. 2015. PET imaging of tau pathology in alzheimer's disease and tauopathies. *Front Neurol.* 9(6):38.

Jeong HE, Shin DH, Lee DC. 2020. Medial temporal atrophy alone is insufficient to predict underlying Alzheimer's disease pathology. *Korean J Fam Med.* 41(5):352–58.

Johnson KA, Fox NC, Sperling RA, Klunk WE. 2012. Brain imaging in Alzheimer disease. *Cold Spring Harb Perspect Med.* 2(4):a006213.

Klunk WE, Engler H, Nordberg A, Wang Y, Blomqvist G, Holt DP, Bergström M, Savitcheva I, Huang GF, Estrada S, Ausén B, Debnath ML, Barletta J, Price JC, Sandell J, Lopresti BJ, Wall A, Koivisto P, Antoni G, Mathis CA, Långström B. 2004. Imaging brain amyloid in Alzheimer's disease with pittsburgh compound-b. *Ann Neurol.* 55:306–19.

Márquez F, Yassa MA. 2019. Neuroimaging biomarkers for Alzheimer's disease. *Mol Neurodegener.* 14(1):21.

Okamura N, Furumoto S, Harada R, Tago T, Yoshikawa T, Fodero-Tavoletti M, Mulligan RS, Villemagne VL, Akatsu H, Yamamoto T, Arai H, Iwata R, Yanai K, Kudo Y. 2013. Novel 18F-labeled arylquinoline derivatives for noninvasive imaging of tau pathology in Alzheimer disease. *J Nucl Med.* 54(8):1420–27.

Pereira JB, Harrison TM, La Joie R, Baker SL, Jagust WJ. 2020. Spatial patterns of tau deposition are associated with amyloid, ApoE, sex, and cognitive decline in older adults. *Eur J Nucl Med Mol Imaging.* 47(9):2155–64.

Risacher SL, Saykin AJ. 2013. Neuroimaging and other biomarkers for Alzheimer's disease: The changing landscape of early detection. *Annu Rev Clin Psychol.* 9:621–48.

Serrano-Pozo A, Frosch MP, Masliah E, Hyman BT. 2011. Neuropathological alterations in alzheimer disease. *Cold Spring Harb Perspect Med.* 1(1):a006189.

Sperling R. 2007. Functional MRI studies of associative encoding in normal aging, mild cognitive impairment, and Alzheimer's disease. *Ann NY Acad Sci.* 1097:146–55.

Suppiah S, Didier MA, Vinjamuri S. 2019. The who, when, why, and how of pet amyloid imaging in management of Alzheimer's disease-review of literature and interesting images. *Diagnostics (Basel).* 9(2):65.

Tiwari S, Atluri V, Kaushik A, Yndart A, Nair M. 2019. Alzheimer's disease: Pathogenesis, diagnostics, and therapeutics. *Int J Nanomedicine.* 19(14):5541–54.

Tom T, Cummings JL, Pollak J. 1998. Posterior cortical atrophy: Unique features. *Neurocase.* 4(1):15–20.

Uzuegbunam BC, Librizzi D, Hooshyar Yousefi B. 2020. PET radiopharmaceuticals for Alzheimer's disease and Parkinson's disease diagnosis, the current and future landscape. *Molecules.* 25(4):977.

Varghese T, Sheelakumari R, James JS, Mathuranath P. 2013. A review of neuroimaging biomarkers of Alzheimer's disease. *Neurol Asia*. 18(3):239–48.
Vemuri P, Jack CR Jr. 2010. Role of structural MRI in Alzheimer's disease. *Alzheimers Res Ther*. 2(4):23.
Wang YT, Edison P. 2019. Tau imaging in neurodegenerative diseases using positron emission tomography. *Curr Neurol Neurosci Rep*. 19(7):45.
Yiannopoulou KG, Papageorgiou SG. 2013. Current and future treatments for Alzheimer's disease. *Ther Adv Neurol Disord*. 6(1):19–33.

10 Therapeutics for Neurodegenerative Diseases

Shubham Kundu[1], *Anupam Maity*[1], *Rajdip Misra*[1], *Uttam Pal*[1,*] *and Nakul C. Maiti*[1,*]
[1]CSIR – Indian Institute of Chemical Biology, Kolkata, India
*Correspondence to uttam2707@gmail.com (U.P.); ncmaiti@iicb.res.in (N.C.M.)

10.1 INTRODUCTION

Neurodegenerative diseases affect a significant population worldwide. Alzheimer's disease (AD) is the most prevalent neurodegenerative disease, followed by Parkinson's disease (PD) (Olanow and Tatton 1999). There are different types of neurodegenerative diseases, for example, Huntington's disease (HD), Lewy body dementia (DLB), frontotemporal dementia (FTD) and spinal muscular atrophy (SMA) (Sun and Roy 2021). These diseases are caused by neuronal damage that may be associated with the formation of protein aggregates (Ross and Poirier 2004). There are different types of aggregated protein species, for example, oligomers, protofibrils and fibrils. It has been claimed that the oligomeric species are more toxic (Cascella et al. 2021). Misfolded protein monomers come together to form a nucleus through nucleation that then grows and forms low molecular weight oligomers followed by high molecular weight oligomers, which then convert into protofibrils and eventually turn into mature fibrils as shown in Figure 10.1. Another type of oligomer is called an off-pathway oligomer that does not directly grow fibrils (Bengoa-Vergniory et al. 2017) but rather breaks down to monomers and then enters the fibril-forming path.

AD is associated with the extracellular deposition of Aβ42, which forms Aβ plaques, and intracellular neurofibrillary tangles that cause cell death, whereas α-synuclein oligomerisation and fibril formation have been claimed to damage the substantia nigra dopaminergic neurons resulting in PD (Bloom 2014; Kalia and Lang 2016). An atomic force microscopy image of a mature α-synuclein fibril is shown in Figure 10.2. Deposition of poly-glutaminated huntingtin protein (mHtt) and aggregation of mHtt are found in HD (Ross and Tabrizi 2011). Disordered aggregates of mHtt in cultured SH-SY5Y cells are also shown in Figure 10.2. Other diseases such as DLB, FTD, SMA, etc. are classified depending on the symptoms and underlying mechanism of cellular deaths. Oxidative stress and neuroinflammation caused by dysregulated immune responses also worsen the disease pathology and may be the trigger that promotes the aggregation of the protein by affecting the protein folding machinery or by producing pro-inflammatory cytokines such as IL-1β, IFN-γ, IL-6, etc., which then produce neuroinflammation resulting in cellular deaths (Dehay et al. 2010; Bae et al. 2014; Mazzulli et al. 2011). Aggregated proteins are also responsible for the promotion of neuroinflammation (Currais et al. 2017).

The origin of protein aggregation has eluded the scientific world for several decades (Pal et al. 2016). Considerable biophysical and bioinformatics studies have been carried out to decipher the sequence complexities of amyloidogenic regions in intrinsically disordered proteins (IDPs) (Das et al. 2014, 2016; Mondal et al. 2018; Pal et al. 2016). Efforts have been made to divulge the characteristic features of α-synuclein, amyloid-β and other IDPs in different microenvironments *in vitro* (Apetri et al. 2006; Maiti et al. 2004, 2008; Huang et al. 2006; Mondal and Maiti 2019). Chaperone proteins such as cyclophilin are often found to help in the refolding of IDPs and thus reduce aggregation propensities in some of them (Supriya Das et al. 2013; Maiti, Mondal, and Das 2015), indicating a possible therapeutic strategy by upregulating specific chaperone protein expression in disease tissue.

DOI: 10.1201/9781003302131-11

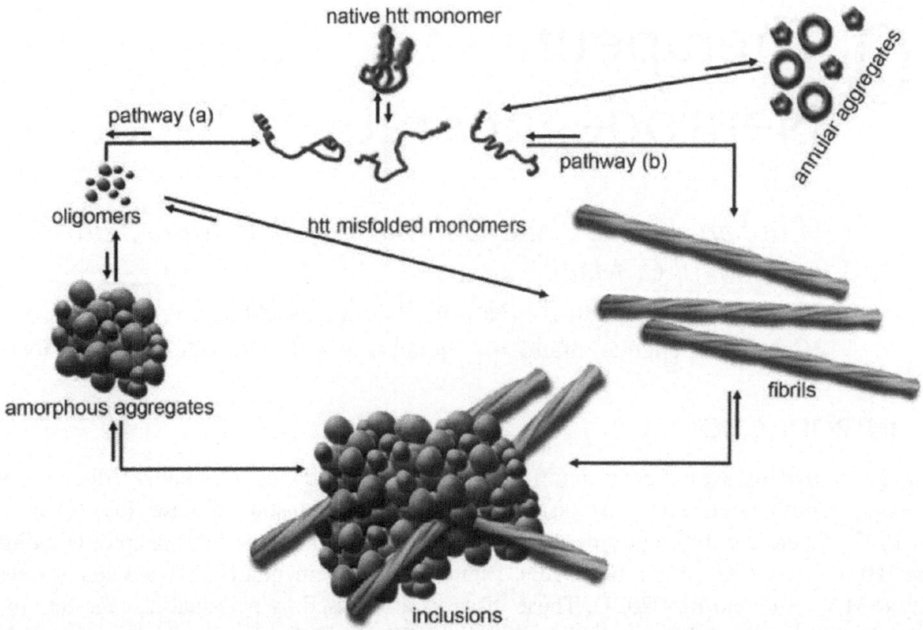

FIGURE 10.1 Schematic representation of the development of oligomers, fibrils and inclusion bodies in the pathogenesis of amyloid diseases with huntingtin as an example. The native structured monomer misfolds and undergoes either oligomerisation (path a) or fibril formation (path b). Oligomers can also directly form fibrils or break down to misfolded monomers and then form fibrils. Finally, the oligomers and fibrils are deposited in inclusion bodies.

Source: Reproduced from Arndt, Chaibva, and Legleiter (2015) under Creative Commons licence CC BY-NC-ND 3.0.

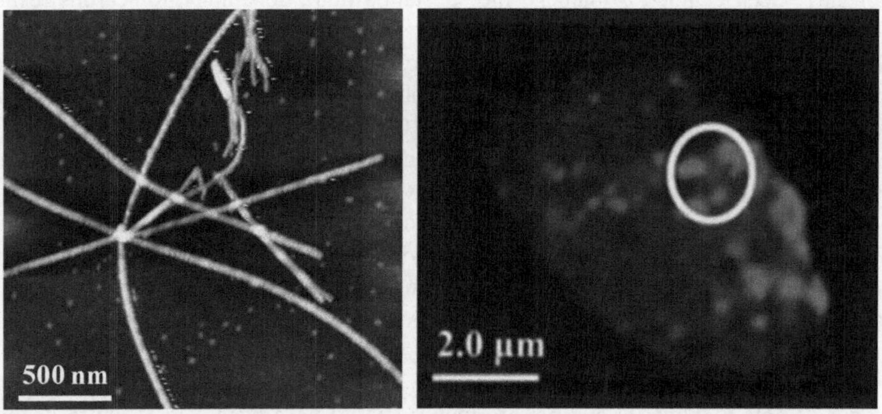

FIGURE 10.2 Ordered and disordered aggregates formed by amyloidogenic proteins α-synuclein and mHtt, respectively. (left) Atomic force microscopy (AFM) image of the mature fibrils of α-synuclein formed *in vitro*. (right) Amorphous/disordered aggregates of mHtt (red) in SH-SY5Y cells transfected with DsRed-HTT-83Q by using Lipofectamine 2000 under a confocal microscope.

Source: (left) Reproduced from Bera et al. (2020) with permission © 2020 American Chemical Society. (right) Reproduced from Bardhan et al. (2021) under a Creative Commons Attribution-NonCommercial 3.0 Unported (CC BY-NC 3.0) licence.

To date, there have been no such potential and efficient therapies that can reverse the aggregation of IDPs and neuronal tissue damage. Present treatments are mainly targeted to alleviate the symptoms or slowing disease progression, which demands intense research to build new therapies with more potential and efficacy. Different strategies are currently being adopted among which gene therapy, stem cell therapy, nanotherapy, small molecule-based therapy and immunotherapy are emerging with promising therapeutic potential. These different strategies deal with different ways to treat these diseases. There are also disadvantages of different types of therapy; thus, combinational therapies are also in focus.

10.2 GENE THERAPY

Gene therapy is the combined application of genetics and molecular biology that has therapeutic effects and is performed by modifying the genetic materials (Kaji and Leiden 2001). It is usually performed by inducing the expression of an important gene that is affected in the diseased condition or by introducing a new gene(s) (Losordo et al. 1998). This is a three-step process consisting of i) an intended gene that will be transfected to the cell, ii) a vector that will deliver the gene into the cell or tissue and iii) a delivery device, for example, a catheter, syringe or stent, to deliver the gene to the cell (Kaji and Leiden 2001). Earlier, gene therapy was used especially in diseases with single-gene defects such as muscular dystrophy and hyperlipidemias; completion of the human genome project, recent developments in vector efficiency and less side effects paved the path for new gene therapy for complex diseases such as diabetes mellitus, PD, AD, etc.

Although PD is a multisystem disorder, many important symptoms, for example, tremors and hypokinesia, are specific to the dopaminergic substantia nigral neuronal death, and the early therapies act by relieving the symptoms, but these therapies have side effects and do not control the disease's progression (Sun and Roy 2021). Levodopa was first used as a therapy in 1960 after findings that the dopamine level is reduced in the striatal region of the brain. The results were inconsistent till 1967 when Cotzias and team reported significant improvement in PD patients with increasing oral levodopa administration over long time periods (Tolosa et al. 1998; Cotzias, Van Woert, and Schiffer 1967; Cotzias, Papavasiliou, and Gellene 1969). L-Dopa therapy is reported to be associated with side effects such as dyskinesia, which is reduced by the co-administration of levodopa with carbidopa, a dopamine decarboxylase inhibitor to inhibit peripheral dopamine level increase. This co-therapy was commercialised in 1975. Adeno-associated virus 2-aromatic amino acid DOPA decarboxylase (AAV2-AADC) therapy has been used, and it showed four years of stable expression and mild symptomatic relief by maintaining the striatal dopamine level (Mittermeyer et al. 2012). Later, real-time MRI-directed accurately targeted AAV2-hAADC was infused into the midbrain region of primates, which ensured tolerability and safety over the long term (Sebastian et al. 2014). Clinical trials on humans have also been performed, and a Phase II study was initiated with 42 patients who were unresponsive to the medications. Similar to Phase I, the primary endpoints in Phase II included the patient rating of motor activity alteration after one year, viral spread of the putamen, alteration in AADC enzyme activity and measures of safety. Changes in activities of quality of life, daily living, global function, non-motor symptoms, and response to levodopa were among the secondary outcomes. The trial was scheduled for the end of 2020, but it was delayed because of the COVID 19 pandemic, and it is has now been halted by the FDA based on the need for more information on MRI abnormalities in study participants.

Recently, gene therapy-based trials for AD have also been performed, and a few trials are ongoing. A nerve growth factor (NGF) gene therapy trial was conducted by Tuszynski et al., and it showed safety and improvement in cognitive decline (Tuszynski et al. 2005). Nagahara et al. showed the efficacy of lentivirus-mediated NGF gene delivery *in vivo* and reported long-term effects by reversing the reduced neuronal size and numbers of cholinergic neurons labelled with p75 without any adverse effects (Nagahara et al. 2009). However, despite showing efficacy and safety in nonhuman primate models, it did not show significant cognitive behaviour progress. Recently, MRI-guided

AAV2-BDNF (brain-derived neurotrophic factor) delivery is being trialled as convection-enhanced therapy to increase the coverage of the gene delivery, and it can be used for further human trials (Nagahara et al. 2018). Antisense oligonucleotide (ASO)-directed microtubule associated protein tau (MAPT) gene silencing that targets the attenuation of tau phosphorylation resulting in the prevention of aggregation is currently under investigation (NCT03186989). Several apolipoprotein variants play different roles in AD pathology; apoE4 is a risk factor, whereas apoE2 is protective (Hudry et al. 2013). After several studies performed by Dodart et al. and Zhao et al., an efficient and safe route of AAV-ApoE2 was investigated by Rosenberg et al. (Dodart et al. 2005; Zhao et al. 2016; Rosenberg et al. 2018). One recent Phase I trial is ongoing (NCT03634007). Different strategies have also been taken to cope with AD by balancing amyloid precursor protein (APP). ASO-mediated exon 17 truncation was reported, but it affects the normal physiological function of APP (Chang et al. 2018). Later, non-homologous end joining (NHEJ)-based CRISPR-CAS9 gene excision was used to produce a YENPTY pentapeptide truncated protein without affecting the normal physiological function of APP (Sun et al. 2019).

HD is caused by a mutated huntingtin protein with unusually long polyglutamine (polyQ) repeats (Ross and Tabrizi 2011). Different strategies have been taken to silence the HTT gene either globally or selectively. HTT_{RX}, a synthetically altered oligomer that is perfectly complementary to a stretch of 20 nucleotides of HTT mRNA, has been administered intrathecally, and it lowered the expression of huntingtin protein by RNAse H1-mediated degradation in cerebrospinal fluid (CSF) without any side effects (Tabrizi et al. 2019). Application of the study has been investigated in HD-deep brain stimulation (DBS) trials (NCT02535884), and this was observed to be a safe and well suited treatment; therefore, a Phase III trial is ongoing (Hartmann et al. 2016). Different RNAi technologies

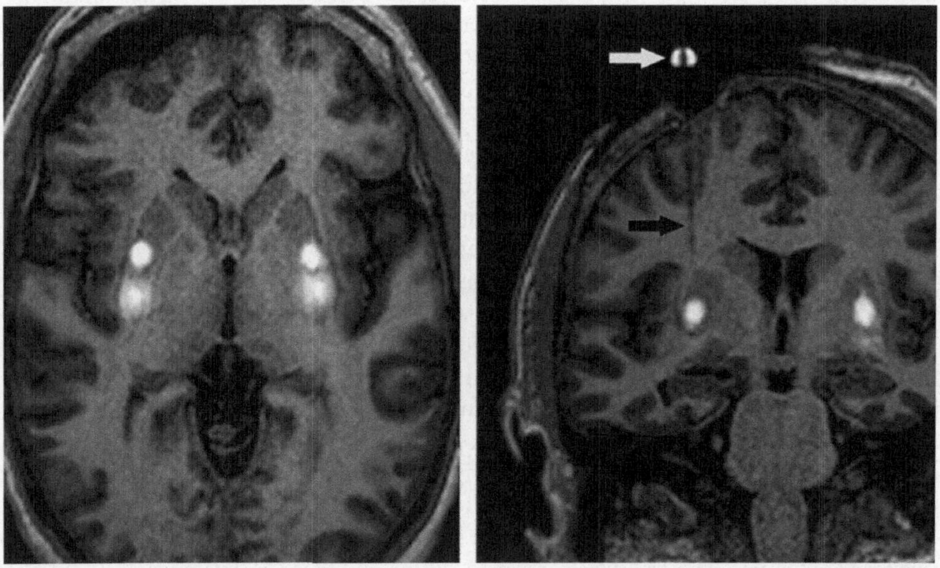

FIGURE 10.3 MRI-guided AAV2-mediated gene therapy. Representative magnetisation prepared rapid gradient echo images during the simultaneous bilateral administration of VY-AADC01 and an AAV2 vector carrying complementary DNA encoding the human aromatic L-amino acid decarboxylase (AADC) enzyme in the putamen. Axial image (left) shows gadoteridol enhancement as a surrogate marker for vector distribution with two administrations in the posterior portion of each putamen. Coronal oblique image (right) in-plane with the right infusion cannula during vector delivery; part of the skull-mounted SmartFrame aiming device is visible above the burr hole (white arrow), as is the SmartFlow cannula (black arrow).

Source: Reproduced from Christine et al. (2019) under Creative Commons Attribution-NonCommercial-NoDerivatives 4.0 International (CC BY-NC-ND 4.0) licence.

have also been performed and are in Phase I/II or III trials, for example, the application of AAV-HTT shRNA into striatum and the intracranial administration of AAV5 with a miRNA against HTT (Boudreau et al. 2009; Keskin et al. 2019). The drawback of global silencing is the defects in the normal physiological function of HTT (Dietrich et al. 2017). To avoid this, the selective silencing of huntingtin genes is ongoing in, for example, AAV-mediated zinc finger protein transcription factors (ZFP-TFs), which targets the CAG repeat region and selectively disrupts the mutant HTT allele (Zeitler et al. 2019).

10.3 STEM CELL THERAPY

Stem cell therapy is now an emerging therapy and is being investigated for curing different diseases such as AD, PD, HD, etc. (Karussis, Petrou, and Kassis 2013). Stem cell therapy can be applied in two different ways to treat neurodegenerative diseases: enriching the environment to support the particular neurons by making different networks of neurons, which produces neurotrophic factors that scavenge reactive oxygen species (ROS), or it can be directly transplanted into the damaged region and/or can produce a particular neuronal subtype and reestablish the connections between neurons that were present before the disease (Karussis, Petrou, and Kassis 2013).

The causes of PD are heterogeneous, and therapies thus apply only dopamine precursors (levodopa) to compensate for the reduced dopamine level and alleviate the symptoms, but the neuronal losses do not reverse (Trombetta-Lima, Sabogal-Guáqueta, and Dolga 2021). Stem cell therapy can resolve this problem by reversing dopaminergic neuron loss by applying embryonic stem cells (ESC) or neuronal stem cells. Many scientists are working with these cells to induce differentiation into dopaminergic neurons (Kim 2011). These studies have been performed in animal models and in clinical trials by transplanting human foetuses' brain cells into PD-diagnosed patients to study their efficacy and side effects (Schwarz and Storch 2010). Takagi et al. and Reddy et al. independently showed recovery after the successful transplantation of differentiated dopaminergic (DA) neurons derived from monkeys' ESC into a mouse brain with PD (Takagi et al. 2005; Reddy, Ravichandran, and Carkaci-Salli 2020). Wang et al. grafted human parthenogenetic embryonic stem cell line-derived DA neurons and showed that there was no tumour formation, which is very important to check if a Phase I/II trial has to be performed (Y.-K. Wang et al. 2018). This study was also helpful for China's first embryonic stem cell-based Phase I/IIa trials (NCT03119636) (Wang et al. 2018). Neuronal stem cells have also been shown to be a successful source of PD treatment by Kim et al. (Kim et al. 2021). Neural stem cells (NSCs) showed neuroprotection against 6-hydroxydopamine neurotoxicity in rat models (Yasuhara et al. 2006). Mesenchymal stem cells (MSC) are another important source of dopaminergic neurons, and intense research is ongoing because of their immune-altering potential, fewer ethical concerns and a low risk of tumour formation. The efficacy of the gene therapy increases if it is co-applied with MSC therapy, as Xiong et al. showed that human umbilical cord mesenchymal stem cells (HUMSCs) combined with vascular endothelial growth factor (VEGF) expression transplantation reduced the loss of dopaminergic neurons in the midbrain substantia nigra region (Xiong et al. 2011). Stem cells communicate by secreting exosomes, and exosomes secreted by human stem cells can cross the blood-brain barrier; stem cells have been shown to be easily taken up by SH-SY5Y cells, resulting in reduced 6-hydroxydopamine toxicity. This approach demonstrated higher neuroprotection and reduced neuronal loss by inducing autophagy that resulted in the upregulation of the dopamine level (Chen et al. 2020). MSCs are also being investigated for mitochondrial delivery in PD because damaged mitochondria are also responsible for PD (Liu et al. 2014). Induced pluripotent stem cell (iPSC)-derived DA neurons can survive and integrate into the patient's neuronal network with functional and behavioural improvements (Cave, Wang, and Baker 2014). The safety and efficacy of iPSC therapy providing long term survival and functionality has also been observed (Doi et al. 2020). Despite some drawbacks, for example, a laborious and expensive process of induction to make iPSC a dopamine-producing neuron, iPSC is currently being used in research for the treatment of neurodegenerative diseases because of its minimal immune

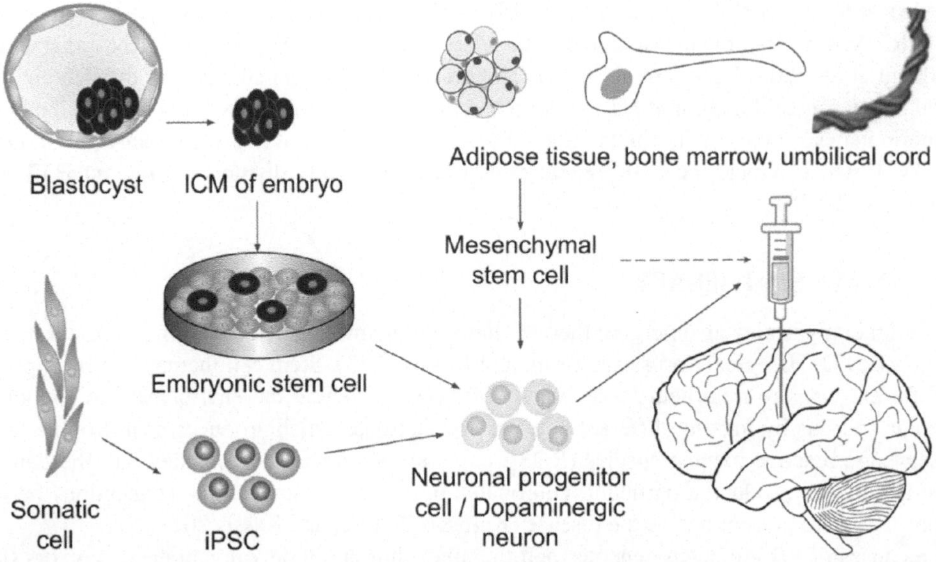

FIGURE 10.4 Schematic representation of various stem cell therapy approaches. Neuronal stem cells, neuronal progenitor cells or dopaminergic neurons are directly administered into the brain. Neuronal progenitor cells can be derived from induced pluripotent stem cells (iPSC), embryonic stem cells or mesenchymal stem cells. Embryonic stem cells are derived from the inner cell mass (ICM) of an embryo (blastocyst); iPSCs can be derived from the somatic cells of the competent donor; mesenchymal stem cells can be derived from various tissues including adipose tissue, bone marrow, umbilical cords, etc.

responses and non-invasive collection of iPSC from blood. Different clinical studies are also ongoing as they have reported efficacy and safety in preclinical studies (Doi et al. 2020). Recently, Hiller et al. showed the maturity and dose dependence of the efficacy of iPSC therapy (Hiller et al. 2022).

Currently, there are no efficient and potential drugs for AD, but research is ongoing to pave a new path in the discovery of AD treatment. Stem cell therapy for AD treatment is in the spotlight. Embryonic stem-derived neural precursor cells (NPC) have shown potential in improving memory loss and in integrating into hippocampal neurons' network after transplantation into an AD rodent model (Moghadam et al. 2009; Bissonnette et al. 2011). As ESC may have a risk of developing uncontrolled cell growth, the application of NSCs has been reported to show increased neuron formation and behavioural function in both rodent and primate models. Many studies have been performed by Zhang et al., Lilja et al. and Ager et al. indicating different strategies that NSCs can take to combat AD such as by reducing inflammation, tau and Aβ pathology and producing new neurons and synapses (Zhang et al. 2016; Lee et al. 2015; Lilja et al. 2015; Ager et al. 2015; Xuan et al. 2009). MSCs are also major players in the treatment of AD. Park et al. showed that adipose-derived MSCs increased physical performance and behavioural function when transplanted into mice (Park et al. 2013).

10.4 NANOTHERAPY

Presently, nanotechnology is a very attractive research field, and intense research is ongoing as there is no potential drug candidate for these genetic diseases. Scientists are trying to apply nanotechnologies to build new therapies for neurodegenerative disorders. Nanomaterials have different properties such as a tiny size (1 nm–900 nm, with at least 1–100 nm in one dimension), chemical properties and biocompatibility that make nanotechnology a field with much more potential in therapeutics (Bera et al. 2020; Poovaiah et al. 2018). Nanomaterials can help in various ways with different strategies to cope with neurodegenerative diseases. ROS and reactive nitrogen species are crucial factors in mediating neurological disorders, and nano enzymes can function by mimicking the antioxidant-protective machinery such as catalase, glutathione peroxidase and superoxide dismutase (SOD) (Li et al. 2020; Abraham et al. 2005). These nano enzymes have many advantages, for example, high stability and multi-enzyme mimicking activity (Liang and Yan 2019). Adhikary et al. showed the glutathione peroxidase (GPx)-mimicking nature of citrate capped manganese based on biocompatible material (c-Mn_3O_4) and validated it preclinically in an HD animal model (Adhikari et al. 2021) by alleviating the disease's symptoms as shown in Figure 10.5. Nanomaterials can also work by indirectly supplying both antioxidant molecules and anti-inflammatory molecules that can alleviate the disease's symptoms. Liu et al. showed the application of ultrasmall copper-based nanoparticles, $Cu_{5.4}O$, in mimicking the actions of SOD, catalase and glutathione peroxidase, resulting in scavenging in the

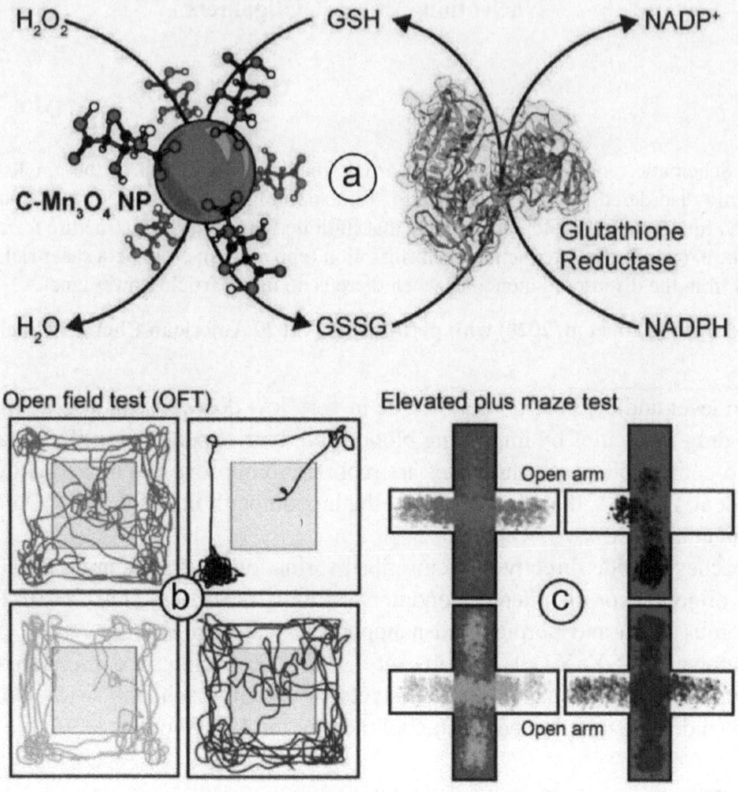

FIGURE 10.5 Nano enzyme-mediated alleviation of symptoms in a 3-Nitropropionic acid (3-NPA)-induced Huntington's model. (a) Citrate capped Mn_3O_4 (C-Mn_3O_4 NP) nanoparticles have been shown to have GPx mimetic activity. C-Mn_3O_4 reduces H_2O_2 by producing H_2O with the help of glutathione (GSH), which becomes oxidised into GSSG. b) Open field test (OFT) to test thigmotactic behaviour; an index of anxiety of the animals shows that the mice spent less time in the central zone of the apparatus, which indicates anxiety, whereas more time spent in the central zone indicates the amelioration of anxiety. 3-NPA-treated animals spent less time in the central region, while C-Mn_3O_4+3-NPA-treated mice spent more time in the central region, and C-Mn_3O_4 (only)-treated mice spent much more time in the central region of the apparatus. (c) Elevated plus maze (EPM) test showing that 3-NPA treated mice spent less time in the open arms of the EPM than other groups, indicating higher anxiety in the 3-NPA treated group. The distance moved by the NPA-treated mice is also less in the case of the 3-NPA-treated mice. The total distance moved by the 3-NPA-treated mice is less in both the OFT and EPM test.

Source: Adopted from (Adhikari et al. 2021) with permission © 2020 Wiley-VCH GmbH.

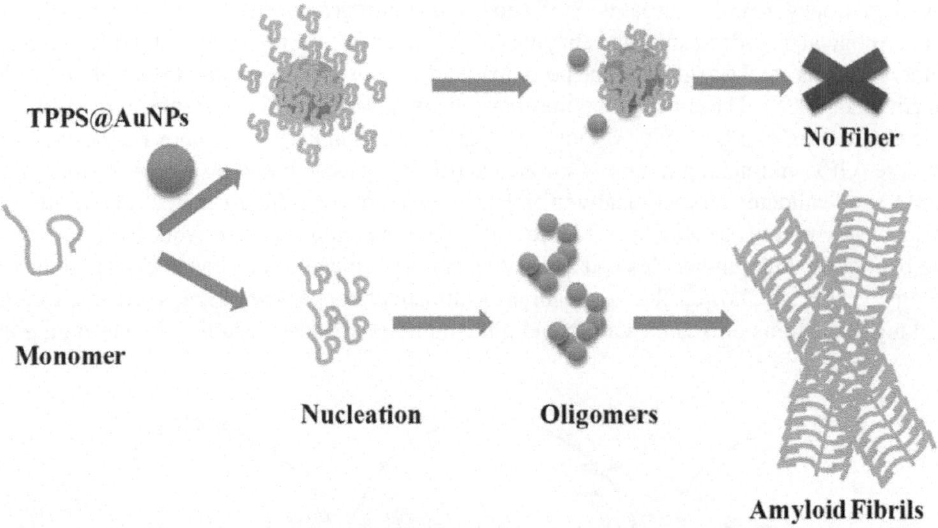

FIGURE 10.6 Schematic representation of nanoparticle-mediated inhibition of the amyloid fibril formation of intrinsically disordered protein (α-synuclein). Tetrasodium salt of meso-tetrakis-(4-sulfonatophenyl) porphyrin (TTPS) functionalised gold nanoparticle that inhibited the beta sheet structure formation from the disordered structure, resulting in a reduction of fibrillisation (upper lane); cross beta sheet-rich fibrillar structures are formed from the disordered monomer when there is no nanoparticle (lower lane).

Source: Adopted from (Bera et al. 2020) with permission © 2020 American Chemical Society.

reactive oxygen level and the protection of tissue in very low doses. It can also be used as an efficient carrier of drug molecules by improving blood-brain barrier penetration (Liu et al. 2020). The half-life and bio-compatibility of nanozymes are reported to improve when nanoparticle capping is performed (Li et al. 2021). Liu et al. determined the biocompatibility of the $Cu_{5.4}O$ nanoparticle in the cell line (Liu et al. 2020).

Other approaches include directly reducing fibril formation by fragmenting the fibres or interaction with the oligomers or disordered monomers (Figure 10.6) (Bera et al. 2020). Bardhan et al. showed that porous silica and porous gold nanoparticles can significantly reduce the huntingtin (mHtt) aggregation in SH-SY5Y cells (Bardhan et al. 2021). Reducing the expression of the mutated protein(s) by supplying siRNA nanomedicine is yet another approach, and si-RNA-mediated gene silencing has been demonstrated to be a potential therapy for AD (Zhou et al. 2020).

10.5 SMALL MOLECULES

Small molecule application in the treatment of neurodegenerative diseases is a relatively new field. Phosphorylation of different proteins, for example, α-synuclein, tau and huntingtin, is a major factor that promotes PD, AD and HD, respectively (Geschwind 2003; Chen et al. 2017; Gupta et al. 2021). Thus, small molecules that target the specific kinases and the enzymes responsible for phosphorylation are a vital choice that can reduce phosphorylation of the proteins and inhibit their aggregation (Krahn et al. 2020). Leucine-rich repeat kinase 2 (LRRK2) and PTEN-induced kinase 1 (PINK1) are found to be mutated in PD, and these are reported as good drug targets (Lin and Farrer 2014). Other kinases such as ABL1 and GSK3β can also be targeted for AD and PD, respectively. Phase III trials have been conducted for Masitinib against AD and ALS, whereas CEP-1347 has been trialed for PD (Shi and Mader 2018; Chico, Van Eldik, and Watterson 2009). Microglial cells mediate the neuroinflammation in uncontrolled conditions, which is seen in neurodegenerative diseases. Toll-like receptors (TLRs) recognise pathogens and activate appropriate responses to combat the pathogen, but it is reported

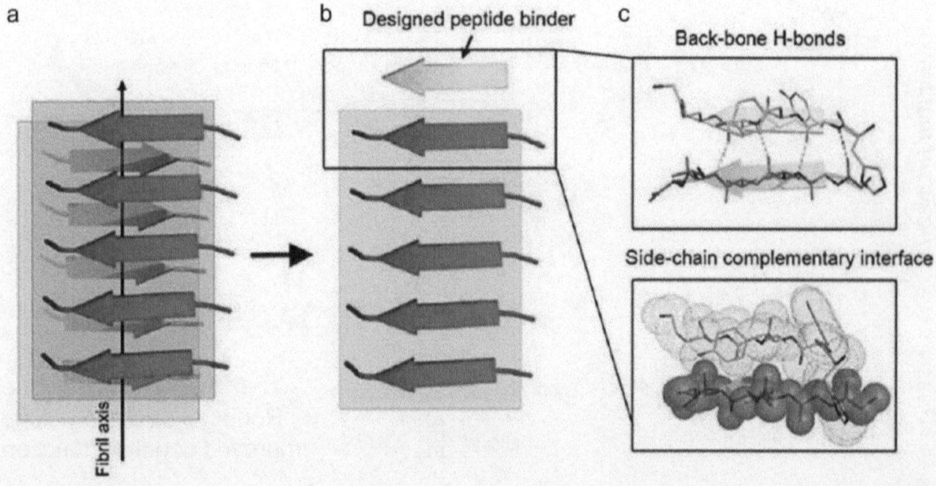

FIGURE 10.7 Peptide-based inhibition of amyloid fibril formation. Aβ fibril formation is inhibited by the application of designed pentapeptides, which block the stacking of beta sheets along the Aβ fibril axis by forming backbone hydrogen bonds and complementary side-chain interactions. (a) Template of the target region extracted from the steric zipper structure. (b) Docked image of designed pentapeptides on one end of the beta sheet of the target sequence. (c) The hydrogen bonding (yellow dashed line) and side-chain complementary interactions (spheres and dots) between the targeted sequence and the designed peptides.

Source: Reproduced from Lu et al. 2019 under Creative Commons licence CC-BY.

that abnormal protein depositions such as α-synuclein and tau also activate TLRs, especially TLR2; thus, pro-inflammatory cytokines IL-1β, TNF-α and IFN-γ are produced that kill the neuronal cells (W.-Y. Wang et al. 2015). Recently, scientists have been trying to develop a potential therapy for these diseases by targeting TLRs. TAK 242, a TLR4-specific inhibitor, has been reported to polarise the microglia into an M2 subset in mice models and reduce neuronal loss (Cui et al. 2020). Dihydrotestosterone, a TLR4 antagonist, showed the reduced secretion of pro-inflammatory cytokines in LPS-induced microglia and a BV2 microglial cell line (Yang et al. 2020). Small molecules/peptides have also been used to inhibit the oligomerisation and aggregation of these proteins, which causes neuronal cell death (Figure 10.7). MM11253 and adapalene have been reported to significantly reduce the initiation of aggregation and elongation of Aβ oligomers *in vitro* (Cohen et al. 2013). As pharmacological chaperones, small molecules having chaperone-like functions that inhibit proteolysis, and unwanted protein-protein interaction can potentially reduce protein aggregation. Oh et al. synthesised and showed the potential of small molecules by inhibiting alpha synuclein aggregation (Oh et al. 2014). Very recently, Wiesen et al. demonstrated that application of L-Dopa with N-acetylcysteine (superDopa and superdopamide) reduced α-synuclein oligomerisation and reversed the oxidative stress-driven phosphorylation of cJUN-N-terminal kinase (JNK) and P38 MAPK (Wiesen and Atlas 2022).

10.6 IMMUNOTHERAPY

Immunotherapy is yet another way to address these neurodegenerative diseases. The main purpose of this therapy is a reduction of the aggregation of different disease-causing proteins such as α-synuclein, tau, Aβ, etc. This is done by either reducing the synthesis or by removing the aggregated fibres or oligomers through phagocytosis (Hanif et al. 2020) as represented by Figure 10.8. The aggregated proteins have been shown to be digested by autophagic/lysosomal machinery; more importantly, a recent report claimed that the degradation of the aggregated proteins can also be executed by a proteasomal pathway when treated with nanoparticles (Liu et al. 2020). Usually, two strategies

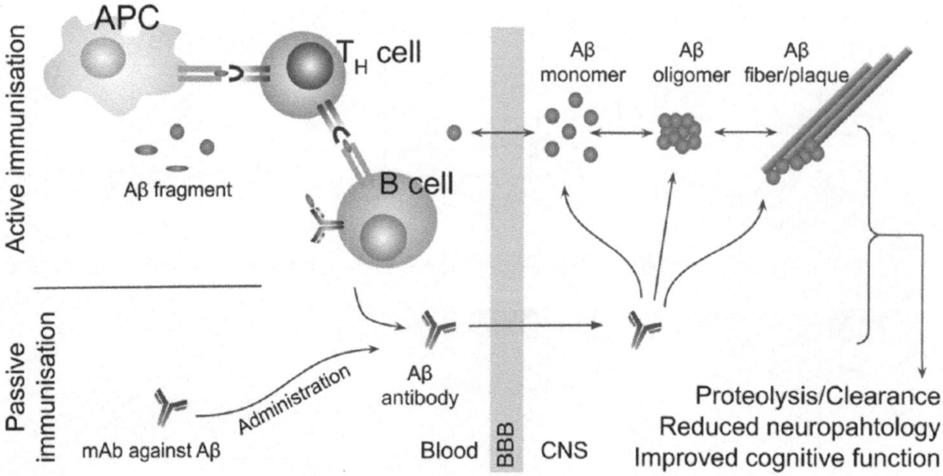

FIGURE 10.8 Schematic presentation of the antibody-mediated treatment of Alzheimer's disease. In the case of active immunisation, antibody production is induced inside the body by injecting antigen, which results in the production of plasma B cells after various complex interactions of antigen-presenting cells (APC), T helper (T_H) cells and B cells. In contrast, in the case of passive immunisation, a monoclonal anti-Aβ antibody is directly injected into the systemic circulation, which crosses the blood-brain barrier (BBB) and reacts with Aβ oligomers and fibrils followed by phagocytosis and proteasomal degradation and reduces the fibril deposition, which results in improvement in cognitive function and a reduction of neuropathology.

are used to treat these diseases with immunotherapy: i) active immunisation and ii) passive immunisation. Both immunisation methods have shown the clearance of Aβ deposition through local microglial activation (Schenk et al. 1999). Active immunisation by immunising with a particular protein antigen, for example, α-synuclein or Aβ, and passive immunisation by directly injecting the monoclonal antibody have been demonstrated to reduce the aggregated proteins either by increasing the autophagy or by activating the macrophage. α-synuclein has been reported to propagate among cells (neural cells and astroglia). This uptake of protein by astroglia results in a considerable production of inflammation; thus, targeting α-synuclein with a region-specific anti-α-synuclein antibody is a good choice (Valera and Masliah 2013). Many studies that immunise with different segments of Aβ have been performed, but most of them have failed either due to side effects or for not meeting the primary goals (Zago et al. 2012; Tariot et al. 2018; Mandler et al. 2019). Different stages of clinical trials are also ongoing that target different species of α-synuclein. Recently, a Phase I trial (NCT02459886) for BIIB054 that targets aggregated α-synuclein concluded that it was safe and tolerable (Brys et al. 2019, 054). Prasinezumab (NCT03100149) is undergoing Phase II trials (Jankovic et al. 2018). A Phase I trial for the safety of AFFITOPE (NCT02270489) was successfully conducted that showed good tolerability (Mandler et al. 2015).

10.7 CONCLUSION

Neurodegenerative diseases are very prevalent worldwide and affect elderly people; proper management and treatment are very important to cope with them. Several investigations have been ongoing for years. There are a few drugs for the management of these diseases, but most of them alleviate the symptoms and cannot reverse the neuronal death or neuroinflammation. Many trials have been performed, but most of these were either withdrawn or failed because of safety issues or side effects. It will be much more plausible if the treatment can reverse the neuronal death, if neuroprotection is built up and if neuroinflammation remains under control. Although stem cell therapy can potentially reverse the neuronal loss, it is far from a direct application in human patients. Recent progress in research has also helped in the investigative progress of this field. Several different strategies have emerged, for example, gene therapy, stem cell therapy, nanotherapy, small molecule therapy and immunotherapy, and intense efforts are underway to unveil the diseases' mechanism and to build new therapies to cure them. In the future, further research is expected to provide more insights into the disease pathology, which will help to design more effective therapeutics.

REFERENCES

Abraham, S., C. C. Soundararajan, S. Vivekanandhan, and M. Behari. 2005. "Erythrocyte Antioxidant Enzymes in Parkinson's Disease." *The Indian Journal of Medical Research* 121 (2): 111–15.

Adhikari, Aniruddha, Susmita Mondal, Monojit Das, Pritam Biswas, Uttam Pal, Soumendra Darbar, Siddhartha Sankar Bhattacharya, et al. 2021. "Incorporation of a Biocompatible Nanozyme in Cellular Antioxidant Enzyme Cascade Reverses Huntington's Like Disorder in Preclinical Model." *Advanced Healthcare Materials* 10 (7): e2001736. https://doi.org/10.1002/adhm.202001736.

Ager, Rahasson R., Joy L. Davis, Andy Agazaryan, Francisca Benavente, Wayne W. Poon, Frank M. LaFerla, and Mathew Blurton-Jones. 2015. "Human Neural Stem Cells Improve Cognition and Promote Synaptic Growth in Two Complementary Transgenic Models of Alzheimer's Disease and Neuronal Loss." *Hippocampus* 25 (7): 813–26. https://doi.org/10.1002/hipo.22405.

Apetri, Mihaela M., Nakul C. Maiti, Michael G. Zagorski, Paul R. Carey, and Vernon E. Anderson. 2006. "Secondary Structure of α-Synuclein Oligomers: Characterization by Raman and Atomic Force Microscopy." *Journal of Molecular Biology* 355 (1): 63–71. https://doi.org/10.1016/j.jmb.2005.10.071.

Arndt, James R., Maxmore Chaibva, and Justin Legleiter. 2015. "The Emerging Role of the First 17 Amino Acids of Huntingtin in Huntington's Disease." *Biomolecular Concepts* 6 (1): 33–46. https://doi.org/10.1515/bmc-2015-0001.

Bae, Eun-Jin, Na-Young Yang, Miyoung Song, Cheol Soon Lee, Jun Sung Lee, Byung Chul Jung, He-Jin Lee, et al. 2014. "Glucocerebrosidase Depletion Enhances Cell-to-Cell Transmission of α-Synuclein." *Nature Communications* 5 (August): 4755. https://doi.org/10.1038/ncomms5755.

Bardhan, Munmun, Sandip Dolui, Siddhi Chaudhuri, Uttam Paul, Gaurav Bhattacharjee, Manorama Ghosal, Nakul C. Maiti, et al. 2021. "Impact of Porous Nanomaterials on Inhibiting Protein Aggregation Behaviour." *RSC Advances* 11 (6): 3354–62. https://doi.org/10.1039/D0RA10927D.

Bengoa-Vergniory, Nora, Rosalind F. Roberts, Richard Wade-Martins, and Javier Alegre-Abarrategui. 2017. "Alpha-Synuclein Oligomers: A New Hope." *Acta Neuropathologica* 134 (6): 819–38. https://doi.org/10.1007/s00401-017-1755-1.

Bera, Kaushik, Animesh Mondal, Uttam Pal, and Nakul C. Maiti. 2020. "Porphyrin-Armored Gold Nanospheres Modulate the Secondary Structure of α-Synuclein and Arrest Its Fibrillation." *The Journal of Physical Chemistry C* 124 (11): 6418–34. https://doi.org/10.1021/acs.jpcc.9b11503.

Bissonnette, Christopher J., Ljuba Lyass, Bula J. Bhattacharyya, Abdelhak Belmadani, Richard J. Miller, and John A. Kessler. 2011. "The Controlled Generation of Functional Basal Forebrain Cholinergic Neurons from Human Embryonic Stem Cells." *Stem Cells (Dayton, Ohio)* 29 (5): 802–11. https://doi.org/10.1002/stem.626.

Bloom, George S. 2014. "Amyloid-β and Tau: The Trigger and Bullet in Alzheimer Disease Pathogenesis." *JAMA Neurology* 71 (4): 505–8. https://doi.org/10.1001/jamaneurol.2013.5847.

Boudreau, Ryan L., Jodi L. McBride, Inês Martins, Shihao Shen, Yi Xing, Barrie J. Carter, and Beverly L. Davidson. 2009. "Nonallele-Specific Silencing of Mutant and Wild-Type Huntingtin Demonstrates Therapeutic Efficacy in Huntington's Disease Mice." *Molecular Therapy: The Journal of the American Society of Gene Therapy* 17 (6): 1053–63. https://doi.org/10.1038/mt.2009.17.

Brys, Miroslaw, Laura Fanning, Serena Hung, Aaron Ellenbogen, Natalia Penner, Minhua Yang, Mackenzie Welch, et al. 2019. "Randomized Phase I Clinical Trial of Anti-α-Synuclein Antibody BIIB054." *Movement Disorders: Official Journal of the Movement Disorder Society* 34 (8): 1154–63. https://doi.org/10.1002/mds.27738.

Cascella, Roberta, Serene W. Chen, Alessandra Bigi, José D. Camino, Catherine K. Xu, Christopher M. Dobson, Fabrizio Chiti, et al. 2021. "The Release of Toxic Oligomers from α-Synuclein Fibrils Induces Dysfunction in Neuronal Cells." *Nature Communications* 12 (1): 1814. https://doi.org/10.1038/s41467-021-21937-3.

Cave, John W., Meng Wang, and Harriet Baker. 2014. "Adult Subventricular Zone Neural Stem Cells as a Potential Source of Dopaminergic Replacement Neurons." *Frontiers in Neuroscience* 8: 16. https://doi.org/10.3389/fnins.2014.00016.

Chang, Jennifer L., Anthony J. Hinrich, Brandon Roman, Michaela Norrbom, Frank Rigo, Robert A. Marr, Eric M. Norstrom, et al. 2018. "Targeting Amyloid-β Precursor Protein, APP, Splicing with Antisense Oligonucleotides Reduces Toxic Amyloid-β Production." *Molecular Therapy: The Journal of the American Society of Gene Therapy* 26 (6): 1539–51. https://doi.org/10.1016/j.ymthe.2018.02.029.

Chen, Hong-Xu, Fu-Chao Liang, Ping Gu, Bian-Ling Xu, Hong-Jun Xu, Wen-Ting Wang, Jia-Yang Hou, et al. 2020. "Exosomes Derived from Mesenchymal Stem Cells Repair a Parkinson's Disease Model by Inducing Autophagy." *Cell Death & Disease* 11 (4): 288. https://doi.org/10.1038/s41419-020-2473-5.

Chen, Zhong-Can, Wei Zhang, Ling-Ling Chua, Chou Chai, Rong Li, Lin Manhua, Zhen Cao, et al. 2017. "Phosphorylation of Amyloid Precursor Protein by Mutant LRRK2 Promotes AICD Activity and Neurotoxicity in Parkinson's Disease." *Science Signaling* 10 (488): eaam6790. https://doi.org/10.1126/scisignal.aam6790.

Chico, Laura K., Linda J. Van Eldik, and D. Martin Watterson. 2009. "Targeting Protein Kinases in Central Nervous System Disorders." *Nature Reviews: Drug Discovery* 8 (11): 892–909. https://doi.org/10.1038/nrd2999.

Christine, Chadwick W., Krystof S. Bankiewicz, Amber D. Van Laar, R. Mark Richardson, Bernard Ravina, Adrian P. Kells, Brendon Boot, et al. 2019. "Magnetic Resonance Imaging–Guided Phase 1 Trial of Putaminal AADC Gene Therapy for Parkinson's Disease." *Annals of Neurology* 85 (5): 704–14. https://doi.org/10.1002/ana.25450.

Cohen, Samuel I. A., Sara Linse, Leila M. Luheshi, Erik Hellstrand, Duncan A. White, Luke Rajah, Daniel E. Otzen, et al. 2013. "Proliferation of Amyloid-B42 Aggregates Occurs Through a Secondary Nucleation Mechanism." *Proceedings of the National Academy of Sciences of the United States of America* 110 (24): 9758–63. https://doi.org/10.1073/pnas.1218402110.

Cotzias, G. C., P. S. Papavasiliou, and R. Gellene. 1969. "Modification of Parkinsonism-Chronic Treatment with L-Dopa." *The New England Journal of Medicine* 280 (7): 337–45. https://doi.org/10.1056/NEJM196902132800701.

Cotzias, G. C., M. H. Van Woert, and L. M. Schiffer. 1967. "Aromatic Amino Acids and Modification of Parkinsonism." *The New England Journal of Medicine* 276 (7): 374–79. https://doi.org/10.1056/NEJM196702162760703.

Cui, Weigang, Chunli Sun, Yuqi Ma, Songtao Wang, Xianwei Wang, and Yinghua Zhang. 2020. "Inhibition of TLR4 Induces M2 Microglial Polarization and Provides Neuroprotection via the NLRP3 Inflammasome in Alzheimer's Disease." *Frontiers in Neuroscience* 14: 444. https://doi.org/10.3389/fnins.2020.00444.

Currais, Antonio, Wolfgang Fischer, Pamela Maher, and David Schubert. 2017. "Intraneuronal Protein Aggregation as a Trigger for Inflammation and Neurodegeneration in the Aging Brain." *FASEB Journal: Official Publication of the Federation of American Societies for Experimental Biology* 31 (1): 5–10. https://doi.org/10.1096/fj.201601184.

Das, Supriya, Uttam Pal, Swagata Das, and Nakul Chandra Maiti. 2013. "Chaperone Action of Cyclophilin on Lysozyme and Its Aggregate." *Journal of Proteins and Proteomics* 4 (2): 129.

Das, Swagata, Supriya Das, Anupam Roy, Uttam Pal, and Nakul Chandra Maiti. 2016. "Orientation of Tyrosine Side Chain in Neurotoxic Aβ Differs in Two Different Secondary Structures of the Peptide." *Royal Society Open Science* 3 (September): 160112. https://doi.org/10.1098/rsos.160112.

Das, Swagata, Uttam Pal, Supriya Das, Khyati Bagga, Anupam Roy, Arpita Mrigwani, and Nakul C. Maiti. 2014. "Sequence Complexity of Amyloidogenic Regions in Intrinsically Disordered Human Proteins." *PLoS One* 9 (3): e89781. https://doi.org/10.1371/journal.pone.0089781.

Dehay, Benjamin, Jordi Bové, Natalia Rodríguez-Muela, Celine Perier, Ariadna Recasens, Patricia Boya, and Miquel Vila. 2010. "Pathogenic Lysosomal Depletion in Parkinson's Disease." *The Journal of Neuroscience: The Official Journal of the Society for Neuroscience* 30 (37): 12535–44. https://doi.org/10.1523/JNEUROSCI.1920-10.2010.

Dietrich, Paula, Irudayam Maria Johnson, Shanta Alli, and Ioannis Dragatsis. 2017. "Elimination of Huntingtin in the Adult Mouse Leads to Progressive Behavioral Deficits, Bilateral Thalamic Calcification, and Altered Brain Iron Homeostasis." *PLoS Genetics* 13 (7): e1006846. https://doi.org/10.1371/journal.pgen.1006846.

Dodart, Jean-Cosme, Robert A. Marr, Milla Koistinaho, Beth M. Gregersen, Seema Malkani, Inder M. Verma, and Steven M. Paul. 2005. "Gene Delivery of Human Apolipoprotein E Alters Brain Abeta Burden in a Mouse Model of Alzheimer's Disease." *Proceedings of the National Academy of Sciences of the United States of America* 102 (4): 1211–16. https://doi.org/10.1073/pnas.0409072102.

Doi, Daisuke, Hiroaki Magotani, Tetsuhiro Kikuchi, Megumi Ikeda, Satoe Hiramatsu, Kenji Yoshida, Naoki Amano, et al. 2020. "Pre-Clinical Study of Induced Pluripotent Stem Cell-Derived Dopaminergic Progenitor Cells for Parkinson's Disease." *Nature Communications* 11 (1): 3369. https://doi.org/10.1038/s41467-020-17165-w.

Geschwind, Daniel H. 2003. "Tau Phosphorylation, Tangles, and Neurodegeneration: The Chicken or the Egg?" *Neuron* 40 (3): 457–60. https://doi.org/10.1016/S0896-6273(03)00681-0.

Gupta, Rohan, Mehar Sahu, Devesh Srivastava, Swati Tiwari, Rashmi K. Ambasta, and Pravir Kumar. 2021. "Post-Translational Modifications: Regulators of Neurodegenerative Proteinopathies." *Ageing Research Reviews* 68 (July): 101336. https://doi.org/10.1016/j.arr.2021.101336.

Hanif, Sumaira, Pir Muhammad, Rose Chesworth, Fawad Ur Rehman, Rong-Jun Qian, Meng Zheng, and Bing-Yang Shi. 2020. "Nanomedicine-Based Immunotherapy for Central Nervous System Disorders." *Acta Pharmacologica Sinica* 41 (7): 936–53. https://doi.org/10.1038/s41401-020-0429-z.

Hartmann, C. J., S. J. Groiss, J. Vesper, A. Schnitzler, and L. Wojtecki. 2016. "Brain Stimulation in Huntington's Disease." *Neurodegenerative Disease Management* 6 (3). https://doi.org/10.2217/nmt-2016-0007.

Hiller, Benjamin M., David J. Marmion, Cayla A. Thompson, Nathaniel A. Elliott, Howard Federoff, Patrik Brundin, Virginia B. Mattis, et al. 2022. "Optimizing Maturity and Dose of IPSC-Derived Dopamine Progenitor Cell Therapy for Parkinson's Disease." *NPJ Regenerative Medicine* 7 (1): 1–15. https://doi.org/10.1038/s41536-022-00221-y.

Huang, Kun, Nakul C. Maiti, Nelson B. Phillips, Paul R. Carey, and Michael A. Weiss. 2006. "Structure-Specific Effects of Protein Topology on Cross-Beta Assembly: Studies of Insulin Fibrillation." *Biochemistry* 45 (34): 10278–93. https://doi.org/10.1021/bi060879g.

Hudry, Eloise, Jonathan Dashkoff, Alysson D. Roe, Shuko Takeda, Robert M. Koffie, Tadafumi Hashimoto, Maria Scheel, et al. 2013. "Gene Transfer of Human Apoe Isoforms Results in Differential Modulation of Amyloid Deposition and Neurotoxicity in Mouse Brain." *Science Translational Medicine* 5 (212): 212ra161. https://doi.org/10.1126/scitranslmed.3007000.

Jankovic, Joseph, Ira Goodman, Beth Safirstein, Tonya K. Marmon, Dale B. Schenk, Martin Koller, Wagner Zago, et al. 2018. "Safety and Tolerability of Multiple Ascending Doses of PRX002/RG7935, an Anti-α-Synuclein Monoclonal Antibody, in Patients with Parkinson Disease: A Randomized Clinical Trial." *JAMA Neurology* 75 (10): 1206–14. https://doi.org/10.1001/jamaneurol.2018.1487.

Kaji, E. H., and J. M. Leiden. 2001. "Gene and Stem Cell Therapies." *JAMA* 285 (5): 545–50. https://doi.org/10.1001/jama.285.5.545.

Kalia, Lorraine V., and Anthony E. Lang. 2016. "Evolving Basic, Pathological and Clinical Concepts in PD." *Nature Reviews Neurology* 12 (2): 65–66. https://doi.org/10.1038/nrneurol.2015.249.

Karussis, Dimitrios, Panayiota Petrou, and Ibrahim Kassis. 2013. "Clinical Experience with Stem Cells and Other Cell Therapies in Neurological Diseases." *Journal of the Neurological Sciences* 324 (1–2): 1–9. https://doi.org/10.1016/j.jns.2012.09.031.

Keskin, Sonay, Cynthia C. Brouwers, Marina Sogorb-Gonzalez, Raygene Martier, Josse A. Depla, Astrid Vallès, Sander J. van Deventer, et al. 2019. "AAV5-MiHTT Lowers Huntingtin MRNA and Protein without Off-Target Effects in Patient-Derived Neuronal Cultures and Astrocytes." *Molecular Therapy: Methods & Clinical Development* 15 (December): 275–84. https://doi.org/10.1016/j.omtm.2019.09.010.

Kim, Hyun-Jung. 2011. "Stem Cell Potential in Parkinson's Disease and Molecular Factors for the Generation of Dopamine Neurons." *Biochimica Et Biophysica Acta* 1812 (1): 1–11. https://doi.org/10.1016/j.bbadis.2010.08.006.

Kim, Seung Won, Hye-Ji Woo, Eun Hee Kim, Hyung Sun Kim, Han Na Suh, Soo-Hyun Kim, Jae-Jin Song, et al. 2021. "Neural Stem Cells Derived from Human Midbrain Organoids as a Stable Source for Treating Parkinson's Disease: Midbrain Organoid-NSCs (Og-NSC) as a Stable Source for PD Treatment." *Progress in Neurobiology* 204 (September): 102086. https://doi.org/10.1016/j.pneurobio.2021.102086.

Krahn, Andrea I., Carrow Wells, David H. Drewry, Lenore K. Beitel, Thomas M. Durcan, and Alison D. Axtman. 2020. "Defining the Neural Kinome: Strategies and Opportunities for Small Molecule Drug Discovery to Target Neurodegenerative Diseases." *ACS Chemical Neuroscience* 11 (13): 1871–86. https://doi.org/10.1021/acschemneuro.0c00176.

Lee, Il-Shin, Kwangsoo Jung, Il-Sun Kim, Haejin Lee, Miri Kim, Seokhwan Yun, Kyujin Hwang, et al. 2015. "Human Neural Stem Cells Alleviate Alzheimer-Like Pathology in a Mouse Model." *Molecular Neurodegeneration* 10 (August): 38. https://doi.org/10.1186/s13024-015-0035-6.

Li, Amanda, Joel Tyson, Shivni Patel, Meer Patel, Sruthi Katakam, Xiaobo Mao, and Weiwei He. 2021. "Emerging Nanotechnology for Treatment of Alzheimer's and Parkinson's Disease." *Frontiers in Bioengineering and Biotechnology* 9: 672594. https://doi.org/10.3389/fbioe.2021.672594.

Li, Haiyun, Jiao Yan, Dejing Meng, Rui Cai, Xinshuang Gao, Yinglu Ji, Liming Wang, et al. 2020. "Gold Nanorod-Based Nanoplatform Catalyzes Constant NO Generation and Protects from Cardiovascular Injury." *ACS Nano* 14 (10): 12854–65. https://doi.org/10.1021/acsnano.0c03629.

Liang, Minmin, and Xiyun Yan. 2019. “Nanozymes: From New Concepts, Mechanisms, and Standards to Applications.” *Accounts of Chemical Research* 52 (8): 2190–2200. https://doi.org/10.1021/acs.accounts.9b00140.

Lilja, Anna M., Linn Malmsten, Jennie Röjdner, Larysa Voytenko, Alexei Verkhratsky, Sven Ove Ögren, Agneta Nordberg, et al. 2015. “Neural Stem Cell Transplant-Induced Effect on Neurogenesis and Cognition in Alzheimer Tg2576 Mice Is Inhibited by Concomitant Treatment with Amyloid-Lowering or Cholinergic A7 Nicotinic Receptor Drugs.” *Neural Plasticity* 2015: 370432. https://doi.org/10.1155/2015/370432.

Lin, Michelle K., and Matthew J. Farrer. 2014. “Genetics and Genomics of Parkinson’s Disease.” *Genome Medicine* 6 (6): 48. https://doi.org/10.1186/gm566.

Liu, Kaiming, Kunqian Ji, Liang Guo, Wei Wu, Huixia Lu, Peiyan Shan, and Chuanzhu Yan. 2014. “Mesenchymal Stem Cells Rescue Injured Endothelial Cells in an in Vitro Ischemia-Reperfusion Model via Tunneling Nanotube like Structure-Mediated Mitochondrial Transfer.” *Microvascular Research* 92 (March): 10–18. https://doi.org/10.1016/j.mvr.2014.01.008.

Liu, Ruiyuan, Jun Yang, Linying Liu, Zhiguo Lu, Zhuyan Shi, Weihong Ji, Jie Shen, and Xin Zhang. 2020. “An ‘Amyloid-β Cleaner’ for the Treatment of Alzheimer’s Disease by Normalizing Microglial Dysfunction.” *Advanced Science (Weinheim, Baden-Wurttemberg, Germany)* 7 (2): 1901555. https://doi.org/10.1002/advs.201901555.

Liu, Tengfei, Bowen Xiao, Fei Xiang, Jianglin Tan, Zhuo Chen, Xiaorong Zhang, Chengzhou Wu, et al. 2020. “Ultrasmall Copper-Based Nanoparticles for Reactive Oxygen Species Scavenging and Alleviation of Inflammation Related Diseases.” *Nature Communications* 11 (1): 2788. https://doi.org/10.1038/s41467-020-16544-7.

Losordo, D. W., P. R. Vale, J. F. Symes, C. H. Dunnington, D. D. Esakof, M. Maysky, A. B. Ashare, et al. 1998. “Gene Therapy for Myocardial Angiogenesis: Initial Clinical Results with Direct Myocardial Injection of PhVEGF165 as Sole Therapy for Myocardial Ischemia.” *Circulation* 98 (25): 2800–2804. https://doi.org/10.1161/01.cir.98.25.2800.

Lu, Jinxia, Qin Cao, Chuchu Wang, Jing Zheng, Feng Luo, Jingfei Xie, Yichen Li, et al. 2019. “Structure-Based Peptide Inhibitor Design of Amyloid-β Aggregation.” *Frontiers in Molecular Neuroscience* 12. www.frontiersin.org/articles/10.3389/fnmol.2019.00054.

Maiti, Nakul C., Mihaela M. Apetri, Michael G. Zagorski, Paul R. Carey, and Vernon E. Anderson. 2004. “Raman Spectroscopic Characterization of Secondary Structure in Natively Unfolded Proteins: α-Synuclein.” *Journal of the American Chemical Society* 126 (8): 2399–2408. https://doi.org/10.1021/ja0356176.

Maiti, Nakul C., Dianlu Jiang, Andrew J. Wain, Sveti Patel, Kim L. Dinh, and Feimeng Zhou. 2008. “Mechanistic Studies of Cu(II) Binding to Amyloid-β Peptides and the Fluorescence and Redox Behaviors of the Resulting Complexes.” *The Journal of Physical Chemistry B* 112 (28): 8406–11. https://doi.org/10.1021/jp802038p.

Maiti, Nakul, Payel Mondal, and Supriya Das. 2015. “Role of Cyclophilin on Amyloid Formation and Protein Disaggregation.” *The FASEB Journal* 29 (S1): 881.7. https://doi.org/10.1096/fasebj.29.1_supplement.881.7.

Mandler, Markus, Edward Rockenstein, Cassia Overk, Michael Mante, Jazmin Florio, Anthony Adame, Changyoun Kim, et al. 2019. “Effects of Single and Combined Immunotherapy Approach Targeting Amyloid β Protein and α-Synuclein in a Dementia with Lewy Bodies-like Model.” *Alzheimer’s & Dementia: The Journal of the Alzheimer’s Association* 15 (9): 1133–48. https://doi.org/10.1016/j.jalz.2019.02.002.

Mandler, Markus, Elvira Valera, Edward Rockenstein, Michael Mante, Harald Weninger, Christina Patrick, Anthony Adame, et al. 2015. “Active Immunization against Alpha-Synuclein Ameliorates the Degenerative Pathology and Prevents Demyelination in a Model of Multiple System Atrophy.” *Molecular Neurodegeneration* 10 (March): 10. https://doi.org/10.1186/s13024-015-0008-9.

Mazzulli, Joseph R., You-Hai Xu, Ying Sun, Adam L. Knight, Pamela J. McLean, Guy A. Caldwell, Ellen Sidransky, et al. 2011. “Gaucher’s Disease Glucocerebrosidase and <alpha>-Synuclein Form a Bidirectional Pathogenic Loop in Synucleinopathies.” *Cell* 146 (1): 37–52. https://doi.org/10.1016/j.cell.2011.06.001.

Mittermeyer, Gabriele, Chadwick W. Christine, Kathryn H. Rosenbluth, Suzanne L. Baker, Philip Starr, Paul Larson, Paul L. Kaplan, et al. 2012. “Long-Term Evaluation of a Phase 1 Study of AADC Gene Therapy for Parkinson’s Disease.” *Human Gene Therapy* 23 (4): 377–81. https://doi.org/10.1089/hum.2011.220.

Moghadam, Farshad Homayouni, Hojatoallah Alaie, Khadije Karbalaie, Somayeh Tanhaei, Mohammad Hossein Nasr Esfahani, and Hossein Baharvand. 2009. “Transplantation of Primed or Unprimed Mouse Embryonic Stem Cell-Derived Neural Precursor Cells Improves Cognitive Function in Alzheimerian Rats.” *Differentiation; Research in Biological Diversity* 78 (2–3): 59–68. https://doi.org/10.1016/j.diff.2009.06.005.

Mondal, Animesh, and Nakul Chandra Maiti. 2019. "Divulging Characteristic Features of the Novel A-Synuclein Oligomers Augmenting Parkinson's Disease." *Biophysical Journal* 116 (3): 185a. https://doi.org/10.1016/j.bpj.2018.11.1027.

Mondal, Animesh, Uttam Pal, Anupam Roy, and Nakul C. Maiti. 2018. "Structural Intricacy of Disordered Regions in Transcription Factors Imparting Colon Cancer." *Journal of Proteins & Proteomics* 9 (3). www.proteomicssociety.in/jpp-journal-master/www.jpp.org.in/index.php/jpp/article/view/302.html.

Nagahara, Alan H., Tim Bernot, Rod Moseanko, Armin Blesch, James M. Conner, Anthony Ramirez, Mehdi Gasmi, and Mark H. Tuszynski. 2009. "Long-Term Reversal of Cholinergic Neuronal Decline in Aged Non-Human Primates by Lentiviral NGF Gene Delivery." *Experimental Neurology* 215 (1): 153–59. https://doi.org/10.1016/j.expneurol.2008.10.004.

Nagahara, Alan H., Bayard R. Wilson, Iryna Ivasyk, Imre Kovacs, Saytam Rawalji, John R. Bringas, Phillip J. Pivirotto, et al. 2018. "MR-Guided Delivery of AAV2-BDNF into the Entorhinal Cortex of Non-Human Primates." *Gene Therapy* 25 (2): 104–14. https://doi.org/10.1038/s41434-018-0010-2.

Oh, Misook, Ji Hoon Lee, Wei Wang, Hui Sun Lee, Woo Sirl Lee, Christopher Burlak, Wonpil Im, et al. 2014. "Potential Pharmacological Chaperones Targeting Cancer-Associated MCL-1 and Parkinson Disease-Associated α-Synuclein." *Proceedings of the National Academy of Sciences of the United States of America* 111 (30): 11007–12. https://doi.org/10.1073/pnas.1320556111.

Olanow, C. W., and W. G. Tatton. 1999. "Etiology and Pathogenesis of Parkinson's Disease." *Annual Review of Neuroscience* 22: 123–44. https://doi.org/10.1146/annurev.neuro.22.1.123.

Pal, Uttam, Anupam Roy, Supriya Das, Swagata Das, Mangaldeep Kundu, Khyati Bagga, and Nakul Chandra Maiti. 2016. "Origin of Protein Aggregation: Identification of Some Characteristic Traits in Structured And Intrinsically Disordered Proteins." *Journal of Proteins and Proteomics* 7 (4): 259–68.

Park, Dongsun, Goeun Yang, Dae Kwon Bae, Sun Hee Lee, Yun-Hui Yang, Jangbeen Kyung, Dajeong Kim, et al. 2013. "Human Adipose Tissue-Derived Mesenchymal Stem Cells Improve Cognitive Function and Physical Activity in Ageing Mice." *Journal of Neuroscience Research* 91 (5): 660–70. https://doi.org/10.1002/jnr.23182.

Poovaiah, N., Z. Davoudi, H. Peng, B. Schlichtmann, S. Mallapragada, B. Narasimhan, and Q. Wang. 2018. "Treatment of Neurodegenerative Disorders through the Blood-Brain Barrier Using Nanocarriers." *Nanoscale* 10 (36): 16962–83. https://doi.org/10.1039/c8nr04073g.

Reddy, Arubala P., Janani Ravichandran, and Nurgul Carkaci-Salli. 2020. "Neural Regeneration Therapies for Alzheimer's and Parkinson's Disease-Related Disorders." *Biochimica Et Biophysica Acta. Molecular Basis of Disease* 1866 (4): 165506. https://doi.org/10.1016/j.bbadis.2019.06.020.

Rosenberg, Jonathan B., Michael G. Kaplitt, Bishnu P. De, Alvin Chen, Thomas Flagiello, Christiana Salami, Eduard Pey, et al. 2018. "AAVrh.10-Mediated APOE2 Central Nervous System Gene Therapy for APOE4-Associated Alzheimer's Disease." *Human Gene Therapy. Clinical Development* 29 (1): 24–47. https://doi.org/10.1089/humc.2017.231.

Ross, Christopher A., and Michelle A. Poirier. 2004. "Protein Aggregation and Neurodegenerative Disease." *Nature Medicine* 10 Suppl (July): S10–17. https://doi.org/10.1038/nm1066.

Ross, Christopher A., and Sarah J. Tabrizi. 2011. "Huntington's Disease: From Molecular Pathogenesis to Clinical Treatment." *The Lancet. Neurology* 10 (1): 83–98. https://doi.org/10.1016/S1474-4422(10)70245-3.

Schenk, D., R. Barbour, W. Dunn, G. Gordon, H. Grajeda, T. Guido, K. Hu, et al. 1999. "Immunization with Amyloid-Beta Attenuates Alzheimer-Disease-like Pathology in the PDAPP Mouse." *Nature* 400 (6740): 173–77. https://doi.org/10.1038/22124.

Schwarz, Johannes, and Alexander Storch. 2010. "Transplantation in Parkinson's Disease: Will Mesenchymal Stem Cells Help to Reenter the Clinical Arena?" *Translational Research: The Journal of Laboratory and Clinical Medicine* 155 (2): 55–56. https://doi.org/10.1016/j.trsl.2009.08.008.

Sebastian, Waldy San, Adrian P Kells, John Bringas, Lluis Samaranch, Piotr Hadaczek, Agnieszka Ciesielska, Michael J Macayan, et al. 2014. "Safety and Tolerability of MRI-Guided Infusion of AAV2-HAADC into the Mid-Brain of Nonhuman Primate." *Molecular Therapy. Methods & Clinical Development* 3 (October): 14049. https://doi.org/10.1038/mtm.2014.49.

Shi, Yuan, and Mary Mader. 2018. "Brain Penetrant Kinase Inhibitors: Learning from Kinase Neuroscience Discovery." *Bioorganic & Medicinal Chemistry Letters* 28 (11): 1981–91. https://doi.org/10.1016/j.bmcl.2018.05.007.

Sun, Jichao, Jared Carlson-Stevermer, Utpal Das, Minjie Shen, Marion Delenclos, Amanda M. Snead, So Yeon Koo, et al. 2019. "CRISPR/Cas9 Editing of APP C-Terminus Attenuates β-Cleavage and Promotes α-Cleavage." *Nature Communications* 10 (1): 53. https://doi.org/10.1038/s41467-018-07971-8.

Sun, Jichao, and Subhojit Roy. 2021. "Gene-Based Therapies for Neurodegenerative Diseases." *Nature Neuroscience* 24 (3): 297–311. https://doi.org/10.1038/s41593-020-00778-1.

Tabrizi, Sarah J., Blair R. Leavitt, G. Bernhard Landwehrmeyer, Edward J. Wild, Carsten Saft, Roger A. Barker, Nick F. Blair, et al. 2019. "Targeting Huntingtin Expression in Patients with Huntington's Disease." *The New England Journal of Medicine* 380 (24): 2307–16. https://doi.org/10.1056/NEJMoa1900907.

Takagi, Yasushi, Jun Takahashi, Hidemoto Saiki, Asuka Morizane, Takuya Hayashi, Yo Kishi, Hitoshi Fukuda, et al. 2005. "Dopaminergic Neurons Generated from Monkey Embryonic Stem Cells Function in a Parkinson Primate Model." *The Journal of Clinical Investigation* 115 (1): 102–09. https://doi.org/10.1172/JCI21137.

Tariot, Pierre N., Francisco Lopera, Jessica B. Langbaum, Ronald G. Thomas, Suzanne Hendrix, Lon S. Schneider, Silvia Rios-Romenets, et al. 2018. "The Alzheimer's Prevention Initiative Autosomal-Dominant Alzheimer's Disease Trial: A Study of Crenezumab versus Placebo in Preclinical PSEN1 E280A Mutation Carriers to Evaluate Efficacy and Safety in the Treatment of Autosomal-Dominant Alzheimer's Disease, Including a Placebo-Treated Noncarrier Cohort." *Alzheimer's & Dementia (New York)* 4: 150–60. https://doi.org/10.1016/j.trci.2018.02.002.

Tolosa, E., M. J. Martí, F. Valldeoriola, and J. L. Molinuevo. 1998. "History of Levodopa and Dopamine Agonists in Parkinson's Disease Treatment." *Neurology* 50 (6 Suppl 6): S2–10; discussion S44–48. https://doi.org/10.1212/wnl.50.6_suppl_6.s2.

Trombetta-Lima, Marina, Angélica María Sabogal-Guáqueta, and Amalia M. Dolga. 2021. "Mitochondrial Dysfunction in Neurodegenerative Diseases: A Focus on IPSC-Derived Neuronal Models." *Cell Calcium* 94 (March): 102362. https://doi.org/10.1016/j.ceca.2021.102362.

Tuszynski, Mark H., Leon Thal, Mary Pay, David P. Salmon, Hoi Sang U, Roy Bakay, Piyush Patel, et al. 2005. "A Phase 1 Clinical Trial of Nerve Growth Factor Gene Therapy for Alzheimer Disease." *Nature Medicine* 11 (5): 551–55. https://doi.org/10.1038/nm1239.

Valera, Elvira, and Eliezer Masliah. 2013. "Immunotherapy for Neurodegenerative Diseases: Focus on α-Synucleinopathies." *Pharmacology & Therapeutics* 138 (3): 311–22. https://doi.org/10.1016/j.pharmthera.2013.01.013.

Wang, Wen-Ying, Meng-Shan Tan, Jin-Tai Yu, and Lan Tan. 2015. "Role of Pro-Inflammatory Cytokines Released from Microglia in Alzheimer's Disease." *Annals of Translational Medicine* 3 (10): 136. https://doi.org/10.3978/j.issn.2305-5839.2015.03.49.

Wang, Yu-Kai, Wan-Wan Zhu, Meng-Hua Wu, Yi-Hui Wu, Zheng-Xin Liu, Ling-Min Liang, Chao Sheng, et al. 2018. "Human Clinical-Grade Parthenogenetic ESC-Derived Dopaminergic Neurons Recover Locomotive Defects of Nonhuman Primate Models of Parkinson's Disease." *Stem Cell Reports* 11 (1): 171–82. https://doi.org/10.1016/j.stemcr.2018.05.010.

Wiesen, Tom, and Daphne Atlas. 2022. "Novel Anti-Apoptotic L-DOPA Precursors SuperDopa and SuperDopamide as Potential Neuroprotective Agents for Halting/Delaying Progression of Parkinson's Disease." *Cell Death & Disease* 13 (3): 227. https://doi.org/10.1038/s41419-022-04667-2.

Xiong, N., Z. Zhang, J. Huang, C. Chen, Z. Zhang, M. Jia, J. Xiong, et al. 2011. "VEGF-Expressing Human Umbilical Cord Mesenchymal Stem Cells, an Improved Therapy Strategy for Parkinson's Disease." *Gene Therapy* 18 (4): 394–402. https://doi.org/10.1038/gt.2010.152.

Xuan, A. G., M. Luo, W. D. Ji, and D. H. Long. 2009. "Effects of Engrafted Neural Stem Cells in Alzheimer's Disease Rats." *Neuroscience Letters* 450 (2): 167–71. https://doi.org/10.1016/j.neulet.2008.12.001.

Yang, Lei, Renyuan Zhou, Yu Tong, Pengfei Chen, Yu Shen, Shuai Miao, and Xiaoqiang Liu. 2020. "Neuroprotection by Dihydrotestosterone in LPS-Induced Neuroinflammation." *Neurobiology of Disease* 140 (July): 104814. https://doi.org/10.1016/j.nbd.2020.104814.

Yasuhara, Takao, Noriyuki Matsukawa, Koichi Hara, Guolong Yu, Lin Xu, Mina Maki, Seung U. Kim, and Cesario V. Borlongan. 2006. "Transplantation of Human Neural Stem Cells Exerts Neuroprotection in a Rat Model of Parkinson's Disease." *The Journal of Neuroscience: The Official Journal of the Society for Neuroscience* 26 (48): 12497–511. https://doi.org/10.1523/JNEUROSCI.3719-06.2006.

Zago, Wagner, Manuel Buttini, Thomas A. Comery, Christopher Nishioka, Shyra J. Gardai, Peter Seubert, Dora Games, et al. 2012. "Neutralization of Soluble, Synaptotoxic Amyloid β Species by Antibodies Is Epitope Specific." *The Journal of Neuroscience: The Official Journal of the Society for Neuroscience* 32 (8): 2696–2702. https://doi.org/10.1523/JNEUROSCI.1676-11.2012.

Zeitler, Bryan, Steven Froelich, Kimberly Marlen, David A. Shivak, Qi Yu, Davis Li, Jocelynn R. Pearl, et al. 2019. "Allele-Selective Transcriptional Repression of Mutant HTT for the Treatment of Huntington's Disease." *Nature Medicine* 25 (7): 1131–42. https://doi.org/10.1038/s41591-019-0478-3.

Zhang, Qi, Hua-Hui Wu, Yuan Wang, Guo-Jun Gu, Wei Zhang, and Rong Xia. 2016. "Neural Stem Cell Transplantation Decreases Neuroinflammation in a Transgenic Mouse Model of Alzheimer's Disease." *Journal of Neurochemistry* 136 (4): 815–25. https://doi.org/10.1111/jnc.13413.

Zhao, Lingzhi, Andrew J. Gottesdiener, Mayur Parmar, Mingjie Li, Stephen M. Kaminsky, Maria J. Chiuchiolo, Dolan Sondhi, et al. 2016. "Intracerebral Adeno-Associated Virus Gene Delivery of Apolipoprotein E2 Markedly Reduces Brain Amyloid Pathology in Alzheimer's Disease Mouse Models." *Neurobiology of Aging* 44 (August): 159–72. https://doi.org/10.1016/j.neurobiolaging.2016.04.020.

Zhou, Yutong, Feiyan Zhu, Yang Liu, Meng Zheng, Yibin Wang, Dongya Zhang, Yasutaka Anraku, et al. 2020. "Blood-Brain Barrier-Penetrating SiRNA Nanomedicine for Alzheimer's Disease Therapy." *Science Advances* 6 (41): eabc7031. https://doi.org/10.1126/sciadv.abc7031.

11 Role of Medical Biotechnology Research in Neurodegenerative Disease Treatment

Nilay Nandi[1,*,#] *and Shashank Shekhar*[1,#]
[1]University of Texas Southwestern Medical Center, Dallas, TX, USA
*Correspondence to nilay.nandi@utsouthwestern.edu, nilaynandi@gmail.com
[#]These authors have contributed equally to this work.

11.1 INTRODUCTION

Alzheimer's disease (AD) is a neurodegenerative disease that leads to impairment in cognitive function and sensory and motor neuron dysfunction. Hallmarks of AD are the presence of neurofibrillary tangles and amyloid beta plaques caused by a hyperphosphorylated Tau protein and amyloid beta protein, respectively. Neurodegenerative disease affects both the central and peripheral nervous system (CNS and PNS) (Citron 2010). According to a WHO report, neurodegenerative disease (mainly AD) is the greatest socioeconomic burden on society. AD is a public health problem among the elderly; since the aging population is increasing at an unprecedentedly rapid rate, the elderly with AD are predicted to increase from 5.5 million to 13.8 million by 2050 (Mike 2018). AD is overwhelming to the people who have it and for their families and caregivers. It is one of the key reasons for disability and dependency among older people worldwide (Yoshida et al. 2012). In looking into these data and reports, there is an urgent need for research on the early detection of the disease and the development of drugs that may fulfill the unmet need of society.

After AD, the most common age-associated neurodegenerative disease is Parkinson's disease (PD). PD is described by the degeneration of dopaminergic neurons in the substantia nigra pars compacta of the midbrain, Lewy bodies and Lewy neuritis (Wakabayashi et al. 2007). Lewy bodies are situated in the cytoplasm of neurons in the brain and are composed of α-synuclein protein. Data from the World Health Organization (WHO) suggest that the total number of PD patients globally is 160 per 100,000 people.

11.2 ALZHEIMER'S DISEASE

Many molecular lesions have been identified in AD, but available data primarily indicate that misfolded protein accumulation in the brain of aging individuals results in inflammatory and oxidative injury, which further leads to synaptic dysfunction and ultimately, to neuronal cell loss (Querfurth and LaFerla 2010). Alterations of lipid metabolism and inflammatory processes are known to play crucial roles in AD and other neurodegenerative diseases. Lipoxygenase (LOX) catalyses the formation of hydroperoxides from fatty acids such as linolenic and arachidonic acid. This process of addition of oxygen to polyunsaturated fatty acids leads to the formation of bioactive molecules that significantly affect the process of neurodegeneration (Czapski et al. 2016).

Previously, it was reported that 5-Lipoxygenase (5-LOX) is found all over the CNS in both glial and neuronal cells. Its expression increases as a factor of aging but is higher in the hippocampus and

DOI: 10.1201/9781003302131-12

cortex, which are vulnerable for neurodegeneration in AD. As aging is one of the key causes of AD, this could be a seeding factor for AD (Chu and Praticò 2016).

In recent years, considerable data emphasise the importance of 5-LOX in the AD brain. The 5-LOX expression gets upregulated in the AD brain; the inhibition of 5-LOX in a transgenic AD mice model reduced the AD-like pathology (Firuzi et al. 2008; Ikonomovic et al. 2008). It is also indicated that 5-LOX can regulate both amyloid beta accumulation and Tau hyperphosphorylation (Vagnozzi, Giannopoulos, and Praticò 2018). The second factor, accumulation of reactive oxygen species (ROS), can cause oxidative damage. The continuous metabolism in the body could lead to the production of ROS, which plays a critical role in the pathophysiology of neurodegenerative diseases such as AD. ROS mainly affects the lipids, proteins in the membrane of the cell and cell organelles such as mitochondria, and the DNA damage can cause synaptic dysfunction and cell death.

Oxidative stress plays a central role in AD pathology by affecting several genes' and proteins' expression/activity. In some recent findings, it is reported that the Repressor element-1silencing transcription factor (REST) regulates healthy aging and the AD pathology through oxidative stress. It is reported that in the AD brain, the level of REST is almost negligible compared to the brain of a control (Lu et al. 2014).

Clinical criteria, neurological investigation and brain imaging are primarily used in diagnosing AD. Invasive CSF Tau and Aβ overload are the only available markers used in clinical practice. Thus far, no potential markers are available that could identify the disease at an early stage. As 500 mL CSF per day diffuses into the blood, this could be a good source for the development of a biomarker. Blood-based protein biomarker identification has many advantages over CSF due to its easy availability from AD patients in both home or clinic visits. Efficient drugs for the treatment of AD are still not available. The only available drug is an inhibitor of acetylcholine esterase, which targets the secondary event in the progression of the disease. To date, there is a lack of available drugs that can target neuroinflammation or oxidative stress, which could be the primary event during disease development. Several traditional natural foods and spices contain high levels of antioxidants and could serve as a better approach towards disease management. These antioxidants may reduce the oxidative load by controlling the genes that are involved in the maintenance of antioxidant enzymes (Olmos et al. 2013). Sirtuin protein directly mediates an oxidative stress response by deacetylating several genes (mainly transcription factors) involved in the regulation of antioxidant genes. Different types of antioxidant compounds present in the plant kingdom can repair the damage by oxygen by-products. Sirtuin proteins are the mediators of the anti-aging effect of calorie restriction (CR) and neuroprotection. Development of senescent cells could be major effect of SIRT1 down regulation or decreased activity, which could be a secondary response to DNA damage mediated by increased chemokines and inflammatory cytokines. This could lead to increased long-lasting inflammation in the aged population, make them vulnerable to age-related diseases and accelerate aging. Hence, SIRT1 could be important in providing protection against age-related diseases and aging (Hayakawa et al. 2015).

11.3 AMYLOID β PRECURSOR PROTEIN (APP)

Some apparently harmless proteins have wicked altered personalities. An important example is amyloid beta precursor protein (APP). APP is a large membrane-spanning protein that has a role in neuronal growth, repair, dendritic arborization and synaptogenesis and needs tightly organized interactions between cells and between cells and substratum (Reinhard, Hébert, and De Strooper 2005). However, in later age, its malfunction leads to neuronal cell death, which can ultimately mean developing AD.

11.4 AMYLOID β (Aβ)

Aβ peptides are of 36–43-mer natural products of metabolism. $A\beta_{40}$ monomers prevail more than $A\beta_{42}$ forms, as Aβ42 forms are susceptible to aggregation. Aβ peptides are created from the proteolysis of APP. Sequential enzymatic cleavage of APP by β-site APP-cleaving enzyme-1 (BACE-1), that is, a β-secretase and γ-secretase with a protein complex with presenilin 1 at its catalytic core

(Haass and Selkoe 2007), generates Aβ peptides. Due to the failure of a balance between clearance and production, Aβ accumulates, and this excessive accumulation could lead to the initiation of AD. This hypothesis is termed the "amyloid hypothesis" (Selkoe 2001; Tanzi and Bertram 2005). However, α-secretase forms a neuroprotective APPα in a non-amyloidgenic pathway.

Aβ has a spontaneous tendency to form self-aggregates into multiple physical forms. One of them comprises a 2- to 6-peptide oligomer that coalesces into intermediary assemblies. Aβ can also form the fibrils, which organise themselves into β-pleated sheets to form insoluble amyloid plaques. The toxicity of amyloid occurs more in cells with soluble oligomers and intermediate amyloids. However, dimers and trimmers of Aβ are toxic to synapse in brain-slice preparation (Klyubin et al. 2008; Walsh et al. 2005). The level of oligomers in the brain of AD patients correlates with the severity of disease that is the cognitive defect, not the total Aβ burden (Lue et al. 1999). Aβ secretions at synapse get increased rapidly upon neuronal activation. The physiological level of synaptic Aβ may prevent neuronal hyperactivity by dampening excitatory transmission (Kamenetz et al. 2003).

Insulin-degrading enzyme and neprilysin maintain the steady state levels of Aβ. Neprilysin belongs to zinc endopeptidase, which destroys Aβ monomers and oligomers (Kanemitsu, Tomiyama, and Mori 2003). Reduction in the level of neprilysin leads to the build-up of cerebral Aβ (Iwata et al. 2001). Insulin-degrading enzyme is a thiolmetalloendopeptidase, and it is known for the degradation of monomeric Aβ, insulin and other small peptides. It is also reported that the over-expression of neprilysin and insulin-degrading enzyme reduces plaque formation (Leissring et al. 2003). Apart from these enzymes, other proteases also play an important role in Aβ removal, such as matrix metalloproteinases, plasminogen activators and angiotensin-converting enzyme. Glial cells mediate the clearance of Aβ by releasing extracellular chaperons, which make their removal possible through blood. α2-macroglobulin, α1-antichymotrypsin and apolipoproteins are the major extracellular chaperons involved in this process. Phagocytosis of Aβ by astrocytes and microglia can also play a vital role in clearance.

The major site of Aβ plaque accumulation and neurodegeneration is the area that metabolizes glucose-6-phosphate into pyruvate and lactate, which is essential for continuous ATP supply to brain cells. This aerobic glycolytic process produces NADH, and NAD^+ is depleted. NAD^+ is an important molecule that plays a crucial role in information transfer and synchronizes events between the nuclear and the mitochondrial genome in the cell. Several drugs under clinical trial primarily target Aβ by either promoting its clearance with antibodies or inhibiting its aggregation with small molecules. Examples of drugs that are under clinical trials targeting Aβ clearance and small molecule-based aggregation blockers are listed in **Table 11.1**.

TABLE 11.1
Drugs Under Clinical Trial Targeting Aβ.

Target type	Name	Therapy type	Status	Company
Aβ clearance	AAB-003	Biological (passive)	Phase I	Janssen Pharmaceutica, Pfizer, Inc.
Aβ clearance	Solanezumab	Biological (passive)	Phase III	Eli Lilly & Co.
Aβ clearance	Crenezumab	Biological (passive)	Phase III	Genentech, Inc.
Aβ clearance	Ganteneruma b	Biological (passive)	Phase III	Chugai Pharmaceutical Co., Ltd., Hoffmann-La Roche
Aβ clearance	BAN2401	Biological (passive)	Phase II	BioArtic Neuroscience AB, Biogen, Inc., Eisai Co., Ltd.
Aβ clearance	Aducanumab	Biological (passive)	Phase III	Biogen, Inc.
γ-secretase inhibitor	EVP-0962	Small molecule	Phase II	FORUM Pharmaceuticals Inc.

11.4.1 Tau

Tau is a microtubule-associated phosphoprotein. Its primary function is to maintain the stability of axons. The phosphorylation status of this protein determines its fate, which is maintained by the homeostatic function of the kinases and phosphatases (Billingsley and Kincaid 1997). Normally, Tau is bound with a microtubule in its phosphorylated state and stabilizes the structure of the neurons and the vesicular transport system in axons. When Tau gets hyperphosphorylated, it detaches from the microtubule and starts forming its disordered structure that is a paired helical filament and consequently forms neurofibrillary tangles. Neurofibrillary tangles are also present in other neurodegenerative diseases and collectively termed "Tauopathies" (Lee, Goedert, and Trojanowski 2001). The number of neurofibrillary tangles reflects the severity of the AD. Normally, Tau is a soluble abundant protein in the axon, but the hyperphosphorylated Tau formed by various kinases such as glycogen synthase kinase 3β (GSK3β), cyclin-dependent kinase 5 (CDK5) and extracellular signal-related kinase 2 (ERK2) tends to form insoluble aggregates (Lee, Goedert, and Trojanowski 2001). Simila to Aβ oligomers, the oligomers of Tau are also toxic to neuronal cells (Khlistunova 1992) and impair cognition (Oddo et al. 2006; Santacruz et al. 2005).

Previously, an association of a high mortality rate in AD with elevated Tau has been observed. There are approximately 79 Ser and Thr phosphorylation sites in Tau (Billingsley and Kincaid 1997). The specific epitope of Tau hyperphosphorylation has been shown to be associated with AD and differs from other types of dementia and neurological conditions. Hyperphosphorylation at threonine Tau181 (p-Tau181) or Tau 231 (p-Tau231) has been positively correlated with neurofibrillary tangles in AD brains (Brunden, Trojanowski, and Lee 2009). High levels of p-Tau181 have also been correlated with a quicker progression of mild cognitive impairment to AD (Blom et al. 2009) and rapid cognitive decline in AD (Sämgård et al. 2010). Hippocampal atrophy has also been reported to be correlated with increased phosphorylation at p-Tau181 (Hampel et al. 2005).

11.5 THE TREATABLE PATHOLOGICAL EVENTS OTHER THAN POST-TRANSLATIONAL MODIFICATION OF TAU

Few major events are related to Tau pathology, which could be a direct therapeutic target rather than post-translational modifications, such as cytoskeletal dysfunction, Tau aggregation and impairment of the protein degradation pathway.

11.5.1 Tau-Targeting Drugs

During the progress of Tau pathology, many events provide opportunities for intervention. For example, the expression of Tau, post-translational modifications, malfunctioning in clearance and aggregation can be key targets.

11.5.2 Decreasing Tau Expression

As Tau not only is involved in neurotoxicity but also is a mediator of Aβ-induced neurotoxicity, (Ittner and Götz 2011), targeting the expression level of Tau seems to be a logical strategy for Tau-based therapeutic development.

11.5.3 Phosphatase Modifier

Memantine improves PP2A activity by blocking inhibitor 2 (Chohan et al. 2006) (also known as I2 PP2A). In moderate to severe AD patients, memantine helped in the short-term progress of cognitive symptoms (Jiang and Jiang 2015; Matsunaga, Kishi, and Iwata 2015). The efficacy of the drug might be improved if it is administered with an acetylcholine esterase inhibitor (Tsoi et al. 2016).

Sodium selenate acts as an activator of PP2A by acting on the regulatory subunit B and thus lowering the phosphorylation of Tau in AD models (Corcoran et al. 2010; van Eersel et al. 2010; Rueli et al. 2017). Sodium selenite has displayed some benefits on diffusion MRI in a Phase II clinical trial for early AD cases. However, there were no differences recognized in CSF Tau, Aβ and cognitive function on MRI (volumetric) and PET imaging (Malpas et al. 2016).

11.5.4 Kinase Inhibitors

Both flavopiridol (alvocidib) (Carlson et al. 1996) and roscovitine (seliciclib) (Meijer et al. 1997) inhibitors act on CDK5 by competing with ATP binding and thereby reduce activity (Mapelli et al. 2005). They can cross the blood-brain barrier and have been examined in various cancer trials, but they have yet been confirmed for AD (Asghar et al. 2015; Khalil et al. 2015).

As an inhibitor of GSK3β, tideglusib acts as an irreversible inhibitor, and it does not compete for an ATP binding site (Domínguez et al. 2012; Martinez et al. 2002). In animal models of AD, this compound showed good results. It has reduced Tau phosphorylation, amyloid plaques and memory impairment. In a Phase II clinical trial, it also showed better results (Lovestone et al. 2015).

Lithium chloride was used for the treatment of bipolar disorder, and now, it has also shown to be capable of inhibiting GSK3β (Klein and Melton 1996; Stambolic, Ruel, and Woodgett 1996). In cell cultures, it reduces Aβ toxicity and Tau phosphorylation. In small clinical studies, it has also shown promising results (Forlenza, De-Paula, and Diniz 2014).

11.5.5 Tau Acetylation Inhibition

A small-molecule, Salsalate, which is an NSAID, prevents Tau acetylation (Lys174) by p300 HAT. In preclinical research, it has shown better results and reduced the activity of p300 HAT (Min et al. 2015). It lowers Tau pathology and maintains hippocampal volume and improved cognitive function in transgenic AD models. Phase I testing of Salsalate was completed in March 2017.

11.5.6 Inhibiting Tau Deglycosylation

MK-8719 is a small molecule that inhibits the O-GlcNAcase (OGA) enzyme. In preclinical studies with different species, a human dose was determined by using oral and intravenous administration. This compound is under consideration for further development (Sandhu et al. 2016).

11.5.7 Tau Aggregation Inhibitors

Methylene blue has been reported to inhibit Tau polymerization in vitro by deceiving the Tau monomers in a conformation that is not competent for forming aggregates (Panza et al. 2016; Wischik et al. 1996). This compound has effectively reduced Tau pathology and improved the cognitive functions in vitro and in vivo (Panza et al. 2016). This compound can cross the blood-brain barrier and seems to be a good drug candidate.

A natural product, curcumin, which is a component of human food, has demonstrated good efficiency in inhibiting aggregation by binding with the β sheet confirmation of Aβ. In animal models of AD, it can decrease Tau and the Aβ burden and improve cognition (Hamaguchi, Ono, and Yamada 2010; Hu et al. 2015).

Several other strategies are also being used that target Tau, such as microtubule stabilizers (Epithilone D), protein degradation modulators (Rolipram) and immunotherapies. Till now, Aβ-based immunotherapies have been almost completely unsuccessful, and the reason might be the late presentation of Aβ during the course of disease pathology. However, hopes remain for early AD patients that they might benefit from Tau-based immunotherapies (Congdon and Sigurdsson 2018).

TABLE 11.2
Drugs Under Clinical Trial Targeting Tau Protein Aggregation and p-Tau Clearance.

Target type	Name	Therapy type	Status	Company
Tau aggregation inhibitor	TRx0237	Small molecule	Phase III	TauRx Therapeutics Ltd.
p-Tau clearance	AADvac-1	Biological (active)	Phase I	Axon Neuroscience SE
p-Tau clearance	ACI-35	Biological (active)	Phase I	AC Immune SA, Janssen Pharmaceutica

Due to its massive role in AD, Tau protein is an important candidate molecule for the development of both biomarker and therapeutic strategies. Some drugs are undergoing clinical trials that target Tau (**Table 11.2**). In a recent scenario, several other hypotheses could explain the sequence and events of disease progression in addition to these two well-established hallmarks of AD.

11.6 5-LIPOXYGENASE: NOVEL LIPID SIGNALING IN AD

An AD brain is defined by extensive neuroinflammation (Giannopoulos, Joshi, and Praticò 2014). 5-Lipoxygenase (5-LOX) is known for its pro-inflammatory enzymatic activity and is distributed widely in the CNS. It has been reported to be increased in AD (Firuzi et al. 2008; Giannopoulos, Joshi, and Praticò 2014).

Lipoxygenases are non-heme iron-containing enzymes that can add oxygen to arachidonic acid. LOX has three isoforms: 5, 12 and 15. 5-Lipoxygenase adds oxygen at the 5th carbon position of arachidonic acid. The 5-LOX final product executes its action through G-protein-coupled receptors (GPCRs) such as BLT1 and BLT2 for LTB4 (Yokomizo et al. 1997, 2000, 2001). LTB4 is well-known for mediating inflammatory responses, as it is a defence against infection in hosts and immune responses (Yokomizo et al. 2001), and it stimulates chemotaxis, degranulation, lysosomal enzyme release and the generation of ROS (Busse 1998; Palmer et al. 1980).

11.6.1 5-LOX in the Aging Brain and AD

There is homogenous distribution of 5-LOX in both the glia and neuronal cells of CNS (Farias et al. 2007). However, expression of 5-LOX in the cortex and hippocampus is high, which are the vulnerable areas for neurodegeneration (Chinnici, Yao, and Praticò 2007; Lammers et al. 1996). In previous studies, it was shown that 5-LOX increased as a factor of aging. Because aging is a key factor of AD, increased 5-LOX may possibly be an early event in the progression of AD (Chu and Praticò 2009). In the post-mortem brain, the immunoreactivity of 5-LOX was elevated in the hippocampus and cortex of AD patients compared to controls. However, there was no difference in the expression when analysed in the cerebellum, an area usually not involved in typical AD pathology (Firuzi et al. 2008; Ikonomovic et al. 2008). In a small pilot study, it was reported that 5-LOX gene polymorphism is associated with the early and late onset of AD. However, a large-scale study is needed to validate and confirm this finding (Qu, Manev, and Manev 2001).

In a recent study, it was shown that Aβ-induced toxicity is mediated by CysLT1 receptor and pharmacological blockade, and the receptor antagonist could be rescued from the Aβ-induced cognitive impairment and hippocampal long-term potentiation (Lai et al. 2014; Tang, Hong, et al. 2014; Tang, Ji, et al. 2014).

11.6.2 5-LOX as a Modulator of Aβ

Chu and Praticò demonstrated in a transgenic mice model of AD that when 5-LOX is genetically knocked down, both the soluble and aggregated forms of Aβ were reduced (Chu and Praticò 2016).

Genetic and pharmacological manipulations further support the idea that 5-LOX may act as a modulator of Aβ development in AD models and in vitro. 5-LOX over-expression in cells (neuroblastoma) that stably express a Swedish mutant of human APP results in a high level of Aβ and APP expression compared to controls that express only APP. This was associated with augmented protein levels of γ-secretase complex components (Chu and Praticò 2011). cAMP-response element binding protein (CREB) is the mediator of 5-LOX response for the regulation of Aβ formation. Treatment of neurons with the 5-LOX metabolite, 5-HETE, resulted in increased Aβ and demonstrated elevated levels and activity of CREB. The effect was nullified by blocking CREB activation. Similar results were obtained in the in vivo experiment. There was a substantial increase in the level of CREB and γ-secretase complex mRNA transcript (Chu et al. 2012).

For approaching a reduction of Aβ, γ-secretase inhibition is one of the highly desirable targets. However, complete inhibition of γ-secretase has a toxic effect because of its involvement in the proteolytic cleavage of various important signaling molecules, such as Notch-1 and cadherin.

In support of this idea, in a recent clinical trial of a γ-secretase inhibitor semagacstat (LY 450139), AD patients experienced functional decline compared to the placebo in addition to several other significant adverse effects (Doody et al. 2013). Therefore, this concept advocates the need for such targets or compounds that can reduce the γ-secretase level without affecting Notch signaling. Chu et al. (2016) have reported that 5-LOX inhibition or knockdown is completely independent of the Notch signaling pathway, but the complete mechanistic details of this effect are still elusive. This finding indicates that 5-LOX activation somehow affects the membrane dynamics and provides greater access to APP than Notch processing, but the neurobiology of this hypothesis is completely unexplored. The new generation inhibitors for γ-secretase are now being investigated, primarily for their effect on Notch signaling. Accordingly, this discovery further emphasizes the significance of the pharmacological blockade of 5-LOX in AD.

11.6.3 5-LOX and γ-Secretase-Activating Protein

γ-secretase and its activating protein, γ-secretase activating protein (GSAP), directly control the formation of Aβ. GSAP is a C-terminal truncated part of 110 kDa precursor protein cleaved by the action of caspase-3 (Chu et al. 2015). Chu et al. (2015) have reported that 5-LOX not only increases the transcription of γ-secretase for augmented Aβ formation but also regulates the levels and availability of the GSAP 16 kDa, which is an active form of the γ-secretase-activating protein. They showed that the knockout or pharmacological blockade of the 5-LOX resulted in a significant decrease of GSAP 16 kDa (Chu et al. 2015).

11.6.4 5-LOX and Tau Hyperphosphorylation

In some recent research, it was shown that 5-LOX not only is involved in Aβ formation but also regulates the phosphorylation status of Tau protein. Tau hyperphosphorylation is mediated by CDK5 by modulating its co-activators p25 and p35. 5-LOX is known to regulate Tau phosphorylation at epitopes Ser202/Thr205, Thr181 and Ser396 (Vagnozzi, Giannopoulos, and Praticò 2018). However, the steady state of total Tau does not change by 5-LOX over-expression. Hence, the 5-LOX pathway seems to have a serious role in AD pathophysiology development including aberrant Aβ accumulation and the hyperphosphorylation of Tau (Giannopoulos, Joshi, and Praticò 2014) (Figure 11.1). Therefore, 5-LOX can be a good candidate with both therapeutic and preventative importance.

11.7 PARKINSON'S DISEASE

Several mechanisms have been proposed for the development of Parkinson's disease (PD), mainly mitochondrial dysfunction, oxidative damage, proteostasis and differentiation of dopaminergic neurons. However, all of these mechanisms together contribute to the death of dopaminergic neurons. PD clinical symptoms appear late in the disease when the dopaminergic neurons in the substantia

TABLE 11.3
Drugs Used in the Management of PD and PD-Associated Symptoms.

Drug (compound)	Mode of action	Implication
Levodopa (L-DOPA)	Dopamine precursor	PD symptom management
Carbidopa	Dopamine decarboxylase inhibitor	Used in combination with L-DOPA
Apomorphin	Dopamine agonist	Used in advanced PD patients
Bromocriptine	Dopamine agonist	To treat PD symptoms and acromegaly
Ropinirole	Dopamine agonist	To treat PD symptoms and RLS
Rotigotine	Dopamine agonist	To treat PD symptoms and RLS
Benztropine	Antagonist of muscarinic receptor	Used as an adjunct in PD treatment (e.g., levodopa); Useful in the control of extrapyramidal disorders (except for tardive dyskinesia) due to neuroleptic drugs (e.g., phenothiazines)
Trihexyphenidyl	Antagonist of muscarinic receptor	Used to treat and prevent Parkinson-like symptoms that are caused by using certain anti-psychotic medications
Biperiden	Antagonist of muscarinic receptor	Used in severe muscle reactions and other side effects from certain medicines (e.g., chlorpromazine)
Rivastigmine	Inhibitor of cholinesterase	Used to manage dementia caused by AD or PD
Amantadine	NMDA receptor antagonist	Treats PD and "Parkinson-like" symptoms such as stiffness or tremors, shaking, and repetitive uncontrolled muscle movements that may be caused by using certain drugs
Istradefylline	Antagonist of Adenosine A_{2A} receptor	Adjunctive treatment for levodopa/carbidopa in adult patients with PD experiencing "OFF" episodes

nigra pars compacta have already been lost (Cheng, Ulane, and Burke 2010). The selective loss of dopaminergic neurons leads to an acute scarcity of dopamine that causes the inadequate activation of dopamine receptors, which, in turn, either under- or overstimulates the parts of the basal ganglia (Rangel-Barajas, Coronel, and Florán 2015). Aging-associated ROS accumulation plays a major role in the loss of dopaminergic neurons (Dias, Junn, and Mouradian 2013). A small proportion of patients (10–15%) shows a familial form of PD caused by a mutation in α-Synuclein and the leucine-rich repeat kinase2 (LRRK2) genes in late-onset cases and parkin (PARK2), ubiquitincarboxy-terminalhydrolase L1 (UCH-L1), PTEN-induced putative kinase1 (PINK1) and oncogene DJ1 (DJ1) in early-onset cases (Grasso et al. 2014). The sporadic form occurs in the majority of patients, and the exact cause is still elusive (Nuytemans et al. 2010). Current therapeutic approaches in PD are mainly based on the dopaminergic and nondopaminergic pharmacological agents for management of the clinical symptoms (Mao et al. 2020) (**Table 11.3**).

11.8 GENE THERAPY IN NEURODEGENERATIVE DISEASES

Gene therapy manipulates the underlying genetic cause of the disease for its treatment or prevention. Gene therapy can aim to replace a mutated disease-causing gene with a non-faulty gene or it can completely switch off an improperly functioning gene. Additionally, a new or altered gene can be introduced to aid in treating the disorder. In recent times, with advances in gene sequencing technology, the understanding of the underlying cause of neurodegenerative diseases has not been restricted only to pathological features; rather, deeper knowledge of genetic causes has started to emerge. Therefore, gene therapy technique is evolving very rapidly as a potent therapeutic strategy to treat a wide array of neurodegenerative diseases, such as AD, PD and Huntington's disease. Although there have been a few unsuccessful clinical trials that could not to find a cure for these diseases, scientists are working relentlessly to enhance the efficacy of gene therapy technique in treating neurodegenerative diseases. The major thrust is to design reliable routes for delivering the transgenes, exploiting new and efficient vectors for this delivery and identifying new therapeutic targets. These new approaches are very closely monitored and evaluated in preclinical and clinical trials for judging improvement in therapeutic efficacy and might ultimately provide a commanding strategy to treat patients with neurodegenerative disease.

Most of these gene-editing tools work by recognizing a precise target DNA sequence in the genome and inducing breaks in the DNA double strand followed by repairing the breaks primarily by non-homologous end joining. This repair mechanism is highly susceptible to introducing mistakes in the DNA and can potentially generate knockout mutation in the gene. However, alternatively, a homology-directed repair could insert a new sequence of DNA in the double-stranded break. Four different nucleases are involved in inducing double-stranded cleavage in specific DNA sequences: meganucleases, zinc finger nucleases Kim et al. 1996), TALENs (Miller et al. 2011) and CRISPR/Cas9 (Doudna and Charpentier 2014). In recent years, CRISPR/Cas9 had evolved as the major gene editing strategy because of its ease and higher efficiency (Jinek et al. 2012). Two new techniques called Base Editor (Komor et al. 2016) and Prime Editor (Anzalone et al. 2019) have recently been reported, which can execute edits on single nucleotides but lack the ability to integrate a new sequence of DNA through homology-directed repair.

Both viral and non-viral vectors can deliver transgenes in gene therapy. Adeno-associated virus-based vectors (AAV) are exclusively used in gene therapy for neurodegenerative diseases. AAV2 is a popular choice for numerous clinical trials due to its non-pathogenicity and its continuous expression in neurons (LeWitt et al. 2011; Marks et al. 2008; Rafii et al. 2018; Marks et al. 2010). Another important factor to consider is the blood-brain barrier, which can prevent AAV vector delivery to the central nervous system. AAV9 and AAVrh.10 have the ability to pass through the blood-brain barrier (Duque et al. 2009; Albright et al. 2018) and are therefore extremely relevant. Viral capsid engineering can enhance the efficacy of these viral vectors. A recently engineered AAV capsid (AAV-PHP.B) can transduce over 50% of neurons and astrocytes and can deliver a forty-fold higher AAV genome in the CNS (Deverman et al. 2016; Allen et al. 2017). Very few scientific investigations have explored adenovirus (Adv)- or lentivirus-based vectors for gene therapy in neurodegenerative diseases, and more clinical data are necessary to pinpoint their importance in this process. However, lentiviral delivery might be advantageous as it can incorporate genetic material within the host genome through reverse transcription and stably express transgenes for a longer time. Conversely, if it is not controlled properly, then lentiviral delivery might provoke a stronger immune response and genotoxicity (Palfi et al. 2014). However, AAV and Adv cannot integrate into the host genome and have a lower probability of introducing insertional mutagenesis and genotoxicity (Barkats et al. 1998). Non-viral vectors can be either lipid- (Yin et al. 2014) or polymer-based. These vectors can be helpful in improving the delivery because of their capabilities of packing larger DNA sequences, controlling release and provoking lower immune responses (Pack et al. 2005; Mintzer and Simanek 2009; Buck et al. 2019). However, associated cytotoxicity (Buck et al. 2019) might be a major disadvantage, and additional clinical trials are needed to completely understand AAV's competency. Local delivery of viral vectors is preferred over systemic administration for delivering AAV in neurodegenerative disease treatment. Intraparenchymal injection is the most efficient route to introduce AAV in the region of interest in brain and neuronal tissues with minimal biodistribution in peripheral tissues (Weinberg, Samulski, and McCown 2013; Tardieu et al. 2017).

11.8.1 Alzheimer's Disease

A distinctive hallmark in AD pathology is the formation of neurofibrillary tangles inside the neurons after tau hyperphosphorylation followed by Aβ accumulation in the CNS. The majority of the gene therapy techniques developed for AD have been primarily concentrated on preventing accumulation of toxic Aβ aggregates. Park et al. efficiently targeted the Bace1 gene in the post-mitotic neurons of adult mice brains using a CRISPR/Cas9 nanocomplex. They demonstrated that Bace1 gene suppression can prevent pathologies related to Aβ in AD knock-in mouse-models (Park et al. 2019). γ-secretase, a large intramembrane protein complex, is also important for Aβ pathology and can be regulated by GSAP. Aβ levels can be dramatically reduced when GSAP is knocked down in AD model mice; therefore, GSAP can be categorized as a potent target for treating AD (He et al. 2010). The role of GSAP in regulating γ-secretase activity is further explored in the CRISPR/Cas9 mediated GSAP knockout of the HEK293 cell line that expresses amyloid precursor protein (APP). These cells that lack GSAP function demonstrate reduced γ-secretase activity, which, in turn, can reduce APP processing and decrease the secretion of Aβ (Wong et al. 2019). György. et al. demonstrated APP Swedish mutation in the amyloid precursor protein gene responsible for inherited AD as a target for gene therapy. In inherited AD, enhanced β-secretase cleavage of the amyloid precursor protein results in high levels of Aβ in both the brain and the peripheral tissues carrying the APP Swedish mutation. However, selective disruption of the APPswe allele by CRISPR/Cas9 successfully reduces the pathogenic levels of Aβ both in and ex vivo. Hence, it might be useful for treating AD caused by mutations in APP (György et al. 2018). Moreover, Nagata et al. edited 3'-UTR of the APP gene by using CRISPR/Cas9 in mouse zygotes and evaluated the protective role of these deletion mutations. Expression of Aβ and APP is measured for the gene-edited mice. Successful deletion of 34-bp in a 52-bp regulatory element near the stop codon prevents Aβ accumulation in the brain due to reduced APP expression (Nagata et al. 2018).

Additionally, genome-wide association studies (GWAS) have identified apolipoprotein E4 (APOE4) as a possible causal element for AD. This E4 allele related to risk differs by a single nucleotide polymorphism in the E3 allele that is not risk-associated. Induced pluripotent stem cells (iPSCs) acquired from AD patients with the E4 allele are edited to E3 by using the CRISPR/Cas9 technique and are differentiated into pure cultures of forebrain excitatory neurons. These edited neurons demonstrate reduced cytotoxicity compared to E4 when treated with ionomycin. Consequently, E4 neurons demonstrate enhanced Tau phosphorylation and an increased release of hyperphosphorylated Tau (Wadhwani et al. 2019).

Chronic neuroinflammation also contributes largely in AD pathogenesis. In different regions of an AD brain, a proinflammatory component, glia maturation factor, remains upregulated. Moreover, reactive glial cells positioned around Aβ plaques are the predominant location for glia maturation factor expression. Activated microglial cells can secrete glia maturation factor and proinflammatory cytokines and can turn on an inflammatory cascade, which ultimately leads to neurodegeneration and AD pathophysiology. Therefore, editing glia maturation factor with CRISPR/Cas9 technology might be an important strategy for the abolition of microglial activation. As a result, the inflammatory signaling pathway is inhibited and suppresses otherwise high MAP kinase activity in AD patients (Lee and Kim 2017). Therefore, editing the glia maturation factor gene using CRISPR/Cas9 is considered an innovative therapeutic approach for AD (Raikwar et al. 2019).

11.8.2 Parkinson's Disease

The available drug-based therapies for PD are not very effective, and gene therapy is thus considered an option for its treatment. One of the major contributors in PD pathology is α-synuclein that is encoded by the SCNA gene. Numerous other genes such as LRRK2, PARK2, DJ-1 and PINK1 have also been demonstrated to be mutated in PD patients (Sundal et al. 2012). Although these gene variants are associated with disease pathology, environmental factors still also play a substantial role.

Another considerable challenge associated with PD is that most cases are sporadic, and only around 3% are familial (Williams-Gray and Worth 2016). Therefore, gene therapy in PD is more focused on promoting expression of the protein required for improving neuronal health and assisting neuronal growth rather than restricting expression of specific genes.

Expressing neurotrophic factors might be key for protecting neurons in the brain. Therefore, researchers are interested in exploring glial cell-derived neurotrophic factors (GDNF) and neurturin (NTN) in developing gene therapy-based treatment for PD. These molecules play an essential part in dopaminergic neuron survival (Lin et al. 2017) by preventing the death of these neurons for both sporadic and familial PD. AAV2-NTN is surgically inserted into the putamen of patients diagnosed with PD, and Phase I and Phase II clinical trials were conducted. The Phase I trial did not show any unpleasant side effects, and 6 months later, patients displayed a significant enhancement in the Unified Parkinson's Disease Rating Scale (UPDRS) motor score. However, in the Phase II clinical trial, compared to the placebo, no substantial improvement in motor function was detected over a 15-month study window in patients who received similar treatment as in Phase I (Warren Olanow et al. 2015). The investigators hypothesized that impaired passage of the vector from the putamen to the substantia nigra might be the cause for the lack of clinical efficiency. Keeping this in mind, they performed follow-up studies and introduced AAV2-NTN directly into the substantia nigra of PD patients in Phase I and in both the putamen and substantia nigra in Phase II, but these modifications still failed to improve efficacy (Bartus et al. 2013). This failure indicates an insufficient and inaccurate delivery of the viral vector that carries the transgene. Although this study failed to treat the symptoms of PD, it provided convincing evidence that AAV2-NTN might hold promise in gene therapy if properly delivered, as none of the patients receiving this therapy suffered weight loss, which can be a significant side effect (Warren Olanow et al. 2015). Therefore, promoting GDNF expression by using gene therapy might offer an alternative strategy for treating PD. In a related research study, an AAV vector bound to a promoter, which can induce the enhanced expression of GDNF, was injected into the brain of a rat model of PD (Tereshchenko et al. 2014). In treated rats, enhanced levels of GDNF improved motor function and prevented the degeneration of dopaminergic neurons. However, the GDNF levels bounced back to baseline within 3 weeks of injection, which indicates that frequent administration of AAV-GDNF is required to maintain the enhanced GDNF level. No significant weight loss was observed in this study, but dramatic weight loss in patients has been reported in other studies where GDNF gene therapy is used to treat PD (Manfredsson et al. 2009). Another preclinical study involved the delivery of GDNF in the striatum and substantia nigra of normal or aged, MPTP-lesioned monkeys (Kordower et al. 2000). The nigrostriatal degeneration was prevented, and no immune response was provoked. After infusion, the monkeys could still continue to express GDNF for 8 months. An altered experimental model was explored where gene therapy was performed after 6 months of PD development that used the delivery of AAV2-GDNF in the putamen of monkeys (Kells et al. 2010). This significantly improved their UPDRS score. Building on this work by using advanced interventional MRI-guided convection-enhanced delivery (iMRI-CED), a Phase I clinical trial was initiated in which advanced-stage PD patients were treated with 4 different doses of AAV2-GDNF (ClinicalTrials.gov: NCT01621581).

Two other AAV2 that mediated gene therapy for PD involved the gene transfer of aromatic L-amino acid decarboxylase (AADC) or glutamic acid decarboxylase (GAD). AADC is the rate-limiting enzyme for dopamine synthesis from levodopa and GAD that catalyses the synthesis of GABA from glutamate. The primary focus of AAV2-GAD-mediated gene therapy is to enhance the local inhibition of GABA by delivering the GAD transgene to the subthalamic nucleus (STN). This can, in turn, rectify pathological over-activity in the basal ganglia node. When AAV2-GAD was delivered to the STN of MPTP-lesioned monkeys, the clinical rating for PD was not lowered (Emborg et al. 2007). However, when AAV2-GAD was delivered via advanced iMRI-CED in the STN of PD patients, it demonstrated improvements in the UPDRS score for up to 12 months post-injection in Phase I. However, in Phase II, although the PD motor defects were improved, it did not yield a better outcome than deep brain stimulation of the STN (LeWitt et al. 2011). Dopamine replacement therapy is effective in PD patients during the early stages. However, with disease progression, increasing levodopa doses fails to deliver a similar therapeutic advantage. AADC gene therapy in patients who are continuing to take the drug levadopa might obtain beneficial effects. hAADC is delivered to the postcommissural (sensorimotor territory) striatum, which can efficiently convert levodopa to dopamine. Preclinical studies involving hAADC delivery to primates with MPTP-induced lesions showed improvement in their behavioral rating scores and substantially lowered the requirements of levodopa and its side effects (Bankiewicz et al. 2006; Hadaczek et al. 2010; Forsayeth et al. 2006; San Sebastian et al. 2012; Bankiewicz et al. 2000). Nevertheless, a Phase I clinical trial did not display much efficacy but demonstrated the safety of this strategy (Eberling et al. 2008; Christine et al. 2009; Mittermeyer et al. 2012). Therefore, a new clinical trial is being conducted that delivers an increased volume of AAV2-hAADC to the putamen by using advanced iMRI-CED technology (ClinicalTrials.gov: NCT01973543). Another trial was conducted with a lentiviral vector, ProSavin, that encodes rate-limiting enzymes for the dopamine biosynthetic pathway, tyrosine hydroxylase and cyclohydrolase 1 in addition to AADC (Palfi et al. 2014). Although lentiviral vectors can carry a larger genetic load, this is the first instance when this strategy has been utilised for clinical trials. The UPDRS motor scores off medication were significantly reduced, but there was no dose-dependent effect. The major aim of this approach is to produce dopamine in a continuous and stable manner in the putamen motor region. However, after ProSavin administration, levodopa uptake is not facilitated, but dopamine binding to the D2 dopamine receptor increases in a dose-dependent manner. Nonetheless, this work demonstrated that a lentiviral-based vector can be safely used for gene delivery in a CNS disorder. To date, L-dopa therapy is considered the major prevailing strategy for treating PD patients, but it treats only the symptoms of the disease and requires frequent administration.

Gene therapy strategies hold considerable promise for treating neurodegenerative disorders. With recent advances in modern technologies, precise and sufficient delivery of the gene might be achieved. This will hold the key for achieving success with clinical trials.

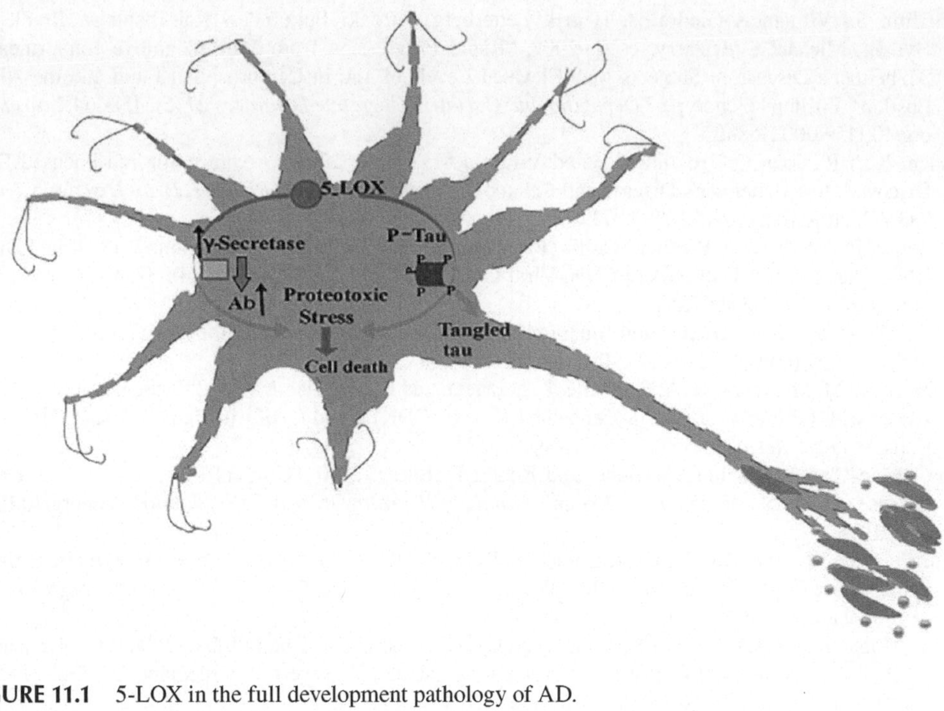

FIGURE 11.1 5-LOX in the full development pathology of AD.

REFERENCES

Albright, Blake H., Claire M. Storey, Giridhar Murlidharan, Ruth M. Castellanos Rivera, Garrett E. Berry, Victoria J. Madigan, and Aravind Asokan. 2018. “Mapping the Structural Determinants Required for AAVrh.10 Transport across the Blood-Brain Barrier.” *Molecular Therapy: The Journal of the American Society of Gene Therapy* 26 (2): 510–23. https://doi.org/10.1016/j.ymthe.2017.10.017.

Allen, William E., Isaac V. Kauvar, Michael Z. Chen, Ethan B. Richman, Samuel J. Yang, Ken Chan, Viviana Gradinaru, et al. 2017. “Global Representations of Goal-Directed Behavior in Distinct Cell Types of Mouse Neocortex.” *Neuron* 94 (4): 891–907.e6. https://doi.org/10.1016/j.neuron.2017.04.017.

Anzalone, Andrew V., Peyton B. Randolph, Jessie R. Davis, Alexander A. Sousa, Luke W. Koblan, Jonathan M. Levy, Peter J. Chen, et al. 2019. “Search-and-Replace Genome Editing without Double-Strand Breaks or Donor DNA.” *Nature* 576 (7785): 149–57. https://doi.org/10.1038/s41586-019-1711-4.

Asghar, Uzma, Agnieszka K. Witkiewicz, Nicholas C. Turner, and Erik S. Knudsen. 2015. “The History and Future of Targeting Cyclin-Dependent Kinases in Cancer Therapy.” *Nature Reviews. Drug Discovery* 14 (2): 130–46. https://doi.org/10.1038/nrd4504.

Bankiewicz, K. S., J. L. Eberling, M. Kohutnicka, W. Jagust, P. Pivirotto, J. Bringas, J. Cunningham, et al. 2000. “Convection-Enhanced Delivery of AAV Vector in Parkinsonian Monkeys; in Vivo Detection of Gene Expression and Restoration of Dopaminergic Function Using pro-Drug Approach.” *Experimental Neurology* 164 (1): 2–14. https://doi.org/10.1006/exnr.2000.7408.

Bankiewicz, Krystof S., John Forsayeth, Jamie L. Eberling, Rosario Sanchez-Pernaute, Philip Pivirotto, John Bringas, Peter Herscovitch, et al. 2006. “Long-Term Clinical Improvement in MPTP-Lesioned Primates after Gene Therapy with AAV-HAADC.” *Molecular Therapy: The Journal of the American Society of Gene Therapy* 14 (4): 564–70. https://doi.org/10.1016/j.ymthe.2006.05.005.

Barkats, M., A. Bilang-Bleuel, M. H. Buc-Caron, M. N. Castel-Barthe, O. Corti, F. Finiels, P. Horellou, et al. 1998. “Adenovirus in the Brain: Recent Advances of Gene Therapy for Neurodegenerative Diseases.” *Progress in Neurobiology* 55 (4): 333–41. https://doi.org/10.1016/s0301-0082(98)00028-8.

Bartus, Raymond T., Tiffany L. Baumann, Joao Siffert, Christopher D. Herzog, Ron Alterman, Nicholas Boulis, Dennis A. Turner, et al. 2013. “Safety/Feasibility of Targeting the Substantia Nigra with AAV2-Neurturin in Parkinson Patients.” *Neurology* 80 (18): 1698–1701. https://doi.org/10.1212/WNL.0b013e3182904faa.

Billingsley, M. L., and R. L. Kincaid. 1997. “Regulated Phosphorylation and Dephosphorylation of Tau Protein: Effects on Microtubule Interaction, Intracellular Trafficking and Neurodegeneration.” *The Biochemical Journal* 323 (Pt 3) (May): 577–91.

Blom, Elin S., Vilmantas Giedraitis, Henrik Zetterberg, Hiroaki Fukumoto, Kaj Blennow, Bradley T. Hyman, Michael C. Irizarry, et al. 2009. “Rapid Progression from Mild Cognitive Impairment to Alzheimer’s Disease in Subjects with Elevated Levels of Tau in Cerebrospinal Fluid and the APOE Epsilon4/Epsilon4 Genotype.” *Dementia and Geriatric Cognitive Disorders* 27 (5): 458–64. https://doi.org/10.1159/000216841.

Brunden, Kurt R., John Q. Trojanowski, and Virginia M.-Y. Lee. 2009. “Advances in Tau-Focused Drug Discovery for Alzheimer’s Disease and Related Tauopathies.” *Nature Reviews. Drug Discovery* 8 (10): 783–93. https://doi.org/10.1038/nrd2959.

Buck, Jonas, Philip Grossen, Pieter R. Cullis, Jörg Huwyler, and Dominik Witzigmann. 2019. “Lipid-Based DNA Therapeutics: Hallmarks of Non-Viral Gene Delivery.” *ACS Nano* 13 (4): 3754–82. https://doi.org/10.1021/acsnano.8b07858.

Busse, W. W. 1998. “Leukotrienes and Inflammation.” *American Journal of Respiratory and Critical Care Medicine* 157 (6 Pt 2): S210–213; discussion S247–248.

Carlson, B. A., M. M. Dubay, E. A. Sausville, L. Brizuela, and P. J. Worland. 1996. “Flavopiridol Induces G1 Arrest with Inhibition of Cyclin-Dependent Kinase (CDK) 2 and CDK4 in Human Breast Carcinoma Cells.” *Cancer Research* 56 (13): 2973–78.

Cheng, Hsiao-Chun, Christina M. Ulane, and Robert E. Burke. 2010. “Clinical Progression in Parkinson Disease and the Neurobiology of Axons.” *Annals of Neurology* 67 (6): 715–25. https://doi.org/10.1002/ana.21995.

Chinnici, Cinzia M., Yuemang Yao, and Domenico Praticò. 2007. “The 5-Lipoxygenase Enzymatic Pathway in the Mouse Brain: Young versus Old.” *Neurobiology of Aging* 28 (9): 1457–62. https://doi.org/10.1016/j.neurobiolaging.2006.06.007.

Chohan, Muhammad Omar, Sabiha Khatoon, Inge-Grundke Iqbal, and Khalid Iqbal. 2006. “Involvement of I2PP2A in the Abnormal Hyperphosphorylation of Tau and Its Reversal by Memantine.” *FEBS Letters* 580 (16): 3973–79. https://doi.org/10.1016/j.febslet.2006.06.021.

Christine, C. W., P. A. Starr, P. S. Larson, J. L. Eberling, W. J. Jagust, R. A. Hawkins, H. F. VanBrocklin, et al. 2009. "Safety and Tolerability of Putaminal AADC Gene Therapy for Parkinson Disease." *Neurology* 73 (20): 1662–69. https://doi.org/10.1212/WNL.0b013e3181c29356.

Chu, Jin, Phillip F. Giannopoulos, Carolina Ceballos-Diaz, Todd E. Golde, and Domenico Pratico. 2012. "Adeno-Associated Virus-Mediated Brain Delivery of 5-Lipoxygenase Modulates the AD-like Phenotype of APP Mice." *Molecular Neurodegeneration* 7 (1): 1. https://doi.org/10.1186/1750-1326-7-1.

Chu, Jin, Jian-Guo Li, Yash B. Joshi, Phillip F. Giannopoulos, Nicholas E. Hoffman, Muniswamy Madesh, and Domenico Praticò. 2015. "Gamma Secretase-Activating Protein Is a Substrate for Caspase-3: Implications for Alzheimer's Disease." *Biological Psychiatry* 77 (8): 720–28. https://doi.org/10.1016/j.biopsych.2014.06.003.

Chu, Jin, and Domenico Praticò. 2009. "The 5-Lipoxygenase as a Common Pathway for Pathological Brain and Vascular Aging." *Cardiovascular Psychiatry and Neurology* 2009: 174657. https://doi.org/10.1155/2009/174657.

———. 2011. "5-Lipoxygenase as an Endogenous Modulator of Amyloid Beta Formation in Vivo." *Annals of Neurology* 69 (1): 34–46. https://doi.org/10.1002/ana.22234.

———. 2016. "The 5-Lipoxygenase as Modulator of Alzheimer's γ-Secretase and Therapeutic Target." *Brain Research Bulletin* 126 (Pt 2): 207–12. https://doi.org/10.1016/j.brainresbull.2016.03.010.

Citron, Martin. 2010. "Alzheimer's Disease: Strategies for Disease Modification." *Nature Reviews. Drug Discovery* 9 (5): 387–98. https://doi.org/10.1038/nrd2896.

Congdon, Erin E., and Einar M. Sigurdsson. 2018. "Tau-Targeting Therapies for Alzheimer Disease." *Nature Reviews. Neurology* 14 (7): 399–415. https://doi.org/10.1038/s41582-018-0013-z.

Corcoran, Niall M., Daniel Martin, Birgit Hutter-Paier, Manfred Windisch, Thanh Nguyen, Lina Nheu, Lars E. Sundstrom, et al. 2010. "Sodium Selenate Specifically Activates PP2A Phosphatase, Dephosphorylates Tau and Reverses Memory Deficits in an Alzheimer's Disease Model." *Journal of Clinical Neuroscience: Official Journal of the Neurosurgical Society of Australasia* 17 (8): 1025–33. https://doi.org/10.1016/j.jocn.2010.04.020.

Czapski, Grzegorz A., Kinga Czubowicz, Joanna B. Strosznajder, and Robert P. Strosznajder. 2016. "The Lipoxygenases: Their Regulation and Implication in Alzheimer's Disease." *Neurochemical Research* 41 (1–2): 243–57. https://doi.org/10.1007/s11064-015-1776-x.

Deverman, Benjamin E., Piers L. Pravdo, Bryan P. Simpson, Sripriya Ravindra Kumar, Ken Y. Chan, Abhik Banerjee, Wei-Li Wu, et al. 2016. "Cre-Dependent Selection Yields AAV Variants for Widespread Gene Transfer to the Adult Brain." *Nature Biotechnology* 34 (2): 204–9. https://doi.org/10.1038/nbt.3440.

Dias, Vera, Eunsung Junn, and M. Maral Mouradian. 2013. "The Role of Oxidative Stress in Parkinson's Disease." *Journal of Parkinson's Disease* 3 (4): 461–91. https://doi.org/10.3233/JPD-130230.

Domínguez, Juan Manuel, Ana Fuertes, Leyre Orozco, María del Monte-Millán, Elena Delgado, and Miguel Medina. 2012. "Evidence for Irreversible Inhibition of Glycogen Synthase Kinase-3β by Tideglusib." *The Journal of Biological Chemistry* 287 (2): 893–904. https://doi.org/10.1074/jbc.M111.306472.

Doody, Rachelle S., Rema Raman, Martin Farlow, Takeshi Iwatsubo, Bruno Vellas, Steven Joffe, Karl Kieburtz, et al. 2013. "A Phase 3 Trial of Semagacestat for Treatment of Alzheimer's Disease." *The New England Journal of Medicine* 369 (4): 341–50. https://doi.org/10.1056/NEJMoa1210951.

Doudna, Jennifer A., and Emmanuelle Charpentier. 2014. "The New Frontier of Genome Engineering with CRISPR-Cas9." *Science* 346 (6213): 1258096. https://doi.org/10.1126/science.1258096.

Duque, Sandra, Béatrice Joussemet, Christel Riviere, Thibaut Marais, Laurence Dubreil, Anne-Marie Douar, John Fyfe, et al. 2009. "Intravenous Administration of Self-Complementary AAV9 Enables Transgene Delivery to Adult Motor Neurons." *Molecular Therapy: The Journal of the American Society of Gene Therapy* 17 (7): 1187–96. https://doi.org/10.1038/mt.2009.71.

Eberling, J. L., W. J. Jagust, C. W. Christine, P. Starr, P. Larson, K. S. Bankiewicz, and M. J. Aminoff. 2008. "Results from a Phase I Safety Trial of HAADC Gene Therapy for Parkinson Disease." *Neurology* 70 (21): 1980–83. https://doi.org/10.1212/01.wnl.0000312381.29287.ff.

Eersel, Janet van, Yazi D. Ke, Xin Liu, Fabien Delerue, Jillian J. Kril, Jürgen Götz, and Lars M. Ittner. 2010. "Sodium Selenate Mitigates Tau Pathology, Neurodegeneration, and Functional Deficits in Alzheimer's Disease Models." *Proceedings of the National Academy of Sciences of the United States of America* 107 (31): 13888–93. https://doi.org/10.1073/pnas.1009038107.

Emborg, Marina E., Maren Carbon, James E. Holden, Matthew J. During, Yilong Ma, Chengke Tang, Jeffrey Moirano, et al. 2007. "Subthalamic Glutamic Acid Decarboxylase Gene Therapy: Changes in Motor Function and Cortical Metabolism." *Journal of Cerebral Blood Flow and Metabolism: Official Journal of the International Society of Cerebral Blood Flow and Metabolism* 27 (3): 501–9. https://doi.org/10.1038/sj.jcbfm.9600364.

Farias, Santiago E., Simona Zarini, Thomas Precht, Robert C. Murphy, and Kim A. Heidenreich. 2007. "Transcellular Biosynthesis of Cysteinyl Leukotrienes in Rat Neuronal and Glial Cells." *Journal of Neurochemistry* 103 (4): 1310–18. https://doi.org/10.1111/j.1471-4159.2007.04830.x.

Firuzi, Omidreza, Jiamin Zhuo, Cinzia M. Chinnici, Thomas Wisniewski, and Domenico Praticò. 2008. "5-Lipoxygenase Gene Disruption Reduces Amyloid-Beta Pathology in a Mouse Model of Alzheimer's Disease." *FASEB Journal: Official Publication of the Federation of American Societies for Experimental Biology* 22 (4): 1169–78. https://doi.org/10.1096/fj.07-9131.com.

Forlenza, O. V., V. J. R. De-Paula, and B. S. O. Diniz. 2014. "Neuroprotective Effects of Lithium: Implications for the Treatment of Alzheimer's Disease and Related Neurodegenerative Disorders." *ACS Chemical Neuroscience* 5 (6): 443–50. https://doi.org/10.1021/cn5000309.

Forsayeth, John R., Jamie L. Eberling, Laura M. Sanftner, Zhu Zhen, Philip Pivirotto, John Bringas, Janet Cunningham, et al. 2006. "A Dose-Ranging Study of AAV-HAADC Therapy in Parkinsonian Monkeys." *Molecular Therapy: The Journal of the American Society of Gene Therapy* 14 (4): 571–77. https://doi.org/10.1016/j.ymthe.2006.04.008.

Giannopoulos, Phillip F., Yash B. Joshi, and Domenico Praticò. 2014. "Novel Lipid Signaling Pathways in Alzheimer's Disease Pathogenesis." *Biochemical Pharmacology* 88 (4): 560–64. https://doi.org/10.1016/j.bcp.2013.11.005.

Grasso, Margherita, Paola Piscopo, Annamaria Confaloni, and Michela A. Denti. 2014. "Circulating MiRNAs as Biomarkers for Neurodegenerative Disorders." *Molecules (Basel, Switzerland)* 19 (5): 6891–6910. https://doi.org/10.3390/molecules19056891.

György, Bence, Camilla Lööv, Mikołaj P. Zaborowski, Shuko Takeda, Benjamin P. Kleinstiver, Caitlin Commins, Ksenia Kastanenka, et al. 2018. "CRISPR/Cas9 Mediated Disruption of the Swedish APP Allele as a Therapeutic Approach for Early-Onset Alzheimer's Disease." *Molecular Therapy. Nucleic Acids* 11 (June): 429–40. https://doi.org/10.1016/j.omtn.2018.03.007.

Haass, Christian, and Dennis J. Selkoe. 2007. "Soluble Protein Oligomers in Neurodegeneration: Lessons from the Alzheimer's Amyloid Beta-Peptide." *Nature Reviews. Molecular Cell Biology* 8 (2): 101–12. https://doi.org/10.1038/nrm2101.

Hadaczek, Piotr, Jamie L. Eberling, Philip Pivirotto, John Bringas, John Forsayeth, and Krystof S. Bankiewicz. 2010. "Eight Years of Clinical Improvement in MPTP-Lesioned Primates after Gene Therapy with AAV2-HAADC." *Molecular Therapy: The Journal of the American Society of Gene Therapy* 18 (8): 1458–61. https://doi.org/10.1038/mt.2010.106.

Hamaguchi, Tsuyoshi, Kenjiro Ono, and Masahito Yamada. 2010. "REVIEW: Curcumin and Alzheimer's Disease." *CNS Neuroscience & Therapeutics* 16 (5): 285–97. https://doi.org/10.1111/j.1755-5949.2010.00147.x.

Hampel, Harald, Katharina Bürger, Jens C. Pruessner, Raymond Zinkowski, John DeBernardis, Daniel Kerkman, Gerda Leinsinger, et al. 2005. "Correlation of Cerebrospinal Fluid Levels of Tau Protein Phosphorylated at Threonine 231 with Rates of Hippocampal Atrophy in Alzheimer Disease." *Archives of Neurology* 62 (5): 770–73. https://doi.org/10.1001/archneur.62.5.770.

Hayakawa, Tomohisa, Mika Iwai, Satoshi Aoki, Koichi Takimoto, Mitsuo Maruyama, Wakako Maruyama, and Noboru Motoyama. 2015. "SIRT1 Suppresses the Senescence-Associated Secretory Phenotype through Epigenetic Gene Regulation." *PloS One* 10 (1): e0116480. https://doi.org/10.1371/journal.pone.0116480.

He, Gen, Wenjie Luo, Peng Li, Christine Remmers, William J. Netzer, Joseph Hendrick, Karima Bettayeb, et al. 2010. "Gamma-Secretase Activating Protein Is a Therapeutic Target for Alzheimer's Disease." *Nature* 467 (7311): 95–98. https://doi.org/10.1038/nature09325.

Hu, Shuxin, Panchanan Maiti, Qiulan Ma, Xiaohong Zuo, Mychica R. Jones, Greg M. Cole, and Sally A. Frautschy. 2015. "Clinical Development of Curcumin in Neurodegenerative Disease." *Expert Review of Neurotherapeutics* 15 (6): 629–37. https://doi.org/10.1586/14737175.2015.1044981.

Ikonomovic, Milos D., Eric E. Abrahamson, Tolga Uz, Hari Manev, and Steven T. Dekosky. 2008. "Increased 5-Lipoxygenase Immunoreactivity in the Hippocampus of Patients with Alzheimer's Disease." *The Journal of Histochemistry and Cytochemistry: Official Journal of the Histochemistry Society* 56 (12): 1065–73. https://doi.org/10.1369/jhc.2008.951855.

Ittner, Lars M., and Jürgen Götz. 2011. "Amyloid-β and Tau-a Toxic Pas de Deux in Alzheimer's Disease." *Nature Reviews. Neuroscience* 12 (2): 65–72. https://doi.org/10.1038/nrn2967.

Iwata, N., S. Tsubuki, Y. Takaki, K. Shirotani, B. Lu, N. P. Gerard, C. Gerard, et al. 2001. "Metabolic Regulation of Brain Abeta by Neprilysin." *Science (New York, N.Y.)* 292 (5521): 1550–52. https://doi.org/10.1126/science.1059946.

Jiang, Jue, and Hong Jiang. 2015. "Efficacy and Adverse Effects of Memantine Treatment for Alzheimer's Disease from Randomized Controlled Trials." *Neurological Sciences: Official Journal of the Italian Neurological Society and of the Italian Society of Clinical Neurophysiology* 36 (9): 1633–41. https://doi.org/10.1007/s10072-015-2221-2.

Jinek, Martin, Krzysztof Chylinski, Ines Fonfara, Michael Hauer, Jennifer A. Doudna, and Emmanuelle Charpentier. 2012. "A Programmable Dual-RNA-Guided DNA Endonuclease in Adaptive Bacterial Immunity." *Science (New York, N.Y.)* 337 (6096): 816–21. https://doi.org/10.1126/science.1225829.

Kamenetz, Flavio, Taisuke Tomita, Helen Hsieh, Guy Seabrook, David Borchelt, Takeshi Iwatsubo, Sangram Sisodia, and Roberto Malinow. 2003. "APP Processing and Synaptic Function." *Neuron* 37 (6): 925–37.

Kanemitsu, Hyoe, Takami Tomiyama, and Hiroshi Mori. 2003. "Human Neprilysin Is Capable of Degrading Amyloid Beta Peptide Not Only in the Monomeric Form but Also the Pathological Oligomeric Form." *Neuroscience Letters* 350 (2): 113–16.

Kells, Adrian P., Jamie Eberling, Xiaomin Su, Philip Pivirotto, John Bringas, Piotr Hadaczek, Wade C. Narrow, et al. 2010. "Regeneration of the MPTP-Lesioned Dopaminergic System after Convection-Enhanced Delivery of AAV2-GDNF." *The Journal of Neuroscience* 30 (28): 9567–77. https://doi.org/10.1523/JNEUROSCI.0942-10.2010.

Khalil, Hilal S., Vanio Mitev, Tatyana Vlaykova, Laura Cavicchi, and Nikolai Zhelev. 2015. "Discovery and Development of Seliciclib. How Systems Biology Approaches Can Lead to Better Drug Performance." *Journal of Biotechnology* 202 (May): 40–49. https://doi.org/10.1016/j.jbiotec.2015.02.032.

Khlistunova, I. 1992. "Diagnosis and Treatment of Cushing's Syndrome. Cushing's Syndrome: Current Clinical Problems, Symposium. Padova, October 19–20, 1990." *Journal of Endocrinological Investigation* 15 (3): 223–44.

Kim, Y. G., J. Cha, and S. Chandrasegaran. 1996. "Hybrid Restriction Enzymes: Zinc Finger Fusions to Fok I Cleavage Domain." *Proceedings of the National Academy of Sciences of the United States of America* 93 (3): 1156–60.

Klein, P. S., and D. A. Melton. 1996. "A Molecular Mechanism for the Effect of Lithium on Development." *Proceedings of the National Academy of Sciences of the United States of America* 93 (16): 8455–59.

Klyubin, Igor, Vicki Betts, Alfred T. Welzel, Kaj Blennow, Henrik Zetterberg, Anders Wallin, Cynthia A. Lemere, et al. 2008. "Amyloid Beta Protein Dimer-Containing Human CSF Disrupts Synaptic Plasticity: Prevention by Systemic Passive Immunization." *The Journal of Neuroscience: The Official Journal of the Society for Neuroscience* 28 (16): 4231–37. https://doi.org/10.1523/JNEUROSCI.5161-07.2008.

Komor, Alexis C., Yongjoo B. Kim, Michael S. Packer, John A. Zuris, and David R. Liu. 2016. "Programmable Editing of a Target Base in Genomic DNA without Double-Stranded DNA Cleavage." *Nature* 533 (7603): 420–24. https://doi.org/10.1038/nature17946.

Kordower, J. H., M. E. Emborg, J. Bloch, S. Y. Ma, Y. Chu, L. Leventhal, J. McBride, et al. 2000. "Neurodegeneration Prevented by Lentiviral Vector Delivery of GDNF in Primate Models of Parkinson's Disease." *Science (New York, N.Y.)* 290 (5492): 767–73. https://doi.org/10.1126/science.290.5492.767.

Lai, Jin'e, Meng Hu, Hao Wang, Mei Hu, Yan Long, Ming-xing Miao, Jia-chang Li, et al. 2014. "Montelukast Targeting the Cysteinyl Leukotriene Receptor 1 Ameliorates Aβ1–42-Induced Memory Impairment and Neuroinflammatory and Apoptotic Responses in Mice." *Neuropharmacology* 79 (April): 707–14. https://doi.org/10.1016/j.neuropharm.2014.01.011.

Lammers, C. H., P. Schweitzer, P. Facchinetti, J. M. Arrang, S. G. Madamba, G. R. Siggins, and D. Piomelli. 1996. "Arachidonate 5-Lipoxygenase and Its Activating Protein: Prominent Hippocampal Expression and Role in Somatostatin Signaling." *Journal of Neurochemistry* 66 (1): 147–52.

Lee, Jong Kil, and Nam-Jung Kim. 2017. "Recent Advances in the Inhibition of P38 MAPK as a Potential Strategy for the Treatment of Alzheimer's Disease." *Molecules (Basel, Switzerland)* 22 (8): E1287. https://doi.org/10.3390/molecules22081287.

Lee, V. M., M. Goedert, and J. Q. Trojanowski. 2001. "Neurodegenerative Tauopathies." *Annual Review of Neuroscience* 24: 1121–59. https://doi.org/10.1146/annurev.neuro.24.1.1121.

Leissring, Malcolm A., Wesley Farris, Alice Y. Chang, Dominic M. Walsh, Xining Wu, Xiaoyan Sun, Matthew P. Frosch, et al. 2003. "Enhanced Proteolysis of Beta-Amyloid in APP Transgenic Mice Prevents Plaque Formation, Secondary Pathology, and Premature Death." *Neuron* 40 (6): 1087–93.

LeWitt, Peter A., Ali R. Rezai, Maureen A. Leehey, Steven G. Ojemann, Alice W. Flaherty, Emad N. Eskandar, Sandra K. Kostyk, et al. 2011. "AAV2-GAD Gene Therapy for Advanced Parkinson's Disease: A Double-Blind, Sham-Surgery Controlled, Randomised Trial." *The Lancet. Neurology* 10 (4): 309–19. https://doi.org/10.1016/S1474-4422(11)70039-4.

Lin, Jing-Ya, Cheng-Long Xie, Su-Fang Zhang, Weien Yuan, and Zhen-Guo Liu. 2017. "Current Experimental Studies of Gene Therapy in Parkinson's Disease." *Frontiers in Aging Neuroscience* 9: 126. https://doi.org/10.3389/fnagi.2017.00126.

Lovestone, Simon, Mercè Boada, Bruno Dubois, Michael Hüll, Juha O. Rinne, Hans-Jürgen Huppertz, Miguel Calero, et al. 2015. "A Phase II Trial of Tideglusib in Alzheimer's Disease." *Journal of Alzheimer's Disease: JAD* 45 (1): 75–88. https://doi.org/10.3233/JAD-141959.

Lu, Tao, Liviu Aron, Joseph Zullo, Ying Pan, Haeyoung Kim, Yiwen Chen, Tun-Hsiang Yang, et al. 2014. "REST and Stress Resistance in Ageing and Alzheimer's Disease." *Nature* 507 (7493): 448–54. https://doi.org/10.1038/nature13163.

Lue, L. F., Y. M. Kuo, A. E. Roher, L. Brachova, Y. Shen, L. Sue, T. Beach, et al. 1999. "Soluble Amyloid Beta Peptide Concentration as a Predictor of Synaptic Change in Alzheimer's Disease." *The American Journal of Pathology* 155 (3): 853–62.

Malpas, Charles B., Lucy Vivash, Sila Genc, Michael M. Saling, Patricia Desmond, Christopher Steward, Rodney J. Hicks, et al. 2016. "A Phase IIa Randomized Control Trial of VEL015 (Sodium Selenate) in Mild-Moderate Alzheimer's Disease." *Journal of Alzheimer's Disease: JAD* 54 (1): 223–32. https://doi.org/10.3233/JAD-160544.

Manfredsson, Fredric P., Corinna Burger, Aaron C. Rising, Kheir Zuobi-Hasona, Layla F. Sullivan, Alfred S. Lewin, Julia Huang, et al. 2009. "Tight Long-Term Dynamic Doxycycline Responsive Nigrostriatal GDNF Using a Single RAAV Vector." *Molecular Therapy: The Journal of the American Society of Gene Therapy* 17 (11): 1857–67. https://doi.org/10.1038/mt.2009.196.

Mao, Qi, Wang-zhi Qin, Ao Zhang, and Na Ye. 2020. "Recent Advances in Dopaminergic Strategies for the Treatment of Parkinson's Disease." *Acta Pharmacologica Sinica* 41 (4): 471–82. https://doi.org/10.1038/s41401-020-0365-y.

Mapelli, Marina, Lucia Massimiliano, Claudia Crovace, Markus A. Seeliger, Li-Huei Tsai, Laurent Meijer, and Andrea Musacchio. 2005. "Mechanism of CDK5/P25 Binding by CDK Inhibitors." *Journal of Medicinal Chemistry* 48 (3): 671–79. https://doi.org/10.1021/jm049323m.

Marks, William J., Raymond T. Bartus, Joao Siffert, Charles S. Davis, Andres Lozano, Nicholas Boulis, Jerrold Vitek, et al. 2010. "Gene Delivery of AAV2-Neurturin for Parkinson's Disease: A Double-Blind, Randomised, Controlled Trial." *The Lancet. Neurology* 9 (12): 1164–72. https://doi.org/10.1016/S1474-4422(10)70254-4.

Marks, William J., Jill L. Ostrem, Leonard Verhagen, Philip A. Starr, Paul S. Larson, Roy Ae Bakay, Robin Taylor, et al. 2008. "Safety and Tolerability of Intraputaminal Delivery of CERE-120 (Adeno-Associated Virus Serotype 2-Neurturin) to Patients with Idiopathic Parkinson's Disease: An Open-Label, Phase I Trial." *The Lancet. Neurology* 7 (5): 400–408. https://doi.org/10.1016/S1474-4422(08)70065-6.

Martinez, Ana, Mercedes Alonso, Ana Castro, Concepción Pérez, and Francisco J. Moreno. 2002. "First Non-ATP Competitive Glycogen Synthase Kinase 3 Beta (GSK-3beta) Inhibitors: Thiadiazolidinones (TDZD) as Potential Drugs for the Treatment of Alzheimer's Disease." *Journal of Medicinal Chemistry* 45 (6): 1292–99.

Matsunaga, Shinji, Taro Kishi, and Nakao Iwata. 2015. "Memantine Monotherapy for Alzheimer's Disease: A Systematic Review and Meta-Analysis." *PloS One* 10 (4): e0123289. https://doi.org/10.1371/journal.pone.0123289.

Meijer, L., A. Borgne, O. Mulner, J. P. Chong, J. J. Blow, N. Inagaki, M. Inagaki, J. G. Delcros, and J. P. Moulinoux. 1997. "Biochemical and Cellular Effects of Roscovitine, a Potent and Selective Inhibitor of the Cyclin-Dependent Kinases Cdc2, Cdk2 and Cdk5." *European Journal of Biochemistry* 243 (1–2): 527–36.

Mike, Lynch. 2018. "New Alzheimer's Association Report Reveals Sharp Increases in Alzheimer's Prevalence, Deaths and Cost of Care." 312.335.4078. Chicago, IL: Alzheimer's Association. www.alz.org/news/2018/new_alzheimer_s_association_report_reveals_sharp_i.

Miller, Jeffrey C., Siyuan Tan, Guijuan Qiao, Kyle A. Barlow, Jianbin Wang, Danny F. Xia, Xiangdong Meng, et al. 2011. "A TALE Nuclease Architecture for Efficient Genome Editing." *Nature Biotechnology* 29 (2): 143–48. https://doi.org/10.1038/nbt.1755.

Min, Sang-Won, Xu Chen, Tara E. Tracy, Yaqiao Li, Yungui Zhou, Chao Wang, Kotaro Shirakawa, et al. 2015. "Critical Role of Acetylation in Tau-Mediated Neurodegeneration and Cognitive Deficits." *Nature Medicine* 21 (10): 1154–62. https://doi.org/10.1038/nm.3951.

Mintzer, Meredith A., and Eric E. Simanek. 2009. "Nonviral Vectors for Gene Delivery." *Chemical Reviews* 109 (2): 259–302. https://doi.org/10.1021/cr800409e.

Mittermeyer, Gabriele, Chadwick W. Christine, Kathryn H. Rosenbluth, Suzanne L. Baker, Philip Starr, Paul Larson, Paul L. Kaplan, et al. 2012. "Long-Term Evaluation of a Phase 1 Study of AADC Gene Therapy for Parkinson's Disease." *Human Gene Therapy* 23 (4): 377–81. https://doi.org/10.1089/hum.2011.220.

Nagata, Kenichi, Mika Takahashi, Yukio Matsuba, Fumi Okuyama-Uchimura, Kaori Sato, Shoko Hashimoto, Takashi Saito, et al. 2018. "Generation of App Knock-in Mice Reveals Deletion Mutations Protective against Alzheimer's Disease-like Pathology." *Nature Communications* 9 (1): 1800. https://doi.org/10.1038/s41467-018-04238-0.

Nuytemans, Karen, Jessie Theuns, Marc Cruts, and Christine Van Broeckhoven. 2010. "Genetic Etiology of Parkinson Disease Associated with Mutations in the SNCA, PARK2, PINK1, PARK7, and LRRK2 Genes: A Mutation Update." *Human Mutation* 31 (7): 763–80. https://doi.org/10.1002/humu.21277.

Oddo, Salvatore, Vitaly Vasilevko, Antonella Caccamo, Masashi Kitazawa, David H. Cribbs, and Frank M. LaFerla. 2006. "Reduction of Soluble Abeta and Tau, but Not Soluble Abeta Alone, Ameliorates Cognitive Decline in Transgenic Mice with Plaques and Tangles." *The Journal of Biological Chemistry* 281 (51): 39413–23. https://doi.org/10.1074/jbc.M608485200.

Olmos, Yolanda, Francisco J. Sánchez-Gómez, Brigitte Wild, Nieves García-Quintans, Sofía Cabezudo, Santiago Lamas, and María Monsalve. 2013. "SirT1 Regulation of Antioxidant Genes Is Dependent on the Formation of a FoxO3a/PGC-1α Complex." *Antioxidants & Redox Signaling* 19 (13): 1507–21. https://doi.org/10.1089/ars.2012.4713.

Pack, Daniel W., Allan S. Hoffman, Suzie Pun, and Patrick S. Stayton. 2005. "Design and Development of Polymers for Gene Delivery." *Nature Reviews Drug Discovery* 4 (7): 581–93. https://doi.org/10.1038/nrd1775.

Palfi, Stéphane, Jean Marc Gurruchaga, G Scott Ralph, Helene Lepetit, Sonia Lavisse, Philip C Buttery, Colin Watts, et al. 2014. "Long-Term Safety and Tolerability of ProSavin, a Lentiviral Vector-Based Gene Therapy for Parkinson's Disease: A Dose Escalation, Open-Label, Phase 1/2 Trial." *The Lancet* 383 (9923): 1138–46. https://doi.org/10.1016/S0140-6736(13)61939-X.

Palmer, R. M., R. J. Stepney, G. A. Higgs, and K. E. Eakins. 1980. "Chemokinetic Activity of Arachidonic and Lipoxygenase Products on Leuocyctes of Different Species." *Prostaglandins* 20 (2): 411–18.

Panza, Francesco, Vincenzo Solfrizzi, Davide Seripa, Bruno P. Imbimbo, Madia Lozupone, Andrea Santamato, Chiara Zecca, et al. 2016. "Tau-Centric Targets and Drugs in Clinical Development for the Treatment of Alzheimer's Disease." *BioMed Research International* 2016: 3245935. https://doi.org/10.1155/2016/3245935.

Park, Hanseul, Jungju Oh, Gayong Shim, Byounggook Cho, Yujung Chang, Siyoung Kim, Soonbong Baek, et al. 2019. "In Vivo Neuronal Gene Editing via CRISPR–Cas9 Amphiphilic Nanocomplexes Alleviates Deficits in Mouse Models of Alzheimer's Disease." *Nature Neuroscience* 22 (4): 524–28. https://doi.org/10.1038/s41593-019-0352-0.

Qu, T., R. Manev, and H. Manev. 2001. "5-Lipoxygenase (5-LOX) Promoter Polymorphism in Patients with Early-Onset and Late-Onset Alzheimer's Disease." *The Journal of Neuropsychiatry and Clinical Neurosciences* 13 (2): 304–5. https://doi.org/10.1176/jnp.13.2.304.

Querfurth, Henry W., and Frank M. LaFerla. 2010. "Alzheimer's Disease." *The New England Journal of Medicine* 362 (4): 329–44. https://doi.org/10.1056/NEJMra0909142.

Rafii, Michael S., Mark H. Tuszynski, Ronald G. Thomas, David Barba, James B. Brewer, Robert A. Rissman, Joao Siffert, et al. 2018. "Adeno-Associated Viral Vector (Serotype 2)-Nerve Growth Factor for Patients with Alzheimer Disease: A Randomized Clinical Trial." *JAMA Neurology* 75 (7): 834–41. https://doi.org/10.1001/jamaneurol.2018.0233.

Raikwar, Sudhanshu P., Ramasamy Thangavel, Iuliia Dubova, Govindhasamy Pushpavathi Selvakumar, Mohammad Ejaz Ahmed, Duraisamy Kempuraj, Smita A. Zaheer, et al. 2019. "Targeted Gene Editing of Glia Maturation Factor in Microglia: A Novel Alzheimer's Disease Therapeutic Target." *Molecular Neurobiology* 56 (1): 378–93. https://doi.org/10.1007/s12035-018-1068-y.

Rangel-Barajas, Claudia, Israel Coronel, and Benjamín Florán. 2015. "Dopamine Receptors and Neurodegeneration." *Aging and Disease* 6 (5): 349–68. https://doi.org/10.14336/AD.2015.0330.

Reinhard, Constanze, Sébastien S. Hébert, and Bart De Strooper. 2005. "The Amyloid-Beta Precursor Protein: Integrating Structure with Biological Function." *The EMBO Journal* 24 (23): 3996–4006. https://doi.org/10.1038/sj.emboj.7600860.

Rueli, Rachel H. L. H., Daniel J. Torres, Andrea S. T. Dewing, Arlene C. Kiyohara, Stephanie M. Barayuga, Miyoko T. Bellinger, Jane H. Uyehara-Lock, et al. 2017. "Selenoprotein S Reduces Endoplasmic Reticulum Stress-Induced Phosphorylation of Tau: Potential Role in Selenate Mitigation of Tau Pathology." *Journal of Alzheimer's Disease: JAD* 55 (2): 749–62. https://doi.org/10.3233/JAD-151208.

Sämgård, Kajsa, Henrik Zetterberg, Kaj Blennow, Oskar Hansson, Lennart Minthon, and Elisabet Londos. 2010. "Cerebrospinal Fluid Total Tau as a Marker of Alzheimer's Disease Intensity." *International Journal of Geriatric Psychiatry* 25 (4): 403–10. https://doi.org/10.1002/gps.2353.

San Sebastian, Waldy, R. Mark Richardson, Adrian P. Kells, Clementine Lamarre, John Bringas, Philip Pivirotto, Ernesto A. Salegio, et al. 2012. "Safety and Tolerability of Magnetic Resonance Imaging-Guided Convection-Enhanced Delivery of AAV2-HAADC with a Novel Delivery Platform in Nonhuman Primate Striatum." *Human Gene Therapy* 23 (2): 210–17. https://doi.org/10.1089/hum.2011.162.

Sandhu, Punam, Junghoon Lee, Jeanine Ballard, Brittany Walker, Joan Ellis, Jacob Marcus, Dawn Toolan, et al. 2016. "Pharmacokinetics Aand Pharmacodynamics to Support Clinical Studies of Mk-8719: An O-Glcnacase Inhibitor for Progressive Supranuclear Palsy." *Alzheimer's & Dementia: The Journal of the Alzheimer's Association* 12 (7): P1028. https://doi.org/10.1016/j.jalz.2016.06.2125.

Santacruz, K., J. Lewis, T. Spires, J. Paulson, L. Kotilinek, M. Ingelsson, A. Guimaraes, et al. 2005. "Tau Suppression in a Neurodegenerative Mouse Model Improves Memory Function." *Science (New York)* 309 (5733): 476–81. https://doi.org/10.1126/science.1113694.

Selkoe, D. J. 2001. "Alzheimer's Disease: Genes, Proteins, and Therapy." *Physiological Reviews* 81 (2): 741–66. https://doi.org/10.1152/physrev.2001.81.2.741.

Stambolic, V., L. Ruel, and J. R. Woodgett. 1996. "Lithium Inhibits Glycogen Synthase Kinase-3 Activity and Mimics Wingless Signalling in Intact Cells." *Current Biology: CB* 6 (12): 1664–68.

Sundal, Christina, Shinsuke Fujioka, Ryan J. Uitti, and Zbigniew K. Wszolek. 2012. "Autosomal Dominant Parkinson's Disease." *Parkinsonism & Related Disorders* 18 Suppl 1 (January): S7–10. https://doi.org/10.1016/S1353-8020(11)70005-0.

Tang, Su-Su, Hao Hong, Lan Chen, Zhen-Lin Mei, Miao-Jin Ji, Guo-Qing Xiang, Ning Li, et al. 2014. "Involvement of Cysteinyl Leukotriene Receptor 1 in Aβ1–42-Induced Neurotoxicity in Vitro and in Vivo." *Neurobiology of Aging* 35 (3): 590–99. https://doi.org/10.1016/j.neurobiolaging.2013.09.036.

Tang, Su-Su, Miao-jin Ji, Lan Chen, Mei Hu, Yan Long, Yong-qi Li, Ming-xing Miao, et al. 2014. "Protective Effect of Pranlukast on $A\beta_{1-42}$-Induced Cognitive Deficits Associated with Downregulation of Cysteinyl Leukotriene Receptor 1." *The International Journal of Neuropsychopharmacology* 17 (4): 581–92. https://doi.org/10.1017/S1461145713001314.

Tanzi, Rudolph E., and Lars Bertram. 2005. "Twenty Years of the Alzheimer's Disease Amyloid Hypothesis: A Genetic Perspective." *Cell* 120 (4): 545–55. https://doi.org/10.1016/j.cell.2005.02.008.

Tardieu, Marc, Michel Zérah, Marie-Lise Gougeon, Jérome Ausseil, Stéphanie de Bournonville, Béatrice Husson, Dimitrios Zafeiriou, et al. 2017. "Intracerebral Gene Therapy in Children with Mucopolysaccharidosis Type IIIB Syndrome: An Uncontrolled Phase 1/2 Clinical Trial." *The Lancet. Neurology* 16 (9): 712–20. https://doi.org/10.1016/S1474-4422(17)30169-2.

Tereshchenko, Julia, Andrea Maddalena, Mathias Bähr, and Sebastian Kügler. 2014. "Pharmacologically Controlled, Discontinuous GDNF Gene Therapy Restores Motor Function in a Rat Model of Parkinson's Disease." *Neurobiology of Disease* 65 (May): 35–42. https://doi.org/10.1016/j.nbd.2014.01.009.

Tsoi, Kelvin K. F., Joyce Y. C. Chan, Nelson W. Y. Leung, Hoyee W. Hirai, Samuel Y. S. Wong, and Timothy C. Y. Kwok. 2016. "Combination Therapy Showed Limited Superiority Over Monotherapy for Alzheimer Disease: A Meta-Analysis of 14 Randomized Trials." *Journal of the American Medical Directors Association* 17 (9): 863.e1–8. https://doi.org/10.1016/j.jamda.2016.05.015.

Vagnozzi, Alana N., Phillip F. Giannopoulos, and Domenico Praticò. 2018. "Brain 5-Lipoxygenase over-Expression Worsens Memory, Synaptic Integrity, and Tau Pathology in the P301S Mice." *Aging Cell* 17 (1). https://doi.org/10.1111/acel.12695.

Wadhwani, Anil R., Amira Affaneh, Stephanie Van Gulden, and John A. Kessler. 2019. "Neuronal Apolipoprotein E4 Increases Cell Death and Phosphorylated Tau Release in Alzheimer Disease." *Annals of Neurology* 85 (5): 726–39. https://doi.org/10.1002/ana.25455.

Wakabayashi, Koichi, Kunikazu Tanji, Fumiaki Mori, and Hitoshi Takahashi. 2007. "The Lewy Body in Parkinson's Disease: Molecules Implicated in the Formation and Degradation of Alpha-Synuclein Aggregates." *Neuropathology: Official Journal of the Japanese Society of Neuropathology* 27 (5): 494–506. https://doi.org/10.1111/j.1440-1789.2007.00803.x.

Walsh, Dominic M., Matthew Townsend, Marcia B. Podlisny, Ganesh M. Shankar, Julia V. Fadeeva, Omar El Agnaf, Dean M. Hartley, et al. 2005. "Certain Inhibitors of Synthetic Amyloid Beta-Peptide (Abeta) Fibrillogenesis Block Oligomerization of Natural Abeta and Thereby Rescue Long-Term Potentiation." *The Journal of Neuroscience: The Official Journal of the Society for Neuroscience* 25 (10): 2455–62. https://doi.org/10.1523/JNEUROSCI.4391-04.2005.

Warren Olanow, C., Raymond T. Bartus, Tiffany L. Baumann, Stewart Factor, Nicholas Boulis, Mark Stacy, Dennis A. Turner, et al. 2015. "Gene Delivery of Neurturin to Putamen and Substantia Nigra in Parkinson Disease: A Double-Blind, Randomized, Controlled Trial." *Annals of Neurology* 78 (2): 248–57. https://doi.org/10.1002/ana.24436.

Weinberg, Marc S., R. Jude Samulski, and Thomas J. McCown. 2013. "Adeno-Associated Virus (AAV) Gene Therapy for Neurological Disease." *Neuropharmacology* 69 (June): 82–88. https://doi.org/10.1016/j.neuropharm.2012.03.004.

Williams-Gray, Caroline H., and Paul F. Worth. 2016. "Parkinson's Disease." *Medicine* 44 (9): 542–46. https://doi.org/10.1016/j.mpmed.2016.06.001.

Wischik, C. M., P. C. Edwards, R. Y. Lai, M. Roth, and C. R. Harrington. 1996. "Selective Inhibition of Alzheimer Disease-like Tau Aggregation by Phenothiazines." *Proceedings of the National Academy of Sciences of the United States of America* 93 (20): 11213–18.

Wong, Eitan, George P. Liao, Jerry C. Chang, Peng Xu, Yue-Ming Li, and Paul Greengard. 2019. "GSAP Modulates γ-Secretase Specificity by Inducing Conformational Change in PS1." *Proceedings of the National Academy of Sciences* 116 (13): 6385–90. https://doi.org/10.1073/pnas.1820160116.

Yin, Hao, Rosemary L. Kanasty, Ahmed A. Eltoukhy, Arturo J. Vegas, J. Robert Dorkin, and Daniel G. Anderson. 2014. "Non-Viral Vectors for Gene-Based Therapy." *Nature Reviews Genetics* 15 (8): 541–55. https://doi.org/10.1038/nrg3763.

Yokomizo, T., T. Izumi, K. Chang, Y. Takuwa, and T. Shimizu. 1997. "A G-Protein-Coupled Receptor for Leukotriene B4 That Mediates Chemotaxis." *Nature* 387 (6633): 620–24. https://doi.org/10.1038/42506.

Yokomizo, T., K. Kato, H. Hagiya, T. Izumi, and T. Shimizu. 2001. "Hydroxyeicosanoids Bind to and Activate the Low Affinity Leukotriene B4 Receptor, BLT2." *The Journal of Biological Chemistry* 276 (15): 12454–59. https://doi.org/10.1074/jbc.M011361200.

Yokomizo, T., K. Kato, K. Terawaki, T. Izumi, and T. Shimizu. 2000. "A Second Leukotriene B(4) Receptor, BLT2. A New Therapeutic Target in Inflammation and Immunological Disorders." *The Journal of Experimental Medicine* 192 (3): 421–32.

Yoshida, Daigo, Toshiharu Ninomiya, Yasufumi Doi, Jun Hata, Masayo Fukuhara, Fumie Ikeda, Naoko Mukai, et al. 2012. "Prevalence and Causes of Functional Disability in an Elderly General Population of Japanese: The Hisayama Study." *Journal of Epidemiology* 22 (3): 222–29.

12 Clinical Manifestation and Molecular Orientation of Huntington's Chorea

Pratik Talukder[1,*], *Anushka Ghosh*[1] *and Arpan Banerjee*[1]
[1]Department of Biotechnology, University of Engineering and Management, Kolkata
*Correpondence to pratik.talukder@uem.edu.in

12.1 INTRODUCTION

Huntington's chorea is a predominantly hereditary somatic neuropathy (Romo et al., 2017). It is caused by a mutation in the HTT gene. It forms an extended polyglutamine chain (CAG) and leads to the death of neurons. It has the ability to penetrate depending on age. The mean CAG length is between 16 and 20 repeats (Krzyzosiak et al., 2012). Age of onset (AO) is also highly dependent on the CAG trinucleotide repeat length. CAG trinucleotides that are 36 and larger in length are known to cause disease, leading to Huntington's disease. Normal genes include CAG repeat lengths from 27–35 (Bates et al., 2015). This intermediate range has some increased risk and can lead to disease (Roos, 2010). A recent study concluded that a normal HTT gene with fewer than 36 repeats and an intermediate or mutant normal gene repeat length from 29–35 would not cause disease (Pringsheim et al., 2010).

George Huntington, a physician from Connecticut, first identified Huntington's Disease (HD, 1872). A neurological disorder, HD is dominantly inherited from generation to generation and is known for causing neuropsychological deficits and excessive, metastable motor movements (Langbehn et al., 2010). As Huntington said, children will have a chance of being affected till their adulthood if one of their parents is affected (Langbehn et al., 2010). However, it is also possible that some offspring will not be affected; then, the string of infection may be discontinued, because in the case of strong infection, it will not likely skip a generation (Reiner, Dragatsis, and Dietrich, 2011). Although it is considered that George Huntington first used the name Huntington, it is also seen that in some European and Latin American journals, the name was first used. George Huntington's description of the particular inheritance pattern started to be taken seriously as science after Mendel was rediscovered in 1900. (McColgan, and Tabrizi, 2018). The Mendelian dominant inheritance pattern described by Huntington was later confirmed by English biologist Reginald Punnett (Krzyzosiak et al., 2012). To make this point, William Bateson, a colleague of Punnett's, first counted the disease in pedigrees. Pedigrees were collected from some families that were observed to have hereditary chorea, and in late 1908, he shared his research with the British Royal Society of Medicine (RSM). Then, in Mendel's Principles of Hereditary, his theory was published (Takahashi et al., 2010). Some researchers continue to study this gene (Gonzalez-Alegre, 2019). They give results on DNA samples that were taken from a region of Venezuela, the area with the highest frequency of this disease and consanguinity. IT15 is now called the Huntingtin (HTT) gene (Bano et al., 2011). It transcripts into an mRNA of approximately 10 kB and translates into approximately 350 kDa (Wanker et al., 2019). A triplet nucleotide repeat cytosine-adenosine-guanine (CAG) is translated into a polyglutamine (poly Q) stretch (Brás, König, and Outeiro, 2019).

 DOI: 10.1201/9781003302131-13

12.2 ETIOLOGY

HD is an autosomal-dominant disease caused by an abnormal repeat of the trinucleotide CAG on the short arm of chromosome 4p16.3 in the HTT gene (Møllersen et al., 2010). This gene encodes the Huntingtin protein and contains the trinucleotide sugar CAG (Paul, and Snyder, 2019). The wild-type contains CAG repeats in the range of 6 to 26 on average (Schultz, Moser, and Nopoulos, 2020). The disease is associated with 36 or more abnormal repeats. Clinical manifestations occur when the number of repetitions exceeds 40. A range of 36 to 39 leads to an incomplete penetration disorder (Zhao et al., 2016). The range from 29 to 35 is known to have intermediate, unstable alleles, which means that these alleles are susceptible to change during gamete formation after reproduction (Matlahov and van der Wel, 2019).

There is an inverse relation between the length of the repeats and the AO that is determined by the first motor manifestation. It has been established that when the CAG repeat is longer, the AO is earlier. If the disease shows its symptoms at an early age such as 21 years or younger, then it is known as Juvenile Huntington's disease (JHD). The repeats, in this case, are more than 55. The repeat length is the cause of 70% variance in the AO without any symptoms at the early stages or during the illness.

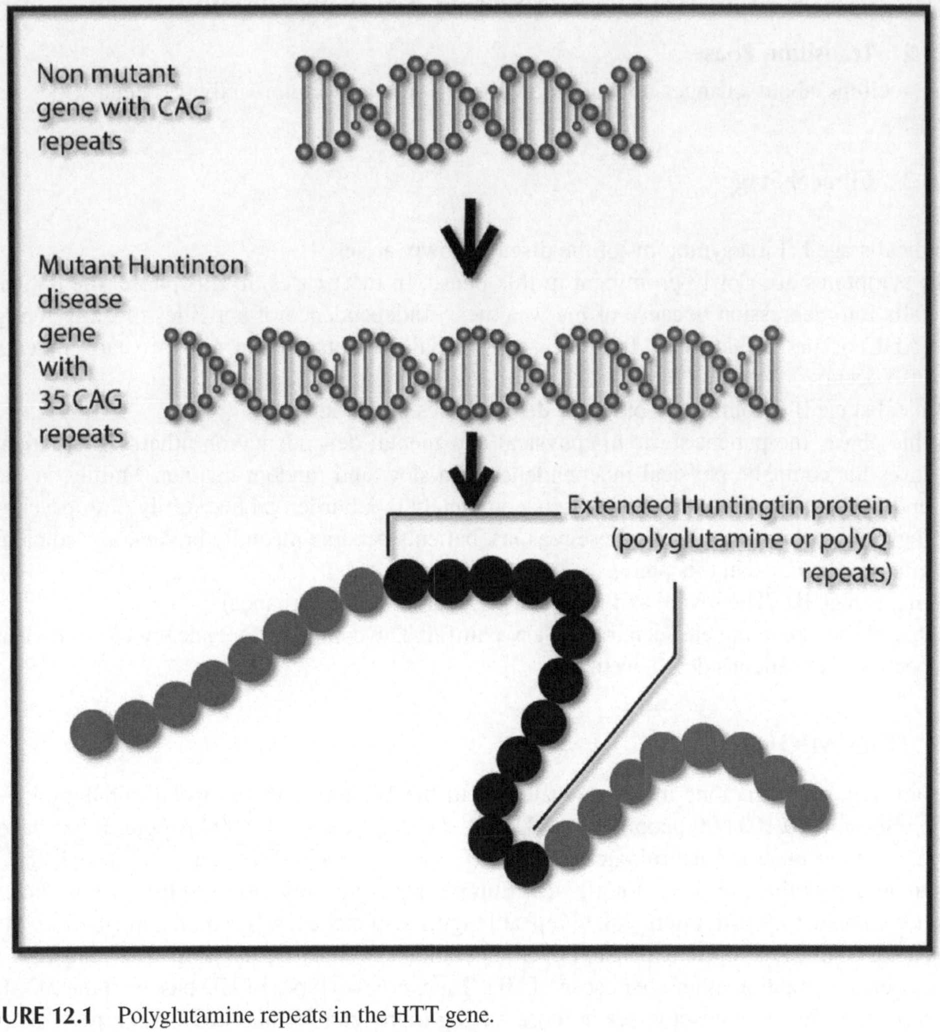

FIGURE 12.1 Polyglutamine repeats in the HTT gene.

A correlation has been established between faster weight loss and longer CAG repeats. Anticipation phenomenon is seen in the paternal inheritance line of families suffering from HD. The function of synapses is significantly influenced by the wild-type gene. In addition to being necessary in the post-embryonic period, it probably also plays an anti-apoptotic role. Some research also shows that the mutant gene causes both a gain in function and a loss in function. Numerous models, including cells, fibroblasts, C. elegans, drosophila, mice, rats, sheep, and monkeys, have been used to investigate the function of the mutant protein. The most studied animal model is mice. However, the role of intracytoplasmic and intranuclear inclusions is not yet understood. Numerous other parts of the brain also have the inclusions. This disease is characterised primarily by severe neuronal loss and striatal brain atrophy.

12.2.1 Stages During the Life of a Huntington's Disease Patient

12.2.1.1 Preclinical Stage

In this stage, the affected parent (at-risk stage 50%) undergoes anxiousness regarding their future of their personal life. Severe carriership uncertainty occurs; affected parents should stay in total care and treatment. This is the phase where the carrying of genes and their expression occurs, which is known as the premanifest stage. The uncertainty about onset is fully renewed in this phase. A treatment and sensitivity are needed for the affected parent and the whole family.

12.2.1.2 Transition Phase

Strong feelings about changes in cognition, behaviour, and motor activity occur. Uncertainty remains.

12.2.1.3 Clinical Stage

Clinical stage I (First symptoms of the disease slowly arise)

The symptoms are slowly prominent in this phase. In most cases, in this phase, the patient falls into depression because of his weakness. Independence of activities of daily living (ADL) occurs in this phase. In rare cases, suicidal attempts happen, and rare natural death also occurs.

Clinical stage II (Prominence of motor disturbances in the body)

In this phase, the patient starts his physical and mental dependency on others. The patient loses his complete physical independence in a slow and random manner. Mainly, at the end of this phase, the patient converts completely to a burden on his family both psychologically and physically. For these reasons, patients become mentally broken, and suicidal attempts increase in this phase.

Clinical stage III (The severe and final/last stage of motor disturbance)

In this phase, the complete scenario is very painful. The complete dependency of the patient occurs. The patient's death occurs.

12.3 EPIDEMIOLOGY

There are regional variations in HD prevalence. In the Western and European populations, it is seen in almost 4–10/100,000 people. In the UK, it is closer to 12.3/100,000 people. It has become a more common inherited neurological disease (Gonzalez-Alegre, 2019). It is believed that HD first spread across the world from northwest Europe. De novo migration resulting from alleles in the general population with intermediate repeat lengths also made a substantial contribution. These alleles arise from CAG repeat expansion over generations. A study on the people of Venezuela discovered genetic mutation as another cause of HD. There are two types of HD based on the AO: JHD and adult-onset. Juvenile onset arises in those having more than 55 inherited CAG repeats that are expressed before an average age of 20 years. Paternal transmission occurs in 90% of cases. Clinical representations of JHD are different from those of adult-onset HD. Bradykinesia, akinesia, and rigidity are predominant in JHD, whereas hyperkinetic presentation is more common in adult-onset. Additionally, 30–50% of individuals experience seizures; this phenotype is referred to as the Westphal or akinetic-rigid variety and is uncommon in cases of adult-onset.

12.3.1 Environmental Factors

Several papers have also investigated certain environmental factors involved in the progression of HD. Among these, the consumption of alcohol and tobacco has been studied in several articles. The results showed no relationship between disease progression and these two factors. In other papers, several dietary factors such as fish, cheese, tea, coffee, fruit juices, and milk consumption were studied of which only tea was associated with the progression of the disease. The development of psychotic symptoms in patients with HD can be caused by a variety of factors, including factors other than the dose-related toxicity of the poly Q-extended Huntingtin protein. It was also found that the length of the CAG repeats do not correlate with cognitive or mental performance (non-motor-related deficits).

The earliest proof of the impact of environmental factors on HD was found in a study of a family from Venezuela. Afterward, it was also studied in mouse models (Gonzalez-Alegre, 2019). Variability of the AO unexplained by the length of the CAG repeats was found to be affected by environmental factors by 60%. Since the time of this study, a considerable number of candidate genes have been found, including polymorphisms in the genes that code for energy metabolism, stress response, apoptosis, synaptic processes, and receptors for neurotransmitters.

Some studied environmental factors and their inferences are discussed in the following.

12.3.1.1 Exercise

Running has been shown to impede the movement of the engine and mental aggregates in R6/1 mouse models. Deliberate practices in R6/2 reduced the movement of circadian mood crashes. Increased exercise has beneficial effects on the cardiovascular, endocrine, metabolic, and other systems. Additionally, it has long been known that exercise increases adult hippocampus neurogenesis and brain-derived neurotrophic factor (BDNF). However, persistent running was unable to reverse the decline in adult hippocampus neurogenesis in transgenic mice. HD mice run less than wild-type mice, and motor dysfunction (developed early in R6/2) is an unavoidable symptom. Chronic voluntary wheel running exacerbated the rat R6/1 BDNF gene expression in the striatum and hippocampus while maintaining protein level abnormalities in the BDNF. However, aggregated HD could not be changed by exercise in R6/1 models.

12.3.1.2 Stress

Stress is a situation in which homeostasis is in danger due to internal or external influences. Prolonged stress can exacerbate the aggressive pathophysiology of neurodegenerative illnesses. In HD mouse models, this disease also seems to be susceptible to the damaging effects of stress. Studies on R6/1 animals showed that wild-types are immune to these alterations, whereas acute stress influences HD development. In particular, compared to male R6/1 mice, R6/1 female rats exhibited depressive-like behaviour and subsequent persistent elevations in stress hormones during 10 min of forced swimming. The main stress hormone released upon activation of the hypothalamic-pituitary-adrenal (HPA) axis is cortisol. Due to excessive stress, HD patients were shown to have an aberrant cortisol diurnal rhythm. Midnight cortisol levels also displayed an abnormal increase.

12.3.1.3 Caloric Restriction

Alternate-day fasting in HD rats showed delayed motor function decline, increased brain nutrient levels, and decreased HTT impurity levels. Research indicates that there is a way to reduce oxidative stress through intermittent fasting. It can also increase trophic factors, induce neurogenesis, and help in gene expression coding for the plasticity and pro-survival proteins found in the striatum, hippocampus, and cortex. These are pathological properties in HD, and the benefits mainly appear in the brain section. However, in the case of the loss of the body, considerable and unnatural weight gain produces highly severe symptoms, so it does not appear that fasting is the key beneficial factor.

12.3.1.4 Antioxidants

Coenzyme Q10, polyphenols, and flavonoids reduce neurodegenerative diseases. They also reduce oxidative damage. In the R6/2 mice model, it was seen that coenzyme Q10 can increase the survival rate, help in motor function, and reduce inclusions and atrophy (Gonzalez-Alegre, 2019). However, this capacity is based on the dose of the coenzyme. Encouraging results have also been seen in human models.

12.3.1.5 Essential Fatty Acids

Moreover, essential fatty acids have been shown to improve cognitive responses in animal models, to boost the BDNF, and to lower oxidative stress. Essential fatty acid delivery to HD mice boosted their survival and exacerbated their motor impairments, while having no effect on striatal atrophy or body weight. Unsaturated fatty acid supplementation in humans also enhances motor functioning. However, a different clinical trial with a pure synthetic fatty acid supplement (Miraxion) demonstrated no improvements in the motor or cognitive function in HD patients in the early stages.

12.4 MANIFESTATION OF HUNTINGTON'S PATHOGENICITY

HTT is a large protein that is anticipated to be mostly made up of 50 amino acid repetitions known as HEAT repeats. HEAT repeats mostly comprise a helical hairpin with two antiparallel helices. With the presence of a hydrophobic core, these features all come together to form a superhelical-shaped structure (Costa and Scorrano, 2012). HTT has many interactive partners, especially at its N terminus. This demonstrates that it functions as a framework for the synchronization of other protein complexes. HTT is also subject to significant changes after translation (Li, Orr, and Li, 2010). HTT's biological functions are still unknown. Poly Q stretches in the mutant protein begin at residue 18, causing conformational alterations and triggering pathogenic cascades (Ross and Tabrizi, 2011). The protein is primarily found in cytoplasm.

12.5 PROGNOSIS

Because of the link with the date of motor onset, pre-diagnosis progression has been described by a combination of CAG repeats with age (Byrne et al., 2018). It has already been recognized that inherent and extrinsic factors influence the age at which symptoms manifest and accumulate. However, there is a need to establish ways to depict the characteristics that predispose mild symptomatology. The duration of disease was selected as a criterion. The availability of young patients made it easy to determine these factors (Zuccato et al., 2011). For researchers, the accurate clinical information was available such as birth rank, sex, presenting neurological signs, dementia, AO, and age of death. These were characterized both qualitatively and quantitatively. As a result, the duration of the disease between the AO and death was discovered to be irregularly distributed. Further regression analyses were performed to determine how the AO and disease duration of affected parents contribute to the variation in the disease duration of patients. None of the variables exert an effect at a 10% significant level, and together, they contributed only 5% (An et al., 2012).

Birth order and sex of the affected parents did not contribute to a significant 10% level of difference in the duration of symptoms, although sex, the manifestation of neurological indications, and the presence of dementia did contribute. Women have worse prognosis than men, living an average of 12.1 years compared with 14.4 years for men. Dementia was the largest component and accounted for 0.8% of the total. For sex, the level is 8.5%, and for neurological signs, the level is

TABLE 12.1

Characteristics of the Disease Based on Patient Gender (Byrne et al., 2018).

Cross-tabulation of the patient frequencies for sex, presenting sign, duration of disease, and time of dementia

Sign	**Sex**	**Short duration**			**Long duration**			**Short and long duration**		
		Early dementia	Late dementia	Total	Early dementia	Late dementia	Total	Early dementia	Late dementia	Total
Rigidity	Male	7	3	10	1	3	4	8	6	14
	Female	6	2	8	3	2	5	9	4	13
Chorea	Male	3	5	8	3	24	27	6	29	35
	Female	8	10	18	4	13	17	12	23	35
Total		24	20	44	11	42	53	35	62	97

7.5%. Patients with an onset of dementia died earlier (9.4 years with 95% confidence intervals of 13.5 and 18.4 years) than patients with a late development of dementia (15.9 years with confidence intervals of 7.4 and 11.7 years). Women with praecox dementia die an average of 10.5 years after onset compared with men with late-onset dementia. The results demonstrate that praecox dementia and possibly rigidity leave side effects (Paulsen, 2010).

Table 12.1 displays the distribution of patients with short (1–12 years) and long (13–55 years) dementia based on dementia onset, presenting symptoms, and sex. Out of 97 patients, 44 passed away within 12 years of the onset of their symptoms, and 35 more displayed dementia symptoms up to two years later.

12.6 SOME SYMPTOMS USED TO CHARACTERIZE THE DISEASE

12.6.1 Motor Symptoms

Motor symptoms comprise compulsory development seen transcendently in early onset to direct, common adult beginning stages and impeded development in the event of the adolescent beginning stages. The TRACK-HD study concentrated on tracked down contrasts in speeded tapping more than three years among control and premanifest patients (10.8 years over from the sickness). Chorea is one more striking component of this sickness. Principally, compulsory unnecessary developments are fleeting. The example of development differs. At first, little height developments of distal limits and faces are noted. Then, the disease spreads slowly to the proximal parts of the body and enhances its amplitude in these parts. It also dramatically changes muscle function, as tongue protrusion, head-turning, and eye closure occur because of facial and neck muscle activation, and in the case of axial muscles, it causes back portion defection. Dystonia is also a common feature that is the main reason of sustained muscle contraction and increased muscle tones leading to torticollis and opisthotonos. With the progression of the disease, hyperkinetic movements decrease, and bradykinesia and akinesia increase more.

12.6.2 Cognitive Features

Cognitive features are universal in HD but vary among patients. Impairments may appear years before onset and range from subtle deficits to dementia in a later stage. HD affects daily functioning independently of motor impairment. In following a group of 800 patients, the PREDICT-HD study found that certain cognitive areas were affected, including verbal learning, attention, information integration, motor speed, and more. The TRACK-HD study showed a decrease in visual attention, psychomotor speed, visual integration, and spatial awareness. There is some overlap with psychiatric symptoms of non-inhibitory and impulsive behaviours.

12.6.3 Psychiatric Features

The psychiatric features are the most distressing symptoms of HD. Among 2,835 patients, the prevalence of depression was found to be 40%. Unresponsiveness is an additional typical and incapacitating element that relates to the illness stage. The TRACK-HD study demonstrated the way that expanded disregard can be identified over three years even in premanifest patients. Other neuropsychiatric side effects are crabbiness and animosity, which are connected to cerebrum brokenness. Hypersexuality is seen in early HD, whereas hyposexuality and psychosis (less common) are seen in late HD.

12.6.4 Peripheral Symptoms

Several effects are seen on the nervous system. Severe weight loss may also lead to cachexia. Plasma samples of HD patients show an increased interleukin 6 (IL6) and IL8, which may also contribute to the peripheral phenotypes.

12.7 CLINICAL DIAGNOSIS

Diagnosis depends on the clinical findings, including obvious motor signs and a positive pedigree history suggesting an autosomal-dominant mode of inheritance (Rawlins et al., 2016). To determine the CAG repeat length in the HD gene, genetic testing confirms the diagnosis (Paulsen et al., 2014). For patients with chorea but no positive family history or other symptoms suggestive of HD, the differential diagnosis, medical, and systemic causes should be excluded. The C9orf72 hexanucleotide repeat expansion mutation has been identified as a major cause of sporadic and familial frontal lobe degeneration and amyotrophic lateral sclerosis. This has also been studied as the most common cause of HD phenocopies. In cases of suspected HD phenocopy, it is indicated that a genetic testing algorithm based on mutation frequency and the presence of other neurological findings should be followed. In most cases, the genetic diagnosis remains unknown.

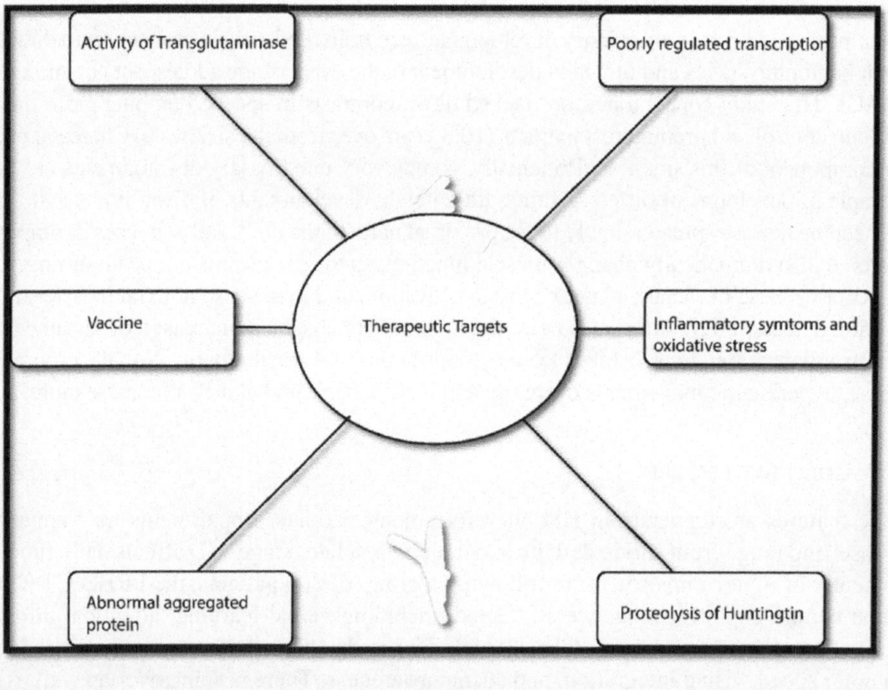

FIGURE 12.2 Therapeutic targets in the diagnosis of HD.

12.8 GENETIC TESTING

A simple blood test that identifies the CAG repeat length of each allele of the HD gene is all that is needed for readily available genetic testing (Novak and Tabrizi, 2011). However, the concerns surrounding genetic testing are very complex. It may be necessary to consult a clinical geneticist.

12.8.1 Diagnostic Testing

Diagnostic testing solely refers to the test performed on persons who exhibit motor symptoms consistent with HD; the test is often recommended to confirm a probable diagnosis. Patients with a positive pedigree but with cognitive or psychiatric symptoms associated with the absence of motor signs should be seen by a clinical geneticist for predictive rather than diagnostic testing (Paulsen, 2010). Individuals should be informed that if they have the mutation, their child has a 50% chance of getting it. Carrier siblings have a 50% chance of inheriting the mutation and may already exhibit symptoms of HD (Ross et al., 2014). Test results must be personally disclosed.

12.8.2 Predictive Analysis

Predictive testing is recommended for asymptomatic people at risk of inheriting the mutation based on a positive family history of HD. According to international standards, a predictive test for HD must be performed at a specialized centre, with at least one counselling session before the test, followed by a period of contemplation and a second counselling session (Sánchez-Castañeda et al., 2013). Written consent must be obtained for testing, and absolute confidentiality must be maintained. In addition, it should be noted that a positive test result indicates that a person will develop HD in the future. It is impossible to predict the AO and rate of progression. Individuals with intermediate repeat length alleles (27–35) should be monitored.

12.8.3 Antenatal Testing

Antenatal testing is another possible route. Chorion between 12 and 13 weeks of gestation are used to test fetal cells for HD gene mutations if the parents are confident that they would terminate the pregnancy if the fetus has the mutation during pregnancy. Villus sampling can also be performed (Scahill, Wild, and Tabrizi, 2012). However, before choosing this course of action, parents should undergo extensive counselling. If the parents decide against termination and continue the pregnancy despite a positive genetic test result, then the child has effectively received a positive predictive test result that was never consented to. Other options that patients can consider include the use of sperm or egg donors or adoption.

12.9 DRUG TREATMENTS

Chorea may not need treatment. Medicine can be considered if the chorea is causing only practical weakness. A review for the suggestive treatment of HD inspected and presumed that as a rule, only tetrabenazine showed clear viability for the control of chorea. Tetrabenazine, which is a dopamine-draining specialist, assists with diminishing choreic developments, but since it can set off sadness, a mental history should be taken before it is endorsed as a medicine (Sánchez-López et al., 2012). Tetrabenazine should not be taken with paroxetine or fluoxetine because these drugs are inhibited by the enzyme cytochrome P450 2D6, which is necessary for metabolism and flexibility. Olanzapine, risperidone, and quetiapine are a few examples of atypical neuroleptics that can be used to treat HD. These are especially useful if patients suffer from mental comorbidities including agitation, grumpiness, or disruption. Haloperidol and sulpiride are like more-seasoned neuroleptic drugs and can also be utilized in HD treatment; however, they have different kinds of hazardous aftereffects and furthermore, upgrade the gamble of this dangerous neuroleptic disorder. The benzodiazepine clonazepam can be used to treat chorea associated with dystonia, inflexibility, or myoclonus. Anticonvulsants, such as sodium valproate and levetiracetam, can be used if myoclonus is a serious side effect. Tizanidine and baclofen help reduce spasticity and flexibility, which are exacerbated in the later stages. Botulinum toxin injections are very effective in treating muscle spasms. Levodopa precursors may help in the development of HD, where akinesia and inflexibility are major problems.

12.10 NONDRUG TREATMENTS

Physiotherapy plays a significant role in helping HD patients improve their gait and balance. Walking assistance might eventually be needed. In the face of advancing dysarthria, communication can be improved by speech and language therapists. In addition, they can be used to help with swallowing difficulties. In instances of cutting-edge HD patients, because of troubles in both swallowing and the upgraded catabolic state, enteral nourishment conveyed through gastrostomy might be required to maintain sufficient sustenance (Sánchez-López et al., 2012). Dieticians can then set up a recommended diet including the GI plot and offer broader guidance on calorie supplementation. Nearby caregivers, support groups and clinicians are fundamental for providing mental and psychologic care nearer to the patient (Rawlins et al., 2016).

12.11 BIOMARKERS AS A DETECTOR

Huntington's chorea is a single genetically driven disease, and informative premanifest predictive genetic testing is the only way to identify it and help patients. Imaging indices of HD progression are therefore being developed for both the manifest disease and the prodromal or premanifest stages. Striatal atrophy is the most studied, with progression beginning in the prodromal phase and continuing through the overt disease stages. Moreover, cross-sectional data from extensive studies, such as the multicentre His PREDICT-HD project and the multicentre European His TRACK-HD study, show that striatum predicted the onset of HD 13–15 years before symptoms manifest. This indicates that body atrophy may begin. In addition, striatal mass is associated with the predicted age of motor onset. Striatal atrophy predicted onset at least 15 years earlier. Interestingly, early cross-sectional and longitudinal data suggest that prodromal white matter damage may also be a highly sensitive part of the disease dynamics. Other dim matter districts (including cortical dim matter) do not have a similar level of volumetric misfortune. Another significant and possibly helpful imaging proportion of Huntington's infection movement is dissemination tensor imaging, which shows white matter decay. In preliminary studies, clinical outcome measures are ordinarily utilized along with cerebrum imaging, which ends up being a heartier biomarker in the premanifest timeframe. Attractive reverberation spectroscopy, particularly with current high-strength field magnets, can likewise provide a profound chance to recognize and follow the original compound biomarkers. This will especially want to have fringe biomarkers of HD movement. Examinations reveal increased resistance to neurotoxic pressure in HD lymphoblasts, although it is unclear the extent to which this can be applied to specific patients. The oxidative physical issue may be shown by the oxidative damage marker (8-hydroxydeoxyguanosine, or 8-OHdG), which is further investigated in a larger study. From bodily fluids, the wet biomarker is isolated, which can measure the source because it reflects the pathobiology of HD. Various mechanisms have been implicated in HD pathology yielding numerous potential molecular biomarkers.

12.12 CURRENT RESEARCH

12.12.1 Cell Models

Though useful for biochemical research, cell lines may not adequately explain the cell biology of neurons (Munoz-Sanjuan, and Bates, 2011). They can be applied to inducible, transient, or stable expression strategies. Many characteristics of neurons in vivo are recapitulated by primary neurons. Mixed cultures are very important in these cases to reproduce and maintain cell interactions (Chang et al., 2015). At present, induced pluripotent undeveloped cells from patients with HD can be used for the investigation of illness pathogenesis and for remedial screening. The point is to research impacted people who have noticeable freak HTT-related aggregates such as harmful impacts or changes in cell digestion. (A model is with the NINDS-subsidized HD infection-prompted pluripotent undifferentiated cell.)

12.12.2 Invertebrate Models (e.g., Caenorhabditiselegans)

Models of invertebrates' disease can show advancing behavioural, physiological, and neuronal degeneration changes. They make it easier to study invertebrate molecular pathogenesis and provide an opportunity to try different treatments at random (Stricker-Shaver et al., 2018). It is unknown, however, how closely their biology resembles that of mammalian HD.

12.12.3 Mouse Models

Clearly understanding disease biology and then accordingly implementing different therapeutic strategies are major issues in the case of HD. Therefore, biological and mouse models of HD have been applied to studies to learn about drugs and their responses. Nuclear reflexes were first discovered in the R6/2 mouse model after they were evident in the postmortem human brain. However, no mouse model exhibits the equivalent giant striatal neuron degeneration seen in humans (Eaton and Wishart, 2017). Various mouse models exhibit N-terminal fragments of the HTT, such as the R6/2 mouse model, which has exon-1 or 90 amino acids in its N-terminal region; another model, the N171–82Q model, has shown that the 171 amino acids are in the N-terminals of its HTT, similar to the N586–82Q model that has additionally shown that the caspase 6 part or 586 amino acids are in its (McBride, Perentos, and Morton, 2016) N-terminal. It appears to have the most irregular and rapidly predominant aggregates, including dyscoordination and mental and other behavioural and physiological abnormalities when suspended by the tail and a significant loss of body weight, leading to early departure and later serving its purpose in preliminary cases (McBride, Perentos, and Morton, 2016). Most of the mice cases show the full-length of the HTT, which normally happens with more conspicuous aggregates than those referenced previously. However, there may be a more specific neurodegeneration in mice. Models consolidating the whole HTT quality that utilise transgenic addition by YACs have been used in investigations of pathogenesis (McBride, Perentos, and Morton, 2016). BAC, YAC, and knock-in models are desperately needed in investigations of the cleavage of full-length HTT and studies of the stages prior to overt behavioural and clinical traits (Morton and Howland, 2013). However, the behavioural phenotypes and pathological phenotypes develop in a very slow manner, and these studies require a stable source of resources. Behavioural tests are very important to standardize accurate yield comparability in the experiment (Tan et al., 2010). The progression of HD models, however small, can be tracked and measured by high-field micro-MRI scans. Preclinical therapeutic trials in human HD patients can be improved with the help of MRI research in mice.

12.12.4 Large Mammalian Models

In behavioural studies and in determining if gene therapy medicines such as viral expression vectors or antisense nucleotides can permeate the entire brain, including subcortical white matter, subcortical grey matter nuclei, and the cortex, models of pigs, lambs, or monkeys may have a substantial advantage (Lin et al., 2011).

12.13 FUTURE PERSPECTIVES OF HD

HD has a deep symptomatic impact on the physiological and psychological health of a patient's body. It is also a socially devastating disorder. A considerable number of studies from recent decades have enormously improved the caring procedure for patients and helped in knowledge-gathering about the disease. For this disease, the main illness is observed at around age 17 years (most cases), which indicates that prior to the onset year is important for treatment; this is known as the premanifest period (Schapira et al., 2014). Since 1983, when the gene was discovered, various procedures, treatments, and therapeutic strategies have been developed. What is the future perspective of this disorder? The research has mainly focused on biomarker-based treatments and pathophysiological pathways. For pathophysiological treatments, a drug needs to be developed for the disease to administer treatment pathologically (Zheng et al., 2018). Drugs that can reverse or stop the beginning of the illness are being examined. Another significant issue is the quest for dependable, timely treatment to distinguish the clinically important biomarkers for the beginning of the end course of the illness. In the quest for biomarkers, a very notable coordinated study is TRACK-HD (Lee et al., 2011). The key thing that this condition needs is broad research. Therefore, after medicine was produced for the human body, the majority of research, including that conducted by the European Huntington's Disease Network (EHDN), has concentrated on planning their main project on a broad scale (Kumar et al., 2020). In addition to these studies, there is also research into various rational ways to treat the disease, with a focus on providing the best possible care for all patients and severe cases. The developments are promising, but the road to solving the problem is long and arduous.

12.14 CONCLUSION

HD is a debilitating and degenerative condition. At both the cellular and macroscopic levels, we have improved over the past ten years in our understanding of the etiology and natural history of HD. In the present scenario, many clinical trials have failed, and few treatments (which are somewhat effective) are available. However, various studies are ongoing, and different therapeutic mechanisms or strategies are being discovered to target the mutated mutant huntingtin protein (mHTT) or its functions. Now more than ever, there is true potential to change and stop HD.

REFERENCES

An, Mahru C., Ningzhe Zhang, Gary Scott, Daniel Montoro, Tobias Wittkop, Sean Mooney, Simon Melov, et al. "Genetic correction of Huntington's disease phenotypes in induced pluripotent stem cells." *Cell Stem Cell* 11, no. 2 (2012): 253–263.

Bano, D., F. Zanetti, Y. Mende, and P. Nicotera. "Neurodegenerative processes in Huntington's disease." *Cell Death & Disease* 2, no. 11 (2011): e228–e228.

Bates, Gillian P., Ray Dorsey, James F. Gusella, Michael R. Hayden, Chris Kay, Blair R. Leavitt, Martha Nance et al. "Huntington disease." *Nature Reviews Disease Primers* 1, no. 1 (2015): 1–21.

Brás, InêsCaldeira, AnnekatrinKönig, and Tiago Fleming Outeiro. "Glycation in Huntington's disease: A possible modifier and target for intervention." *Journal of Huntington's Disease* 8, no. 3 (2019): 245–256.

Byrne, Lauren M., Filipe B. Rodrigues, Eileanor B. Johnson, Peter A. Wijeratne, Enrico De Vita, Daniel C. Alexander, Giuseppe Palermo et al. "Evaluation of mutant huntingtin and neurofilament proteins as potential markers in Huntington's disease." *Science Translational Medicine* 10, no. 458 (2018): eaat7108.

Chang, Renbao, Xudong Liu, Shihua Li, and Xiao-Jiang Li. "Transgenic animal models for study of the pathogenesis of Huntington's disease and therapy." *Drug Design, Development and Therapy* 9 (2015): 2179.

Costa, Veronica, and Luca Scorrano. "Shaping the role of mitochondria in the pathogenesis of Huntington's disease." *The EMBO Journal* 31, no. 8 (2012): 1853–1864.

Eaton, S. L., and T. M. Wishart. "Bridging the gap: Large animal models in neurodegenerative research." *Mammalian Genome* 28, no. 7 (2017): 324–337.

Gonzalez-Alegre, Pedro. "Recent advances in molecular therapies for neurological disease: Triplet repeat disorders." *Human Molecular Genetics* 28, no. R1 (2019): R80–R87.

Krzyzosiak, Wlodzimierz J., Krzysztof Sobczak, MarzenaWojciechowska, AgnieszkaFiszer, AgnieszkaMykowska, and Piotr Kozlowski. "Triplet repeat RNA structure and its role as pathogenic agent and therapeutic target." *Nucleic Acids Research* 40, no. 1 (2012): 11–26.

Kumar, Ashok, Vijay Kumar, Kritanjali Singh, Sukesh Kumar, You-Sam Kim, Yun-Mi Lee, and Jong-Joo Kim. "Therapeutic advances for Huntington's disease." *Brain Sciences* 10, no. 1 (2020): 43.

Langbehn, Douglas R., Michael R. Hayden, Jane S. Paulsen, and PREDICT-HD Investigators of the Huntington Study Group. "CAG-repeat length and the age of onset in Huntington disease (HD): A review and validation study of statistical approaches." *American Journal of Medical Genetics Part B: Neuropsychiatric Genetics* 153, no. 2 (2010): 397–408.

Lee, Seung-Jae, Hee-Sun Lim, Eliezer Masliah, and He-Jin Lee. "Protein aggregate spreading in neurodegenerative diseases: Problems and perspectives." *Neuroscience Research* 70, no. 4 (2011): 339–348.

Li, Xiao-Jiang, Adam L. Orr, and Shihua Li. "Impaired mitochondrial trafficking in Huntington's disease." *BiochimicaetBiophysicaActa (BBA)-Molecular Basis of Disease* 1802, no. 1 (2010): 62–65.

Lin, Yow-Sien, Chiung-Mei Chen, Bing-wen Soong, Yih-Ru Wu, Hui-Mei Chen, Wen-Ying Yeh, Dai-Rong Wu et al. "Dysregulated brain creatine kinase is associated with hearing impairment in mouse models of Huntington disease." *The Journal of Clinical Investigation* 121, no. 4 (2011): 1519–1523.

Matlahov, Irina, and Patrick CA van der Wel. "Conformational studies of pathogenic expanded polyglutamine protein deposits from Huntington's disease." *Experimental Biology and Medicine* 244, no. 17 (2019): 1584–1595.

McBride, Sebastian D., Nicholas Perentos, and A. Jennifer Morton. "A mobile, high-throughput semi-automated system for testing cognition in large non-primate animal models of Huntington disease." *Journal of Neuroscience Methods* 265 (2016): 25–33.

McColgan, Peter, and Sarah J. Tabrizi. "Huntington's disease: A clinical review." *European Journal of Neurology* 25, no. 1 (2018): 24–34.

Møllersen, Linda, Alexander D. Rowe, Elisabeth Larsen, TorbjørnRognes, and Arne Klungland. "Continuous and periodic expansion of CAG repeats in Huntington's disease R6/1 mice." *PLoS Genetics* 6, no. 12 (2010): e1001242.

Morton, A. Jennifer, and David S. Howland. "Large genetic animal models of Huntington's disease." *Journal of Huntington's Disease* 2, no. 1 (2013): 3–19.

Munoz-Sanjuan, Ignacio, and Gillian P. Bates. "The importance of integrating basic and clinical research toward the development of new therapies for Huntington disease." *The Journal of Clinical Investigation* 121, no. 2 (2011): 476–483.

Novak, Marianne JU, and Sarah J. Tabrizi. "Huntington's disease: Clinical presentation and treatment." *International Review of Neurobiology* 98 (2011): 297–323.

Paul, Bindu D., and Solomon H. Snyder. "Impaired redox signaling in Huntington's disease: Therapeutic implications." *Frontiers in Molecular Neuroscience* 12 (2019): 68.

Paulsen, Jane S. "Early detection of Huntington's disease." *Future Neurology* 5, no. 1 (2010): 85–104.

Paulsen, Jane S., Jeffrey D. Long, Christopher A. Ross, Deborah L. Harrington, Cheryl J. Erwin, Janet K. Williams, Holly James Westervelt et al. "Prediction of manifest Huntington's disease with clinical and imaging measures: A prospective observational study." *The Lancet Neurology* 13, no. 12 (2014): 1193–1201.

Pringsheim, Tamara, Katie Wiltshire, Lundy Day, Jonathan Dykeman, Thomas Steeves, and Nathalie Jette. "The incidence and prevalence of Huntington's disease: A systematic review and meta-analysis." *Movement Disorders* 27, no. 9 (2010): 1083–1091.

Rawlins, Michael D., Nancy S. Wexler, Alice R. Wexler, Sarah J. Tabrizi, Ian Douglas, Stephen JW Evans, and Liam Smeeth. "The prevalence of Huntington's disease." *Neuroepidemiology* 46, no. 2 (2016): 144–153.

Reiner, Anton, IoannisDragatsis, and Paula Dietrich. "Genetics and neuropathology of Huntington's disease." *International Review of Neurobiology* 98 (2011): 325–372.

Romo, Lindsay, Ami Ashar-Patel, Edith Pfister, and Neil Aronin. "Alterations in mRNA 3′ UTR isoform abundance accompany gene expression changes in human Huntington's disease brains." *Cell Reports* 20, no. 13 (2017): 3057–3070.

Roos, Raymund AC. "Huntington's disease: A clinical review." *Orphanet Journal of Rare Diseases* 5, no. 1 (2010): 1–8.

Ross, Christopher A., Elizabeth H. Aylward, Edward J. Wild, Douglas R. Langbehn, Jeffrey D. Long, John H. Warner, Rachael I. Scahill et al. "Huntington disease: Natural history, biomarkers and prospects for therapeutics." *Nature Reviews Neurology* 10, no. 4 (2014): 204–216.

Ross, Christopher A., and Sarah J. Tabrizi. "Huntington's disease: From molecular pathogenesis to clinical treatment." *The Lancet Neurology* 10, no. 1 (2011): 83–98.

Sánchez-Castañeda, Cristina, Andrea Cherubini, Francesca Elifani, Patrice Peran, Sara Orobello, Giovanni Capelli, Umberto Sabatini, et al. "Seeking Huntington disease biomarkers by multimodal, cross-sectional basal ganglia imaging." *Human Brain Mapping* 34, no. 7 (2013): 1625–1635.

Sánchez-López, Fernando, InmaculadaTasset, Eduardo Agüera, Montserrat Feijóo, Ricardo Fernández-Bolaños, Francisco M. Sánchez, María C. Ruiz, et al. "Oxidative stress and inflammation biomarkers in the blood of patients with Huntington's disease." *Neurological Research* 34, no. 7 (2012): 721–724.

Scahill, Rachael I., Ed J. Wild, and Sarah J. Tabrizi. "Biomarkers for Huntington's disease: An update." *Expert Opinion on Medical Diagnostics* 6, no. 5 (2012): 371–375.

Schapira, Anthony HV, C. Warren Olanow, J. Timothy Greenamyre, and ErwanBezard. "Slowing of neurodegeneration in Parkinson's disease and Huntington's disease: Future therapeutic perspectives." *The Lancet* 384, no. 9942 (2014): 545–555.

Schultz, Jordan L., Amelia D. Moser, and Peg C. Nopoulos. "The association between CAG repeat length and age of onset of juvenile-onset Huntington's disease." *Brain Sciences* 10, no. 9 (2020): 575.

Stricker-Shaver, J., A. Novati, L. Yu-Taeger, and H. P. Nguyen. "Genetic rodent models of Huntington disease." *Polyglutamine Disorders* (2018): 29–57.

Takahashi, Masaki, Shoko Watanabe, Miho Murata, HirokazuFuruya, Ichiro Kanazawa, Keiji Wada, and HirohikoHohjoh. "Tailor-made RNAi knockdown against triplet repeat disease-causing alleles." *Proceedings of the National Academy of Sciences* 107, no. 50 (2010): 21731–21736.

Tan, Sonny KH, RinskeVlamings, LeeWei Lim, ThibaultSesia, Marcus LF Janssen, Harry WM Steinbusch, VeerleVisser-Vandewalle, et al. "Experimental deep brain stimulation in animal models." *Neurosurgery* 67, no. 4 (2010): 1073–1080.

Wanker, Erich E., Anne Ast, Franziska Schindler, Philipp Trepte, and Sigrid Schnoegl. "The pathobiology of perturbed mutant huntingtin protein–protein interactions in Huntington's disease." *Journal of Neurochemistry* 151, no. 4 (2019): 507–519.

Zhao, Ting, Yan Hong, Xiao-Jiang Li, and Shi-Hua Li. "Subcellular clearance and accumulation of Huntington disease protein: A mini-review." *Frontiers in Molecular Neuroscience* 9 (2016): 27.

Zheng, Ju, JorisWinderickx, Vanessa Franssens, and Beidong Liu. "A mitochondria-associated oxidative stress perspective on Huntington's disease." *Frontiers in Molecular Neuroscience* 11 (2018): 329.

Zuccato, Chiara, Manuela Marullo, Barbara Vitali, AlessiaTarditi, CaterinaMariotti, Marta Valenza, NayanaLahiri et al. "Brain-derived neurotrophic factor in patients with Huntington's disease." *PLoS One* 6, no. 8 (2011): e22966.

13 Clinical Manifestation and Molecular Orientation of Leishmaniasis

Pratik Talukder[1,*], *Sounak Chanda*[1] *and Suravi Datta*[1]
[1]Department of Biotechnology, University of Engineering and Management, Kolkata
*Correspondence to: pratik.talukder@uem.edu.in

13.1 INTRODUCTION

Neglected tropical diseases (NTDs) are a group of infectious illnesses that mostly affect the world's poor. They have been consistently ignored for decades, initially because of the developing world's overall tolerance, and more recently, as a result of the focus on HIV/AIDS, tuberculosis, and malaria. In low-income communities in Africa, Asia, and the Americas, NTDs rank among the most prevalent chronic infections (de Menezes et al., 2016).

A biologically varied set of diseases that are to blame includes protozoa carried by vectors (such as Trypanosoma brucei), bacteria (Chlamydia trachomatis ocular serovars), and filarial parasites (such as *Onchocerca volvulus*). Two types of non-tuberculosis mycobacteria cause leprosy and Buruli ulcer, for which the method of infection is not yet fully known, and soil-transmitted helminths (STHs).

13.1.1 Leishmaniasis

Leishmaniases are a diverse set of NTDs with endemic distribution that affect people in 98 countries in the tropics, subtropics, and Mediterranean basin. All of these diseases are caused by an estimated 20 species of the protozoan parasite Leishmania.

Self-healing and widespread cutaneous leishmaniasis (CL), mutilating mucosal/mucocutaneous leishmaniasis (ML), and fatal visceral leishmaniasis (VL) are among the disease's forms.

Skin ulcers are the primary symptom of the cutaneous form, while mouth, nose, and skin ulcers are the primary symptom of the mucocutaneous form (Chappuis et al., 2007). Skin ulcers and fever of the visceral form are followed by enlargement of the spleen and liver, a relatively low red blood cell level, and skin ulcers.

Leishmaniasis is essentially a set of parasitic obligatory protistan diseases that are transmitted through the biting of phlebotomine sandflies, specifically, Phlebotomus and Lutzomyia.

13.1.2 Source

There are over 1.3 million cases annually that cause 20,000 to 30,000 deaths. It is spread through the sting of female sandflies; over 90 species of sandflies are known to transmit the Leishmaniasis parasite.

Sandflies become susceptible to infection when they bite a Leishmania-infected person or animal. A wide variety of animals can be infected with Leishmania including rodents, dogs, and foxes. A few uncommon cases of transmissions include blood transfusions, used needles or syringes, or from a mother to the fetus. Not everyone infected with the Leishmania parasite develops the disease, and some people can also be infected but be asymptomatic (Strazzulla et al., 2013).

DOI: 10.1201/9781003302131-14

Kala Azar, also known as VL, is categorized into two groups. One is prompted by the parasitic organism *Leishmania infantum* and impacts both dogs and humans. The other is caused by the pathogen *Leishmania donovani*, which impacts Asia and Africa and occurs in Southern Europe, the Middle East, and Brazil. Both of these parasites are transmitted in the same way, that is, by using a vector (intermediary) in the form of a ting blood-sucking fly barely a few millimeters long, namely, the female *Phlebotomus* sp.

13.2 MODE OF MECHANISM OF THE PATHOGEN

After the sandfly has transmitted the parasite to a new victim (for example, a human), inside the human body, there are immune cells (macrophages) whose job it is to destroy pathogens and swallow the parasites. If the immune system succeeds in fighting the infection, then the victim is immunized against any future attacks, but if it fails, the victim falls seriously ill very rapidly.

Inside the macrophages, the parasite changes form, multiplies and invades all the organs with a distinct preference for the spleen and liver.

Fever, anemia, and an enlarged spleen and liver are the initial indicators of the disease, and if left untreated, the patient will die. In Asia, Leishmaniasis is nick-named black fever as the skin often darkens. Leishmaniasis weakens the immune system, leading patients vulnerable to so-called opportunistic diseases such as tuberculosis, malaria, or dysentery (Aliaga et al., 2003). Leishmaniasis itself is an opportunistic disease. People living with HIV are more susceptible to the parasite growing rapidly in their bodies, and when a Phlebotomus bites someone who is co-infected, it ingests a higher parasite mood that it will transmit when it bites again; this makes co-infected patients an important reservoir for the transmission of Kala Azar.

13.3 TYPES OF LEISHMANIASIS

13.3.1 Cutaneous Leishmaniasis

CL is the type of leishmaniasis that is the least harmful. *Leishmania tropica*, *Leishmania major*, and *Leishmania mexicana* are the three organisms that cause it.

When a person infected with cutaneous Leishmania is examined, the following are noticed:

> Hyperkeratosis, focal pseudo – epitheliomatous hyperplasia in the epidermis, dermis becomes heavy, mixed inflammatory infiltrate with few epithelioid granuloma without caseation necrosis, multinucleated giant cell wall seen, numerous neutrophils, lymphocytes, plasma cells and histiocytes were seen, H&E and Giemsa section shows leishmania-like parasite within macrophages, parasite were small round structures with eccentrically located kinetoplast.
>
> (Castro et al., 2007)

The protozoan parasite Leishmania, which is carried by infected female sandflies, causes CL. Although CL is rarely rapidly fatal, it can be fatal to surrounding communities.

Symptoms include the following:

> Breathing difficulty, skin sores which in future may become skin ulcer with slow healing rate, stuffy nose, runny nose, and nosebleeds.
>
> The diagnosis is made by light microscopic analysis of stained slides using a molecular technique to identify Leishmania parasites (or DNA) in cultured cells, such as those from skin lesions or bone marrow. In almost all situations, the healing process takes between three and eighteen months. From two weeks to a few months pass during the incubation stage.
>
> The skin lesions may get smaller, but they typically get bigger and don't get better. They are often painless, only affect one area of the body, and might be wet and leak fluid or have a dried-up crust.
>
> (Goto and Lindoso, 2010)

13.3.2 Visceral Leishmaniasis

Tropical and subtropical regions are home to this species. Internal organs are generally affected, including the spleen, liver, and bone marrow. It primarily affects young people under the age of five years. It is linked to malnutrition and other immune-suppressive disorders such as HIV/AIDS. It can cause no symptoms in some people, but in others, it might cause fever, weight loss, and spleen or liver swelling.

The most common contributing factors of VL are *Leishmania donovanii* and *Leishmania infantum*. If the parasite is not treated, then it nearly always will cause the host to die because it spreads to inner organs such as the liver, spleen, and bone marrow. VL affects residents in impoverished rural parts of Bangladesh, Ethiopia, India, Nepal, South Sudan, Sudan, and Brazil; 200,000–400,000 cases have been documented (Guimarães et al., 2016).

Infected individuals may have abdominal pain, fever, decreased appetite, decreased blood counts, lethargy, night sweats, abnormal thinness, skin discoloration, diarrhea, and swollen lymph nodes.

The incubation period usually lasts between two and six months. Fatigue and weakness can be made worse by anemia, which is caused by an ongoing inflammatory condition, hypersplenism (the outer disintegration of red cells in the spleen), and bleeding. VL has a similar clinical appearance in different endemic locations, but there are notable variances. For instance, enlarged lymph nodes are more typical in Sudanese VL victims than in Indian VL patients. Another feature unique to Indian VL patients is hyperpigmentation of the skin, which is why it is also called "Kala Azar" in Hindi or black fever. As the illness worsens, splenomegaly can cause abdominal pain and cramping, which is occasionally made worse by coexisting hepatomegaly. The initial diagnosis may be skewed by the symptoms and signs of bacterial co-infections including pneumonia, diarrhea, or tuberculosis. Before seeking medical attention or passing away from bacterial co-infections, heavy bleeding, or severe anemia, patients with VL generally exhibit symptoms for weeks or months (Camuset et al., 2007).

The people who should be treated must be identified via confirmatory testing because the exact pathophysiology of VL lacks specificity. Because VL is a deadly illness, tests must be not only extremely sensitive (>95%) but also extremely specific (>95%) because the present medications used to treat it are toxic. A test that can discriminate between acute illness and silent infection is ideal because none of the currently available drugs are safe enough to just treat symptomless infections. Moreover, such examinations should be straightforward and inexpensive.

13.3.3 Mucosal Leishmaniasis

A metastatic variant of New World cutaneous illness, ML, also known as espundia, is brought on by a parasite spreading from the epidermal to the naso-oropharyngeal membranes. The species of the *Viannia subgenus* are to blame. Depending on the type of pathogens and the complexity of the pathogenesis, the clinical manifestations of ML might significantly change (Seed, 1996).

Traditional mucosal lesions that develop during American tegumentary leishmaniasis (ATL) are extremely dangerous, painfully disfiguring, and even lethal. Most commonly, ulcerous lesions cause septum perforation. In 5–20% of patients, cutaneous lesions come before ML. Before mucosal invasion, these lesions may clinically present or be treated days to decades in advance. Risk factors for the advancement of ML include wrong diagnosis and mistreatment and a significant number of skin lesions. However, isolated or solitary ML caused by the complex species of *L. braziliensis* has been seen in subjects with frequent mucosal involvement (17–18%). The most common initial sites of ML lesions are the upper respiratory system, nose, and oral cavity, with pharynx and larynx involvement occurring later as the disease progresses. As the disease spreads to the lower respiratory tract and the soft tissues of the nose, mouth, and throat degrade, ML's ambiguous inflammatory symptoms – such as nasal congestion, erythema, edoema, serous rhinorrhea, and epistaxis – get worse over time. ML was described as having ocular involvement and presenting with exophthalmos in a case report. Due to incomplete clinical information, a number of contagious diseases, including leprosy, schistosomiasis, blastomycosis, and mycobacteria infections, can enter the differential diagnosis (Waki et al., 2007)

Based on the definition of Leishmania amastigotes inside the mucosal lesions, ML is diagnosed. The most common technique to see Leishmania bodies is through histological diagnosis, which results from the appearance of amastigotes in mucosal tissues stained with Giemsa or hematoxylin-eosin. However, this seems to work better in Old World ML than in New Frontier ML. In fact, ML histology in *L. infantum* has a high specificity (varying from 50–70% to about 100%), whereas *in L. braziliensis* ML, it drops to 35–70%. Both infections have a high level of specificity (>95%). Because *L. infantum* cells are more common in mucosal lesions than *L. braziliensis* cells, which are somewhat scarce, there is a difference in sensitivity. The liquid cultures from Novy-McNeal-Nicolle (NNN) and Schneider verified the presence of Leishmania promastigotes in the medium.

13.4 EPIDEMIOLOGY OF THE PATHOGEN

Leishmania is a protozoan pathogen. Protozoan creatures may spend their whole lives in human hosts, although many spend a part of their breeding cycle in insects and other hosts. Protozoal infection causes tissue damage, which results in disease. The main contributors to tissue damage in persistent infections are changes in the cytokine profiles and immune system responses to pathogen and/or host antigens. It could also be the result of harmful protozoal products or mechanical injury. The protozoa adhere themselves to the small intestine lining and prevent the host from adequately absorbing nutrients. They cause diarrhea, stomach pains, and fever. These symptoms have been linked to the protozoa's hyperproliferation, regardless of whether the parasites live intracellularly or extracellularly in their respective host cells (Adams et al., 2012).

The species once referred to as "protozoa" is only superficially similar and not closely related (eukaryotic, unicellular, and motile, although with exceptions). The term "protozoa" (also "protist") is often avoided in modern biosciences. However, this expression is still employed in medicine. This is a result of both the traditional nature of medical classification and the requirement to identify organisms based on morphology.

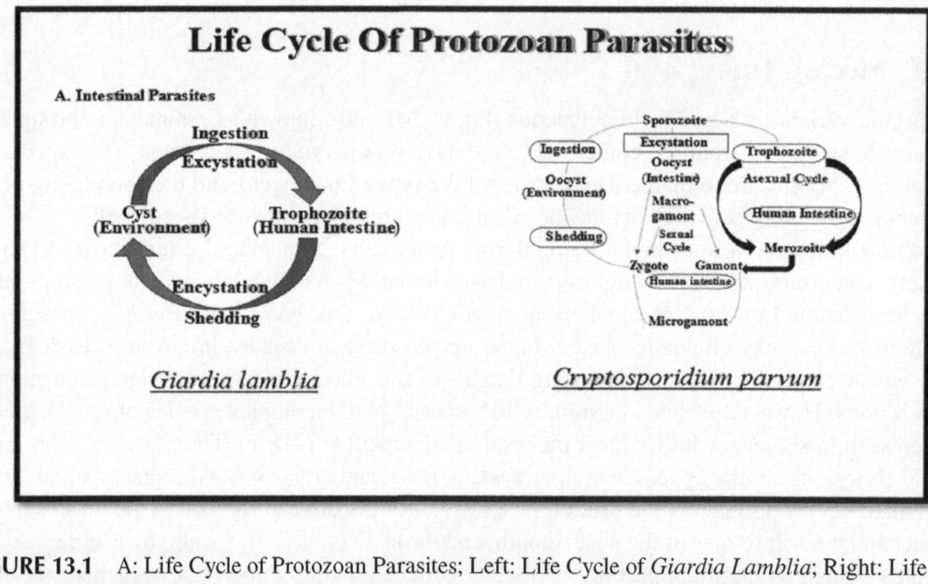

FIGURE 13.1 A: Life Cycle of Protozoan Parasites; Left: Life Cycle of *Giardia Lamblia*; Right: Life Cycle of *Cryptosporidium parvum*.

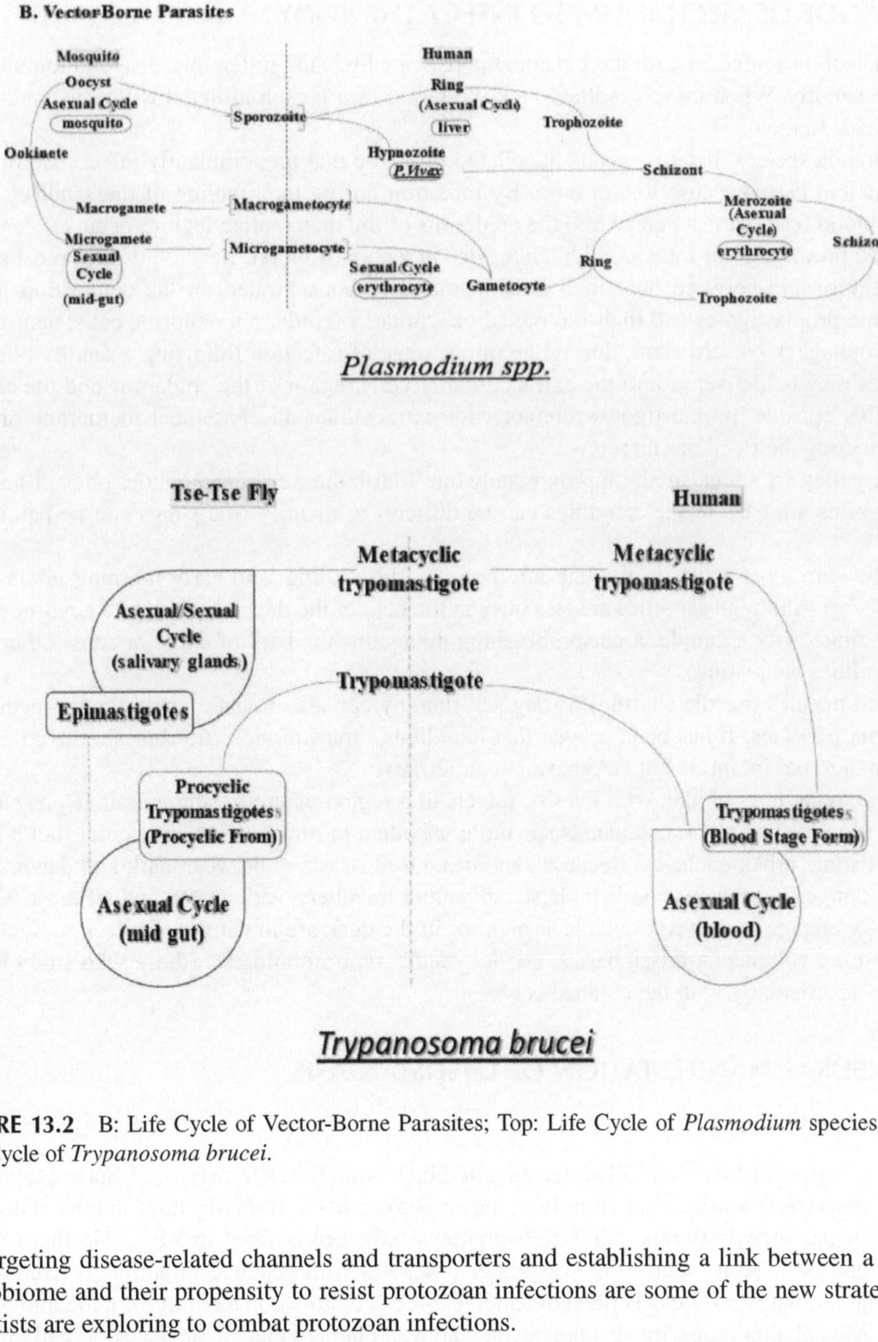

FIGURE 13.2 B: Life Cycle of Vector-Borne Parasites; Top: Life Cycle of *Plasmodium* species; Bottom: Life Cycle of *Trypanosoma brucei*.

Targeting disease-related channels and transporters and establishing a link between a person's microbiome and their propensity to resist protozoan infections are some of the new strategies that scientists are exploring to combat protozoan infections.

As chemoorganotrophic protists, protozoa have three different mechanisms for obtaining nutrition. Saprotrophic nutrition is the initial way to obtain nutrients. In saprotrophic feeding, nutrients are extracted from decomposing organic matter through enzymatic processes. The second approach for acquiring nutrients is osmotrophic feeding, where nutrients are gained by the absorption of soluble products (Akhavan et al., 2010). Holozoic feeding is the third way to obtain nutrients in which solid nutrients are taken through phagocytosis. Photoautotrophic protists include protozoa. Strict aerobes are among these protists, and they use photosystems I and II to carry out photosynthesis, which produces oxygen. Organic and inorganic carbon compounds are used by mixotrophic protists to obtain nutrition.

13.5 MODE OF MECHANISM TO INFECT THE BODY

A person becomes infected with the Leishmania parasite by being bitten by a diseased female phlebotomine sandfly. When these sandflies suck the blood of a person afflicted with Leishmania, the virus spreads to them.

Leishmania species, heterogeneous flagellated protozoa that predominantly infect macrophages in mammalian tissues, cause this disease. By infection during the nibbling of the sandfly vectors during a blood feed, parasite entry into the epidermis of the mammalian legions occurs.

Injected promastigotes interact with leukocytes in a pool of blood. Because of this, most studies on the relationship between hosts and Leishmania have concentrated on the connection linking Leishmania promastigotes and their biological benchmarks (epidermal dendritic cells, neutrophils, and macrophages). Nevertheless, during the initial stage of infection following a sandfly bite, promastigotes may be delivered into the extracellular (ECM) region of the epidermis and the circulation. In this episode, promastigotes interact with extracellular and basement membrane proteins before infecting their cellular targets.

Because they are silent, small (approximately one-fourth the size of mosquitoes or even smaller), and their bites are hard to see, sandflies can be difficult to identify (the stings can be painless or painful).

Sandflies are most active in the late afternoon, early evening, and early morning hours (from dusk to dawn). Although sandflies are less busy in the heat of the day, they may bite anytime if they become agitated (for example, a person brushing up against the bark of a tree or some other place where sandflies are resting).

Infected needles (needle sharing) or targeted therapy can also transfer some types (species) of Leishmania parasites. It has been proven that hereditable transmission (transmission from a pregnant woman to her infant) occurs (Akhavan et al., 2010)

Anyone, regardless of age, who lives or travels in a region where Leishmaniasis is present is at risk of getting the disease. Leishmaniasis is more prevalent in rural areas than in cities, but it is still present in some urban enclaves. Because sandflies are most active between dark and dawn, this is when the danger of transmission is highest. Adventure travellers, ecotourists, and all those who do research (or engage in activity) outside at night or in the dusk are in danger of infection, including Army Reserve volunteers, missionaries, service members, ornithologists (those who study birds), and others (particularly with the cutaneous type).

13.6 NEURAL MANIFESTATION OF LEISHMANIASIS

13.6.1 Case Reports

On gross examination, VL-infected individuals from East Africa typically have extensively dispersed parasites, particularly in the cerebro spinal fluid (CSF) flow and the endothelia of meningeal arteries. Paralytic symptoms have been linked to VL in Northern Africa and have been found to improve after treatment with pentavalent antimony. Neurologic abnormalities were described in patients infected with VL in a report from the mid-1960s. Mental alterations are also prevalent. Carswell claimed that "patients had a really unique mental melancholy and were utterly apathetic" although they may walk 20 kilometers to the dispensary. In a group of clinical cases from Ethiopia, "definite mental abnormalities" were described in three of 18 individuals.

A general perineural inflammatory cell infiltrate consisting of lymphocytes or a mixture of lymphocytes, plasma cells, and macrophages was identified in cutaneous lesion biopsy specimens in 14 impacted biopsies, according to a study seeking to explore affiliated distal nerve participation. This study described four cases of inflammatory cell infiltration of the nerve (neuritis), one of which included granulomatous irritation and nerve injury. Amastigotes were found within the perineural sheath of two people. Two patients with CL had reduced feeling in the location of their CL lesions, according to sensory tests of 50 patients with CL (Akhoundi et al., 2013)

Based on more recent reports over the past 20 years, neurological issues are widespread in patients with Leishmania infection. In a study of 111 individuals in Sudan, neurological abnormalities were reported in 46% of the patients, with foot drop and the feeling of burning feet being the most prevalent, along with less common to extremely rare deafness and cranial nerve abnormalities. After starting anti-Leishmania therapy, most patients' neurological symptoms faded or disappeared completely.

13.6.2 Peripheral Neural Involvement

The most common neurological manifestation of Leishmaniasis, independent of the host's immune condition, is peripheral neuropathy; central nervous system (CNS) signs are uncommon, regardless of the host's immune status, but have been recorded in immunocompromised persons, generally in association with widespread illness. CNS involvement is assumed to be caused by a contiguous infection that spreads to the CNS, usually through the nasal cavities; neurogenic impairment and meningitis have already been reported. Following the progression of conjunctival and sinus infection in the orbital apex, with invasion and erosion of nearby bone structures, optic nerve involvement has been documented.

In CL, peripheral neurological involvement has recently been discovered. Several tender hyperesthetic cutaneous nodules were shown by a male patient. Histopathological examination proved that the identification of CL was correct. The appearance of a remarkable lymph histiocytic inflammatory infiltrate around and inside cutaneous nerves and visible Leishmania organisms in the perineural region were of particular interest. Following this discovery, a complete histopathologic examination of Leishmanial skin lesions formed in 13 mice following subcutaneous injection of major Leishmania promastigotes was conducted. Nine animals' cutaneous lesions showed varying degrees of neurological involvement. Organisms were visible within the likely Schwann cells in certain lesions, in addition to inflammatory cells in the nerves (Cannet et al., 2016).

13.6.3 Evidence in Animal Models

The sandfly is the primary vector of VL transmission, with canines and humans as natural hosts. The main source of zoonotic VL in endemic regions is sick dogs, which also provide the greatest risk of human infection. There are more locations where people are infected because of the expanding range of canine VL transmitted via vectors. Chemotherapy for dogs is frequently ineffective, resulting in the risk of infection spreading from infected dogs to humans. Clinical indicators in dogs frequently match those observed in humans, and neurological manifestations are no exception.

In a study of Leishmaniasis in U.S. guard dogs, the dog's caregiver observed neurological symptoms such as "altered states of awareness," spasms, hind limb immobility, and impairment of scenting skills. These symptoms are comparable to descriptions of dullness or lethargy in human patients.

These Leishmania-infected dogs showed a widespread non-suppressive meningoencephalitis with gliosis in both the white and grey matter, sporadic regions of necrosis in the grey matter, and diffused vacuolation in the white matter. Previous research on Foxhounds in the early 2000s discovered that dogs infected with VL frequently displayed aberrant behavior, including temporary aggressiveness, seizures, hyperesthesia, paralytic episodes, and photophobia. After a post-mortem inspection, all of the dogs tested negative for rabies because they had all had their rabies vaccinations.

Rodents are also naturally infected with Leishmaniasis, primarily cutaneous forms, while BALB/c mice and golden Syrian hamsters have a variation of visceralizing illness. In hamsters, studies dating back to the 1920s showed cerebral symptoms of VL infection.

CNS-tropic VL has also been discovered in BALB/c mice.

13.6.4 Disruption of the Blood-Brain Barrier

An interruption of the blood-brain barrier (BBB) is suggested by the existence of immunoglobulin in aqueous humor and cerebrospinal fluid (CSF) and in reports of neuro-inflammation linked to both innate and experimental VL infection. The blood-cerebrospinal barrier underlies the CNS manifestation of the disease. The BBB can reject both large and small hydrophilic molecules due to endothelial cell specializations including strong cell-to-cell contacts, a particular basal lamina surrounding the brain capillaries, and an ensheathment of the veins made of astrocytic end foot (Utzinger et al., 2012). Under typical physiological circumstances, as shown in VL, the BBB would stop internalized amastigotes and following inflammatory cytokines from entering the system.

13.6.5 Immunology

Leishmanial infections must be resolved by the proliferation of certain type 1 helper T cells that create or display on their surfaces lymphokines that can rouse the macrophages that house such parasites to a microbicidal state. Due to their capacity to generate interferon-, CD8+ T cells that are stimulated during infections appear to also play a part in protective immunity. An effect that appears to be caused by the capacity of certain type 2 T helper-derived lymphokines to deactivate macrophages and restrict the manufacturing of boosting cytokines by type 1 helper T cells is the expansion of specific type 2 helper T cells, which worsens the illness in a mouse model of a sizable Leishmania infestation.

13.6.5.1 Gender's Role in the Spread of Infection

One risk factor for VL is the male sex, according to population-based research and not just as a socio-cultural determinant. The pathophysiology of this disease is thought to be influenced by biological sex differences.

It is believed that physiological differences in host-pathogen interactions and the immune response after infection account for the male prevalence in VL. Sexual hormones particularly have the power to change the immune system and the immune response to pathogen infection. The idea that this physiologic variation also plays a significant role in Leishmaniasis is supported by a number of animal studies that demonstrate higher susceptibility to illness and worse associated diseases in male animals following Leishmania infection. However, no research has examined the causes of the male dominance in the reported VL on the Indian subcontinent. The idea that gender-based differences in health-seeking behavior and access to care account for the gender disparity in reported cases was put to the test by looking at the gender ratio of VL findings in population-based studies carried out in the same area as those for which medical system data are available (Roberts et al., 2000). In these long-term population-based studies, incident VL was meticulously documented over several years for all individuals residing within designated zones through door-to-door household surveys. Because of this, we may assume that the population statistics on male and female VL cases were gathered with equal validity and that any potential variations in health-care-seeking behavior or access to care had a negligible effect on these findings.

13.7 TREATMENTS

Miltefosine, sodium stibogluconate, or amphotericin B lipid complex are the therapies of choice for VL, although miltefosine is not accessible in the United States. Many therapies, most of which are parenteral, might have serious side effects. For HIV-infected patients, amphotericin B lipid complex has been utilized as first-line therapy. Relapse occurs in up to 30% of HIV-infected persons with Leishmaniasis within 6 months of treatment, regardless of the treatment administered; persistent suppressive therapy is therefore indicated (Sundar and Chakravarty, 2017). Treatment failure and resistance to basic drugs are common in HIV patients, necessitating the use of combination therapy. Concurrent HAART treatment reduces the occurrence of symptomatic VL.

13.8 FUTURE ASPECTS

The entire genomic sequences of humans and *Leishmania major* would be finished over the next few generations with ongoing financial support. This additional data will allow researchers to look at illness development from two different angles. First, the human genomic sequencing will hasten the discovery of genes that control host vulnerability and resistance. Second, the Leishmania gene that controls virulence will be discovered. There will be answers to questions such as how the parasite survives in the hostile atmosphere of the macrophage phagolysosome and which pathogen and host parameters affect the type of helper T cell response (Feasey et al., 2010). More significantly, these biological discoveries will lead to novel therapy targets. Knowing all of the Leishmania genes will considerably expand the vaccine design options. This promising future extends beyond Leishmaniasis research. Numerous significant infectious illnesses, such as tuberculosis, malaria, schistosomiasis, filariasis, and trypanosomiasis, as well as many harmful microorganisms, are undergoing genome projects.

13.9 CONCLUSION

Associations seen between host and pathogen, which are mediated by the host and parasite genomes, dictate the outcome of Leishmaniasis. The fact that the genomes of the host and the virus have both been selected for sequencing is positive. With the aim of ultimately sequencing the entire genome, the Leishmania genome project was started as a massive undertaking to sequence a fragment of each transcript for every gene in the organism. The introduction of new techniques such as microarray technology and genetic data from both humans and parasites will help us to gain a deeper understanding of the interactions between parasite virulence traits and the elements of the host response. The ability to target novel treatments using the biochemical understanding of the host-parasite relationship will be aided.

REFERENCES

A. A. Akhavan, H. Mirhendi, A. Khamesipour, M. H. Alimohammadian, Y. Rassi, P. Bates, S. Kamhawi, J. G. Valenzuela, M. H. Arandian, H. Abdoli, N. Jalali-zand, R. Jafari, N. Shareghi, M. Ghanei, M. R. Yaghoobi-Ershadi. (2010). Leishmania species: Detection and identification by nested PCR assay from skin samples of rodent reservoirs. *Experimental Parasitology.* 126:552–55 https://doi.org/10.1016/j.exppara.2010.06.003.

E. R. Adams, D. Jacquet, G. Schoone, K. Gidwani, M. Boelaert, J. Cunningham. (2012). Leishmaniasis direct agglutination test: Using pictorials as training materials to reduce inter-reader variability and improve accuracy. *PLOS Neglected Tropical Diseases.* 6:e194. https://doi.org/10.1371/journal.pntd.0001946

M. Akhoundi, H. Hajjaran, A. Baghaei, M. Mohebali. (2013). Geographical distribution of leishmania species of human cutaneous leishmaniasis in Fars province, southern Iran. *Journal of Parasitology.* 8:85–91.

L. Aliaga, F. Cobo, J. D. Mediavilla, J. Bravo, A. Osuna, J. M. Amador, J. Martin-Sanchez, E. Cordero, J. Navarro. (2003). Localized mucosal leishmaniasis due to Leishmania (Leishmania) infantum clinical and microbiologic findings in 31 patients. *Medicine.* 82:147–58. https://doi.org/10.1097/00005792–200305000–00001.

G. Camuset, V. Remy, Y. Hansmann, D. Christmann, C. Gomes de Albuquerque, G.-A. Sena Casseb. (2007). Mucocutaneous leishmaniasis in Brazilian Amazonia. *Medecine et Maladies Infectieuses*. 37(6):343–46. https://doi.org/10.1016/j.medmal.2007.03.012

A. Cannet, M. Akhoundi, G. Michel, P. Marty, P. Delaunay. (2016). Experimental infection of Phlebotomus perniciosus by bioluminescent Leishmania infantum using murine model and artificial feeder. *Memórias do Instituto Oswaldo Cruz*. 111:495–500.

E. A. Castro, V. Thomaz-Soccol, C. Augur, E. Luz. (2007). Leishmania (Viannia) braziliensis: Epidemiology of canine cutaneous leishmaniasis in the State of Parana (Brazil). *Experimental Parasitology*. 117(1):13–21.

F. Chappuis, S. Sundar, A. Hailu, H. Ghalib, S. Rijal, R. W. Peeling, J. Alvar, M. Boelaert. (2007). Visceral leishmaniasis: What are the needs for diagnosis, treatment and control? *Nature Reviews Microbiology*. 5(11):873–82. https://doi.org/10.1038/nrmicro1748.

J. P. de Menezes, E. M. Saraiva, B. da Rocha-Azevedo. (2016). The site of the bite: Leishmania interaction with macrophages, neutrophils and the extracellular matrix in the dermis. *Parasites Vectors*. 9:264. https://doi.org/10.1186/s13071-016-1540-3.

Nick Feasey, Mark Wansbrough-Jones, David C. W. Mabey, Anthony W. Solomon. (March 2010). Neglected tropical diseases, *British Medical Bulletin*. 93(1):179–200, https://doi.org/10.1093/bmb/ldp046

H. Goto, J. A. L. Lindoso. (2010). Current diagnosis and treatment of cutaneous and mucocutaneous leishmaniasis. *Expert Review of Anti-Infective Therapy*. 8(4):419–33. https://doi.org/10.1586/eri.10.19.

L. H. Guimarães, A. Queiroz, J. A. Silva, S. C. Silva, V. Magalhaes, E. L. Lago, P. R. Machado, O. Bacellar, M. E. Wilson, S. M. Beverley, E. M. Carvalho. (2016). Atypical manifestations of cutaneous leishmaniasis in a region endemic for Leishmania braziliensis: Clinical, immunological and parasitological aspects. *PLoS Neglected Tropical Diseases*. 10(12):e0005100.

L. J. Roberts, E. Handman, S. J. Foote. (2000). Science, medicine, and the future: Leishmaniasis. *BMJ (Clinical Research Ed.)*. 321(7264):801–04. https://doi.org/10.1136/bmj.321.7264.801

J. R. Seed. (1996). Chapter 78. Protozoa: Pathogenesis and defenses. In: Baron, S. (Ed.), *Medical Microbiology* (4th ed.). Galveston, TX: University of Texas Medical Branch at Galveston.

A. Strazzulla, S. Cocuzza, M. R. Pinzone, M. C. Postorino, S. Cosentino, A. Serra, B. Cacopardo, G. Nunnari. (2013). Mucosal leishmaniasis: An underestimated presentation of a neglected disease. *BioMed Research International*. https://doi.org/10.1155/2013/805108.

S. Sundar, J. Chakravarty. (2017). Drug resistance in leishmaniasis. In: Mayers, D., Sobel, J., Ouellette, M., Kaye, K., Marchaim, D. (Eds.), *Antimicrobial Drug Resistance*. Cham: Springer. https://doi.org/10.1007/978-3-319-47266-9_29.

J. Utzinger, S. L. Becker, S. Knopp, J. Blum, A. L. Neumayr, J. Keiser, C. F. Hatz. (2012, November 22). Neglected tropical diseases: Diagnosis, clinical management, treatment and control. *Swiss Medical Weekly*. https://doi.org/10.4414/smw.2012.13727.

K. Waki, S. Dutta, D. Ray, K. B. Kolli, L. Akman, S. Kawazu, C. P. Lin., K. P. Chang. (2007, February 1). Transmembrane Molecules for Phylogenetic Analyses of Pathogenic Protists: Leishmania-Specific informative sites in hydrophilic loops of trans-endoplasmic reticulum n-acetylglucosamine-1-phosphate transferase. *ASM Journals Eukaryotic Cell*. 6(2). https://doi.org/10.1128/EC.00282-06

14 Molecular Pathogenesis of the Ebola Virus

Pratik Talukder[1,], Rajarshi Ray[1], Mainak Sarkar[1], Anirban Das[1], Baishakhi Sinha[1], Arundhuti Mukherjee[1] and Srishti Chakraborty[1]*

[1]Department of Biotechnology, University of Engineering and Management, Kolkata, India

*Correpondence to pratik.talukder@uem.edu.in

Since its discovery in 1976, the Ebola virus has posed a threat to human health due to its dangerous, highly lethal, and infectious behaviour. Ebola fever has emerged as one of the most lethal identified forms of haemorrhagic fever, with no specific treatment available. Humans spread the virus primarily through the exchange of blood and bodily secretions. Other obvious modes of transmission include hospital-acquired infection and poor hygiene practises. There is an urgent need for community information dissemination and training sessions for doctors, nurses, and other hospital personnel. Initiatives should be organised to raise awareness and knowledge about the disease for its eradication. Research must focus on the development of rapid and simple diagnostic kits for dealing with Ebola infection. A proper effort with a perfect strategy is required to transport effective drugs and vaccines from the lab to clinical trials for initiating the ultimate treatment for Ebola patients.

14.1 INTRODUCTION

Ebola virus disease (EVD) can be a fatal illness that occurs chiefly within the African continent. EVD most typically infects humans and non-human primates such as monkeys, gorillas, chimpanzees, etc. It is caused by an infection from the Ebola virus family.

Filovirus was initially brought to light in 1976 close to the Ebola fever watercourse, which is currently in the Democratic Republic of the Congo (DRC). Afterwards, the virus has sometimes infected humans and caused eruptions in many countries of Africa. Scientists are not aware from where the filovirus evolved. Supported by the same type of viruses, they believe that EVD is transmitted by animals, along with non-human primates appearing to be the most important source of infection. Infected animals carrying the virus transmit it to a wide variety of animals including apes, monkeys, duikers, and humans. Viruses first infect humans through direct contact with animal blood, body fluids, or tissues.

Filovirus is then spread to others by direct contact with the bodily fluids of individuals. This will occur once someone touches these infected bodily fluids or objects contaminated with them. The virus then enters the body through cuts in the skin and the mucous membranes of the eyes, nose, and mouth. The virus is also transmitted through sexual contact with somebody who has or has recovered from filovirus illness. The virus will continue in the bodily fluids of individuals when they are convalescent from associated ill-health.

EVD initially affects the macrophages and immune cells of the nerve fibres. The innate response is characterized by a protein flow that releases an oversized range of pro-inflammatory cytokines, inflicting an oversized range of contradictory signals and damaging immune cells and different

DOI: 10.1201/9781003302131-15

tissues. Different extremely unhealthful viruses also cause protein storms, but filoviruses are considered to be significantly deadly because they affect good types of tissues. Additionally, in the system, Ebola virus attacks the spleen and kidneys, which serve to balance fluids and chemicals, and kills cells that build proteins that facilitate blood coagulation. In addition, EVD causes not only the liver, lungs, and kidneys to stop functioning but also fluid to leak from blood vessels into surrounding tissues.

14.2 ETIOLOGY

Filovirus enters the body through surface contact, skin injuries, and wounds within the skin or by direct transfer. The dimensions of this virus's particles are a regular diameter of 80 nm, but this varies in range from 600–1,400 miles (Goeijenbier et al., 2014). Contact of the viral agent and cellular membrane is mediated by GP2, which ends up from the chemical process cleaving GP1 by using endosomal proteases cathepsin B and L. Cells defective for HOPS complicated or primary fibroblasts derived from human NPC1 patients are proof against Ebola fever infection virus (Kuhn et al., 2014). Ebola virus takes advantage of a non-specific engulfing method referred to as macropinocytosis, which permits the virus to be "eaten" by a wave-type motion of the semipermeable membrane (Hoenen et al., 2010). Once within the cell, the virus manipulates the cell's system to multiply itself and proliferate. Ebola, like several viruses, works partly by inhibiting interferon – a variety of molecules that cells use to hinder additional infective agent replications (Beniac et al., 2012). With defective nerve fiber cells that fail to allow the proper signal, the T cells do not respond to the infection. Ebola virus targets specific cell types together with liver cells, cells within the system, and epithelium cells that line the blood vessels (Iampietro et al., 2017). By infecting vas cells, the virus causes the leakage of vessels, resulting in hemorrhage and internal injuries.

14.3 PATHOGENESIS

Ebola virus causes both internal and external bleeding. Although rare, it is a highly fatal virus that leads to the destruction of the immune system and organs and spreads throughout the body resulting in severe bleeding, thus leading to haemorrhagic fever with almost a 90% fatality rate. The glycoprotein (GP) of Ebola virus is obtained as a dimeric form/secreted form-sGP, which has a complete length transmembrane form with specific biochemical and biological attributes for every gene. Examples include GP-forming trimeric complex and binding to endothelial cells, which sGP cannot do. In certain studies, it has been observed that viral GP plays a prominent role in Ebola manifestation. The transmembranic forms of GP contribute to the formation of cells from viruses that are optimal for pathogenesis. More precisely, GP makes the virus release its contents into monocytes and macrophages while viral exposure causes cytokine action and release leading to damage to vascular integrity (Alvarez et al., 2002). Thus, sGP can modify the immune system by inhibiting the activation of neutrophils. Meanwhile, Transmembranic GP leads to a fever and hemorrhagic condition by leading the virus towards cells having a reticuloendothelial network and blood vessel linings.

Host immune responses to filovirus and cell harm due to monocyte and macrophage infection cause a cytokine discharge related to inflammation and fever. Infection of the epithelium cells conjointly causes cytopathic impact and harm to the epithelium barrier, which along with protein effects, end up losing integrity. Transient expression of filovirus MD in human venous blood vessel epithelium cells or 293T cells causes a discount of specific integrins (primary molecules to blame for cell attachment to the animate matrix) and immune molecules on the cell surface. Protein dysregulation and infection might synergize at the epithelium surface, causing hemorrhage and constriction collapse.

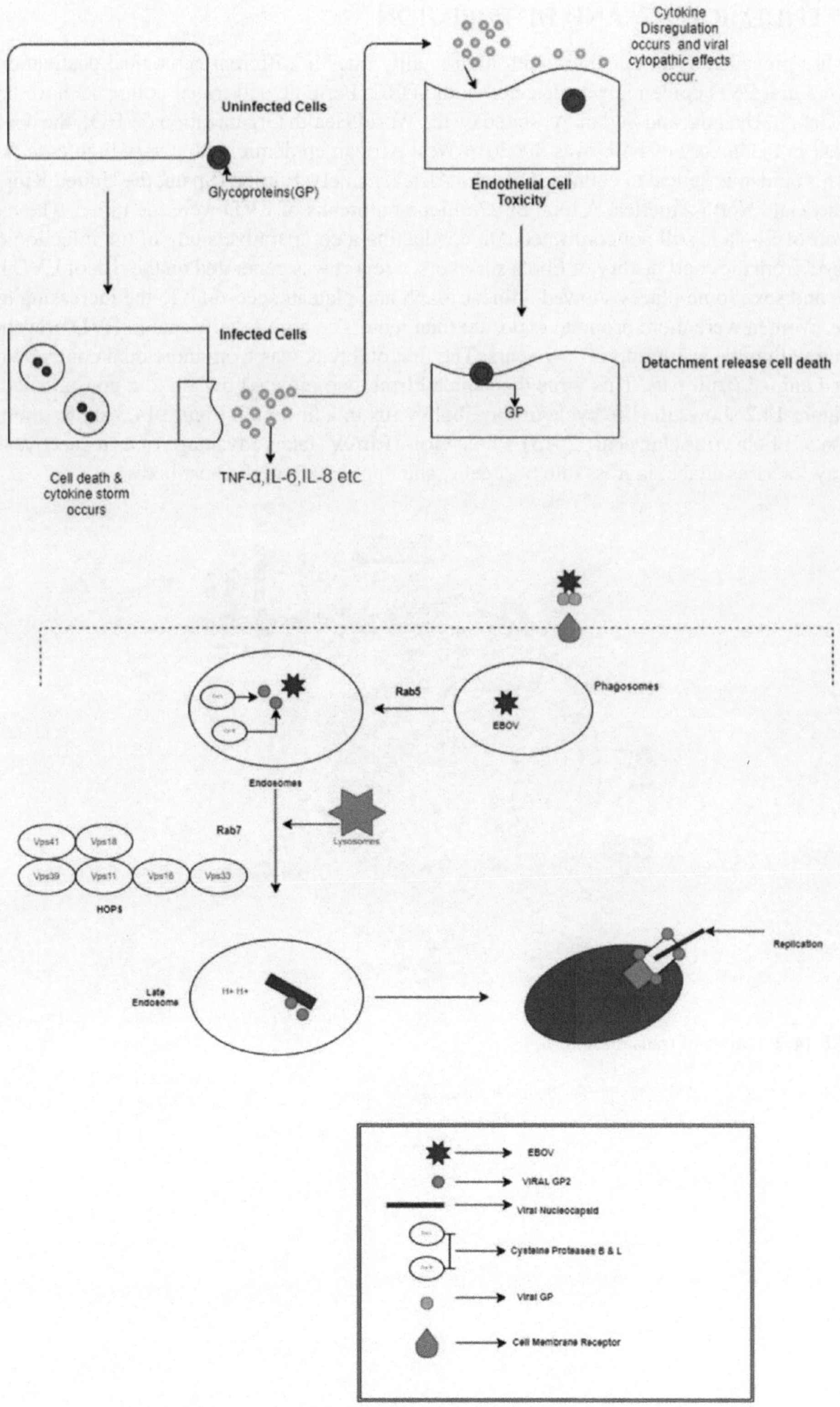

FIGURE 14.1 Diagrammatic representation of EVD etiology and pathogenesis.

14.4 EPIDEMIOLOGY AND DISTRIBUTION

Ebola has prevailed as an epidemic with high fatality rates in different ecological destinations. In 1976, the first EVD epidemic was detected in the DRC. Periodic and trivial outbreaks have broken out in Gabon, Uganda, and Sudan. As stated by the World Health Organization (WHO), the deadliest outbreak in the history of EVD was the 2014 West African epidemic with a very high case fatality rate. This epidemic spread to countries beyond Africa, namely, Europe (Spain, the United Kingdom, and Italy) and North America. A total of 22 unique outbreaks of EVD were identified. The natural reservoir of Ebola is still not confirmed. On conducting a comparative study of the infection exposure level from the chart history of Ebola survivors, a report was generated on the risk of EVD based on age and sex. Some places showed a linear graph and plateau according to the increasing age of people. Women were more prone to exposure than men. The high fatality rate in EVD prevailed in children within the age group of 1–5 years. The first outbreak was from individual contact with an infected animal (fruit bat). This virus then spread from one infected person to a non-infected one, etc. Figure 14.2 shows the life cycle of the Ebola virus in a host cell. Figure 14.3 shows the transmission of Ebola virus (Jun et al., 2015). Ebola virus (EBOV) takes advantage of its respective host's pathway for virus attack, ingress into host cells, and duplication in the host body.

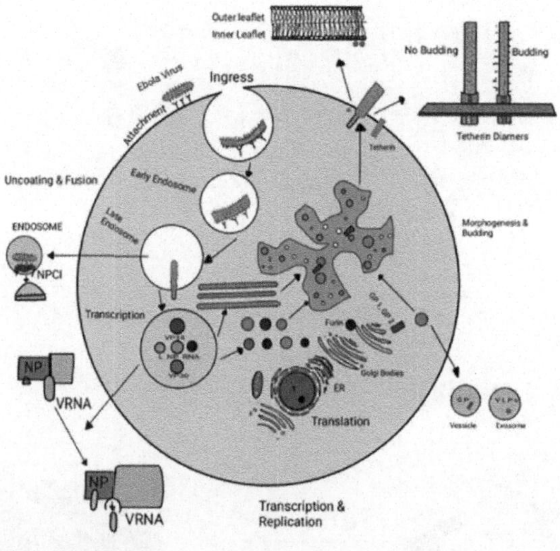

FIGURE 14.2 Mode of transmission.

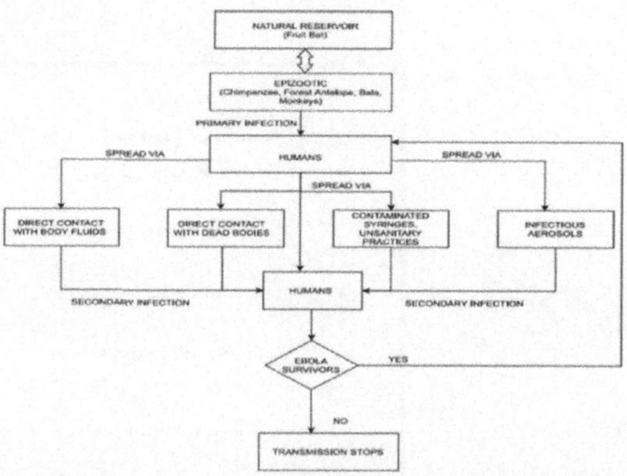

FIGURE 14.3 Signal transduction associated with disease progression.

14.5 DIAGNOSIS

Ebola is a relatively recent viral disease that appeared in 2014–2015 and highlighted the need to develop more equipped and potent diagnosis tools. Since the disease is highly transmissible with an even higher mortality rate, the diagnosis and associated biosafety measures need to be taken care of. A proper diagnostic approach depends on a number of factors, such as the characteristics of the population, financial infrastructure, disease prevalence, and training equipment. As for the structure of EBOV, the virus contains single-stranded RNA encoded by seven viral proteins, namely, nucleoproteins, glycoproteins, polymerase L, viral protein 24, viral protein 30, viral protein 35, and viral protein 40 (Bixler et al., 2017). Out of every effort made to detect the viral protein, all of the approaches fall under three main categories (Broadhurst et al., 2016):

I. Detect antibodies released by the host against the virus;
II. Antigen tests; and
III. Molecular approach to test viral RNA.

The most conventional detecting method is cell culture propagation. The propagated viruses are then visualized under immunofluorescence microscopy. Due to the close relation between the Marburg virus and Ebola virus, an indirect fluorescent antibody detection test (IFAT) is carried out. Ebola virus-specific antibodies such as IgM and IgG are detected by an enzyme-linked immunosorbent assay (ELISA); this technique is a faster and more effective approach. Viral protein antigen circulates in the blood stream, whose detection also became a diagnostic tool for Ebola virus.

Moreover, reverse transcription polymerase chain reaction (RT-PCR) has become a pivotal tool in detecting the virus (Broadhurst et al., 2016). This technique amplifies the L, GP, and NP genes of the virus, which then undergo gel electrophoresis. The conventional RT-PCR detection test is also carried out to test the presence of the virus in non-blood bodily fluids, such as saliva, seminal fluids, etc. Real-time RT-PCR has detected the presence of the virus from the serum samples of patients (Broadhurst et al., 2016). Compared to conventional RT-PCR detection tests, real-time RT-PCR detection tests are more distinct, accurate, and less time-consuming, but both techniques require developed infrastructure, high-tech laboratories, technical expertise, temperature-dependent reagents, etc.

14.6 TREATMENT

The outbreak of Ebola virus as recent as 2014–2015 in West Africa and its alarming rate of mortality in humans as high as 90% triggered researchers to discover drugs to suppress and subdue this deadly virus. One of the major approaches is antivirals that target the viral life cycle, and they often comprise small interfering RNAs that facilitate the degradation of mRNA transcripts phosphodiamidate morpholino oligomer (PMO), which often interferes in the translation process, thereby inhibiting the replication of the virus (Bixler et al., 2017). TKM-100802 and its derivative TKM-130803 primarily target the West African strain of EBOV, and they function by inhibiting viral proteins such as L, VP35, and VP24. Cathepsin-like CatB and CatL constitute cysteine proteases in the endosome that cleave EBOV glycoprotein before its entry. Ebola virus is responsible for viral hemorrhagic fever, and because of its anti-coagulants, recombinant human activated protein C (rhAPc) and recombinant nematode anticoagulant protein c2 (rNAPc2) are manifested.

The most promising of these passive immunotherapy techniques is ZMapp, which contains three humanised monoclonal antibodies (Yazdanpanah et al., 2015). This product's efficacy at preventing deadly disease was demonstrated in a recent trial in non-human primates (NHPs) [8]. Several health personnel, mainly in northern nations, have received ZMapp doses according to these findings. Although this is based on a limited number of individuals, it is possible that some of these people benefited from the product. This drug's supply is running out, and scaling up production may take

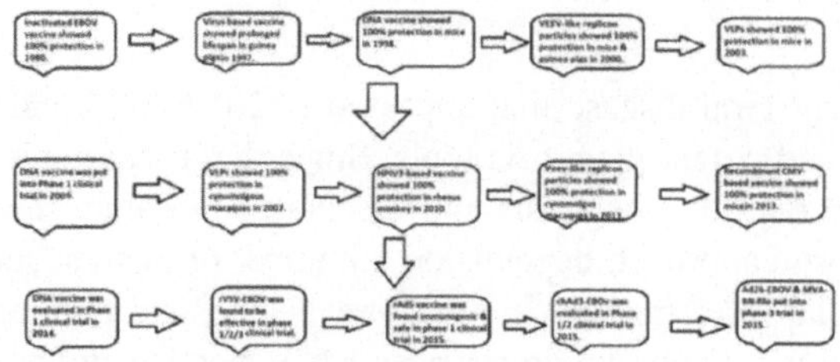

FIGURE 14.4 Mechanism of proposed treatment.

months, which limits access. The initial trial used formalin-fixed or heat-inactivated virions for the development of the vaccine, but the results of the vaccine on some animal models were cumbersome; hence, its efficacy was not tested on human beings. DNA vaccination with genes that code for the glycoprotein and nucleoprotein have proved to be somewhat successful (Feldmann et al., 2003). To date, the most effective mechanism has been a DNA primer followed by an adenovirus booster. The effectivity of this strategy was tested on cynomolgus monkeys, and the results demonstrated that T-memory cells and antibody response helped in protection from further exposure. All laboratory studies and immunological assays fall under the biosafety level 4, so proper precautions must be taken to address studies regarding this virus. This is the reason why research related to Ebola virus requires not only a laboratory with high-tech infrastructure to prevent the spread of the virus but also technical expertise and ample funding to aid this research.

14.7 CURRENT RESEARCH – VACCINE AND DRUG DEVELOPMENT

14.7.1 Vaccines

Five of the thirteen EBOV vaccine candidates have advanced to clinical testing after Phase I clinical trials in humans.

The most cutting-edge vaccines, licensed in both the United States and Europe, include

1. Ervebo and
2. Zabdeno/Mvabea.

Manufacturing challenges, the presence of difficulties, and supply chain adjustments to the new viral varieties are some basic drawbacks of the vectors of virus vaccines.

14.7.2 Current Experimental Therapeutics Against EVD

The techniques identified for fighting EBOV include the following:

(i) at once concentrated on the virus;
(ii) inflect the host elements or an immunological reaction; and
(iii) management of the ailment.

The current remedy of EVD is centered on controlling coagulation and maintaining oxygen ranges in EVD patients.

14.7.3 Small Molecules: Direct EBOV Inhibitors

The creation of powerfully small molecules aimed towards the RNA-established RNA polymerase L, which is appropriate for viral reproduction, is the most promising medical practice. These molecules have detailed antiviral impacts because of intracellular change to their related nucleoside triphosphate for connecting to the viral genome and the blockage of viral polymerase.

Among numerous FDA screens, a couple of compounds, inclusive of amodiaquine, diphenylpyraline, ketotifen, and diphenoxylate, have been identified to block EBOV reproduction, even as others, inclusive of verapamil, dronedarone, sertraline, toremifene, chloroquine, teicoplanin, and amiodarone, have been considered EBOV access inhibitors (Mirza et al., 2019).

REFERENCES

Alvarez CP, Lasala F, Carrillo J, Corbi AL, Delgado R. C-type lectins DC-SIGN and L-SIGN mediate cellular entry by Ebola virus in *cis* and in *trans*. *J. Virol.* (2002);76:6841–6844.

Beniac DR, Melito PL, Devarennes SL, Hiebert SL, Rabb MJ, Lamboo LL, et al. The organisation of Ebola virus reveals a capacity for extensive, modular polyploidy. *PLoS One.* (2012);7(1):e29608.

Bixler, SL, Duplantier, AJ, Bavari S. Discovering Drugs for the Treatment of Ebola Virus. *Curr Treat Options Infect Dis.* (2017);9:299–317.

Broadhurst MJ, Brooks TJ, Pollock NR. Diagnosis of Ebola Virus Disease: Past, Present, and Future. *Clin Microbiol Rev.* (2016);29(4):773–793.

Feldmann H, Jones S, Klenk HD, Schnittler HJ. Ebola virus: from discovery to vaccine. *Nat Rev Immunol.* (2003);3:677–685.

Goeijenbier M, van Kampen JJ, Reusken CB, Koopmans MP, van Gorp EC. Ebola virus disease: a review on epidemiology, symptoms, treatment and pathogenesis. *Neth J Med.* (2014);72:442–448.

Hoenen T, Biedenkopf N, Zielecki F, Jung S, Groseth A, Feldmann H, et al. Oligomerization of Ebola virus VP40 is essential for particle morphogenesis and regulation of viral transcription. *J Virol.* (2010);84:7053–7063.

Iampietro M, Younan P, Nishida A, Dutta M, Lubaki NM, Santos RI, et al. Ebola virus glycoprotein directly triggers T lymphocyte death despite of the lack of infection. *PLoS Pathog.* (2017);13(5):e1006397.

Jun SR, Leuze MR, Nookaew I, Uberbacher EC, Land M, Zhang Q, et al. Ebola virus comparative genomics. *FEMS Microbiol Rev.* (2015);39(5):764–778.

Kuhn JH, Andersen KG, Baize S, Bào Y, Bavari S, Berthet N, et al. Nomenclature- and database-compatible names for the two ebola virus variants that emerged in Guinea and the democratic republic of the Congo in 2014. *Viruses.* (2014);6:4760–4799.

Mirza MU, Vanmeert M, Ali A, Iman K, Froeyen M, Idrees M. Perspectives towards antiviral drug discovery against Ebola virus. *J Med Virol.* (2019);91(12):2029–2048.

Yazdanpanah Y, Arribas, JR, Malvy, D. Treatment of Ebola virus disease. *Intensive Care Med.* (2015);41:115–117.

15 Importance of Plant-Derived Antioxidants in the Field of Medical Biotechnology

Tista Basu[1] *and Suranjana Sarkar*[2,*]
[1]CRIJAF, Nilganj, Barrackpore, India
[2]Surendranath College, 24/2 M. G. Road, Kolkata, India
*Correspondence to: suranjanasarkar@surendranathcollege.ac.in

15.1 INTRODUCTION

A wide variety of extremely beneficial chemical compounds have been found in plants that are popularly named antioxidants. These compounds have a repository of highly pivotal characteristics that range from preventive to therapeutic uses. Plants' antioxidant system consists of both enzymatic and non-enzymatic components that are responsible for scavenging reactive oxygen species (ROS) that are generated when 5% or more oxygen is inhaled. This ROS, in turn, is responsible for various types of physiological disorders (Trachootham et al. 2009). A free radical (FR) consists of an unpaired electron, which may be positively, negatively, or neutrally charged. These substances often result in the generation of various types of cardiovascular diseases, such as diabetes, cancer, inflammation, and other neurological disorders, as well as the ageing process. Antioxidants play a role by offering an electron to this FR, resulting in a harmless molecule. This is performed by preventing it from radical formation, reducing the energy of the FR, or repairing damage by a reconstitution of membranes (Bagchi et al. 2000). The human body has its own multifaceted mechanism system to fight these toxic ROS; this system is commonly called antioxidants and comprises enzymes, proteins, vitamins A, C, and E, flavonoids, carotenoids, polyphenols, and trace elements, such as selenium. Various types of plant-derived antioxidants can be categorised in this manner (see Figure 15.1) (Kurutas 2015).

15.1.1 Different Types of Plant-Derived Antioxidants

15.1.1.1 Vitamins

15.1.1.1.1 Vitamin A

Vitamin A or retinol was first discovered in 1912 by Frederick Gowland Hopkins when he identified some unknown factors that are a dietary essential for rats for which he was awarded Nobel Prize in 1929. Vitamin A is fat-soluble, and the most common dietary sources of vitamin A include beef liver, carrots, fish, eggs, pumpkin, sweet potatoes, spinach and other green leafy vegetables, tomatoes, and dairy products. Vitamin A not only plays a crucial role in vision, cell division, and immunity but also has a great deal of antioxidant properties, for example, a higher intake of vitamin A can lower ischemic heart disease (Palgi 1981). Carotenoids, which are also a precursor of vitamin A, further provide antioxidant properties due their strong and highly stable chemical structure that helps in fighting FRs. Consumption of carotenoids help can lower the chances of neurophysiological disorders in the aged population (Kesse-Guyot et al. 2014).

DOI: 10.1201/9781003302131-16

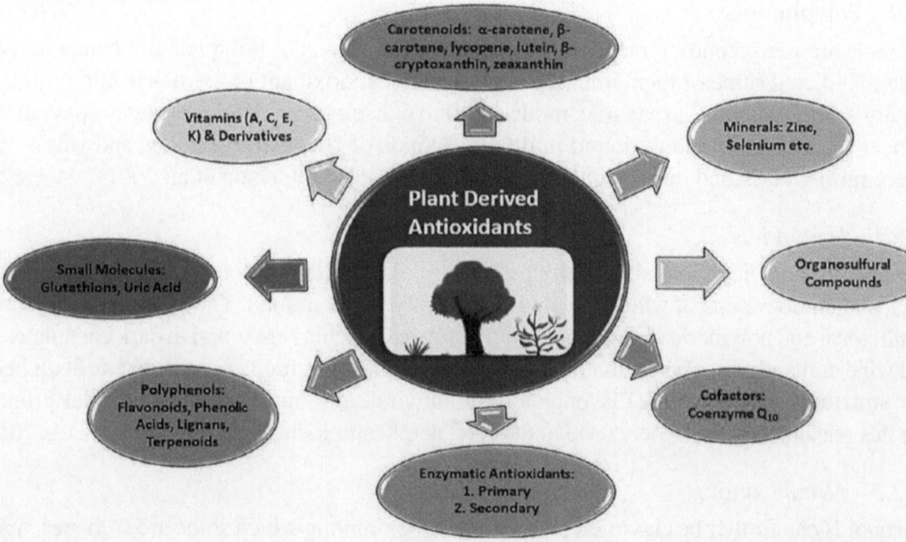

FIGURE 15.1 Different Types of Plant-Derived Antioxidants.

15.1.1.1.2 Vitamin E

Vitamin E was first isolated by Evan and Bishop as an essential dietary resource for rats in 1922. (Evans et al. 1922). It is fat-soluble, and the most commonly form used for human consumption is α tocopherol, which has several antioxidant properties ranging from strengthening the immune and cardiac systems and anti-ageing to the prevention of cancer. Some plant sources that have rich sources of vitamin E include almonds, asparagus, avocados, mangos, red bell peppers, peanuts, spinach, sunflower seeds, and wheat germ oil. Although there is no direct involvement of vitamin E in combating heart disease, it has been recently found that a regular intake of direct supplements markedly reduced the heart disease rate for people who have type II diabetes (Milman et al. 2008). A recent trial shows that vitamin E in combination with beta carotene, zinc, and vitamin C are effective in the reduction of age-related macular degeneration (AMD) in people who had a chance of contracting this disease (AREDS report no. 8 2001).

15.1.1.1.3 Vitamin C

Vitamin C or ascorbic acid was first isolated by Szent-Gyorgyi in 1928. It acts as a good FR scavenger mainly due to its free ascorbyl radical that is relatively unreactive; this makes it a good antioxidant that can interact with a reactive harmful FR and reduce it. Thus, it serves as a well-known reducing agent (Bielski et al. 1975). It is water-soluble, and common dietary sources include berries, broccoli, cabbage, cauliflower, citrus fruits, papayas, tomatoes, watermelons and red bell peppers. It is needed to make collagen, helps in healing wounds, fights against infection, and thus serves an important function in the immune system and in the fight against cancer. It also helps the human body to absorb and stock iron. Evidence from some experimental results also shows that it plays a pivotal role in the prevention of hypertension and vascular disease (Padayatty et al. 2003). Occurrence of the common cold is significantly reduced if vitamin C supplements are regularly consumed. Vitamin C with other vitamin supplements are also found to lower the risk of Acid Mine Drainage (AMD) in aged persons. It further acts as a cofactor and thus plays a part in hypoxia-inducible factor (HIF) pathway inhibition, which may offer an alternative and added approach in controlling tumor progression, inflammation, and infections (Traber et al. 2011).

15.1.1.2 Polyphenols

Polyphenols are the secondary metabolites in plants. Thousands of polyphenolic compounds have been identified, and some of them impart distinguishable antioxidant properties resulting in protection against cardiovascular diseases, neurodegenerative diseases, skeletal muscle atrophy, diabetes, and cancer. They are commonly found in different kinds of berries, vegetables, and whole grains. These are further classified into flavonoids and nonflavonoids (Anderson et al. 2001).

15.1.1.2.1 Flavonoids

Flavonoids are subdivided into six subgroups, namely, flavanols, isoflavanols, flavanones, flavonols, flavones, and anthocyanins of which flavonols are mostly found in food. They exist in both monomer (catechin) form and polymeric (proanthocyanidins) form. Catechins are found in dark chocolate, green tea, red wine and apricots. The estimation of proanthocyanidins in food is somewhat difficult because of their structural arrangements. Flavonoids have antiviral, antifungal, and antibacterial properties, and for this reason, they are widely used in the food and health industries (Del Río and Luis 2015).

15.1.1.2.2 Nonflavonoids

Nonflavonoids can further be classified into various types, among which some are stilbenes, hydroxycinnamic acids, tannins, and phenolic acids that comprise four molecules of ellagic acid, gallic acid, syringic acid, and vanillic acid. Major dietary sources include tea, wine, strawberries, raspberries, blackberries, persimmon, peanuts, pistachios, and mangos. All of these compounds exhibit some antioxidant properties. Resveratrol is one of the most extensively studied stilbenes as it has been found to regulate many different processes such as cell proliferation, inflammation, senescence, and apoptosis (Bellavia et al. 2021).

15.1.1.3 Small Molecules

15.1.1.3.1 Glutathione

Glutathione is a multifunctional intracellular antioxidant that is found in nuclei, mitochondria, chloroplasts, endoplasmic reticulum, vacuoles, and cytosol. Major dietary sources include avocados, asparagus, green beans, spinach, and strawberries. Glutathione plays a pivotal role in different biological processes, namely, cell division, cellular ageing in humans, signal transduction, protein synthesis, synthesis of phyochelatins for metal chelation, and the expression of stress-responsive genes. It serves as an FR scavenger as it can react chemically with oxygen and hydrogen peroxide (Demirkol et al. 2004).

15.1.1.3.2 Ubiquinone

Ubiquinone, commonly known as coenzymeQ10, is a vitamin-like lipophilic element that serves an important function in energy metabolism and aerobic cellular respiration. It plays a crucial role in the prevention of cell damage that is associated with FRs, which, in turn, is an important antioxidant property. Major dietary sources of CoQ10 include beef, whole grains, organ meat, salmon, and tuna (fatty fishes). Supplements of ubiquinone are reportedly said to prevent gastric ulcers, mitochondrial diseases, radiation injury, kidney failure, ageing, Parkinson's disease, migraine headaches, and heart disease.

15.1.1.4 Minerals

15.1.1.4.1 Selenium

Selenium is a micronutrient, and almost 25 selenoproteins are found in humans (Kryukov et al. 2003). Major dietary sources include seafood, meat products, and some cereals. Mostly depending upon soil selenium levels, some food may or may not contain this trace element. It has been found that a deficiency of selenium leads to Keshan disease (Chen 2012). It has strong antioxidant properties that can prevent oxidative damage, heart disease, and cancer, mainly in advanced-age prostate cancer (Yoshizawa et al. 1998). Another study suggested that a diet supplemented with selenized yeast massively reduced cancer morbidity (Clark et al. 1996)

15.2 WIDELY USED METHODOLOGIES FOR THE IDENTIFICATION OF ANTIOXIDANTS

A thorough study to confirm the presence of a specific antioxidant compound and its endogenous portions in each food would necessitate the use of highly sophisticated and expensive identity verification and quantification techniques and a sequence of protocols and methods to prepare the samples for analysis. High Performance Liquid Chromatography (HPL) with photometric, electrochemical, or fluorescence detection and mass spectrometry (MS-GC) or liquid chromatography (LC) are the most extensively used separation, identification, and quantification methods (MS-LC). However, knowing a food's potential antioxidant components does not always imply that it has antioxidant activity. Furthermore, because of the synergistic impact that can occur among multiple antioxidants, the whole antioxidant effect could be greater than the total of the separate antioxidant activities, and isolation of one chemical may not accurately reflect the entire activity (Poeggeler et al. 1995, 1999). A single plant might have extremely complicated profiles of these chemicals, which are occasionally labile to heat, air, and light and can exist in very low quantities. As a result, separating and detecting these antioxidant phytochemicals is difficult (Tsao et al. 2004). Moreover, unrecognised or unquantified substances in the sample might have a similar impact. As a result, for many years, ways to measure the so-called total antioxidant activity (TAA) have been pursued (Halliwell et al. 1990; Stanley and Mogg 1995). TAA is a metric that measures a sample's ability to scavenge FRs in aqueous media (whether natural or treated). Antioxidant activity is viewed as a global property of a product, regardless of its specific makeup.

15.3 DISCUSSION

Within plant cells, chloroplasts and mitochondria are the two primary powerhouses and sources of ROS. These materials also play a role in maintaining a delicate balance between energy-related operations and ROS production management. Within plant cells, peroxisomes, which are solitary, are a major source of ROS such as hydrogen peroxide (H_2O_2), superoxide (O_2-), and nitric oxide (NO). Catalase (CAT)- and hydrogen peroxide (H_2O_2)-producing flavin oxidases (Del Río et al. 2006) are fundamental enzymatic components found in peroxisomes. ROS are produced in the chloroplasts' photosystems I and II (PS I and PS II), the peroxisome's membrane and matrix, and the mitochondrial electron transport chain's complex I, ubiquinone, and complex III (ETC) (Gill et al. 2010). Under normal physiological circumstances, electrons escape from the chloroplast PS I and PS II membranes, the mitochondrial ETC membrane, and the peroxisome. These electrons are then combined with molecular oxygen to form the superoxide radical (O_2-). The superoxide radical is next transformed to the hydroperoxyl radical (HO_2) and then to H_2O_2 (Alscher et al. 1997; Zhao et al. 2005; Carocho et al. 2013).

Reactive nitrogen species (RNS) such as nitric oxide radical (NO•) and peroxinitrite (ONOO-) are produced in many cell compartments, including chloroplasts, mitochondria, and peroxisomes (Møller et al. 2007). The interaction of thiols with ROS is thought to create the third type of FR: reactive sulphur species (RSS) (Lü et al. 2010). These FRs are continually created in live cells' subcellular organelles. Because FRs operate as signalling molecules, their production is usually genetically programmed (Zhao et al. 2005) (Mittler et al. 2002). Overproduction of FRs, however, can occasionally damage biomolecules such as DNA, proteins, and lipids. To escape the damaging effects of FRs, plants have effective complicated enzymatic and non-enzymatic antioxidant defence mechanisms. SOD, catalase (CAT), glutathione peroxidase (GPx), and glutathione reductase (GR) are examples of enzymatic systems, whereas non-enzymatic systems include low molecular weight antioxidants (ascorbic acid, glutathione, proline, carotenoids, phenolic acids, flavonoids, etc.) and high molecular weight compounds such as tannins.

The production and accumulation of these non-enzymatic antioxidants by plants might be due to one of two factors. Plants have a natural capacity to produce a broad array of phytochemicals to execute their regular physiological tasks, which act as a protector against microbial diseases and animal herbivores, because of their genetic make-up. Moreover, the production of reductant phytochemicals is often associated with these tasks, as the plants have a natural predisposition to respond to stress.

The preservation of redox equilibrium is essential for good health and disease prevention. An imbalance between ROS and antioxidants causes oxidative stress. Excess ROS causes lipid, protein, and nucleic acid breakdown, which imparts oxidative damage to cells and as a result, oncogene overexpression, mutagen creation, initiation of atherogenic activity, or inflammation. The development of cardiovascular diseases, neurodegeneration, malignancies, immunological disorders, diabetes, and other illnesses, as well as ageing, is often assumed to be influenced by oxidative stress. Antioxidants are abundant in plants, particularly in fruits and vegetables. Antioxidants are thought to provide health advantages via reducing oxidative stress directly. The antioxidant network in the body functions in concert through a variety of methods, including ROS scavenging, lipid peroxidation termination, and metal chelation (Szymanska et al. 2016). Although considerable information on antioxidant structures, characteristics, and biological effects has been obtained, there are still many questions that need to be answered, and more research needs to be done. The biological processes underlying their therapeutic potential, interactions with other drugs, optimal dose, and therapy success are all unknown (especially their effect in randomised clinical trials). A substantial number of natural chemicals are also unknown in terms of their bioactivity. Antioxidants, both well-known and freshly discovered, have raised expectations for their utility in the prevention and treatment of the illnesses indicated previously.

15.3.1 Plant-Derived Antioxidant's Role Against Cancer

ROS and NOS have a crucial role in carcinogenesis. The presence of FRs often results in mutation, transformation, and carcinogenesis, which ultimately leads to cancer (Lobo et al. 2010). Cancer cells are often subjected to increased ROS cell production and sometimes altered redox regulation. As known to all, ROS promotes tumour development by inducing genomic instability, DNA damage, and DNA mutations, but sometimes, high levels of ROS become lethal to cancer cells, which ultimately results in cell death. This unique antagonistic dual role of ROS has the power to stop tumour progression, thus making it a perfect candidate for antioxidant cancer therapy. Evidence of tumorigenesis through dietary sources of antioxidants has been well implicated in many articles but the failure of therapeutic antioxidants might be due to ROS or how it can be controlled to actually make this antioxidant-rich component useful (Trachootham et al. 2009). Over the years, almost all of the review articles have repeatedly stated that plant-derived antioxidants are an enormous therapeutic that not only prevents cancer or induces tumorigenesis (Bandy and Davison 1990) but also reduces oxidative stress-induced carcinogenesis. Several groups of antioxidant-rich fruit and vegetable consumption are said to be beneficial for the prevention of any type of cancer progression (Owen et al. 2004). Olive oil, which consists of phenolic compounds, exhibits highly beneficial antioxidant properties that can lower cancer risk. Upper aerodigestive, urinary, overall gastrointestinal, and breast cancer are some of these types of cancer. Omega-3 fatty acids also reduce inflammatory conditions by producing superoxide anions. Positive effects against FRs, oxidative stress, and tissue injury have been seen in lycopene. Grape seed proanthocyanidin extract (GSPE) has shown similar effects where it exhibits great protection against these DNA damage- and FR-induced lipid peroxidation than beta-carotene and vitamins C and E (Bagchi et al. 2000). Beta-carotene also has an anticarcinogenic effect (Van Poppel et al. 1995). Green tea, which is full of catechins, has shown both antioxidative and pro-oxidative effects thus inhibiting the progression of tumorigenesis (Lambert et al. 2010). Vitamin C prevents cancer in multiple ways, sometimes by enhancing immune response, blocking nitrosamines formation, or accelerating the detoxification of liver enzymes (Glatthaar et al. 1986). Vitamin E obstructs carcinogenesis by being immunocompetent, for example, by creating resistance to bacterial infections, inhibiting mutagen formation, blocking micro cell line formation, repairing DNA membranes, and increasing humoral antibody production. Curcumin has always received great accolades for its antioxidant, anti-cancerous, and anti-inflammatory properties. Its effect on cancer has been thoroughly studied in different types of cancer including breast cancer, colorectal cancer, pancreatic cancer, prostate cancer, and head and neck cancer both in vitro and in vivo (Tomeh et al. 2019). Sesame oil, which is widely used, is also rich in antioxidant properties that can prevent carcinogenesis (Sinha et al. 2021).

15.3.2 Plant-Derived Antioxidant's Role Against Alzheimer's Disease

Alzheimer's disease (AD) is one of the fastest growing cognitive diseases in the world. Plant-derived antioxidants such as phenolic acids, flavonoids, vitamins, and some active peptides help to scavenge the FRs and reduce damage to the nerve cells, thus protecting the brain and nerve cells against ageing (Del Río and Luis 2015). Tocopherols and carotenoids are often reported in improving memory levels. Resveratrol also has a protective effect on AD. Vitamin C can reduce metal-like iron and copper in the body to thus scavenge reactive species, peroxyl, and hydroxyl resulting in neuroprotection. Vitamin E has the abilities to inhibit neuronal deaths occurring from inflammation, eliminate FRs in red blood cell membranes, and stop the spread of lipoperoxidation, thus preventing hemolysis (Sinyor et al. 2020). Coffee or its antioxidative derivative caffeine has been found to dramatically decrease the cholesterol-induced phosphorylation of tau proteins and amyloid accumulation in rabbits (Prasanthi et al. 2010). Furthermore, curcumin inhibits peroxidases by crossing the blood-brain barrier and therefore reduces the release of ROS by stimulating neutrophils resulting in decreased cytokine inflammation (Reddy et al. 2018).

15.3.3 Plant-Derived Antioxidant's Role Against Ageing

FRs often damage cells, and the human body is constantly finding ways to combat this damage. A reduction of radicals may lead to a delay in ageing. Almost all plant-derived antioxidants have some type of anti-ageing property. Vitamin C or ascorbic acid, the most popular among all of them, has multiple roles ranging from depigmentation and stimulating blood flow to epidermal differentiation (Boyce et al. 2002). Both α-Tocopherol and γ-Tocopherol suppress melanogenesis in which it is seen that γ-Tocopherol has more ROS scavenging than α-Tocopherol (Yoshida et al. 2006). They also act as a cofactor for enzymatic activity that hydroxylates prolyl residues in elastic and procollagen (Myllylä et al. 1984). Resveratrol prevents skin ageing and increases cell survival. It also inhibits tyrosinase activity, which, in turn, decreases pigmentation (Cao et al. 2009). Ubiquinone or CoQ10 further increases epidermal basement membrane components to thus result in anti-ageing effects (Muta-Takada et al. 2009). Lycopene concentration contributes to reducing skin roughness (Darvin et al. 2008).

15.3.4 Plant-Derived Antioxidant's Role Against Cardiovascular Disease

The American Heart Association (AHA) has dietary guidelines that advise that antioxidants rich in fruits, vegetables, lean meat, and low-fat dairy products may reduce the risk of cardiovascular diseases (Kris-Etherton et al. 2004). Polyunsaturated fatty acids (PUFA) are present in a major part of low-density lipoproteins (LDL). High levels of oxidized lipids often damage the blood vessels and lead to the formation of foam cells, which cause atherosclerosis. This oxidized LDL unfailingly damages endothelial cells. Antioxidants such as vitamin E or beta-carotene help in the prevention of these cardiovascular diseases (Lobo et al. 2010). Some studies have reported that selenoprotein P that is found in plasma not only plays a pivotal role in the prevention of atherosclerosis (Traulsen et al. 2004) but also regulates plasma cholesterol levels. It has further been found that the selenium level in humans helps to lower the risk of heart disease (Flores-Mateo et al. 2006). Both vitamins C and E have been found to impart valuable and positive effects in the functioning of the vascular endothelial in patients with cardiovascular disease (Brown et al. 2001). Patients suffering from hypertension are often suggested to include pomegranate juice in their diet as it significantly reduces systolic blood pressure by protecting against lipid peroxidation to thus inhibit serum angiotensin (Aviram et al. 2000). A clinical study on carotid artery stenosis patients who were subjected to having pomegranate juice for 3 consecutive years reported reduced blood pressure, LDL oxidation, and carotid intima-media thickness (Aviram et al. 2004). Polyphenol-rich walnuts are reported to lower LDL cholesterol in vitro. Another research article concluded that the consumption of the right number of nuts and walnuts can prominently reduce the risk of coronary artery disease (Feldman 2002).

15.3.5 Plant-Derived Antioxidant's Role Against Diabetes

Oxidative stress and inflammation have been identified as important factors in the development and progression of diabetes and its consequences over time. As a result, many studies have been conducted on the effect of antioxidants in diabetes. The negative effects of beta cells cause type-2 diabetes. Many studies show that a high glucose level can result in the production of ROS, which can contribute to oxidative stress (Suresh et al. 2021). Antioxidants have been explored in various experimental studies to see if they might lessen diabetes problems to a greater extent. Antioxidants can be employed as a pharmacological substrate and a combination medicine in the treatment of diabetes because of their action (Koya et al. 2003).

Antioxidants can protect the function of beta cells by resisting oxidative stress, which attacks beta cells. As a result, it can help to reduce diabetes-related problems and restore insulin sensitivity. Antioxidant consumption in the diet can also be used as a technique in the treatment of diabetes (Marrazzo et al. 2014). Various antioxidants are found in nature that can help lessen the chance of developing this condition. These include vitamin E, vitamin C, alpha lipoic acid, and selenium. These antioxidants function in lowering the risk of diabetes if consumed in a specified amount (Cao et al. 2015). For example, several studies in recent years have shown that astaxanthin may be effective in the treatment and control of diabetes. Moreover, astaxanthin has been found to protect T-cells, neurons, and various organs, including the eyes, kidneys, and liver, against oxidative damage caused by diabetes (Kanwugu et al. 2021). In both T1DM and T2DM, epidemiological studies have found a deterioration in cognitive function and a high chance of dementia (Biessels and Despa 2018; Feng et al. 2018). More than a dozen studies have found that astaxanthin improves cognitive performance. Supplementing with astaxanthin has been observed to reduce cognitive impairment in animal diabetes models, mostly by reducing inflammation and oxidative stress (Feng et al. 2018; Li et al. 2016; X. Zhou et al. 2015). However, the increase in cognitive function is not statistically significant in the limited human trials that are available (Hayashi et al. 2018; Katagiri et al. 2011). Depression is another well-known diabetes comorbidity that affects around 20% of diabetic people (Zhou et al. 2017). Given these factors, astaxanthin has a promising future in the management of diabetes and its complications, especially since foods and natural edible products possessing biological and pharmacological properties are becoming popular in the 21st-century food and pharmaceutical sector (Kanwugu et al. 2021). Resveratrol reduces endothelial nitric oxide synthase production (Yar et al. 2012), oxidative biomarker expression, and superoxide dismutase activity in diabetic rats' retinas and blood (Soufi et al. 2012). In diabetic rats, carotenoids such as lutein and zeaxanthin appear to reduce Diabetic Retinopathy (DR). Zeaxanthin has anti-inflammatory properties and prevents oxidative damage (Kowluru et al. 2008). In addition, lutein reduces apoptosis indicators (caspase-3) and normalises retinal oxidative stress markers (Sasaki et al. 2010). Oral antioxidant supplementation may be an effective long-term supplementary therapy for non-proliferative DR. Another study found that supplementing lutein and zeaxanthin for three months improved visual acuity, contrast sensitivity, and foveal thickness in type 1 diabetic mellitus (T1DM) and type 2 diabetic mellitus (T2DM) patients (Hu et al. 2011).

15.4 FUTURE STRATEGIC ROADMAP FOR THE PATHOPHYSIOLOGICAL USES OF PLANT-DERIVED ANTIOXIDANTS

Plant antioxidants' therapeutic relevance cannot be determined using in vitro antioxidant potential evaluation techniques. Furthermore, multiple variables impact the capacity of plant-derived antioxidants and the existence of co-antioxidants along with transition metal ions. As a result, the findings of in vivo plant antioxidant evaluation studies are inconsistent (Antolovich et al. 2002). Consequently, a comprehensive research strategy involving a variety of in vitro and in vivo trials is required to offer more precise therapeutic values for plant antioxidants. To impart therapeutic antioxidant properties amongst plants and their constituents, one often-advised technique is to undertake both in vitro and in vivo antioxidant evaluation strategies at the same time. In vitro plant antioxidant evaluation studies should be driven by in vivo results, not the other way around, according to Holst and Williamson (Holst et al. 2008). They also proposed that after a phytochemical has been proved to have an impact in vivo, the mechanisms of this effect may be studied in vitro to minimise failures when evaluating in vitro principles in vivo. Because flavonoids and lignans are converted to low molecular antioxidants in the body, the suggested experimental research studies are thought to be very appropriate for investigating the antioxidant activity of these phytochemicals (Sinha et al. 2021). The majority of dietary flavonoids consumed by humans are metabolised into different phenolic acids, which may have radical scavenging properties (Pietta et al. 2000). In the same way, the lignan secoisolariciresinol diglucoside is metabolised in the body to produce stronger antioxidants such as secoisolariciresinol, enterodiol, and enterolactone (Prasad et al. 2000). To attribute a more accurate therapeutic value to specific or mixed-plant antioxidant entities, a disease pathophysiology that targets a coupled in vitro and in vivo antioxidant activity assessment as a study strategy is recommended. To use this technique, researchers must first identify the primary target of FR-related disease pathophysiology, such as mitochondrial malfunction, which is a contributing factor in the development of various degenerative illnesses. Following this, screening is necessary for low molecular antioxidants with effective radical scavenging activity in vitro and the ability to cross the blood-brain barrier and reach the target by using in silico methods, upon comparing them to proposed or known structural analogues of mitochondria-targeted antioxidants. Therefore, in vivo research must be used to validate the findings.

15.5 CONCLUSION

Antioxidant molecules derived from plant sources that are pharmacologically strong and have few or no adverse effects are now being sought for application in protective medicine and the food sector. The use of various chemicals, pesticides, pollution, smoking, alcohol consumption, and even certain synthetic medicines all raise the risk of illness caused by FRs. Increasing our understanding of antioxidant phytoconstituents and incorporating them into our everyday lives and diets can provide the human body with enough assistance to combat these illnesses. Pathological states of illness produced by FRs can also be managed with phytoconstituents and herbal medication. As a result, greater scientific interest in this topic is critical not only to obtain a better understanding of existing antioxidant molecules but also to identify novel natural antioxidants. As a result, commercially important items for human health can be ensured. Biofortification techniques aimed at increasing plants' nutraceutical and/or medicinal value will allow for the development of novel, enhanced plant-derived foods and goods. These might be beneficial in the preservation of human health and in the prevention and treatment of human diseases. Integrating this scientific information into ex situ value-added measures underlines the need for the in situ protection of local species, which must be included in future natural resource management plans.

15.6 ACKNOWLEDGEMENT

The authors want to express their gratitude to Sriranjani Sarkar of Auxilium Convent School, Dum Dum for the digital drawing.

REFERENCES

Age-Related Eye Disease Study Research Group. "A randomized, placebo-controlled, clinical trial of high-dose supplementation with vitamins C and E, beta carotene, and zinc for age-related macular degeneration and vision loss: AREDS report no. 8." *Archives of Ophthalmology* 119, no. 10 (2001): 1417–1436.

Alscher, Ruth G., Janet L. Donahue, and Carole L. Cramer. "Reactive oxygen species and antioxidants: Relationships in green cells." *Physiologia Plantarum* 100, no. 2 (1997): 224–233.

Anderson, Koren J., Suzanne S. Teuber, Alayne Gobeille, Peader Cremin, Andrew L. Waterhouse, and Francene M. Steinberg. "Walnut polyphenolics inhibit in vitro human plasma and LDL oxidation." *The Journal of Nutrition* 131, no. 11 (2001): 2837–2842.

Antolovich, Michael, Paul D. Prenzler, Emilios Patsalides, Suzanne McDonald, and Kevin Robards. "Methods for testing antioxidant activity." *Analyst* 127, no. 1 (2002): 183–198.

Aviram, Michael, Leslie Dornfeld, Mira Rosenblat, Nina Volkova, Marielle Kaplan, Raymond Coleman, Tony Hayek, Dita Presser, and Bianca Fuhrman. "Pomegranate juice consumption reduces oxidative stress, atherogenic modifications to LDL, and platelet aggregation: Studies in humans and in atherosclerotic apolipoprotein E–deficient mice." *The American Journal of Clinical Nutrition* 71, no. 5 (2000): 1062–1076.

Aviram, Michael, Mira Rosenblat, Diana Gaitini, Samy Nitecki, Aaron Hoffman, Leslie Dornfeld, Nina Volkova, et al. "Pomegranate juice consumption for 3 years by patients with carotid artery stenosis reduces common carotid intima-media thickness, blood pressure and LDL oxidation." *Clinical Nutrition* 23, no. 3 (2004): 423–433.

Bagchi, Debasis, Manashi Bagchi, Sidney J. Stohs, Dipak K. Das, Sidhartha D. Ray, Charles A. Kuszynski, Shantaram S. Joshi, et al. "Free radicals and grape seed proanthocyanidin extract: Importance in human health and disease prevention." *Toxicology* 148, no. 2–3 (2000): 187–197.

Bandy, Brian, and Allan J. Davison. "Mitochondrial mutations may increase oxidative stress: Implications for carcinogenesis and aging?." *Free Radical Biology and Medicine* 8, no. 6 (1990): 523–539.

Bielski, Benon H. J., Helen W. Richter, and Phillip C. Chan. "Some properties of the ascorbate free radical." *Annals of the New York Academy of Sciences* 258, no. 1 (1975): 231–237.

Biessels, Geert Jan, and Florin Despa. "Cognitive decline and dementia in diabetes mellitus: Mechanisms and clinical implications." *Nature Reviews Endocrinology* 14, no. 10 (2018): 591–604.

Boyce, Steven T., Andrew P. Supp, Viki B. Swope, and Glenn D. Warden. "Vitamin C regulates keratinocyte viability, epidermal barrier, and basement membrane in vitro, and reduces wound contraction after grafting of cultured skin substitutes." *Journal of Investigative Dermatology* 118, no. 4 (2002): 565–572.

Brown, Allison A., and Frank B. Hu. "Dietary modulation of endothelial function: Implications for cardiovascular disease." *The American Journal of Clinical Nutrition* 73, no. 4 (2001): 673–686.

Cao, Cong, Shan Lu, Rebecca Kivlin, Brittany Wallin, Elizabeth Card, Andrew Bagdasarian, Tyrone Tamakloe, et al. "SIRT1 confers protection against UVB-and H_2O_2-induced cell death via modulation of p53 and JNK in cultured skin keratinocytes." *Journal of Cellular and Molecular Medicine* 13, no. 9b (2009): 3632–3643.

Cao, Hui, Yixi Xie, and Xiaoqing Chen. "Type 2 diabetes diminishes the benefits of dietary antioxidants: Evidence from the different free radical scavenging potential." *Food Chemistry* 186 (2015): 106–112.

Carocho, Márcio, and Isabel CFR Ferreira. "A review on antioxidants, prooxidants and related controversy: Natural and synthetic compounds, screening and analysis methodologies and future perspectives." *Food and Chemical Toxicology* 51 (2013): 15–25.

Chen, Junshi. "An original discovery: Selenium deficiency and Keshan disease (an endemic heart disease)." *Asia Pacific Journal of Clinical Nutrition* 21, no. 3 (2012): 320–326.

Clark, Larry C., Gerald F. Combs, Bruce W. Turnbull, Elizabeth H. Slate, Dan K. Chalker, James Chow, Loretta S. Davis, et al. "Effects of selenium supplementation for cancer prevention in patients with carcinoma of the skin: A randomized controlled trial." *Jama* 276, no. 24 (1996): 1957–1963.

Darvin, Maxim, Alexa Patzelt, Saskia Gehse, Sabine Schanzer, Christian Benderoth, Wolfram Sterry, and Juergen Lademann. "Cutaneous concentration of lycopene correlates significantly with the roughness of the skin." *European Journal of Pharmaceutics and Biopharmaceutics* 69, no. 3 (2008): 943–947.

Del Río, Luis A. "ROS and RNS in plant physiology: An overview." *Journal of Experimental Botany* 66, no. 10 (2015): 2827–2837.

Del Río, Luis A., Luisa M. Sandalio, Francisco J. Corpas, José M. Palma, and Juan B. Barroso. "Reactive oxygen species and reactive nitrogen species in peroxisomes. Production, scavenging, and role in cell signaling." *Plant Physiology* 141, no. 2 (2006): 330–335.

Demirkol, Omca, Craig Adams, and Nuran Ercal. "Biologically important thiols in various vegetables and fruits." *Journal of Agricultural and Food Chemistry* 52, no. 26 (2004): 8151–8154.

Evans, Herbert M., and K. Scott Bishop. "On the existence of a hitherto unrecognized dietary factor essential for reproduction." *Science* 56, no. 1458 (1922): 650–651.

Feldman, Elaine B. "The scientific evidence for a beneficial health relationship between walnuts and coronary heart disease." *The Journal of Nutrition* 132, no. 5 (2002): 1062S–1101S.

Feng, Yonghao, Aiqun Chu, Qiong Luo, Men Wu, Xiaohong Shi, and Yinghui Chen. "The protective effect of astaxanthin on cognitive function via inhibition of oxidative stress and inflammation in the brains of chronic T2DM rats." *Frontiers in Pharmacology* 9 (2018): 748.

Flores-Mateo, Gemma, Ana Navas-Acien, Roberto Pastor-Barriuso, and Eliseo Guallar. "Selenium and coronary heart disease: A meta-analysis." *The American Journal of Clinical Nutrition* 84, no. 4 (2006): 762–773.

Gill, Sarvajeet Singh, and Narendra Tuteja. "Reactive oxygen species and antioxidant machinery in abiotic stress tolerance in crop plants." *Plant Physiology and Biochemistry* 48, no. 12 (2010): 909–930.

Glatthaar, Beat E., Dietrich H. Hornig, and Ulrich Moser. "The role of ascorbic acid in carcinogenesis." In *Essential Nutrients in Carcinogenesis*, 357–377. Boston, MA: Springer, 1986.

Halliwell, Barry, and John MC Gutteridge. "Role of free radicals and catalytic metal ions in human disease: an overview." *Methods in Enzymology* 186 (1990): 1–85.

Hayashi, Masahiro, Takashi Ishibashi, and Takashi Maoka. "Effect of astaxanthin-rich extract derived from Paracoccus carotinifaciens on cognitive function in middle-aged and older individuals." *Journal of Clinical Biochemistry and Nutrition* 62, no. 2 (2018): 195–205.

Holst, Birgit, and Gary Williamson. "Nutrients and phytochemicals: From bioavailability to bioefficacy beyond antioxidants." *Current Opinion in Biotechnology* 19, no. 2 (2008): 73–82.

Hu, Bo-Jie, Ya-Nan Hu, Song Lin, Wen-Jiang Ma, and Xiao-Rong Li. "Application of Lutein and Zeaxanthin in nonproliferative diabetic retinopathy." *International Journal of Ophthalmology* 4, no. 3 (2011): 303–306.

Kanwugu, Osman N., Tatiana V. Glukhareva, Irina G. Danilova, and Elena G. Kovaleva. "Natural antioxidants in diabetes treatment and management: Prospects of astaxanthin." *Critical Reviews in Food Science and Nutrition* (2021): 1–24.

Katagiri, Mikiyuki, Akira Satoh, Shinji Tsuji, and Takuji Shirasawa. "Effects of astaxanthin-rich Haematococcus pluvialis extract on cognitive function: A randomised, double-blind, placebo-controlled study." *Journal of Clinical Biochemistry and Nutrition* (2011): 1203270150–1203270150.

Kesse-Guyot, Emmanuelle, Valentina A. Andreeva, Véronique Ducros, Claude Jeandel, Chantal Julia, Serge Hercberg, and Pilar Galan. "Carotenoid-rich dietary patterns during midlife and subsequent cognitive function." *British Journal of Nutrition* 111, no. 5 (2014): 915–923.

Kowluru, Renu A., Bindu Menon, and Dennis L. Gierhart. "Beneficial effect of zeaxanthin on retinal metabolic abnormalities in diabetic rats." *Investigative Ophthalmology & Visual Science* 49, no. 4 (2008): 1645–1651.

Koya, Daisuke, Kazuyuki Hayashi, Munehiro Kitada, Atsunori Kashiwagi, Ryuichi Kikkawa, and Masakazu Haneda. "Effects of antioxidants in diabetes-induced oxidative stress in the glomeruli of diabetic rats." *Journal of the American Society of Nephrology* 14, no. suppl 3 (2003): S250-S253.

Kris-Etherton, Penny M., Alice H. Lichtenstein, Barbara V. Howard, Daniel Steinberg, and Joseph L. Witztum. "Antioxidant vitamin supplements and cardiovascular disease." *Circulation* 110, no. 5 (2004): 637–641.

Kryukov, Gregory V., Sergi Castellano, Sergey V. Novoselov, Alexey V. Lobanov, Omid Zehtab, Roderic Guigó, and Vadim N. Gladyshev. "Characterization of mammalian selenoproteomes." *Science* 300, no. 5624 (2003): 1439–1443.

Kurutas, Ergul Belge. "The importance of antioxidants which play the role in cellular response against oxidative/nitrosative stress: Current state." *Nutrition Journal* 15, no. 1 (2015): 1–22.

Lambert, Joshua D., and Ryan J. Elias. "The antioxidant and pro-oxidant activities of green tea polyphenols: A role in cancer prevention." *Archives of Biochemistry and Biophysics* 501, no. 1 (2010): 65–72.

Li, Xiaobin, Zhonghua Qi, Longshan Zhao, and Zhan Yu. "Astaxanthin reduces type 2 diabetic-associated cognitive decline in rats via activation of PI3K/Akt and attenuation of oxidative stress." *Molecular Medicine Reports* 13, no. 1 (2016): 973–979.

Lobo, Vijaya, Avinash Patil, A. Phatak, and Naresh Chandra. "Free radicals, antioxidants and functional foods: Impact on human health." *Pharmacognosy Reviews* 4, no. 8 (2010): 118.

Lü, Jian-Ming, Peter H. Lin, Qizhi Yao, and Changyi Chen. "Chemical and molecular mechanisms of antioxidants: Experimental approaches and model systems." *Journal of Cellular and Molecular Medicine* 14, no. 4 (2010): 840–860.

Marrazzo, Giuseppina, Ignazio Barbagallo, Fabio Galvano, Michele Malaguarnera, Diego Gazzolo, Alessandro Frigiola, Nicolantonio D'Orazio et al. "Role of dietary and endogenous antioxidants in diabetes." *Critical Reviews in Food Science and Nutrition* 54, no. 12 (2014): 1599–1616.

Milman, Uzi, Shany Blum, Chen Shapira, Doron Aronson, Rachel Miller-Lotan, Yefim Anbinder, Junia Alshiek, et al. "Vitamin E supplementation reduces cardiovascular events in a subgroup of middle-aged individuals with both type 2 diabetes mellitus and the haptoglobin 2–2 genotype: A prospective double-blinded clinical trial." *Arteriosclerosis, Thrombosis, and Vascular Biology* 28, no. 2 (2008): 341–347.

Mittler, Ron. "Oxidative stress, antioxidants and stress tolerance." *Trends in Plant Science* 7, no. 9 (2002): 405–410.

Møller, Ian M., Poul Erik Jensen, and Andreas Hansson. "Oxidative modifications to cellular components in plants." *Annual Review of Plant Biology* 58 (2007): 459–481.

Muta-Takada, Keiko, Takatomo Terada, Haruyo Yamanishi, Yutaka Ashida, Shinji Inomata, Toshio Nishiyama, and Satoshi Amano. "Coenzyme Q10 protects against oxidative stress-induced cell death and enhances the synthesis of basement membrane components in dermal and epidermal cells." *Biofactors* 35, no. 5 (2009): 435–441.

Myllylä, R., K. Majamaa, V. Günzler, H. M. Hanauske-Abel, and K. I. Kivirikko. "Ascorbate is consumed stoichiometrically in the uncoupled reactions catalyzed by prolyl 4-hydroxylase and lysyl hydroxylase." *Journal of Biological Chemistry* 259, no. 9 (1984): 5403–5405.

Owen, R. W., R. Haubner, Gl Würtele, W. E. Hull, B. Spiegelhaider, and H. Bartsch. "Olives and olive oil in cancer prevention." *European Journal of Cancer Prevention* (2004): 319–326.

Padayatty, Sebastian J., Arie Katz, Yaohui Wang, Peter Eck, Oran Kwon, Je-Hyuk Lee, Shenglin Chen, et al. "Vitamin C as an antioxidant: Evaluation of its role in disease prevention." *Journal of the American College of Nutrition* 22, no. 1 (2003): 18–35.

Palgi, Aviva. "Association between dietary changes and mortality rates: Israel 1949 to 1977; a trend-free regression model." *The American Journal of Clinical Nutrition* 34, no. 8 (1981): 1569–1583.

Pietta, Pier-Giorgio. "Flavonoids as antioxidants." *Journal of Natural Products* 63, no. 7 (2000): 1035–1042.

Poeggeler, Burkhard, Miguel A. Pappolla, Rüdiger Hardeland, Arash Rassoulpour, Paul S. Hodgkins, Paolo Guidetti, and Robert Schwarcz. "Indole-3-propionate: A potent hydroxyl radical scavenger in rat brain." *Brain Research* 815, no. 2 (1999): 382–388.

Poeggeler, B., R. J. Reiter, R. Hardeland, E. Sewerynek, D. Melchiorri, and L. R. Barlow-Walden. "Melatonin, a mediator of electron transfer and repair reactions, acts synergistically with the chain-breaking antioxidants ascorbate, trolox and glutathione." *Neuroendocrinology Letters* 17, no. 2 (1995): 87–92.

Prasad, Kailash. "Antioxidant activity of secoisolariciresinol diglucoside-derived metabolites, secoisolariciresinol, enterodiol, and enterolactone." *International Journal of Angiology* 9, no. 04 (2000): 220–225.

Prasanthi, Jaya RP., Bhanu Dasari, Gurdeep Marwarha, Tyler Larson, Xuesong Chen, Jonathan D. Geiger, and Othman Ghribi. "Caffeine protects against oxidative stress and Alzheimer's disease-like pathology in rabbit hippocampus induced by cholesterol-enriched diet." *Free Radical Biology and Medicine* 49, no. 7 (2010): 1212–1220.

Reddy, P. Hemachandra, Maria Manczak, Xiangling Yin, Mary Catherine Grady, Andrew Mitchell, Sahil Tonk, Chandra Sekhar Kuruva, et al. "Protective effects of Indian spice curcumin against amyloid-β in Alzheimer's disease." *Journal of Alzheimer's Disease* 61, no. 3 (2018): 843–866.

Sasaki, M., Yoko Ozawa, T. Kurihara, S. Kubota, Kenya Yuki, K. Noda, S. Kobayashi, S. Ishida, and Kazuo Tsubota. "Neurodegenerative influence of oxidative stress in the retina of a murine model of diabetes." *Diabetologia* 53, no. 5 (2010): 971–979.

Sinha, Aritrika, Das, Abhishek, Chakraborty, Rishita and Suranjana, Sarkar. 2021. "An Overview of the Properties and Pharmaceutical uses of Bioactive Compounds from Sesame (*Sesamum indicum* L.)." In: Tathagata Ray Chaudhuri, Aditi Sarbajna Sandip Poddar and Amiya Bhaumik (Eds.), *Therapeutics: Key to Well Being.* Lincoln Research and Publications Limited. (2021): 18–27. doi:10.31674/book.2021.tkwb

Sinyor, Benjamin, Jocelyn Mineo, and Christopher Ochner. "Alzheimer's disease, inflammation, and the role of antioxidants." *Journal of Alzheimer's Disease Reports* 4, no. 1 (2020): 175–183.

Soufi, Farhad Ghadiri, Mina Vardyani, Roghayeh Sheervalilou, Mustafa Mohammadi, and Mohammad Hussein Somi. "Long-term treatment with resveratrol attenuates oxidative stress pro-inflammatory mediators and apoptosis in streptozotocin-nicotinamide-induced diabetic rats." *General Physiology and Biophysics* 31, no. 4 (2012): 431–438.

Stanley, J., and A. Mogg. "MAFF antioxidant research programme." *The Biochemist* (1995): 22–24.

Suresh, Varuna, Amala Reddy, Pavithra Muthukumar, and Thendarl Selvam. "Antioxidants: Pharmacothearapeutic boon for diabetes." In: Viduranga, Y., Waisundara (Ed.), *Antioxidants-Benefits, Sources, Mechanisms of Action*. London: IntechOpen, 2021.

Szymanska, Renata, Pavel Pospisil, and Jerzy Kruk. "Plant-derived antioxidants in disease prevention." *Oxidative Medicine and Cellular Longevity* 2016 (2016).

Tomeh, Mhd Anas, Roja Hadianamrei, and Xiubo Zhao. "A review of curcumin and its derivatives as anticancer agents." *International Journal of Molecular Sciences* 20, no. 5 (2019): 1033.

Traber, Maret G., and Jan F. Stevens. "Vitamins C and E: beneficial effects from a mechanistic perspective." *Free Radical Biology and Medicine* 51, no. 5 (2011): 1000–1013.

Trachootham, Dunyaporn, Jerome Alexandre, and Peng Huang. "Targeting cancer cells by ROS-mediated mechanisms: A radical therapeutic approach?." *Nature Reviews Drug Discovery* 8, no. 7 (2009): 579–591.

Traulsen, Henrik, Holger Steinbrenner, Darius P. Buchczyk, Lars-Oliver Klotz, and Helmut Sies. "Selenoprotein P protects low-density lipoprotein against oxidation." *Free Radical Research* 38, no. 2 (2004): 123–128.

Tsao, Rong, and Zeyuan Deng. "Separation procedures for naturally occurring antioxidant phytochemicals." *Journal of Chromatography B* 812, no. 1–2 (2004): 85–99.

Van Poppel, Geert, and R. Alexandra Goldbohm. "Epidemiologic evidence for beta-carotene and cancer prevention." *The American Journal of Clinical Nutrition* 62, no. 6 (1995): 1393S-1402S.

Yar, Atiye Seda, Sevda Menevse, Irem Dogan, Ebru Alp, Volkan Ergin, Ahmet Cumaoglu, Aysel Aricioglu, Abdullah Ekmekci, and Adnan Menevse. "Investigation of ocular neovascularization–related genes and oxidative stress in diabetic rat eye tissues after resveratrol treatment." *Journal of Medicinal Food* 15, no. 4 (2012): 391–398.

Yoshida, Emiko, Tatsuya Watanabe, Jiro Takata, Akihiko Yamazaki, Yoshiharu Karube, and Shizuko Kobayashi. "Topical application of a novel, hydrophilic γ-tocopherol derivative reduces photo-inflammation in mice skin." *Journal of Investigative Dermatology* 126, no. 7 (2006): 1633–1640.

Yoshizawa, Kazuko, Walter C. Willett, Steven J. Morris, Meir J. Stampfer, Donna Spiegelman, Eric B. Rimm, and Edward Giovannucci. "Study of prediagnostic selenium level in toenails and the risk of advanced prostate cancer." *JNCI: Journal of the National Cancer Institute* 90, no. 16 (1998): 1219–1224.

Zhao, Jian, Lawrence C. Davis, and Robert Verpoorte. "Elicitor signal transduction leading to production of plant secondary metabolites." *Biotechnology Advances* 23, no. 4 (2005): 283–333.

Zhou, Xiao-yan, Fang Zhang, Xiao-tong Hu, Jing Chen, Ren-xian Tang, Kui-yang Zheng, and Yuan-jian Song. "Depression can be prevented by astaxanthin through inhibition of hippocampal inflammation in diabetic mice." *Brain Research* 1657 (2017): 262–268.

Zhou, Xiaoyan, Fang Zhang, Xiaotong Hu, Jing Chen, Xiangru Wen, Ying Sun, Yonghai Liu, et al. "Inhibition of inflammation by astaxanthin alleviates cognition deficits in diabetic mice." *Physiology & Behavior* 151 (2015): 412–420.

16 In Vitro Genetic Approach to Identify Disease Gene Mutations

Pratik Talukder[1,], Anushka Dutta[1] and Suravi Datta[1]*
[1]Department of Biotechnology, University of Engineering and Management, Kolkata, India
*Correspondence to: pratik.talukder@uem.edu.in

16.1 INTRODUCTION

It has long been recognised that new mutations might result in genetic disease, but it is only recently that it has become possible to assess the true impact of these diseases by utilizing family-based whole-genome or whole-exome scanning methods. Inherited genetic variation is the main focus of conventional genetic techniques such as linkage studies and genome-wide association studies. We can now examine the function of de novo mutations in heath and illness for the first time because of impartial whole-genome and whole-exome sequencing (Veltman and Brunner 2012).

Our genome experiences an average of 74 de novo single-nucleotide variations (SNVs), three unique minor insertions or deletions (indels), and 0.02 larger copy number variants (CNVs) per generation (per individual). Advanced paternal age during conception, a geographic genomic architecture rich in segmental duplications, and unidentified genetic variation are risk factors that raise this de novo mutation rate. It can be difficult to predict the toxicity of uncommon de novo missense mutations in new genes (Mahdieh and Rabbani 2013). However, this is made much easier by the discovery of recurring mutations in people with comparable characteristics, which enables in-depth genotype-phenotype research. As the reoccurring mutations will be uncommon for genetically varied disorders, the discovery of these mutations needs international cooperation.

16.2 TYPES OF GENETIC MUTATIONS

An alteration in one or many genes is known as a gene mutation. Certain mutations can result in genetic diseases or disorders.

16.2.1 Point Mutation

When one single base pair is introduced, removed, or altered in a genome, it is referred to as a point mutation. Although the majority of point mutations are harmless, they can also have a variety of functional effects, such as modifications to encoded proteins or changes in gene expression (Bais et al. 2004).

Single point mutations in particular genes, which result in the generation of non-functional proteins, are frequently the cause of monogenic disorders. Point mutations are typically the root cause of numerous blood illnesses, including ß-thalassemia, sickle cell disease, hereditary spherocytosis, Fanconi anaemia, and haemophilia A and B.

 DOI: 10.1201/9781003302131-17

16.2.2 Insertion Mutation

In terms of genomics, an insertion mutation is a form of mutation in which one or more nucleotides are inserted into a DNA sequence. Any number of nucleotides can be added during an insertion, ranging from one nucleotide to an entire chromosome (Feuk 2010).

Insertional mutations are the cause of a variety of illnesses. For instance, a few conditions brought on by insertional mutations are cystic fibrosis, Huntington's disease, Fragile X syndrome, and myotonic dystrophy.

16.2.3 Deletion Mutation

In terms of genomics, a deletion is a sort of mutation that happens when one or even more nucleotides are lost from a section of DNA. Any number of nucleotides can be lost during a deletion, ranging from one nucleotide to the loss of an entire section of a chromosome.

Many genetic disorders, including two-thirds of the instances of cystic fibrosis and Cat's Cry syndrome – so named because babies with this syndrome frequently have a cry that sounds like a cat meowing – are really caused by deletion mutations.

16.2.4 Substitution Mutation

In terms of genomics, substitution is a type of mutation where one nucleotide is changed for a different nucleotide. Substitution mutation can also be used to describe switching out one amino acid for another in a protein.

It only takes a single substitution mutation to develop the blood condition sickle-cell anaemia. The DNA region that codes for one unit of haemoglobin has a single nucleotide replaced because of the mutation. Colour blindness can result from a substitution mutation in the gene encoding for one of these proteins (Davisson et al. 2012).

16.2.5 Forward Mutation

A forward mutation is a genetic change from a wild-type to a mutant allele or the generation of a new mutation from a wild-type allele. New qualities in the population evolve as a result of a forward mutation.

Family-based studies and genome-wide association (GWA) studies are examples of forward genetic methods used in humans. For instance, family investigations have shown that mutations in the clock genes PERIOD2 (PER2) and CASEIN KINASE 1 DELTA (CK1) are linked to the circadian condition of familial advanced sleep phase syndrome (FASPS).

16.2.6 Backward Mutation

A mutation is often a one-way process, but occasionally, a mutation can return the original (wild-type) allele to the population; this is known as a backward or back mutation. Back mutation is an extremely odd and infrequent occurrence in nature.

Reversion assays have been used mostly to study stress-induced mutagenesis, which is the increase in the mutation rate in response to cellular stress, frequently under the direction of stress responses.

16.2.7 Silent Mutation

A silent mutation is non-expressive. In a silent mutation, a new codon codes for the same amino acid as the wild-type amino acid (Bond et al. 2020).

In genes considered to be involved in hereditary illnesses such Laron dwarfism, Crouzon syndrome, +-thalassemia, and phenylalanine hydroxylase deficiency (phenylketonuria (PKU)), silent mutations that result in such skipping of exon excision have been discovered.

16.2.8 Mis-Sense Mutation

In a mis-sense mutation, a codon is created as a result of a nucleotide mutation that codes for various amino acids. It may cause a protein's function to be altered or lost.

At the protein level, this triggers the leucine to take the place of arginine at position 527. This makes the salt bridge collapse, and the structure becomes unstable. This shows up as overlapping mandibuloacral dysplasia and progeria syndrome at the phenotypic level.

16.2.9 Nonsense Mutation

A premature protein gains a stop codon. Since a stop codon ends protein synthesis and produces an early or truncated protein, it halts protein synthesis.

Nonsense mutations can result in both common diseases, such as malignancies, metabolic problems, and neurological disorders, and rare genetic diseases, such as Duchenne muscular dystrophy, cystic fibrosis, and haemophilia (Yan et al. 2000).

16.2.10 Frameshift Mutation

A base pair change results in an aberrant reading frame, which, in turn, leads to the production of a defective protein. There is a start codon and a stop codon for each reading frame. A distinct coding sequence can be found in the space between both codons. A frameshift mutation causes this reading frame to move from one location in the genome to another because of a change in the DNA. As a result, the start or stop codon's position shifts. Frameshift mutations are responsible for Crohn's disease, cystic fibrosis, and certain cancers (Nakamura 2009).

16.2.11 Spontaneous Mutation

Unknown causes, such as metabolic mistakes, replication faults, or development errors, might result in spontaneous mutations. These are rare and random mutations: it was born that way. Because larger genes have a higher possibility of experiencing replication errors, they are more susceptible to spontaneous mutation. During replication, there are 10–5 spontaneous mutations per gene per generation (Rosenthal 1994).

Throughout a person's life, these inherited mutations are present in practically every cell of their body. Cystic fibrosis, haemophilia, and sickle cell disease are examples of hereditary mutations. Throughout a person's life, further mutations may occur on their own. Spontaneous, sporadic, or novel mutations are the terms used to describe these types of mutations.

16.2.12 Induced Mutation

When an organism is exposed to mutagenic substances, induced mutation occurs. Chemicals, UV light, and radiation are the three main mutagenic agents. Skin cancer and xeroderma pigmentation are both brought on by UV light's ability to penetrate the skin and trigger unchecked cell division. The lower intensity of UV radiation prevents it from penetrating other human tissues, but it can penetrate skin cells, where it activates oncogenes. Chemical mutagens include substances such as base analogues and alkylating agents. However, our DNA is somewhat mutagenic to every chemical. Furthermore, accumulating mutations are significantly influenced by lifestyle choices and climate change (Eichenlaub-Ritter 1996).

A change in the sex cell genes is an example of an induced mutation. The sex cell genes become mutated when the gonads are exposed to mutagens such as radiation.

16.3 GENETIC APPROACHES

Genomic disorders use biochemical and morphological assertions. Some of the approaches could be used to identify single base pair alterations overall; this entails locating each single nucleotide sequence and contrasting it with the order of the healthy gene, which is done by direct sequencing. Direct DNA sequencing, which can correctly discover and characterise the majority of pattern changes, is, in principle, the method for alteration identification that is most efficient in such a Target DNA. Capillary array electrophoresis with laser-induced fluorescence detection is used in this category for Sanger DNA sequencing (Rosenstein and Zeitlin 1998).

Before the late 1970s, only one anomaly was significant enough to be noticed by the cytogenetics technique of looking at a collection of cells under a microscope – where only DNA variations were observed. Medical biochemical findings might be used to imply alterations that caused absent or changed protein expression, such as phenylketonuria (PKU). Ever since, advances in DNA modification have increased the range of techniques for identifying chromosomal mutations, with the most suitable technique varying according to the amount and type of the mutant. Certain methods are used on the DNA itself, others on the RNAs made by the transcribed operative DNA, and still others on the particular genetic polypeptide output. All of them use more than one fundamental DNA or DNA-interacting enzymatic characteristic.

16.3.1 HOW SINGLE BASE PAIR MUTATIONS ARE IDENTIFIED

- **Direct DNA Sequencing:** Locating each single nucleotide sequence and contrasting it with the order of the healthy gene is done by direct sequencing. When other approaches fail to discover the condition that is medically indicated, this typically labour-intensive procedure is used to look for unknown abnormalities or uncommon variations of a prevalent illness (cystic fibrosis).
- **DNA hybridization methods:** Techniques for separating DNA with abnormalities from DNA with a sequential manner include DNA hybridization, which relies on the strong binding of a single-stranded DNA to another strand whose pattern is the exact counterpart of the initial one. A hybridization-based technique can pinpoint the 31 most prevalent cystic fibrosis mutations, which account for almost 90% of the condition's white sufferers (Van Ommen et al.1999).
- **Restriction enzyme digestion or hybridization:** Specialised biomolecules known as restriction enzymes detect and break the DNA double helix when they come across a specific, short segment. A few of these segments may be eliminated or conversely added to by nucleotide sequence alterations, which will change the lengths of the resulting DNA pieces.

16.3.2 WHAT LARGER MUTATION APPROACHES INVOLVE

The elimination, reorganization, extension, or multiplication of portions of genomes, complete genetic mutations, or many genes constitute larger genetic variations.

- By using a tagged "probe" carrying the genetic sequence, hybridization could be used to identify whether a genome or a significant portion of a genome is present or absent. There would be a signalling if the genes are present; otherwise, there will be no signalling. Significant alterations modify the distinct appearance of breaks in the genomes formed by restriction endonucleases by altering the spacing among domains or by deleting regions altogether. For instance, the amplification of genes producing one of the proteins to form the myelin membranes results in the hereditary neuropathy Charcot-Marie-Tooth disease type 1A. Classifying the structure created when endonucleases cleave the mutant with an initial chromosome and analysing the results are necessary for diagnosis (Aitman 2001).

- **Polymerase chain reaction** is one method that is used to increase the precise domain containing the mutation. Other tests look for changed proteins rather than the actual alteration. Present-day tests for BRCA1 and BRCA2 mutations rely on the identification of the curtailed protein generated by the affected genes.

16.3.3 Other Approaches by Which Mutations are Identified

- **Gene therapy:** With the aim to counteract a variation that results in a genetic disorder, a healthy gene is inserted into a person's genome during genetic engineering. A normal gene would most probably assimilate into a chromosomal position distinct from the deficient allele once it is put into a different nucleus; while this may correct the variation, a further mutation could arise if the healthy gene incorporates into yet another gene sequence. There is a possibility that the changed cells will multiply and create sufficient regular genetic material to substitute the mutation gene, reverting the full body to its healthy state. Human gene therapy for conditions such as cancer and severe combined immunodeficiency syndrome has only been tried on somatic (body) cells thus far (SCIDS). Although somatic cells altered with gene therapy can alleviate a patient's clinical manifestations, the alteration is not carried onto the patient's progeny. The goal of germline gene therapy is to introduce improved cells into the germ line (e.g., cells of the ovary or testis). If successful, then these individuals will proceed through meiosis and contribute normally to the following century's gametes. Artificial germline gene therapy has proved to be successful in mammals but not in people (King and Sinha 2001).
- **Reverse genetics:** Reverse genetics is a branch of genomics facilitated by recombinant DNA techniques. Typically, genetic study begins with a mutant allele, which may then be linked to a particular genetic variant by Mendel crossover analysis. The path of genetic manipulation is in totally the reverse direction. Starting with a genome whose role is unclear, investigators use microscopic examination to ascertain its phenotypic expression. Gene elimination is a critical tool in genetic manipulation. The aberrant phenotype that results from altering the replicated gene with arbitrary function and utilising it to substitute the local copies will reveal whatever bioactivity this gene typically regulates.
- It is also possible to test for many mutations at once via **DNA chips or microarrays**: Screening for numerous abnormalities is a challenge that can be solved by DNA "chips" or microarrays. For each genome and every array's location, single-stranded DNA strands from various human gene sequences are glued to a sturdy substrate (such a glass slide) in a DNA chip. The microchip is again coated with denatured DNA generated from biopsies or scientific samples and prelabelled with fluorescent tags. According to this, it is possible to create DNA that symbolises every activated expression in the unit by (enzymatically) creating a DNA duplicate of the mRNA molecule existing in the cell. Any DNA in the material that is complimentary to DNA on the chip bonds to the permanent DNA strands that are matched during incubation at the correct temperature (Sadée 1999). The existence of fluorescence is then determined by using a laser; signals at any point on the microarray denote the availability of DNA for the gene fastened to the microchip at that location in the specimen. The ability to continuously study several genes gives this approach its potency. Today's microchip may store sections of thousands and thousands of distinct genes, which is sufficient to describe all the genetic traits that are active in a specific cell, such as a lymphocyte, for example. The development of chips with all genetic mutations is almost complete. The use of DNA microchips in healthcare situations still

seems to be investigational. The sequence of expression of genes detected by microarrays has nonetheless been enough to distinguish among subgroups of leukaemia, lymphoma, melanoma, and breast carcinoma in current papers, with a focus on changes in gene expression that signal cancer development and therapy tolerance (Fridovich-Keil 2019). It is not necessary to identify the genes responsible for these events in order to gain relevant knowledge.

- **Genetic testing** is another promising option, especially for disorders that may originate from various potential gene mutations. Human genotyping, which analyses numerous disease markers concurrently to ascertain disorder susceptibility, or the developing field of pharmacogenomics, which forecasts reactions to treatments based on variations in intended binding site types or in enzymatic reactions used during absorption or metabolic activity of drugs, are typical, similar applications. To quickly identify and subtype microorganisms, particularly those that are resistant to antibiotics in contagious diseases, DNA chips can be utilised (Boileau et al. 2005).
- ***In vitro* gene editing CRISPR-Cas9:** Gene editing allows for extremely precise genetic alterations to be made to biota, thereby altering their genetic composition. To remove current DNA and install substitute DNA, this is carried out by utilising enzymes, notably, endonuclease or exonuclease, that have been specifically intended to attack a certain sequence of DNA. CRISPR-Cas9, a potent developed application, is crucial amongst genomic tools. Scientists can delete and introduce DNA inside the right places because of CRISPR-precise Cas9's operation. Arguments regarding moral and cultural ramifications of human gene editing have been ongoing for a while, but this major advancement has given them new importance. In some way, a number of questions, including if genetic modification must be utilised to heal medical sickness or to change qualities such as intellect or attractiveness, had already been raised for generations. Nevertheless, such problems are not any more hypothetical with the advent of simple and effective gene-editing techniques, especially CRISPR-Cas9, and the solutions promise to have serious repercussions in medicine and healthcare.

16.3.4 How Unknown Mutations Are Identified

- **Single-Strand Conformational Polymorphism (SSCP)** is among the most straightforward testing methods for identifying unidentified changes (microlesions), including such unidentified single-base replacements, minor losses, small insertions, or micro inversions (Hunt and Hassold 2008). When damaged fragments of DNA migrate by gel electrophoresis, a DNA variant alters their shape. The reasoning is based on an analysis of the different gel electrophoresis migratory patterns of denatured wild-type and mutant components. In this method, ss DNA fragments can develop structural shapes by being denatured and then renatured under specific circumstances that prohibit the production of ds DNA. The initial nucleotide sequence gives rise to the particular shape. Nondenaturing polyacrylamide gels are used to differentiate the movement of these pieces, and their structural configurations provide the basis for the identification of changes.
- **Denaturing Gradient Gel Electrophoresis (DGGE)** has been utilised to check for unidentified point mutations. It is predicated on variations in how tiny fragments of DNA dissolve; even just a single nucleotide mutation can result in just such a variation. In this method, DNA will first be isolated and electrophoresed on a denatured proteins gel. The component entirely breaks into single strands as the denatured proteins' state worsens. In acrylamide gels, the physical form of the particle affects how quickly it moves. DNA segment melting patterns on denaturing gradient gels could be compared to identify mutant fragments. The detection rate of point mutations with DGGE is a little less than 100%. By using this approach, a segment of up to roughly 1,000 bp can be examined.

- **Characterization of Heteroduplexes:** Heteroduplexes are formed by denatured and renatured mixtures of wild-type and mutant DNA molecules. Through nondenaturing polyacrylamide gels, homoduplexes and heteroduplexes exhibit various electrophoretic mobilities. Heteroduplex analysis is thought to be able to find about 80% of point mutations. Point mutations can affect the recognition sites in DNA by changing how restriction enzymes cleave the DNA and produce pieces of different sizes. This phenomenon is known as restriction fragment length polymorphism (RFLP). To find mutations at restriction sites, RFLP is performed.

16.4 CONCLUSION

Any alteration to the DNA sequence may be harmful if it abnormally affects cellular biological systems. For a definitive diagnosis, the hereditary foundation of the illness must be characterised. A very appropriate and sensitive method, polymerase chain reaction (PCR) may replicate minuscule quantities of DNA. This method is used to diagnose hereditary illnesses at the DNA level and has numerous applications in various fields of biology. DNA analysis would show the potential disease-causing gene alterations. Pathogenic abnormalities in DNA and/or samples obtained are found during the scientific analysis of genetic illnesses. It might make the diagnosis, prognosis, and treatment of illnesses easier. Because most inherited diseases manifest in children, it is crucial for paediatricians to be knowledgeable about the methods used in genetic analysis and the clinical uses of these requirements with the aim of providing a precise diagnosis. To provide an effective lab test for illness diagnosis, practitioners should be able to identify and classify genetic diseases and impacted individuals based on clinical signs to a subclass of chromosome or specific gene diseases. A clinical geneticist could be consulted and used as a source of discussion (Richette et al. 2008).

16.4.1 ACKNOWLEDGEMENT

The authors sincerely acknowledge the University of Engineering and Management, Kolkata.

REFERENCES

Aitman TJ. DNA microarrays in medical practice. *BMJ.* 2001 Sep 15;323(7313):611–615. doi: 10.1136/bmj.323.7313.611.

Bais HP, Park SW, Weir TL, Callaway RM, Vivanco JM. How plants communicate using the underground information superhighway. *Trends Plant Sci.* 2004;9:26–32.

Boileau C, Jondeau G, Mizuguchi T, Matsumoto N. Molecular genetics of marfan syndrome. *Curr Opin Cardiol.* 2005 May;20(3):194–200. doi: 10.1097/01.hco.0000162398.21972.

Bond TA, Karhunen V, Wielscher M, Auvinen J, Männikkö M, Keinänen-Kiukaanniemi S, Gunter MJ, Felix JF, Prokopenko I, Yang J, Visscher PM, Evans DM, Sebert S, Lewin A, O'Reilly PF, Lawlor DA, Jarvelin M-R. Exploring the role of genetic confounding in the association between maternal and offspring body mass index: Evidence from three birth cohorts. *Int J Epidemiol.* 2020;49(1):233–243, doi:10.1093/ije/dyz095

Davisson MT, Bergstrom DE, Reinholdt LG, Donahue LR. Discovery genetics-the history and future of spontaneous mutation research. *Curr Protoc Mouse Biol.* 2012; 1(2):103–118.

Eichenlaub-Ritter U. Parental age-related aneuploidy in human germ cells and offspring: A story of past and present. *Environ Mol Mutagen.* 1996;28(3):211–236. doi: 10.1002/(SICI)1098-2280(1996)28:3.

Feuk, L. Inversion variants in the human genome: role in disease and genome architecture. *Genome Med.* 2010;2:11.

Fridovich-Keil, Judith L. "Gene editing." *Encyclopedia Britannica.* 2019 Jun 4. Accessed September 15, 2022. www.britannica.com/science/gene-editing.

Hunt PA, Hassold TJ. Human female meiosis: what makes a good egg go bad? *Trends Genet.* 2008 Feb;24(2):86–93. doi: 10.1016/j.tig.2007.11.010.

King HC, Sinha AA. Gene expression profile analysis by DNA microarrays: Promise and pitfalls. *JAMA.* 2001;286(18):2280–2288. doi: 10.1001/jama.286.18.2280

Mahdieh N, Rabbani B. An overview of mutation detection methods in genetic disorders. *Iran J Pediatr.* 2013 Aug;23(4):375–388.

Nakamura, Y. DNA variations in human and medical genetics: 25 years of my experience. *J Hum Genet.* 2009;54:1–8. https://doi.org/10.1038/jhg.2008.6.

Richette P, Bardin T, Stheneur C. Achondroplasia: from genotype to phenotype. *Joint Bone Spine.* 2008 Mar;75(2):125–130. doi: 10.1016/j.jbspin.2007.06.007.

Rosenstein BJ, Zeitlin PL. Cystic fibrosis. *Lancet.* 1998 Jan 24;351(9098):277–282. doi: 10.1016/S0140-6736(97)09174-5.

Rosenthal N. Tools of the trade-recombinant DNA. *N Engl J Med.* 1994 Aug;331(5):315–317. doi: 10.1056/NEJM199408043310508.

Sadée W. Pharmacogenomics. *BMJ.* 1999;319:1286. doi: 10.1136/bmj.319.7220.1286.

Van Ommen GJ, Bakker E, den Dunnen JT. The human genome project and the future of diagnostics, treatment and prevention. *Lancet.* 1999;354. doi: 10.1016/S0140-6736(99)90241-6

Veltman J, Brunner H. De novo mutations in human genetic disease. *Nat Rev Genet.* 2012;13:565–575.

Yan H, Kinzler KW, Vogelstein B. Tech.sight. Genetic testing-present and future. *Science.* 2000 Sep15;289(5486):1890–1892. doi: 10.1126/science.289.5486.1890

17 In Vitro Models for Studying Cancer Progression

Pratik Talukder[1,], Sounak Chanda[1], Meghna Mishra[1], Baishakhi Sinha[1], Anushka Dutta[1] and Ayan Adhikary[1]*
[1]Department of Biotechnology, University of Engineering and Management, Kolkata, India
*Correspondence to: pratik.talukder@uem.edu.in

17.1 INTRODUCTION

To analyze the metabolic pathways of cancer progression, efforts are made to artificially recreate the microenvironments via the use of several models. One such notable attempt made is in vitro models. Our key focus is utilizing in vitro models to study cancer progression. Various dimensions of cancer need to be extensively studied for potent drug development. For example, in vivo models help us to study cancer metastasis, angiogenesis, chemokine secretion, extravasation, and the negative feedback of drugs. Metastasis is the key characteristic feature of a cancer cell. It is mediated in steps through the stromal, endothelial blood connective tissue, vascular system, and even tissues located at a secondary site (Chambers et al., 2002; Fidler, 2003; Steeg, 2006). The process of metastasis is successfully carried out through the interactions occurring between the micro-environment and the progressing cancer cell. In vitro models aid us in efficiently studying these negotiations. This, in turn, helps in better analyzing cancer cell behaviors.

The leading cause of mortality in recent years has been cancer. The reason is the abrupt occurrence of cancer fuelled by heredity, mutations at the molecular level, and carcinogens such as UV light, gamma radiation, and mutagens. An anomaly in the quiescent phase (G0) during cell division often lays the foundation for cancer growth. Cancer is not specific in its origin, which makes drug development tedious work. Hence, in vitro models come into play, which are initiated with cancer stem cells (CSCs). CSCs serve as the functional basis of cancer-induced tumor growth. Characteristic features of CSCs such as self-renewal abilities, a high differentiation rate, and fewer nutrient requirements make them ideal for in vitro models by efficiently mimicking a mature cancer cell. Therefore, in vitro models play an important role in studying the behavior of cancer cells and developing effective drugs (Al-Hajj et al., 2003; Allegra et al., 2014; Ciurea et al., 2014).

17.2 BIOLOGICAL EVENTS THAT LEAD TO CANCER

Circulating tumor cells (CTCs) are the primary cause of tumors. They are known to secrete cells from the primary tumor into the capillaries and continue to circulate in the bloodstream of cancer patients. Researchers have tried to elucidate their nature and meaning. Even in the early stages of the disease, CTCs circulate in the bodily fluids before spreading to different regions of the body (Yang et al., 2019). However, their amount is very small, so it is difficult to identify them. CTCs first develop from the primary tumor and stay in the bloodstream for a while before manifesting in new tissue. CTCs can also adhere to capillary walls and provide force to penetrate within the new unaffected tissue. Due to their large size, they may act as an obstruction to the capillaries while in the bloodstream. Different tumors can discharge various kinds of CTCs in multiple sites of action. CTCs secreted from many tumor sites may show variations of the same tumor. Researchers continue to make attempts, but they have not yet determined the significance of CTCs' behavior.

 DOI: 10.1201/9781003302131-18

Metastasis is one of the most devastating aspects of cancer, and it is caused by tumor cells that have broken out from the primary tumor and entered the bloodstream. Even in the early stages of cancer, tumor cells block arteries and reside within platelets and leukocytes. External conditions cause a large number of CTCs to perish in circulation, and only a tiny percentage of CTCs may spread. These metastasis-initiating CTCs have the following characteristics: stemlessness, a hybrid epithelial-mesenchymal transition (EMT) state, evasion of immune surveillance, and resistance to drugs (Hu et al., 2020). The presence of CTCs in the blood circulation and their ability to spread inside the host are triggered by their interactions with other formed elements of blood such as neutrophils and platelets. Metastasis broadly includes the following five stages: (1) tumor cell invasion; (2) intravasation within the closed circulatory network; (3) communication and interaction with the immunity system; (4) indulging in the secondary tissue and adherence to the vascular endothelium; and (5) mushrooming of micro- and macro-metastasis (Menyailo et al., 2021). Each stage of the metastatic cascade is important and may be addressed to decrease the likelihood of metastasis. Anti-cancer treatment mainly focuses on combating the primary tumor. In addition, there is a risk remaining of building-up metastasis following the successful removal of the primary tumor; this is due to the release of tumor cells into the bloodstream during surgery or the presence of disseminated tumor cells (DTCs).

The cell-matrix adhesion of tumor cells is frequently reduced. There is growing evidence that a reduction in cell adhesion corresponds with tumor invasion and metastasis. Cadherins and CAMs, two major families of adhesion molecules, have been linked to tumor aggressiveness (Horwitz, 2012). The relationship between cell adhesion molecule loss and tumor growth and the direct movement of these molecules in certain signal transduction pathways are investigated, with a focus on how these pathways change in transformed cells. When tumor cells travel from their original place to distant organs, the majority of cases have been described as active participation in the loss of cell adhesion and the development of unfavorable features. Tumor invasion and metastasis are frequently associated with a loss of E-cadherin-mediated cell adhesion (Horwitz, 2012). The absence of intercellular connections may likely facilitate tumor cell migration since E-cadherin, the most prevalent adhesion molecule in the epithelial tight junction, is expressed.

Tumors form when cells in the body divide and expand too quickly. Tumor vascularization differs not only by the tumor type and anatomic location but also by the cancer tissue type. Many cancers, however, can get the blood flowing through vascular arteries without the necessity for tumor angiogenesis. Tumor start is the result of a genetic change that causes aberrant cell proliferation in a single cell. A population of clonally produced tumor cells originates as a result of cell proliferation. Tumor development continues as a new mutation occurs inside the tumor population's cells. This biological phenomenon is referred to as "clonal selection" because a new copy of tumor cells emerges due to its higher growth rate (Jiong Yan et al. 2022). Clonal selection occurs at every stage of tumor growth, which results in tumors that grow more quickly and become more aggressive.

17.3 MIMICKING TUMOR MICROENVIRONMENTS IN IN VITRO MODELS

Revolutionary methods such as tissue engineering and stem-cell culturing have come to the fore in recent years, which have enabled the methodical characterization of cancer cells and thereby aided in the development of potent treatment strategies. All of these techniques employ mechanical generation of the tumor microenvironment, which is heterogenic and comprises a wide array of components ranging from stromal fibroblast, extracellular matrix, and immune cells (Wu & Swartz, 2014). Tumour microenvironments play a pivotal role in studying cancer progression as they maneuver the rate at which tumors grow and control metastasis, host immune response to the tumor (chemokine secretion), and the mediation of chemical signaling among neighboring cells. Thus, the study of the tumor microenvironment explicates the malignancy-promoting contributors, whose knowledge when implemented may result in a sophisticated cancer curative approach.

As discussed previously, in vitro models are engineered for categorical cancer-related studies and associated oncologic diagnosis. In vitro models thus require cells that mimic or possess tumor-like properties to assimilate a better understanding of the biochemical events (extravasation, angiogenesis, invasion, etc.) that take place in cancer-positive patients. These cells are referred to as CSCs or tumor-initiating cells (Franco et al., 2016). CSCs are recognized as progenitors of tumor cells and exhibit properties such as unregulated differentiation, self-renewal, generation of new tumors, resistance to cell death, and specific tolerance to the immune reaction of the host.

17.4 ISOLATION OF CSCs

The twentieth-century marked the pioneering effort made by Bonnet and Dick to isolate CSCs. This effort was done by isolating $CD34^{+}CD38^{-}$ cells, which can lead to acute myeloid leukemia in non-obese diabetic\severe immunodeficient mice (Abbaszadegan et al., 2017). CSCs can be cultured from primary tumors by reprogramming the cell lines to pluripotent forms, which helps to unravel the mechanisms involved in cancer progression (Franco et al., 2016). A rather uncomplicated isolation of CSCs can be performed via specific cellular markers, for example, CD44 and CD133 assist in the recognition and isolation of CSCs (Abbaszadegan et al., 2017). Cellular markers generally operate by enhancing cell signaling pathways, which ultimately leads to increased cell division. Aldehyde dehydrogenase has also been identified as an intracellular marker of CSCs.

The most well-known approach to CSC isolation is by culturing them from tumor tissues in serum-free conditions. CSCs have the potential to form non-adherent tumor spheroids when they are grown in media containing specific growth factors such as epithelial growth factors, fibroblast growth factors, etc. Although this approach is acknowledged as comparatively easier, it has downsides, for example, the low efficiency of cells, the requirement of appropriate media, employing different techniques for different types of CSCs, and sphere heterogeneity leading to the lesser self-renewal ability of CSCs (Abbaszadegan et al., 2017). However, it is also necessary to recognize that CSC-based therapies have come to the fore, wherein the elimination of CSCs may effectively escalate the lifespan of cancer patients.

17.5 THE EFFICIENCY OF 3D IN VITRO MODELS IN ANALYZING CANCER CELL BEHAVIOR

Successful recreation of tumor microenvironments artificially in 3D in vitro models allows the interpretation and examination of biological events that occur during tumor progression. 3D spheroid tumor cell cultures show that phenotypic distinctions result in structural and functional differences between normal mammary cells and carcinoma cells that often pave the way to tumorigenesis (Wu & Swartz, 2014).

The growth of tumors is mainly regulated by oxygen diffusion and consumption rates. Therefore, the study of these factors helps develop methods to inhibit angiogenesis. Microfluid models provide well-developed microenvironments for proper studies of blood and lymphatic angiogenesis.

Cancer metastasis is usually propagated by downregulating cell-cell adhesion followed by the invasion of the extracellular matrix (Wu & Swartz, 2014).

Chemokine-driven tumor invasion occurs via chemokine receptors that play roles in the metastasis of tumor cells. The presence of chemokine gradients in tumor microenvironments due to extracellular matrix heterogeneity, cell-cell signaling, metabolic stresses, etc., often acts as the driving force of angiogenesis through chemotaxis and hence propels the rate of tumor survival and proliferation (Wu & Swartz, 2014).

The presence of CTCs in patients' blood serves in the early identification and diagnosis of cancer. Microfluid models have the advantage of entrapping these CTCs and thereby assisting in the identification of CTCs in patients suffering from prostate, breast, lung, and colon cancer.

17.6 TYPES OF MODELS

Culturing is a commonly utilized in vitro approach for bettering our knowledge of biological sciences, body architecture, illness causes, medication activity, and gene expression. In vitro assays involving (2D) cultured cells account for the majority of cancer research, but this method has a lot of shortcomings. For these drawbacks, prototypes that can better imitate in vivo settings have been developed. Three-dimensional culture is, however, another way (3D). The research and targeted medicines may well be aided by optimizing conditions (Rolver et al., 2019).

17.6.1 2D Model

The benefits of adherence to 2D models include inexpensive and easy culturing upkeep and the ability to conduct the testing process. But 2D culture has a slew of drawbacks. 2D cultivated cells do not have the same rounded appearance as lumps. Cell discernment, multiplication, stimulus-response, and certain other cell components are all governed by these relationships. The shape of the lymphocytes and the method of cell replication are modified once they are isolated first from tissue and transferred to 2D settings. 2D culture also results in a decrease in phenotypic diversity. Changes in morphology can have an impact on their performance, internal structural organization, release, and signal transduction. Cells developing innately mislay their polarization because of disruptions in connections with the extraneous conditions, which changes their responsiveness to numerous events including cell lysis. A disadvantage of this is that single-layer cells have unrestricted access to selected compositions. Compared to 3D cultivation, cells are more responsive to drugs (Barbosa et al., 2021).

17.6.2 3D Model

3D models can be split into four types based on the technique used to prepare them: i) suspension cultures on non-adherent plates; ii) suspension cultures on adherent plates; iii) cultures in concentrated media or gel-like substances; and iv) cultures on scaffolds (Yamada & Cukierman, 2007). When the patient's tissue cells are cultivated in multilayers, a 3D structure can closely resemble the design of the original material than other models. This property is the outcome of correct cell signaling pathways, which were established to accomplish tissue pattern replication. In the same way that cells get signals from their surroundings in vivo, tissues can also receive inducement from their surroundings in the lab. Furthermore, the shape and polarization of the tissues are preserved, and they may be reinstated to cells already cultured in adherent. Distinct internal structures with lumen creation have been seen in several 3D models. This is the outcome of apoptotic cell death in the core region. Cell growth is influenced by the subcellular level and is greater near the 3D construction's periphery. Cellular architecture, gene function, transduction, and metabolic activity are other essential features of 3D cultures in their resemblance to developing in vivo cells (Amirghasemi et al., 2021).

17.6.2.1 Behavioral Analysis of 3D Cancer Models

17.6.2.1.1 Response to Viability Assay

3D cell cultures constitute the 3D models used for cancer analysis. Special culture systems, namely, tumor organoids and tumor spheroids, developed using multicellular structures are seen to be remarkably useful for studies of cancer progression, drug development, screening, testing, and precision medicine (Dominijanni et al., 2021). Cell viability assays aid in developing an understanding of the effects of various biological reagents on the patient's health at cellular levels. These assays are key to the in vitro study of various cellular models including those of cancer. With the growing expansion of the usage of 3D cell models, even 2D assays are used, outside their determined operation on 3D constructs. The reason is purely commercial (Dominijanni et al., 2021). The use of a 3D cell viability assay, specifically, CellTiter-Glo®, on a breast cancer cell line with a triple-negative character, named MDA-MB-231, revealed crucial information about the cells' structure and growth characteristics (Huang et al., 2020).

17.6.2.1.2 3D Models that Aid in Enhancing Drug Screening

Biologists have long overcome the challenge of creating microenvironments by utilizing advanced tissue engineering and developed biomaterials. 3D cancer models developed through 3D cell cultures of CSCs fulfill the qualities needed to achieve high-throughput and high-contrast screening results (Brancato et al., 2020). Traditional 2D in vitro models are dependent on monolayer cell cultures that yield inaccurate and unproductive data. Drug screening is a stepwise procedure to test the reactions and consequences of the selected drugs on the sorted cell groups. This includes using various assays to examine diverse candidate drugs and analyze their physiological effects on the cell cultures. To obtain a single test-approved drug, a significant variety of drugs are tested (Kondo & Inoue, 2019). 3D models have been shown to help improve the predictability of the health and toxic effects of various drugs on cell lines. Two-dimensional cell cultures, aka flat biology, do not certainly represent the animal model conditions in natural environments. A certain level of stiffness applied to the cultures in a Petri dish alters the gene expression and the cells' sensitivity to certain drugs (Kondo & Inoue, 2019).

The wide range of cancer models available at present demands a greater and more efficient way of categorizing them based on certain characteristics. One such classification is to categorize them based on the technology used for development.

1. Scaffold-free systems
2. Scaffold-based systems
3. Hydrogels
4. Bioreactors
5. Microcarriers
6. Tumor-on-chip

The listed models have individual pros and cons to be applied and utilized according to the specific needs. Individual companies focus on producing each kind of model with a specific use in mind (Brancato et al., 2020).

17.7 COMPARATIVE STUDY OF THE DRUGS IN VARIOUS CANCER MODELS

In this context, the various results of drug screening and how a comparative study of drug screening analyses is conducted deserve mention. Statin drugs were tested against models of human breast cancer cell groups, namely, the MDA-MB-231 and MCF-7. Statin drugs are prescribed to lower cholesterol levels in blood. Recently, these drugs have proved effective against cancer cell lines and hence have the potential of being used extensively in cancer therapy (Bytautaite & Petrikaite, 2020).

In the performed test, four statin drugs, specifically, lovastatin, mevastatin, pitavastatin, and simvastatin, were introduced for comparative study. The findings were examined using both 2D and 3D cell line models. Pitavastatin had the greatest impact on these cancer cell lines' viability, according to the results of the 2D model, whereas pitavastatin and metavastatin were able to reduce cell colony formation. All the drugs in specific amounts had suppression effects on the cell lines. However, the results for the 3D cell spheroids proved to be divergent. MDA-MB-231 cell spheroids grew slower with lovastatin and simvastatin, while MCF-7 spheroids grew slower with mevastatin and pitavastatin. This revealed that the drugs' activities differ respective to the cell lines and the type of culture (Bytautaite & Petrikaite, 2020).

17.8 DIFFERENCES IN DRUG SENSITIVITIES IN THE MODELS

The study revealed the varied sensitivities of drugs on the different types of cell cultures. It has been reported that these varying results are caused by the differential gene expression of genes associated with signal transduction, varying protein expression patterns, cell-cell signaling mechanisms, varied growth of the cells, and morphology of the cells (Hongisto et al., 2013).

A detailed study utilizing several drugs on various 2D and 3D cell lines revealed the key reason for such differences in sensitivities is differential gene expression. The purpose of the study was to evaluate different 3D models for high-outturn drug screening. It showed that the most drug-sensitive cells were those grown in the Matrigel matrix. When cultivated on a 3D Matrigel matrix as opposed to a 2D culture, the JIMT1 cells were found to be more responsive to medicines. The gene expression patterns in Matrigel cultures resemble xenografts (Hongisto et al., 2013).

17.9 THE WARBURG EFFECT

The Warburg effect refers to the increased aerobic glycolysis in tumor cells that leads to the over-exhaustion of glucose to produce the energy needed for cell proliferation (Diaz-Ruiz et al., 2011). Using this feature of the tumor, increased knowledge of drug efficacy has been exposed by biologists. Glycolysis, by nature, is an anaerobic reaction. Cancer cells have the luxury of carrying out this reaction in aerobic conditions. This has proved to be significantly beneficial for the survival of these cancer cells. The region surrounding the tumor cells is characterized by hypoxia. With an increase in oxygen sources, rapid exhaustion of glucose occurs to feed into the low levels of energy (Kim et al., 2009). The increased need for glucose by tumor cells can be targeted via target-specific drugs to limit the glucose sources and to let the tumor cells starve of the required energy (Bhattacharya et al., 2016).

17.10 TUMOR MODELS FOR ANGIOGENESIS

Cancer cells can recruit blood vessels through the process of angiogenesis to carry out and promote growth. Artificial microenvironments aim at reproducing such circumstances to visualize and analyze this important phenomenon. Various studies have led to the conclusion that the type of cell lines and cell culture are the controlling factors for angiogenic behaviors of the cells. In Chrisey's lab, a model for studying the angiogenic abilities of cancer cells was developed from the ex vivo mesentery tissue of rats. These cells had breast cancer cells bioprinted on them. This made it possible to analyze how cancer cells and endothelial cells interact. Unsurprisingly, the outcome favored the argument. Angiogenesis was observed to be more active when the bioprinted cells were present (Phamduy et al., 2015).

17.11 CONCLUSION

Three-dimensional cancer cell models have opened several doorways to developing an efficient way to cure cancer and developing anti-cancer drugs. As we learn more about the growth, development, and multiplication of cancer cells within the body through in vitro recreations of their natural habitats, the likelihood of curing cancer and developing anti-cancer drugs will increase. The inefficiency and inaccuracy of 2D cell lines have been somewhat compensated by the 3D models. The only thing holding back the further expanded usage of 3D cancer models is their lack of standardization in pre-clinical trials. In the near future, we can expect to generate 3D models from specific patients and thus have more specific and personalized therapy accompanied by individual protocols for treatment.

REFERENCES

Abbaszadegan MR, Bagheri V, Razavi MS, Momtazi AA, Sahebkar A, Gholamin M. Isolation, identification, and characterization of cancer stem cells: A review. J Cell Physiol. (2017);232(8):2008–2018.

Al-Hajj M, Wicha MS, Benito-Hernandez A, Morrison SJ, Clarke MF. Prospective identification of tumorigenic breast cancer cells. Proc Natl Acad Sci U S A. (2003);100(7):3983–3988.

Allegra A, Alonci A, Penna G, Innao V, Gerace D, Rotondo F, Musolino C. The cancer stem cell hypothesis: A guide to potential molecular targets. Cancer Invest. (2014);32(9):470–495.

Amirghasemi F, Adjei-Sowah E, Pockaj BA, Nikkhah M. Microengineered 3D tumor models for anti-cancer drug discovery in female-related cancers. Ann Biomed Eng. (2021);49(8):1943–1972.

Barbosa MAG, Xavier CPR, Pereira RF, Petrikaitė V, Vasconcelos MH. 3D cell culture models as recapitulators of the tumor microenvironment for the screening of anti-cancer drugs. Cancers (Basel). (2021);14(1):190.

Bhattacharya B, Omar MFM, Soong R. The Warburg effect and drug resistance. British J Pharmacol. (2016);173(6):970–979.

Brancato V, Oliveira JM, Correlo VM, Reis RL, Kundu SC. Could 3D models of cancer enhance drug screening?. Biomaterials. (2020);232:119744.

Bytautaite M, Petrikaite V. Comparative study of lipophilic statin activity in 2D and 3D in vitro models of human breast cancer cell lines MDA-MB-231 and MCF-7. Onco Targets Ther. (2020);13:13201–13209.

Chambers AF, Groom AC, MacDonald IC. Dissemination and growth of cancer cells in metastatic sites. Nat Rev Cancer. (2002);2(8):563–572.

Ciurea ME, Georgescu AM, Purcaru SO, Artene S-A, Emami GH, Boldeanu MV, Tache DE, Dricu A. Cancer stem cells: Biological functions and therapeutically targeting. Int J Mol Sci. (2014);15(5):8169–8185.

Diaz-Ruiz R, Rigoulet M, Devin A. The Warburg and Crabtree effects: On the origin of cancer cell energy metabolism and yeast glucose repression. Biochim Biophys Acta Bioenerg. (2011);1807(6):568–576.

Dominijanni A, Devarasetty M, Forsythe S, Votanopoulos KI, Votanopoulos K, Soker S. Cell viability assays in three-dimensional hydrogels: A comparative study of accuracy. Tissue Eng. Part C Methods. (2021);27(7):401–410.

Fidler IJ. The pathogenesis of cancer metastasis: The 'seed and soil' hypothesis revisited. Nat Rev Cancer. (2003);3(6):453–458.

Franco SS, Szczesna K, Iliou MS, Al-Qahtani M, Mobasheri A, Kobolák J, Dinnyés A. In vitro models of cancer stem cells and clinical applications. BMC Cancer. (2016);16(Suppl 2):738.

Hongisto V, Jernstrom S, Fey V, Mpindi J-P, Sahlberg KK, Kallioniemi O, Perala M. High-throughput 3D screening reveals differences in drug sensitivities between culture models of JIMT1 breast cancer cells. PLoS One. (2013);8(10):e77232.

Horwitz AR. The origins of the molecular era of adhesion research. Nat Rev Mol Cell Biol. (2012);13(12): 805–811.

Hu B, Tian X, Li Y, Liu Y, Yang T, Han Z, An J, Kong L, Li Y. Epithelial-mesenchymal transition may be involved in the immune evasion of circulating gastric tumor cells via downregulation of ULBP1. Cancer Med. (2020);9(8):2686–2697.

Huang Z, Yu P, Tang J. Characterization of triple-negative breast cancer MDA-MB-231 cell spheroid model. Onco Targets Ther. (2020);13:5395–5405.

Kim MY, Oskarsson T, Acharyya S, Nguyen DX, Zhang XH, Norton L, Massagué J. Tumor self-seeding by circulating cancer cells. Cell. (2009);139(7):1315–1326.

Kondo J, Inoue M. Application of cancer organoid model for drug screening and personalized therapy. Cells. (2019);8(5):470.

Menyailo ME, Bokova UA, Ivanyuk EE, Khozyainova AA, Denisov EV. Metastasis prevention: Focus on metastatic circulating tumor cells. Mol Diagn Ther. (2021);25(5):549–562.

Phamduy TB, Sweat RS, Azimi MS, Burow ME, Murfee WL, Chrisey DB. Printing cancer cells into intact microvascular networks: A model for investigating cancer cell dynamics during angiogenesis. Integr Biol (Camb). (2015);7(9):1068–1078.

Rolver MG, Elingaard-Larsen LO, Pedersen SF. Assessing cell viability and death in 3D spheroid cultures of cancer cells. J Vis Exp. (2019);(148).

Steeg PS. Tumor metastasis: Mechanistic insights and clinical challenges. Nat Med. (2006);12(8):895–904.

Wu M, Swartz MA. Modeling tumor microenvironments in vitro. J Biomech Eng. (2014);136(2):021011.

Yamada KM, Cukierman E. Modeling tissue morphogenesis and cancer in 3D. Cell. (2007);130(4): 601–610.
Yan J, Wan P, Choksi S, Liu ZG. Necroptosis and tumor progression. Trends Cancer. (2022);8(1):21–27.
Yang C, Xia BR, Jin WL, Lou G. Circulating tumor cells in precision oncology: Clinical applications in liquid biopsy and 3D organoid model. Cancer Cell Int. (2019);19:341.

18 Immunotherapy
Adding a New Dimension to Cancer Therapy

Susmita Mukherjee[1], Shreya Banerjee[2], Chiranjeeb Dey[2], and Sonali Paul[1,]*

[1]Department of Biotechnology, University of Engineering and Management, Kolkata

[2]Applied Immunology and Parasitology Unit, Department of Zoology, Serampore College, West Bengal

*Correpondence to: sonali.paul@uem.edu.in

18.1 INTRODUCTION

Immunotherapy is not very old. A major concept behind it involves the medical field in which challenges are faced during the application of radiotherapy and chemotherapy, such as anemia, blood clotting, fatigue, bladder problems, flu-like symptoms, and many other difficulties (Chen et al. 2019).

Therefore, to overcome all of these problems, the new concept of immunotherapy comes in where in some new, unique way, we can apply this as a new hope in the field of cancer biology (Crommelin et al. 2020).

Immunotherapy is basically a process where various therapeutic procedures are conducted by either suppressing or stimulating our innate immune system and communicating with some passive immunotherapeutic procedure.

The main target is to reduce the fatality rate of diseases along with reducing the adverse side effects of traditional methods of treatment in a simple and economic procedure.

18.2 IMMUNOTHERAPEUTIC STRATEGIES

18.2.1 Monoclonal Antibodies

Naked and conjugated (radiolabelled, chemotoxin, and immunotoxin), monoclonal antibodies are used for therapeutic purposes. They basically involve a passive immunotherapeutic process where some monoclonal antibodies are fused with other oncolytic cells and employed in immunotherapy via injecting the fused cell after keeping it in a culture medium (Martin et al. 2020).

18.2.2 Adoptive Immunotherapy

Adoptive immunotherapeutic applications are also used. Adoptive immunotherapy is the process of passive immunotherapy where it is injected into the patient's body, for example, therapeutic cancer vaccines (Martin et al. 2020).

18.2.3 Therapeutic Cancer Vaccines

Cancer vaccines often require additional substances, such as adjuvants, for optimal effectiveness, for example, ex-DNA vaccines, antigen vaccines, and Dendritic cell vaccines. This is also a type of adaptive and passive immunotherapy. Here, acquired immunity is being established (Kranz et al. 2016).

DOI: 10.1201/9781003302131-19

18.2.4 Checkpoint Inhibitors

Checkpoint inhibitors are also a major type of therapy. In this mechanism, T cell receptors are helped by MHC antigen processing and presentation and rely on cell-mediated immunity mechanisms. Some examples include ex-complement mediated systems, phagocytosis, cytotoxic-mediated killing producing perforin, granzyme, etc. (Rakoff et al. 2009).

18.2.5 Oncolytic Virus Therapy

These viruses can directly kill cancer cells and activate immune cells such as Dendritic cells and T cells to fight cancer cells. They are genetically modified and are combined with other immunotherapeutic processes such as monoclonal antibodies and cancer vaccines.

18.2.6 Cytokines

Interferons and interleukins are involved in therapeutic processes for the quick division of immune cells and for attacking cancer cells. Cytokines are a messenger molecule that helps in the growth and control of immune system cells (Zheng et al. 2013).

18.3 ADVANCED IMMUNOTHERAPEUTIC PROTOCOLS

18.3.1 Surface-enhanced Raman Spectroscopy for Cancer Immunotherapy and the Use of Cancer Nanomedicine

To know cancer patients' immunotherapy responses, very appropriately multiplexed immunoassays are used in sensing the multiple immunomarkers of tumors (Linkov et al. 2008). One of the best examples is Surface-enhanced Raman spectroscopy (SERS) in multiplexed immunoassays, which is very well used in the field of cancer biology. SERS' active nanomaterials include in vivo and in vitro biomarker sensing and drug screening and delivery (Lane et al. 2015).

When cancer nanomedicine meets immunotherapy techniques, this boosts immunity and activates antigen processing, antigen presentation, and immune cell-mediated killing. There are mainly three mechanisms involved – targeting cancer-affected cells, targeting the tumor immune microenvironment, and targeting the peripheral immune system. For example, recently, this has been performed in a Phase III trial where paclitaxel was combined with atezolizumzb for the treatment of patients suffering from triple-negative breast cancer (He et al. 2019).

18.3.2 The Role of the CRISPER-CAS 9 Genome Editing Tool in Cancer Biology

Clustered Regularly Interspaced Short Palindromic Repeats (CRISPER)-associated protein 9 (CRISPER-Cas9) has become mainly a part of adaptive immune systems (Matsumura and Maeda 1986). It helps in the modification and correction of cancer-causing gene mutations and deletions and boosts immune cells, for example, chimeric antigen receptor (CAR T) cells, which are widely used in the field of antitumor therapy.

18.3.3 Overcoming HIV with Immunotherapeutic Strategies and RNA Nanotechnology in Cancer

In this field, very pristine therapies that have been used, such as antiretroviral therapies, are currently outdated. Thus, modern immunotherapeutic strategies that utilize siRNA and MicroRNA are now widely used for treating malignancy, which can activate both the innate and adaptive immune response via silencing or upregulating immune-relevant gene mechanisms (Wu et al. 2014). mRNA vaccines are also useful in the treatment of cancer patients. Furthermore, it has been seen that versatile nanomaterials collaborate with these RNAs to amplify their delivery process to tumor-affected cells.

18.3.4 Lymphocytic Immunotherapy for Pregnancy and Cancer Immunotherapy in Pediatric Patients

An embryo basically generates paternal antigens, and for this, it is recognized as a semi-allograft that causes repeated implantation failure (RIF) and recurrent pregnancy loss (RPL). To overcome this, lymphocyte immunotherapy (LIT) is widely used to prevent pregnancy loss following various methods such as progesterone-induced blocking factor (PIBF), anti-idiotypic antibodies (Ab2), mixed lymphocyte reaction blocking antibodies (MLR-Bf), and anti-paternal cytotoxic antibodies (APCA). The ratio between helper T cells 1 and 2 is reduced.

Therapies, specifically, adaptive cell therapy, bispecific T cell engagers, checkpoint inhibition, and oncolytic virotherapy, are major mechanisms seen in treating paediatric patients. CD 19 chimeric antigen receptor T cells for pre-B cell acute lymphoblastic leukemia has recently been studied (Tang et al. 2018).

18.3.5 Immunotherapy in the Case of Glioblastoma

For promoting tumor eradication, enhancing innate and adaptive immune responses, and overcoming tumor resistance, immunotherapy is broadly used in glioblastoma, which is the most common malignant brain tumor. Mechanisms such as vaccines, oncolytic viruses, checkpoint inhibitors, and adaptive cell transfer are being used. However, in these techniques, some challenges are faced, for example, the intracranial location and heterogeneity of the tumor and a lack of appropriate tumor models (Galluzi et al. 2017).

18.3.6 Role of Immune Checkpoint Signaling in Cancer Immunotherapy

Various therapeutic applications used to block checkpoints have become the major herald in checkpoint biology (Chen and Mellman 2013) and involve processes such as surface delivery, internalization, recycling, and degradation to regulate surface expression levels in the case of checkpoint proteins. Checkpoint proteins induce signaling and regulate immune responses by engaging conventional trans and cis interactions.

18.4 AFTERMATH OF CANCER IMMUNOTHERAPY: A NEW CHALLENGE TO FIGHT AND CONQUER

Developing newer, more efficient cancer treatment strategies has currently become one of the major issues where very old and pristine therapies such as radiotherapy and chemotherapy are shown to be very difficult for patients due to their adverse side effects. Accordingly, scientists have carried out research on the concept of immunotherapy for the overall benefit of patients. But recently, some drawbacks and side effects are being discovered, namely, its overall efficiency is not high, and it is resistant to drugs (Peer et al. 2007). Therefore, scientists have developed the idea of combination therapy that can effectively reduce drug resistance. To mitigate the side effects, effective mechanisms such as activating ferroptosis, utilizing thioredoxin, FATP 2, and the reprogramming of tumor blood vessels are being considered

18.5 PRESENT AND FUTURE IMMUNOTHERAPY: CHALLENGES AND PERSPECTIVES

By using the applications of monoclonal antibodies, interferons, and natural products, a new approach in the field of immunotherapy is overcoming limitations such as the significant cost of treatment and adverse side effects of treatment (Feng et al. 2019). Presently, we can also see that not only for cancer but also for diseases such as allergies, infections, hyperlipidemia, osteoporosis, and neurodegenerative diseases, applications of immunotherapies are used.

In certain types of malignancies, for example, melanoma, immunotherapy response rates often vary. In this regard, various modal approaches are undertaken, such as studying the role of gut microbiota in improving the efficacy and the response rates for patients undergoing cancer immunotherapy.

The etoposide-platinum (EP) combination is one of the major heralds in small lung cancer treatment. Thus, searching for some important new approaches in this field has become very important. Programmed cell death protein (PD1) and Cytotoxic T lymphocyte-associated protein 4 (CTLA4) are the immune checkpoint inhibitors that have been used in this treatment procedure in recent years. In fact, some chemotherapeutic drugs such as temozolomide and transcription inhibitor lurbinectedin are also used for their immunomodulatory effects.

Nanoparticle drug delivery systems have become a new approach to boost cancer immunotherapy, which reduces the toxicity and enhances the effectiveness of immunotherapeutic drugs, specifically, immunogenic cell death (ICD)-inducing drugs, cytokines, and adjuvants (Wang et al. 2017).

The concept of immunoradiotherapy is also very much accepted in this field where a combination of radiotherapy with some immunomodulators, for example, checkpoint inhibitors such as anti PD1/PD -L1 and anti CTLA 4, are used for amplifying the antitumor response (Saeed et al. 2019).

18.6 CONCLUSION

Advanced research in cancer therapy, for example, the use of monoclonal antibodies, interferons, and other active immunological molecules, are now multimodal approaches in the field of immunotherapy. Some drawbacks in this respect, such as adverse effects of therapy and the high cost of treatment, will surely be overcome in the future by further research to make it the therapy of future clinical oncology.

REFERENCES

Chen DS, Mellman I. 2013. Oncology meets immunology: the cancer-immunity cycle. *Immunity.* 39:1–10.

Chen Q, Wang C, Zhang X, Chen G, Hu Q, Li H. 2019. In situ sprayed bioresponsive immunotherapeutic gel for post-surgical cancer treatment. *Nat Nanotechnol.* 14:89–97.

Crommelin DJ, van Hoogevest P, Storm G. 2020. The role of liposomes in clinical nanomedicine development. What now? Now what? *J Control Release.* 318:256–63.

Feng X, Xu W, Li Z, Song W, Ding J, Chen X. 2019. Disease immunotherapy: immunomodulatorynanosystems. *Adv Sci.* 6:1970100.

Galluzzi L, Buqué A, Kepp O, Zitvogel L, Kroemer G. 2017. Immunogenic cell death in cancer and infectious disease. *Nat Rev Immunol.* 17:97–111.

He H, Liu L, Morin EE, Liu M, Schwendeman A. 2019. Survey of clinical translation of cancer nanomedicines-lessons learned from successes and failures. *Acc. Chem Res.* 52:2445–61.

Kranz LM, Diken M, Haas H, Kreiter S, Loquai C, Reuter KC. 2016. Systemic RNA delivery to dendritic cells exploits antiviral defence for cancer immunotherapy. *Nature.* 534:396–401.

Lane LA, Qian X, Smith AM, Nie S. 2015. Physical chemistry of nanomedicine: understanding the complex behaviors of nanoparticles in vivo. *Annu Rev Phys Chem.* 66:521–47.

Linkov I, Satterstrom FK, Corey LM. 2008. Nanotoxicology and nanomedicine: making hard decisions. *Nanomedicine.* 4:167–71.

Martin JD, Cabral H, Stylianopoulos T, Jain RK. 2020. Improving cancer immunotherapy using nanomedicines: progress, opportunities and challenges. *Nat Rev Clin Oncol.* 17:251–66.

Matsumura Y, Maeda H. 1986. A new concept for macromolecular therapeutics in cancer chemotherapy: mechanism of tumoritropic accumulation of proteins and the antitumor agent smancs. *Cancer Res.* 46:6387–92.

Peer D, Karp JM, Hong S, Farokhzad OC, Margalit R, Langer R. 2007. Nanocarriers as an emerging platform for cancer therapy. *Nat Nanotechnol.* 2:751–60.

Rakoff-Nahoum, Seth, Ruslan Medzhitov. 2009. Toll-like receptors and cancer. *Nature reviews cancer.* 9(1): 57–63.

Saeed M, Gao J, Shi Y, Lammers T, Yu H. 2019. Engineering nanoparticles to reprogram the tumor immune microenvironment for improved cancer immunotherapy. *Theranostics*. 9:7981–8000.

Tang J, Yu JX, Hubbard-Lucey VM, Neftelinov ST, Hodge JP, Lin Y. 2018. The clinical trial landscape for PD1/PDL1 immune checkpoint inhibitors. *Nat Rev Drug Discov*. 17:854–5.

Wang C, Wen D, Gu Z. 2017. Cellular bioparticulates with therapeutics for cancer immunotherapy. *Bioconj Chem*. 29:702–8.

Wu TYH, Singh M, Miller AT, De Gregorio E, Doro F, D'Oro U. 2014. Rational design of small molecules as vaccine adjuvants. *SciTransl Med*. 6:263ra160.

Zheng Y, Stephan MT, Gai SA, Abraham W, Shearer A, Irvine DJ. 2013. In vivo targeting of adoptively transferred T-cells with antibody-and cytokine-conjugated liposomes. *J Control Release*. 172:426–35.

19 Analytical Tools to Identify Cancer Biomarkers from Extracellular Vesicles

S. Saroj[1,#], Debashish Paul[1,#], U. S. Poornima[1,#] and Tatini Rakshit[1,]*
[1]Department of Chemistry, Shiv Nadar Institution of Eminence, Delhi-NCR, India
*Correspondence to: tatini.rakshit@snu.edu.in
[#]These authors contributed equally to this work.

19.1 INTRODUCTION

Extracellular vesicles (EVs) refer to the sub-micron- to nanometer-sized bi-layer phospholipid membrane-bound vesicles released by cells into the extracellular matrix. These vesicles enter bodily fluids (for example, blood, saliva, breast milk, synovial fluid, urine and faeces) and serve as an intercellular communication system (Datta et al. 2021). EVs are mainly classified into various subtypes, such as exosomes, microvesicles and apoptotic bodies, on the basis of their size, release and biogenesis pathway (**Figure 19.1**). EVs have been found to be mainly involved in intercellular crosstalk involving RNA. They carry different biomolecules including proteins, nucleic acids and carbohydrates (Kalluri and LeBleu 2020). Exosomal content has not yet been fully revealed, and it is dependent on a variety of physiological processes (Datta et al. 2022).

The EV platform provides a non-invasive method of sample collection, quick analysis and early detection. The link between cancer cell EVs and cancerous cells is mainly valued in cases of tissue biopsy sample scarcity, for example, brain tumours (Shi et al. 2015). The traditional methods used for EV isolation from body fluids include ultracentrifugation (gold standard), microfiltration and density gradient separation (Liangsupree, Multia, and Riekkola 2021). In this chapter, we discuss

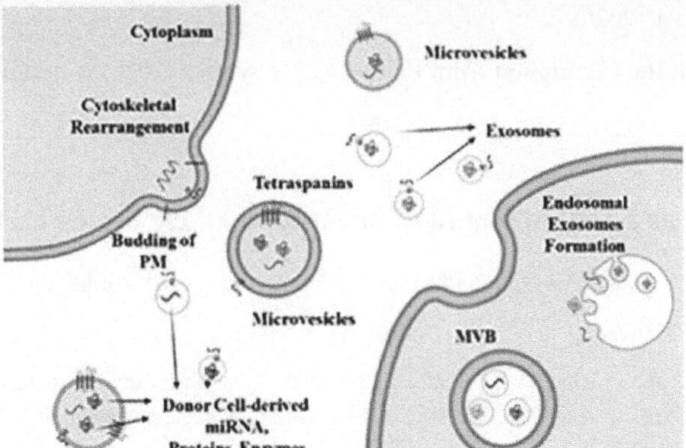

FIGURE 19.1 Exosomes are endogenously formed inside multivesicular bodies (MVBs). MVBs fuse into the plasma membrane and release exosomes into the extracellular matrix (ECM). Microvesicles are formed by the direct budding of the plasma membrane. Both exosomes and microvesicles are selectively incorporated with membrane-bound specific proteins such as tetraspanins. EVs carry specific proteins, DNA, sugar and miRNA molecules across the ECM as mediators of cell communication.

DOI: 10.1201/9781003302131-20

the uses of infrared (IR), Raman spectroscopy, mass spectroscopy (MS), nuclear magnetic resonance (NMR) and atomic force microscopy and spectroscopy (AFM/AFS) in EV research.

19.2 ANALYTICAL TOOLS

19.2.1 IR Spectroscopy

IR spectra are mainly used to determine the molecular fingerprint by identifying the functional groups. IR's high sensitivity helps acquire the small chemical changes (i.e., showing the addition or absence of spectral peaks) in biological samples. Fourier transform (FT)-IR is the easiest technique since it requires a minute amount of sample, can be operated with liquid samples and is label-free. FT-IR spectrophotometer consists of an IR source, interferometer, sample chamber, detector, amplifier, convertor and computer. Radiation generated from an IR source passes across the interferometer and hits the detector. The interferometer measures the energy, which is transmitted to the sample. As a result, an interferogram is generated in which the wavelengths of specific energy are absorbed. Then, the signal is transferred to generate a Fourier transformation of the energy signals by using a computer (Schmitt and Flemming 1998) (**Figure 19.2A**). Recently, we examined HCT116 colon cancer cell-derived EVs for identifying colon cancer biomarker Hyaluronan using FT-IR spectra with respect to normal colon epithelial cell CCD-18CO EVs. We showed distinct differences in the spectral signatures between cancer and normal cell-derived EVs (**Figure 19.2B**) (Paul et al. 2020). More examples are listed in Table 19.1.

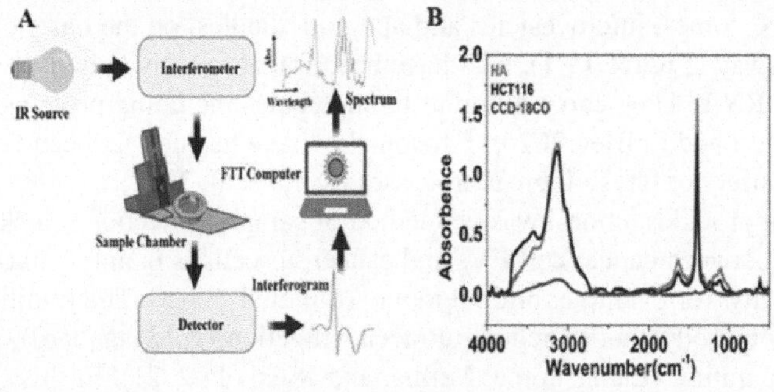

FIGURE 19.2 (A) FT-IR setup. (B) FT-IR spectra of Hyaluronic acid (HA) along with HCT116 EV and CCD-18CO EV (Paul et al. 2020).

Source: Reprinted with the permission from Paul et al. Copyright (2020) American Chemical Society.

TABLE 19.1
FT-IR Spectroscopy as a Tool to Identify Cancer Biomarkers on the EV Platform

Sample Name	Samples Obtained From		IR Findings	Ref.
	Cancer	Normal		
Salivary EVs	Oral cancer patient	Healthy individuals	Significant differences were observed in spectral signatures at 1072, 1543, 2924 and 2854 cm	(Zlotgorski-Hurvitz et al. 2019)
Serum EVs	Cancer patient	Healthy Individuals	Significant differences were observed in amide I, II, IIIand amide A bands	(Krafft et al. 2017)

TABLE 19.1 (*Continued*)
FT-IR Spectroscopy as a Tool to Identify Cancer Biomarkers on the EV Platform

Sample Name	Samples Obtained From		IR Findings	Ref.
	Cancer	Normal		
Urinary EVs	Prostate cancer patient	Healthy individuals	Significant differences were observed at 1640, 1120 and 967 cm	(Yap et al. 2019)
Colorectal EVs	HCT116 cell line	CCD-18CO cell line	Significant differences were observed at 622, 1113, 1407, 1639, 3005 and 3137 cm^{-1}	(Paul et al. 2020)

TABLE 19.2
Assignment of FT-IR Absorption Bands to EVs

Serial No.	Wavelength Range (cm^{-1})	Functional Group
1	700–1000	C-H out of plane bending vibration (Schulz and Baranska 2007)
2	900–1200	Phosphodiester group, phospholipids of nucleic acid and C-O absorption of carbohydrates (Wood et al. 1998)
3	1480–1700	Amide I region of proteins carrying information on the secondary structure of proteins: in-plane stretching of the C-O bonds, weakly coupled to stretching of the C-N and in-plane bending of the N-H bond (Barth 2007)
4	2800–3000	Symmetric and asymmetric stretching of lipids (Andrus and Strickland 1998)
5	3000–3500	Amide A of proteins; symmetric and asymmetric N-H stretching (Yang et al. 2005)

19.2.2 Raman Spectroscopy

Raman spectroscopy delivers a molecular structural fingerprint by detecting the vibrational, rotational and some low-frequency states. The main struggle in acquiring a Raman spectrum is separating the feeble inelastic scattering from the Rayleigh intense scattering (Raman Setup: **Figure 19.3A**) (Downes and Elfick 2010). IR absorption and Raman scattering spectroscopy enable us to acquire complementary bond-specific information of any biological sample. Surface-enhanced Raman scattering is a technique that improves Raman scattering by using molecules that are adsorbed on rough metal surfaces or nanostructures where the enhancement factor reaches 10^{10} to 10^{11} (Langer et al. 2020). This surface-sensitive Raman analysis of colon cancer and normal EVs was reported in 2012. Their results indicate that tumour EVs have crucial RNA content compared to healthy EVs (Tirinato et al. 2012). Recently, we showed that HCT 116-derived EVs are coated with Hyaluronan molecules compared to their healthy counterparts by using Raman spectroscopy (**Figure 19.3B**) (Paul et al. 2020). Nina et. al. reported that ovarian cancer cell EVs were distinguishable by their Surface-enhanced Raman spectroscopy (SERS) spectral signatures, and subsequently, cancer staging was studied (Ćulum et al. 2021). More examples can be found in Table 19.3.

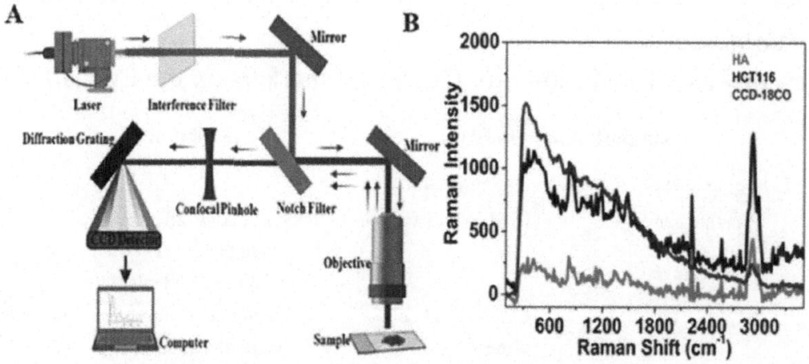

FIGURE 19.3 (A) Raman Setup. (B) Raman Spectra HA, HCT116 and CCD-18CO cell-derived EV (Paul et al. 2020).

Source: Reprinted with permission from Paul et al. Copyright (2020) American Chemical Society.

TABLE 19.3
SERS as a Tool to Identify Cancer Biomarkers on the EV Platform

Sample Name	Samples Obtained From		SERS Findings	Ref.
	Cancer	Normal		
Breast cancer EVs	MCF7 and MDA-MB-231 cell line	MCF-10A cell line	1221–1573 cm (Low intensity) and 735 cm (High intensity)	(P Zhang et al. 2019)
Hepatoma EV	HepG2 cell lines	L02 cell line	729, 735, 964, 1318, 1376, 1470 and 1580 cm	(P Zhang et al. 2019)
EVs	EC109, Kysel50 and EC9706 cell lines	L02 cell line	700–750 cm and 1200–1600 cm	(P Zhang et al. 2019)
Lung cancer EVs	PC9, HCC827 and H197 cell line	L929, NL-20 and Beas-2b cell line	Cancer EVs show the strong spectral bands only at protein regions	(Sivashanmugan et al. 2017)
Melanoma EVs	B16F10 cell line	RBC EV (Obtained from healthy individuals)	Intense peak at 900–1100 cm	(Stremersch et al. 2016)
Colorectal EVs	HCT116 cell line	CCD-84CoN cell line	Normal EVs show a high presence of lipids, whereas cancer EVs show a high presence of RNA	(Tirinato et al. 2012)
Colorectal EVs	HCT116 cell line	CCD-18CO cell line	Significant differences were observed at 1049, 1132, 1170, 1339, 1488, 2909 and 2975 cm	(Paul et al. 2020)
Ovarian EVs	OV-90, NIH:0VCAR3 cancer cell line EOC6 (High grade cancer cell line) and EOC18 (Low grade cancer cell line)	hIQSE(OSE364) cell line	Normal EVs can be different from cancer EVs by positive peaks at 782, 1181 and 1483 cm, whereas high-grade cancer also differentiates or predicts based on the positive peaks at 940, 1060 and 1170 cm	(Ćulum et al. 2021)

TABLE 19.4
Assignment of the Raman Absorption Band of EVs

Serial No.	Wavelength (cm)	Functional Group	Ref.
1	505–550	Disulfides	(Van Wart et al. 1973)
2	1230–1300	Amide III	(Notingher et al. 2004)
3	1550	Amide II	(Schulz and Baranska 2007)
4	1630–1650	Amide I	(Notingher al. 2004)
6	800–1050	-CH bending	(Schulz and Baranska 2007)
7	1050–1200	C-C and P-O stretching	(Stone et al. 2004)
8	1200–1300	CH deformation	(Notingher et al. 2004)
9	1400–1500	CH2CH3 bending and scissoring	(Schulz and Baranska 2007)
10	1600–1800	C=C and C=O stretching	(Schulz and Baranska 2007)
11	2800–3100	CH/CH2/CH3 stretching	(Schulz and Baranska 2007)

TABLE 19.5
AFM and AFS as Tools to Identify Cancer Biomarkers on the EV Platform

Sample Name	Sample Obtained From		AFM and AFS Finding	Ref.
	Cancer	Normal		
Urothelial EV	T24 cell line (malignant non-metastatic) FL3 cell line (malignant metastatic)	HCV-29 cell line (Non-malignant)	Malignant (metastatic and non-metastatic) cell-derived EVs show a lower stiffness value compared to non-malignant cell-derived EVs	(Whitehead et al. 2015)
Serum EV	Obtained from patients with hereditary spherocytosis	Obtained from healthy individuals	Patient EVs show lower bending modulus than donor EVs, which means that patient EVs are more rigid than normal EVs	(Vorselen et al. 2018)
Saliva EV	Obtained from oral cancer patient	Obtained from healthy individuals	Highly specific detection of CD63 protein on EV surface using AFS	(Sharma et al. 2010)
Saliva EV	Obtained from oral cancer patient	Obtained from healthy individuals	The population of cancer EVs is high, and they also have irregular morphology compared to normal EVs. The density of CD63 protein is high in cancer EVs.	(Sharma et al. 2011)

19.2.3 Mass Spectrometry (MS)

MS separates atoms or molecules (ionized form) by using the mass/charge ratio (m/z). Since molecules have distinct fragmentation patterns, MS is effective for quantifying atoms/molecules and providing information to recognize and characterize the structural constituents (Kaklamanos, Aprea, and Theodoridis 2012). A mass spectrometer is made up of three parts: an ion generator, an analyzer (mass-selective) and a detector (for ions) (**Figure 19.4**). Since the analyzer and ion detector (also

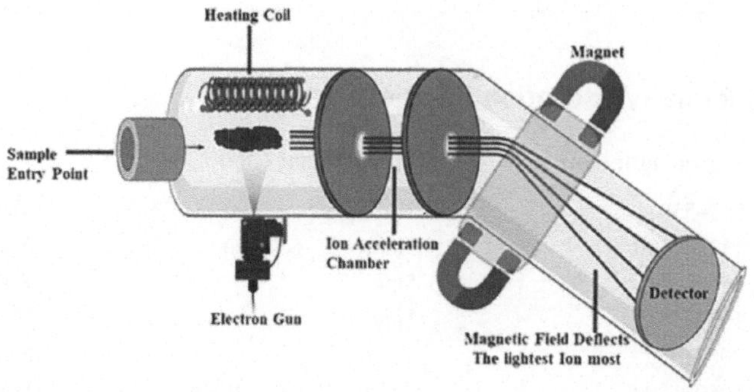

FIGURE 19.4 Schematic of an MS setup.

including ion sources) operate at a small amount of pressure, the system also requires a pump. Moreover, separation devices are frequently employed in conjunction with mass spectrometers to improve the analytical method's sensitivity, dynamic range and selectivity (Matta et al. 2010). MS offers the potential to identify and analyze the molecular composition of EVs; nevertheless, there are currently only a few (but a growing number of) papers that apply MS procedures in the exosome field.

19.2.3.1 MS in EV Proteomic and Lipidomics Analysis

The most popular and adaptable analytical approach for studying the chemical composition of EVs is liquid chromatography-electrospray ionization tandem mass spectrometry; it particularly offers high sensitivity, where hundreds of proteins can be detected, identified, characterized and quantified from a single EV sample (Erozenci et al. 2019). Several studies have been performed with recently developed data-independent acquisition MS, which allows for high-throughput, in-depth studies of exosome proteomes derived from cancer patients with improved quantitative accuracy (Collins et al. 2017; Guo et al. 2015).

Exosomes possess a notable enrichment of different lipids, indicating a high level of lipid discrimination from the parent cell (Skotland, Sandvig, and Llorente 2017). GC-MS is an established lipidomics analysis platform that provides excellent temporal resolution for lipids having a low molecular weight (500 Da) of diverse types and allows for the separation of positional isomers and diastereomers before mass estimations by MS. Llorente et al. discovered possible biomarkers within PC-3-derived exosomes in their study. The study provides a comprehensive lipid analysis of cells and exosomes based on the detection of roughly 280 molecular lipid species (Llorente et al. 2013). Another study highlighted the role of lipidomics analysis of urinary exosomes to detect prostate cancer from cancer patients and healthy individuals using high-throughput mass spectrometry (Skotland et al. 2017). The study suggested the changes in particular sphingolipids, which corresponded to the known cancer-related aberrant metabolism of sphingolipid. Recently, EV metabolite biomarker candidates were found in a variety of disorders, including liver disease and prostate and pancreatic malignancies (Salciccia et al. 2021; Kdadra et al. 2019).

19.2.4 Nuclear Magnetic Resonance Spectroscopy

NMR spectroscopy is extensively employed for studying the structure of organic molecules in solution. Magnetic resonance imaging (MRI) is also routinely used as an advanced medical imaging technique. The most commonly used nuclei are ^{1}H and ^{13}C. EV-based biomarker research is relatively new, and NMR has been used for only a couple of reports (Table 19.7). Shao et al. created a microfluidic micro-nuclear magnetic resonance biosensor that integrated microfiltration and NMR, allowing for the concurrent assessment of exosome quantity and exosomal protein expression. The NMR biosensor demonstrated great repeatability and exceptional detection sensitivity than the nanoparticle tracking analyzer (NTA) characterization system and other biochemical methods (Céline M. Dubéa, Amy L. Brewstera 2012).

TABLE 19.6
Mass Spectrometry-Based EV Analysis

SL. No	Source of EVs	Strategies Used and Biomarker	Ref.
1.	Colon cancer cells	Proteomic profiling, statistical analysis were used to find distinct protein statistical patterns	(Stübiger et al. 2018)
2.	Melanoma cancer cells	Protein fingerprint analysis was used to find distinct protein fingerprint peaks	(Zhu et al. 2019)
3.	U-87MG glioblastoma cell	Peptide mass fingerprinting was used to find proteins including hCG and annexin A5 under temperature stress	(Chun et al. 2016)|
4.	Oviduct fluid	Lipidomic profiling, statistical analysis were used to find phosphatidylcholines, phosphatidylethanolamine, sphingomyelins	(Banliat et al. 2020)
5.	Plasma from osteosarcoma patient	Statistical analysis was used to find exosomal proteins	(Han et al. 2021)
6.	Serum obtained from children with coronary artery aneurysms caused by Kawasaki disease	Proteomic profiling was used to find proteins as biomarkers	(Xie et al. 2019)
7.	Serum obtained from patients with coronary artery aneurysms caused by Kawasaki disease	Proteomic profiling was used to find proteins as biomarkers	(L. Zhang et al. 2016)
8.	Tear	Fingerprinting analysis, statistical analysis were used to find distinct fingerprints and characteristic peaks	(T. Zhang et al. 2020)
9.	Urine	Lipidomic profiling was used to find sphingolipids	(Singhto. Vinaiphat. and Thongboonkerd 2019)

TABLE 19.7
NMR-Based EV Analysis

SL. No	Source of EVs	Strategies Used and Biomarker	Ref.
1.	Liver-derived exosomes	Aptasensor-based strategies used to detect the CD63	(Zhou et al. 2016)
2.	Plasma	Microfluidic biochip combined microfiltration with NMR-based strategies used to detect CD63, EGFR, EGFRvIII, PDPN, IDH1 and R132H biomarkers	(Shao et al. 2012)
3.	Serum from hepatocellular carcinoma	1H-NMR-based serum metabolomics changes in serum glucose, lactate, lipids and several amino acids	(Casadei-Gardini et al. 2020)

19.2.5 Atomic Force Microscopy and Spectroscopy

In AFM/AFS, the cantilever/tip assembly probes the sample with raster scanning, monitored by a photo-detector (PSPD) that captures the reflected laser from the back of the cantilever in a position-sensitive manner (**Figure 19.5**). The probe's motion then forms a 3D image of the surface.

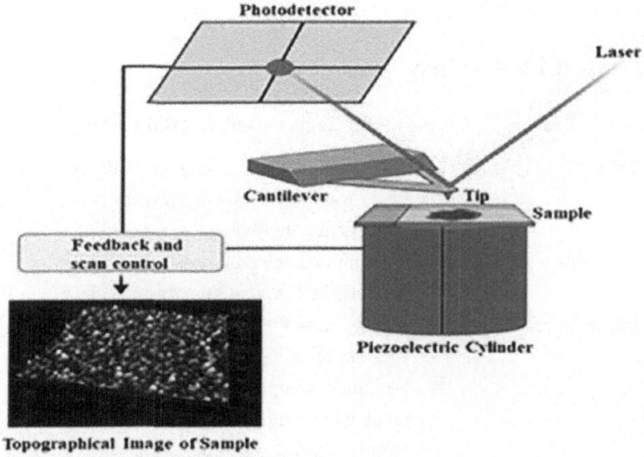

FIGURE 19.5 Schematic of an atomic force microscopy setup.

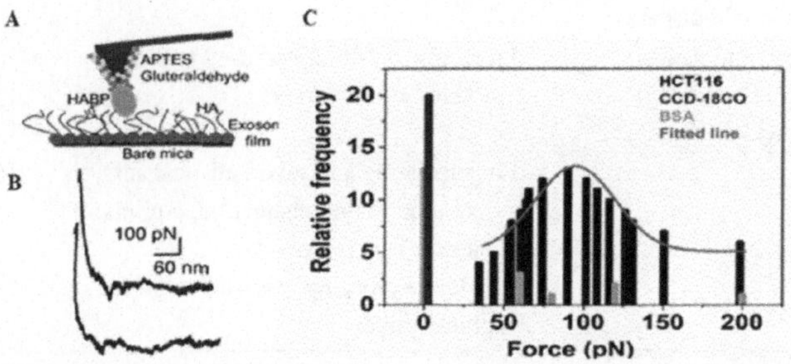

FIGURE 19.6 (A) AFS schematic showing that HCT116 EVs are interacting with a HABP functionalized tip. (B) Force (pN) vs. separation (nm) for a single pulling experiment. (C) Histogram of unbinding force between HCT 116 EVs and a HABP coated tip (Paul et al. 2020).

Source: Reprinted with the permission from Paul et al. Copyright (2020) American Chemical Society.

Forces between the probe and surface can be in the pico-Newton regime, and very sharp probes with a diameter of a few nanometers can be used. Thus, horizontal resolutions of a few nanometers are achievable by AFM imaging.

Along with topographic imaging (the intermittent contact mode is most suitable for visualizing EVs under buffer), force-distance curves are taken by moving the probe onto the sample vertically towards the substrate as a function of the distance. Nano-indentation and adhesion measurements using AFM have been employed in a few reports for EV biomarker identification (see Table 19.6). Recently, we visualized colon cancer and normal EVs (HCT 116 and CCD-18 Co cells) using high-resolution AFM imaging and noticed a distinguishable change in cancer EV morphology: they were larger, had network arrangements and were irregularly shaped compared to the normal epithelial cell-EVs. We probed tetraspanin CD9 on cancer EVs and identified a colon cancer biomarker Hyaluronan using AFM unbinding experiments (Paul et al. 2020) (**Figure 19.6**). Despite being tedious and time-intensive, we believe that single molecule-based highly sensitive and reproducible AFM/AFS investigations on cancer EVs are important aspects of identifying EV-based cancer biomarkers, especially for early detection.

19.2.5.1 Acknowledgements

DST/INSPIRE/04/2017/002375, CRG/2019/007013 and Shiv Nadar Institution of Eminence are gratefully acknowledged for their generous funding.

BIBLIOGRAPHY

Akers, Johnny C., Valya Ramakrishnan, Ryan Kim, Johan Skog, Ichiro Nakano, Sandeep Pingle, Juliya Kalinina, et al. 2013. "MiR-21 in the Extracellular Vesicles (EVs) of Cerebrospinal Fluid (CSF): A Platform for Glioblastoma Biomarker Development." *PLoS ONE* 8 (10): 1–13. https://doi.org/10.1371/journal.pone.0078115.

Andrus, Paul G. L., and Robert D. Strickland. 1998. "Cancer Grading by Fourier Transform Infrared Spectroscopy." *Biospectroscopy* 4 (1): 37–46. https://doi.org/10.1002/(SICI)1520-6343(1998)4:1<37::AID-BSPY4>3.0.CO;2-P.

Banliat, Charles, Daniel Le Bourhis, Ophélie Bernardi, Daniel Tomas, Valérie Labas, Pascal Salvetti, Benoît Guyonnet, Pascal Mermillod, and Marie Saint-Dizier. 2020. "Oviduct Fluid Extracellular Vesicles Change the Phospholipid Composition of Bovine Embryos Developed in Vitro." *International Journal of Molecular Sciences* 21 (15): 1–13. https://doi.org/10.3390/ijms21155326.

Barth, Andreas. 2007. "Infrared Spectroscopy of Proteins." *Biochimica et Biophysica Acta – Bioenergetics* 1767 (9): 1073–1101. https://doi.org/10.1016/j.bbabio.2007.06.004.

Casadei-Gardini, Andrea, Laura Del Coco, Giorgia Marisi, Fabio Conti, Giulia Rovesti, Paola Ulivi, Matteo Canale, et al. 2020. "1H-NMR Based Serum Metabolomics Highlights Different Specific Biomarkers between Early and Advanced Hepatocellular Carcinoma Stages." *Cancers* 12 (1). https://doi.org/10.3390/cancers12010241.

Céline M. Dubéa, Amy L. Brewstera, and Tallie Z. Baram. 2012. "Protein Typing of Ci." *Bone* 23 (1): 1–7. https://doi.org/10.1038/nm.2994.Protein.

Chun, Sohyun, Seunghyun Ahn, Chang Hwan Yeom, and Seyeon Park. 2016. "Exosome Proteome of U-87MG Glioblastoma Cells." *Biology* 5 (4): 1–11. https://doi.org/10.3390/biology5040050.

Collins, Ben C., Christie L. Hunter, Yansheng Liu, Birgit Schilling, George Rosenberger, Samuel L. Bader, Daniel W. Chan, et al. 2017. "Multi-Laboratory Assessment of Reproducibility, Qualitative and Quantitative Performance of SWATH-Mass Spectrometry." *Nature Communications* 8 (1): 1–11. https://doi.org/10.1038/s41467-017-00249-5.

Ćulum, Nina M., Tyler T. Cooper, Gilles A. Lajoie, Thamara Dayarathna, Stephen H. Pasternak, Jiahui Liu, Yangxin Fu, Lynne Marie Postovit, and François Lagugné-Labarthet. 2021. "Characterization of Ovarian Cancer-Derived Extracellular Vesicles by Surface-Enhanced Raman Spectroscopy." *Analyst* 146 (23): 7194–7206. https://doi.org/10.1039/d1an01586a.

Datta, Brateen, Debashish Paul, Tina Dey, Suchetan Pal, and Tatini Rakshit. 2022. "Importance of Extracellular Vesicle Derived RNAs as Critical Colorectal Cancer Biomarkers." *ACS Bio & Med Chem Au*. https://doi.org/10.1021/acsbiomedchemau.1c00043.

Datta, Brateen, Debashish Paul, Uttam Pal, and Tatini Rakshit. 2021. "Intriguing Biomedical Applications of Synthetic and Natural Cell-Derived Vesicles: A Comparative Overview." *ACS Applied Bio Materials* 4 (4): 2863–85. https://doi.org/10.1021/acsabm.0c01480.

Downes, Andrew, and Alistair Elfick. 2010. "Raman Spectroscopy and Related Techniques in Biomedicine." *Sensors* 10 (3): 1871–89. https://doi.org/10.3390/s100301871.

Erozenci, Leyla Ayse, Franziska Böttger, Irene V. Bijnsdorp, and Connie R. Jimenez. 2019. "Urinary Exosomal Proteins as (Pan-)Cancer Biomarkers: Insights from the Proteome." *FEBS Letters* 593 (13): 1580–97. https://doi.org/10.1002/1873-3468.13487.

Guo, Tiannan, Petri Kouvonen, Ching Chiek Koh, Ludovic C. Gillet, Witold E. Wolski, Hannes L. Röst, George Rosenberger, et al. 2015. "Rapid Mass Spectrometric Conversion of Tissue Biopsy Samples into Permanent Quantitative Digital Proteome Maps." *Nature Medicine* 21 (4): 407–13. https://doi.org/10.1038/nm.3807.

Han, Zhenzhen, Cheng Peng, Jia Yi, Yiwen Wang, Qi Liu, Yi Yang, Shuping Long, Liang Qiao, and Yuhui Shen. 2021. "Matrix-Assisted Laser Desorption Ionization Mass Spectrometry Profiling of Plasma Exosomes Evaluates Osteosarcoma Metastasis." *IScience* 24 (8): 102906. https://doi.org/10.1016/j.isci.2021.102906.

Jeppesen, Dennis K., Aidan M. Fenix, Jeffrey L. Franklin, James N. Higginbotham, Qin Zhang, Lisa J. Zimmerman, Daniel C. Liebler, et al. 2019. "Reassessment of Exosome Composition." *Cell* 177 (2): 428–445.e18. https://doi.org/10.1016/j.cell.2019.02.029.

Kaklamanos, George, Eugenio Aprea, and Georgios Theodoridis. 2012. Mass Spectrometry. In Chemical Analysis of Food: Techniques and Applications. https://doi.org/10.1016/B978-0-12-384862-8.00009-1.

Kalluri, Raghu, and Valerie S. LeBleu. 2020. "The Biology, Function, and Biomedical Applications of Exosomes." *Science* 367 (6478). https://doi.org/10.1126/science.aau6977.

Kdadra, Marouane, Sebastian Höckner, Hing Leung, Werner Kremer, and Eric Schiffer. 2019. "Metabolomics Biomarkers of Prostate Cancer: A Systematic Review." *Diagnostics* 9 (1): 1–44. https://doi.org/10.3390/diagnostics9010021.

Krafft, Christoph, Konrad Wilhelm, Anna Eremin, Sigrun Nestel, Nikolas von Bubnoff, Wolfgang Schultze-Seemann, Juergen Popp, and Irina Nazarenko. 2017. "A Specific Spectral Signature of Serum and Plasma-Derived Extracellular Vesicles for Cancer Screening." *Nanomedicine: Nanotechnology, Biology, and Medicine* 13 (3): 835–41. https://doi.org/10.1016/j.nano.2016.11.016.

Lane, R. E., D. Korbie, M. M. Hill, and M. Trau. 2018. "Extracellular Vesicles as Circulating Cancer Biomarkers: Opportunities and Challenges." *Clinical and Translational Medicine* 7 (1). https://doi.org/10.1186/s40169-018-0192-7.

Langer, J., D. Jimenez de Aberasturi, J. Aizpurua, R. A. Alvarez-Puebla, B. Auguié, J. J. Baumberg, G. C. Bazan, S. E. J. Bell, A. Boisen, A. G. Brolo, J. Choo, D. Cialla-May, V. Deckert, L. Fabris, K. Faulds, F. J. García de Abajo, R. Goodacre, D. Graham, A. J. Haes, C. L. Haynes, C. Huck, T. Itoh, M. Käll, J. Kneipp, N. A. Kotov, H. Kuang, E. C. Le Ru, H. K. Lee, J. F. Li, X. Y. Ling, S. A. Maier, T. Mayerhöfer, M. Moskovits, K. Murakoshi, J. M. Nam, S. Nie, Y. Ozaki, I. Pastoriza-Santos, J. Perez-Juste, J. Popp, A. Pucci, S. Reich, B. Ren, G. C. Schatz, T. Shegai, S. Schlücker, L. L. Tay, K. G. Thomas, Z. Q. Tian, R. P. Van Duyne, T. Vo-Dinh, Y. Wang, K. A. Willets, C. Xu, H. Xu, Y. Xu, Y. S. Yamamoto, B. Zhao, and L. M. Liz-Marzán. 2020. "Present and Future of Surface-Enhanced Raman Scattering." *ACS Nano* 14 (1): 28–117. https://doi.org/10.1021/acsnano.9b04224.

Liangsupree, Thanaporn, Evgen Multia, and Marja Liisa Riekkola. 2021. "Modern Isolation and Separation Techniques for Extracellular Vesicles." *Journal of Chromatography A* 1636: 461773. https://doi.org/10.1016/j.chroma.2020.461773.

Lin, Shih Yi, Chao Hsiang Chang, His Chin Wu, Ching Chan Lin, Kai Po Chang, Chi Rei Yang, Chi Ping Huang, Wu Huei Hsu, Chiz Tzung Chang, and Chao Jung Chen. 2016. "Proteome Profiling of Urinary Exosomes Identifies Alpha 1-Antitrypsin and H2B1K as Diagnostic and Prognostic Biomarkers for Urothelial Carcinoma." *Scientific Reports* 6 (September): 1–12. https://doi.org/10.1038/srep34446.

Llorente, Alicia, Tore Skotland, Tuulia Sylvänne, Dimple Kauhanen, Tomasz Róg, Adam Orłowski, Ilpo Vattulainen, Kim Ekroos, and Kirsten Sandvig. 2013. "Molecular Lipidomics of Exosomes Released by PC-3 Prostate Cancer Cells." *Biochimica et Biophysica Acta – Molecular and Cell Biology of Lipids* 1831 (7): 1302–9. https://doi.org/10.1016/j.bbalip.2013.04.011.

Matta, Ajay, Ranju Ralhan, Leroi V. Desouza, and K. W. Michael Siu. 2010. "Mass Spectrometry-Based Clinical Proteomics: Head-and-Neck Cancer Biomarkers and Drug-Targets Discovery." *Mass Spectrometry Reviews* 29. https://doi.org/10.1002/mas.20296.

Notingher, Ioan, Chris Green, Chris Dyer, Elaine Perkins, Neil Hopkins, Chris Lindsay, and Larry L. Hench. 2004. "Discrimination between Ricin and Sulphur Mustard Toxicity in Vitro Using Raman Spectroscopy." *Journal of the Royal Society Interface* 1 (1): 79–90. https://doi.org/10.1098/rsif.2004.0008.

Paul, Debashish, Anuradha Roy, Arpita Nandy, Brateen Datta, Prateeka Borar, Samir Kumar Pal, Dulal Senapati, and Tatini Rakshit. 2020. "Identification of Biomarker Hyaluronan on Colon Cancer Extracellular Vesicles Using Correlative AFM and Spectroscopy." *Journal of Physical Chemistry Letters* 11 (14): 5569–76. https://doi.org/10.1021/acs.jpclett.0c01018.

Salciccia, S., A. L. Capriotti, A. Laganà, S. Fais, M. Logozzi, E. De Berardinis, G. M. Busetto, G. B. Di Pierro, G. P. Ricciuti, F. Del Giudice, A. Sciarra, P. R. Carroll, M. R. Cooperberg, B. Sciarra, and M. Maggi. 2021. "Biomarkers in Prostate Cancer Diagnosis: From Current Knowledge to the Role of Metabolomics and Exosomes." *International Journal of Molecular Sciences* 22 (9). https://doi.org/10.3390/ijms22094367.

Schmitt, Jürgen, and Hans Curt Flemming. 1998. "FTIR-Spectroscopy in Microbial and Material Analysis." *International Biodeterioration and Biodegradation* 41 (1): 1–11. https://doi.org/10.1016/S0964-8305(98)80002-4.

Schulz, Hartwig, and Malgorzata Baranska. 2007. "Identification and Quantification of Valuable Plant Substances by IR and Raman Spectroscopy." *Vibrational Spectroscopy* 43 (1): 13–25. https://doi.org/10.1016/j.vibspec.2006.06.001.

Shao, Huilin, Jaehoon Chung, Leonora Balaj, Alain Charest, Darell D. Bigner, Bob S. Carter, Fred H. Hochberg, Xandra O. Breakefield, Ralph Weissleder, and Hakho Lee. 2012. "Protein Typing of Circulating Microvesicles Allows Real-Time Monitoring of Glioblastoma Therapy." *Nature Medicine* 18 (12): 1835–40. https://doi.org/10.1038/nm.2994.

Sharma, Shivani, Boyd M. Gillespie, Viswanathan Palanisamy, and James K. Gimzewski. 2011. "Quantitative Nanostructural and Single-Molecule Force Spectroscopy Biomolecular Analysis of Human-Saliva-Derived Exosomes." *Langmuir* 27 (23): 14394–400. https://doi.org/10.1021/la2038763.

Sharma, Shivani, Haider I. Rasool, Viswanathan Palanisamy, Cliff Mathisen, Michael Schmidt, David T. Wong, and James K. Gimzewski. "Structural-mechanical characterization of nanoparticle exosomes in human saliva, using correlative AFM, FESEM, and force spectroscopy." *ACS Nano* 4, no. 4 (2010): 1921–1926.

Shi, R., P. Y. Wang, X. Y. Li, J. X. Chen, Y. Li, X. Z. Zhang, C. G. Zhang, T. Jiang, W. B. Li, W. Ding, and S. J. Cheng. 2015. "Exosomal Levels of MiRNA-21 from Cerebrospinal Fluids Associated with Poor Prognosis and Tumor Recurrence of Glioma Patients." *Oncotarget* 6 (29): 26971–81. https://doi.org/10.18632/oncotarget.4699.

Singhto, Nilubon, Arada Vinaiphat, and Visith Thongboonkerd. 2019. "Discrimination of Urinary Exosomes from Microvesicles by Lipidomics Using Thin Layer Liquid Chromatography (TLC) Coupled with MALDI-TOF Mass Spectrometry." *Scientific Reports* 9 (1): 1–11. https://doi.org/10.1038/s41598-019-50195-z.

Sivashanmugan, Kundan, Wei Lun Huang, Chen Hsueh Lin, Jiunn Der Liao, Chien Chung Lin, Wu Chou Su, and Ten Chin Wen. 2017. "Bimetallic Nanoplasmonic Gap-Mode SERS Substrate for Lung Normal and Cancer-Derived Exosomes Detection." *Journal of the Taiwan Institute of Chemical Engineers* 80: 149–55. https://doi.org/10.1016/j.jtice.2017.09.026.

Skotland, Tore, Kim Ekroos, Dimple Kauhanen, Helena Simolin, Therese Seierstad, Viktor Berge, Kirsten Sandvig, and Alicia Llorente. 2017. "Molecular Lipid Species in Urinary Exosomes as Potential Prostate Cancer Biomarkers." *European Journal of Cancer* 70: 122–32. https://doi.org/10.1016/j.ejca.2016.10.011.

Skotland, Tore, Kirsten Sandvig, and Alicia Llorente. 2017. "Lipids in Exosomes: Current Knowledge and the Way Forward." *Progress in Lipid Research* 66: 30–41. https://doi.org/10.1016/j.plipres.2017.03.001.

Stone, Nicholas, Catherine Kendall, Jenny Smith, Paul Crow, and Hugh Barr. 2004. "Raman Spectroscopy for Identification of Epithelial Cancers." *Faraday Discussions* 126 (1): 141–57. https://doi.org/10.1039/b304992b.

Stremersch, Stephan, Monica Marro, Bat El Pinchasik, Pieter Baatsen, An Hendrix, Stefaan C. De Smedt, Pablo Loza-Alvarez, Andre G. Skirtach, Koen Raemdonck, and Kevin Braeckmans. 2016. "Identification of Individual Exosome-like Vesicles by Surface Enhanced Raman Spectroscopy." *Small* 12 (24): 3292–3301. https://doi.org/10.1002/smll.201600393.

Stübiger, Gerald, Michael D. Nairn, Tom K. Abban, Matthew E. Openshaw, Luis Mancera, Barbara Herzig, Michael Wuczkowski, Daniel Senfter, and Robert M. Mader. 2018. "MALDI-MS Protein Profiling of Chemoresistance in Extracellular Vesicles of Cancer Cells." *Analytical Chemistry* 90 (22): 13178–82. https://doi.org/10.1021/acs.analchem.8b03756.

Tirinato, L., F. Gentile, D. Di Mascolo, M. L. Coluccio, G. Das, C. Liberale, S. A. Pullano, G. Perozziello, M. Francardi, A. Accardo, F. De Angelis, P. Candeloro, and E. Di Fabrizio. 2012. "SERS Analysis on Exosomes Using Super-Hydrophobic Surfaces." *Microelectronic Engineering* 97: 337–40. https://doi.org/10.1016/j.mee.2012.03.022.

Vorselen, Daan, Susan M. van Dommelen, Raya Sorkin, Melissa C. Piontek, Jürgen Schiller, Sander T. Döpp, Sander A. A. Kooijmans, et al. 2018. "The Fluid Membrane Determines Mechanics of Erythrocyte Extracellular Vesicles and Is Softened in Hereditary Spherocytosis." *Nature Communications* 9 (1). https://doi.org/10.1038/s41467-018-07445-x.

Wart, H. E. Van, A. Lewis, H. A. Scheraga, and F. D. Saeva. 1973. "Disulfide Bond Dihedral Angles from Raman Spectroscopy." *Proceedings of the National Academy of Sciences of the United States of America* 70 (9): 2619–23. https://doi.org/10.1073/pnas.70.9.2619.

Whitehead, Bradley, Lin Ping Wu, Michael Lykke Hvam, Husnu Aslan, Mingdong Dong, Lars DyrskjØt, Marie Stampe Ostenfeld, Seyed Moein Moghimi, and Kenneth Alan Howard. 2015. "Tumour Exosomes Display Differential Mechanical and Complement Activation Properties Dependent on Malignant State: Implications in Endothelial Leakiness." *Journal of Extracellular Vesicles* 4 (1). https://doi.org/10.3402/jev.v4.29685.

Wood, Bayden R., Michael A. Quinn, Brian Tait, Martin Ashdown, Tracy Hislop, Melissa Romeo, and Don McNaughton. 1998. "FTIR Microspectroscopic Study of Cell Types and Potential Confounding Variables in Screening for Cervical Malignancies." *Biospectroscopy* 4 (2): 75–91. https://doi.org/10.1002/(sici)1520-6343(1998)4:2<75::aid-bspy1>3.0.co;2-r.

Xie, Xiao Fei, Hong Juan Chu, Yu Fen Xu, Liang Hua, Zhou Ping Wang, Ping Huang, Hong Ling Jia, and Li Zhang. 2019. "Proteomics Study of Serum Exosomes in Kawasaki Disease Patients with Coronary Artery Aneurysms." *Cardiology Journal* 26 (5): 584–93. https://doi.org/10.5603/CJ.a2018.0032.

Yang, Ying, Josep Sulé-Suso, Ganesh D. Sockalingum, Gregory Kegelaer, Michel Manfait, and Alicia J. El Haj. 2005. “Study of Tumor Cell Invasion by Fourier Transform Infrared Microspectroscopy.” *Biopolymers* 78 (6): 311–17. https://doi.org/10.1002/bip.20297.

Yap, Xin-Le, Teng-Aik Ong, Jasmine Lim, Bayden Wood, and Wai-Leng Lee. 2019. “Study of Prostate Cancer-Derived Extracellular Vesicles in Urine Using IR Spectroscopy.” *Progress in Drug Discovery & Biomedical Science* 2 (1): a0000026. https://doi.org/10.36877/pddbs.a0000026

Zhang, Li, Wei Wang, Jun Bai, Yu Fen Xu, Lai Qing Li, Liang Hua, Li Deng, and Hong Ling Jia. 2016. “Proteomic Analysis Associated with Coronary Artery Dilatation Caused by Kawasaki Disease Using Serum Exosomes.” *Revista Portuguesa de Cardiologia* 35 (5): 265–73. https://doi.org/10.1016/j.repc.2015.11.016.

Zhang, Ping, Limin Wang, Yaping Fang, Dawei Zheng, Taifeng Lin, and Huiqin Wang. 2019. “Label-Free Exosomal Detection and Classification in Rapid Discriminating Different Cancer Types Based on Specific Raman Phenotypes and Multivariate Statistical Analysis.” *Molecules* 24 (16). https://doi.org/10.3390/molecules24162947.

Zhang, Ting, Liang Hu, Huixiang Ma, Fangfang Ni, Fei Liu, and Hao Chen. 2020. “Detection of Tear Components Using Matrix-Assisted Laser Desorption Ionization/Time-of-Flight Mass Spectrometry for Rapid Dry Eye Diagnosis.” *Journal of Proteome Research* 19 (9): 3644–51. https://doi.org/10.1021/acs.jproteome.0c00138.

Zhou, Qing, Ali Rahimian, Kyungjin Son, Dong Sik Shin, Tushar Patel, and Alexander Revzin. 2016. “Development of an Aptasensor for Electrochemical Detection of Exosomes.” *Methods* 97: 88–93. https://doi.org/10.1016/j.ymeth.2015.10.012.

Zhu, Yingdi, Horst Pick, Natalia Gasilova, Xiaoyun Li, Tzu En Lin, Heinz Philipp Laeubli, Alfred Zippelius, Ping Chih Ho, and Hubert H. Girault. 2019. “MALDI Detection of Exosomes: A Potential Tool for Cancer Studies.” *Chem* 5 (5): 1318–36. https://doi.org/10.1016/j.chempr.2019.04.007.

Zlotogorski-Hurvitz, Ayelet, Ben Zion Dekel, Dov Malonek, Ran Yahalom, and Marilena Vered. 2019. “FTIR-Based Spectrum of Salivary Exosomes Coupled with Computational-Aided Discriminating Analysis in the Diagnosis of Oral Cancer.” *Journal of Cancer Research and Clinical Oncology* 145 (3): 685–94. https://doi.org/10.1007/s00432-018-02827-6.

20 RNA Interference (RNAi)
A Boon to Medical Biotechnology

Shreya Das[1]*, Chandrani Dey*[1]*, Santanu Chakraborty*[2] *and Arunima Sengupta*[1,*]

[1]Department of Life Science and Biotechnology, Jadavpur University, Kolkata, India, shreyadas1991@gmail.com)

[2]Department of Life Sciences, Presidency University, Kolkata, India, santanu.dbs@presiuniv.ac.in)

*Correpondence to arunimasengupta2013@gmail.com

20.1 INTRODUCTION

Ever since its inception, medical biotechnology has undergone a vast transformation in its approach to develop potent drugs in order to increase the longevity of the human race. One such Nobel Prize-winning discovery in the field of medical biotechnology is RNA interference (RNAi) by Fire and Mello in 2006 (Figure 20.1). RNAi technology is utilised by artificially introducing short synthetic double-stranded RNA (dsRNA) molecules to target specific genes post-transcriptionally. The silencing of genes by the administration of transgenes or dsRNA has now been used as a fundamental procedure in the protection against microorganism-causing diseases in both the plant and animal kingdoms (Cogoni and Macino 2000). In this technology, the mRNA of the target gene is cleaved, which results in decreased abundance of the respective proteins. In 2018, the first ground-breaking development in the field of RNAi was achieved when the first small interfering RNA (siRNA)-based drug was approved by the U.S. Food and Drug Administration (FDA). It was named Patisiran and used for the treatment of transthyretin-mediated amyloidosis (hATTR) in adults (Adams et al. 2018). In 2020, another siRNA drug that received approval was Givosiran to treat acute hepatic porphyria (AHP) (Scott 2020). The field of RNAi-based therapeutics is undergoing evolution with each passing day with multiple drug candidates already being in various phases of clinical trials (Figure 20.1).

In this chapter, we elucidate the molecular mechanisms of RNAi. We highlight the numerous developing modes of delivery of RNAi therapeutics for efficient gene silencing. We briefly

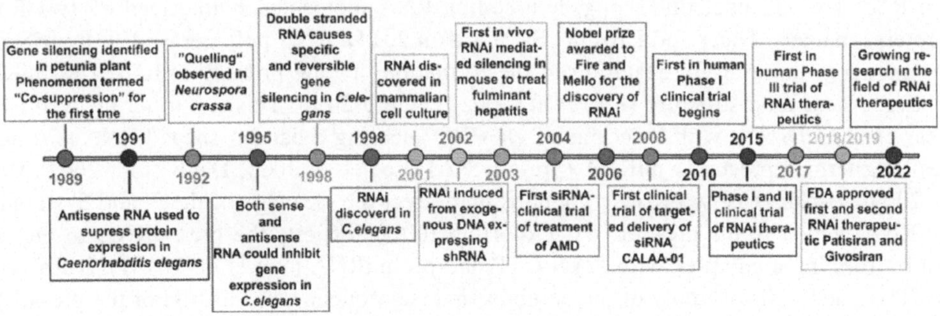

FIGURE 20.1 Timeline for milestones in RNAi and RNAi-based therapeutics. A timeline of the major historical developments leading to the discovery of RNA interference (RNAi) and the progress of RNAi therapeutics.

DOI: 10.1201/9781003302131-21

outline the various hurdles faced for the targeted delivery of RNAi therapeutics such as stability, off-target effects and immune elicitation. The chapter brings forth comprehensive knowledge regarding the available RNAi therapeutics for the cure of diseases such as cancer, cardiovascular diseases, neurological disorders, ocular disorders, etc., which have brought a revolutionary change in the arena of medical biotechnology. Lastly, the chapter delineates the numerous RNAi therapeutics that are currently under Phase I, II and III clinical trials with the hope that these drugs will reveal promising results in the treatment of various life-threating disorders similar to that of the FDA-approved drugs Patisiran and Givosiran, which have revolutionised medical biotechnology.

20.2 BIOGENESIS AND MECHANISTIC INSIGHTS OF RNA INTERFERENCE (RNAi)

20.2.1 Small Interfering RNA (siRNA)

siRNAs are 21–23 nucleotide-long double-stranded non-coding regulatory RNA having 3′ two-nucleotide overhangs that silence genes at the post-transcriptional level (Fougerolles et al. 2005). The precursors (long double-stranded RNA) for siRNA can be either introduced into the system exogenously or they may be of endogenous origin (transcribed from a host genome or infecting pathogens). In the cytoplasm, the dsRNA is cleaved by Dicer (an RNase III-like enzyme) to form siRNA (Figure 20.2). The siRNA is then loaded to an RNA-induced silencing complex (RISC), which is a ribonuclease that comprises a multi-protein complex by Dicer in association with TAR-RNA binding protein (TRBP) (Hammond et al. 2000). Once the siRNA is loaded, the endonucleolytic slicing protein of the RISC complex Ago2 comes into play. The siRNA duplex inside the RISC complex is separated by Ago2. The "passenger strand" is cleaved and released by Ago2, resulting in the formation of an activated RISC complex harbouring the "guide strand". This guide strand directs the specificity of the targeted cleavage of the desired mRNA (Tang 2005). The selectivity of the strand loading in the RISC complex is guided by the thermodynamic stability of the siRNA ends. The strand having relatively low 5'stability is retained in the RISC and loaded into Ago selectively. This strand is referred to as the "guide strand" (Schwarz et al. 2003). For a more potent and efficient functioning of siRNA, a 5'phosphate group must be present in the antisense or guide strand (Martinez et al. 2002; Nykänen, Haley, and Zamore 2001). The guide RNA then binds to a specific mRNA with perfect or near perfect complementarity, and the mRNA is cleaved by Ago2.

20.2.2 Micro RNA (miRNA)

Micro RNA are ~22 nucleotide-long, non-coding RNA molecules synthesized by two RNase III proteins, namely, Dicer and Drosha (Ha and Kim 2014). RNA polymerase II transcribes the genes encoding the miRNA, resulting in the formation of long primary miRNA (pri-miRNA) that have a stem loop structure (Lee et al. 2002) (Figure 20.2). This is further cleaved by the nuclear enzyme Drosha with its co-factor DGCR8 releasing a hairpin-shaped RNA ~70 nucleotides long termed precursor miRNA (pre-miRNA) (Lee et al. 2003; Denli et al. 2004; Martin et al. 2009). Drosha has two RNase domains that direct the cleavage of the 3' and 5' strands of pri-miRNA, which forms pre-miRNA. Exportin5 further exports the pre-miRNA to the cytoplasm by forming a complex with RAN-GTP and pre-miRNA (Wu et al. 2018). Hydrolysis of the GTP results in disassembly of the complex and subsequent translocation of the pre-miRNA to the cytosol. Inside the cytosol, it is cleaved by the RNase III enzyme Dicer, which binds to the pre-miRNA, has a two-nucleotide-long 3′ overhang generated by Drosha and cleaves near the terminal loop to thus form the miRNA duplex (Bernstein et al. 2001; Ketting et al. 2001). Dicer cleaves the pre-miRNA in combination with cofactors TRBP and protein kinase RNA activator

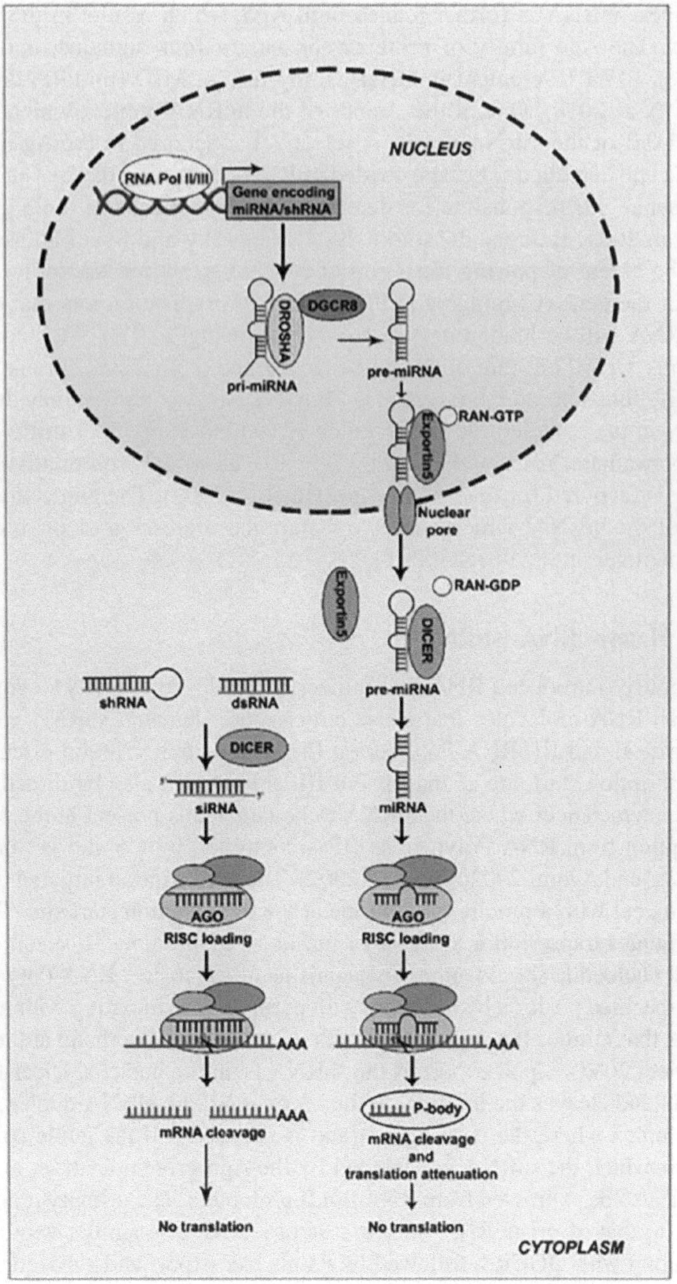

FIGURE 20.2 Biogenesis and mechanistic insights of RNAi in mammals. Inside the nucleus, RNA pol II drives the transcription of the primary miRNA (pri-miRNA) transcript, which is a hairpin structure. The pri-miRNA is processed by the RNase III enzyme Drosha and DGCR8, which cleaves the hairpin structure and results in the formation of precursor miRNA (pre-miRNA). The pre-miRNA is exported out of the nucleus to the cytoplasm by Exportin5 with the subsequent conversion of RAN-GTP to RAN-GDP. Inside the cytoplasm, Dicer cleaves the pre-miRNA. Following strand separation, the guide strand is loaded into the RNA-induced silencing complex (RISC), which cleaves the target mRNA by either Argonaut (Ago) or guide translational repression. Endogenously transcribed shRNA from plasmid or viral vectors or exogenously administered dsRNA is converted to siRNA by a cytoplasmic RNase III enzyme Dicer. siRNA is then loaded into an RISC complex where it is unwound by a helicase. The siRNA guide strand then binds to the complementary target mRNA, and the Ago protein causes the degradation of target mRNA.

(PACT). The duplex miRNA is further loaded onto Ago, which results in the formation of the miRISC complex. The Ago family of proteins consists of four domains, namely, PAZ (Piwi/Argonaute/Zwille), PIWI (P-element induced wimpy testis), MID (middle) domain and an N-terminal domain (Niaz 2018). One of the strands of the miRNA duplex, which has complementarity with the 3'UTR of the mRNA target, is selectively anchored to the Ago protein. Another strand is unwound and degraded. The Ago-loaded miRNA containing the 2–7 nucleotides (known as the "seed sequence") is responsible for determining the sequence in the 3'UTR of the target mRNA (Grün et al. 2005; Haley and Zamore 2004; Rajewsky and Socci 2004). Apart from the seed sequence, the extent of pairing the 3'end of miRNA to its mRNA target also contributes in the silencing of the gene (Chipman and Pasquinelli 2019). The factors that determine which strand of the miRNA will be loaded in Ago are the following. a) The 5'nucleotide of the guide strand is commonly Uracil but that of the passenger strand is Cytosine (Frank, Sonenberg, and Nagar 2010; Warf, Johnson, and Bass 2011). Moreover, it was shown that the existence of a phosphate moiety in the 5'nucleotide of the guide strand is required for miRNA loading in the RISC complex (Kawamata, Yoda, and Tomari 2011). b) The strand with relatively lower thermodynamic stability is favoured for loading in Ago (Hu et al. 2009). The guide strand further binds with the 3'UTR of the mRNA, which causes translational repression or in some cases, mRNA degradation (Huntzinger and Izaurralde 2011).

20.2.3 Short Hairpin RNA (shRNA)

shRNA is exogenously introduced RNA that utilizes the endogenic miRNA synthesis machinery to synthesize small RNA molecules that cause target gene silencing. shRNA can be transcribed by RNA Polymerase II and III. RNA Polymerase II-mediated transcription often does not have a well-defined transcription start site as that of Pol III. shRNA can also be directly expressed as a substrate for the enzyme Dicer where the siRNA to be targeted is present at the 5' or 3' end of the shRNA. Transcription from RNA Polymerase III is controlled by U6 and H1 promoters (Brummelkamp, Bernards, and Agami 2002; Sui et al. 2002). The transcription initiated under the U6 promoter is more stringent with a proclivity to purine at the transcription start site. The H1 promoter, to the contrary, initiates transcription with either purine or pyrimidine. To specifically express the shRNA, the initial nucleotide should either be guanosine or adenosine. RNA Polymerase III begins transcription at a specified position by initiating with purine and terminating with two extra uracil to synthesize shRNA that mimics the pre-miRNA with a 2 nucleotide overhang at the 3' end (Braglia, Percudani, and Dieci 2005). Upon export of the shRNA from the nucleus, Dicer further processes the shRNA hairpin and cleaves the terminal to thus form a ~22 bp siRNA duplex, which is loaded into the RISC complex where the passenger strand is released and the guide strand binds to the target mRNA after which the mRNA is degraded by the Ago protein (Park et al. 2011; Gu et al. 2012). shRNA can also be expressed from RNA Pol II promoters. The primary transcript containing shRNA is similar to that of pri-miRNA; hence, it is processed in a similar way by the RNase III enzyme Drosha along with DGCR8, followed by its nuclear export and subsequent processing by the RISC complex.

20.3 THE DISCOVERY OF RNA INTERFERENCE (RNAi)

This novel technique of post-transcriptional silencing of a gene was first identified in the Petunia flower in 1990 (Napoli, Lemieux, and Jorgensen 1990). On overexpressing chalcone synthase (CHS), a gene that increases anthocyanin expression to thus give a deeper violet colour to a petunia, most of the flowers were found to lose their colour. They termed this phenomenon "co-suppression" because they observed an abrogated expression of an endogenous CHS gene and the exogenously administered transgenic copy of the gene. A similar phenomenon was found in *Neurospora crassa*. They termed this "quelling" of the endogenous gene by the introduction of exogenous homologous RNA sequences (Romano and Macino 1992; Hammond, Caudy, and Hannon 2001).

In 1998, Fire and colleagues explained the terms co-suppression, quelling and sense mRNA through the endogenously expressed genes of *Caenorhabditis elegans*. They used both single-stranded RNAs (ssRNAs) and dsRNAs to target the UNC-22 gene. The dsRNA was shown to be more potent and its efficacy of gene downregulation was higher than ssRNA. The effect of the interference was shown to persist in the progeny of the injected animals. They also showed that only a few molecules of dsRNA were enough against each cell, thus negating stoichiometric interference with endogenous mRNA and highlighting the presence of a catalytic or amplification component during the interference process (Fire et al. 1998).

In the mammalian system, 21 nucleotide siRNA duplexes were shown to silence genes in cultured mammalian cells (Elbashir et al. 2001). In this study, a synthetic siRNA duplex containing a 2-nt 3' overhang against reporter gene coding for a sea pansy (Renilla reniformis, RL) and two sequence variants of firefly (Photinus pyralis, GL2 and GL3) luciferases were used. siRNA duplex transfection caused significant downregulation in the expression of the entire silenced gene.

The first human evidence of siRNA silencing was published in 2011 (Davis 2009). The nanoparticle formulation was denoted as CALAA-01, which was under a Phase I trial (NCT00689065). However, the work is now terminated due to toxicity issues.

The first ever RNAi drug approved by the FDA was Alnylam's Onpattro (Patisiran) (Adams et al. 2018). It is used to treat patients with the rare hereditary transthyretin amyloidosis (hATTR) after its invigorating result in the Phase III clinical trial "Apollo" (NCT01960348). Treatment with Patisiran halted or reversed disease progression. Givosiran (Givlaari™) is another siRNA-mediated therapeutic drug that was recently approved in the USA for treatment of acute hepatic porphyria (AHP) in adults after a positive response from the Phase III ENVISION trial. In the EU, the Givosiran drug received its approval for treatment of AHP in adults and others aged 12 years and over in 2020. Givosiran targets aminolevulinate synthase 1 (*alas1*) mRNA and prevents the cumulation of neurotoxic δ-aminolevulinic acid and porphobilinogen that cause AHP. Givosiran is developed by Alnylam Pharmaceuticals (Scott 2020).

20.4 HURDLES TO THE DELIVERY SYSTEM OF RNAI THERAPEUTICS

20.4.1 Stability

The RNAi-based therapeutic molecules require the crossing of various physiological barriers before their uptake in the desired organ system. These therapeutic molecules are generally large, hydrophilic and polyanionic and are unable to pass in the naked state due to repulsion from the cell membrane that is negatively charged (de Fougerolles et al. 2007; Kuninty et al. 2016). Intravenous administration of naked RNAi-inducing molecules such as siRNA and miRNA was shown to cause poor distribution due to renal clearance, uptake by mononuclear phagocytes and destruction by serum RNases (Kuninty et al. 2016). Delivery through nanoparticles has certain disadvantages as they do not provide a uniform delivery of the molecules to the site of solid tumors in a sufficient amount (Jain and Stylianopoulos 2010; Kumari, Kumar, and Yadav 2011). Liposome-mediated delivery also poses a threat as the positive charge present on them results in conglomeration with red blood cells. The RNAi-inducing molecules are also destructed by reticuloendothelial cells. The liposomes often interact with immunoglobulins present in blood, which results in their conglomeration and subsequent phagocytosis by the Reticuloendothelial System (RES) system in the liver and spleen (Gilmore et al. 2006). For these reasons, several protective vehicles and chemical modifications of siRNA and miRNA are made to alter its pharmacokinetic properties to increase their half-life and to protect them from RNases in vivo (Choung et al. 2006; Amarzguioui et al. 2003).

20.4.2 Immune Activation

Several studies have shown that the structure of the siRNA molecule and its delivery system often trigger the innate immunity resulting in the induction of interleukin-6 (IL-6) and tumor necrosis factor-α (TNF-α) in adherent peripheral blood mononuclear cells (PBMCs). First, it was shown that immune elicitation by siRNA is sequence-dependent (Sioud 2005). The immune response elicited by certain sequences of the siRNA occurs via the toll-like receptors (TLRs) (Meng and Lu 2017). TLR7 was shown to act in mice, whereas TLR7/8 acts in humans. Some siRNAs have been shown to use immunostimulatory motifs such as 5′-UGUGU-3′, 5′-GUCCUUCAA-3′, AU, UGGC and GU to act through the TLR7/8 pathways and activate IFN-β signaling (Forsbach et al. 2008; Judge et al. 2005; Gantier et al. 2008; Song et al. 2003; Lin et al. 2005; Hornung et al. 2005). Blunt-ended synthetic siRNA were shown to induce Interferon-induced protein with tetratricopeptide repeats 1 (IFIT1) gene expression via a Retinoic acid-inducible gene-I (RIG-I) pathway (Marques et al. 2006). Replacement of the immunostimulatory marks with adenosines was shown to reduce the activation of the immune system. Exchanging the 2′-hydroxyl uridines with 2′-fluoro, 2′-O-methyl or 2′-deoxy uridines was shown to abrogate the immune activation. Thus, modifications of the 2′-ribose group of uridines could solve the RNAi-mediated immune elicitation phenomenon. However, such modification does not always confer full protection against immune activation.

20.4.3 Off-Target Effect

Another hurdle in the delivery of RNAi molecules is the off-target effect. In addition to binding to their complementary mRNA sequence, siRNA often bind target sites that resemble miRNA in the 3'UTR of other mRNAs by using their seed sequence, thus resulting in gene silencing. The miRNA-like off-target activities are often reduced by employing two strategies (Jackson and Linsley 2010; Seok et al. 2018). The first is modification of the guide siRNA at position 2 from the 5' end within their seed sequence. The modifications that are generally used are an addition of 2′-O-methylations or locked nucleic acid. Another modification is the pooling of multiple siRNAs, in which the individual siRNA is directed against the same target gene, albeit at different positions, but each siRNA has a unique off-target mark (Neumeier and Meister 2021).

20.5 DELIVERY SYSTEM OF RNAi THERAPEUTICS

20.5.1 Viral Vectors

Adeno-associated virus (AAV) vector encoding of siRNA, shRNA and miRNA has been successfully used for target gene silencing. AAV-encoding miR-26a was shown to be effective against hepatocellular carcinoma (Kota et al. 2009). AAV miR-196a was reported to reduce spinal and bulbar muscular atrophy (Miyazaki et al. 2012). The AAV siRNA vector-mediated silencing of neuronal nitric oxide synthase (nNOS) was shown to boost the functional outcome following traumatic brain injury (TBI) (Boone et al. 2017). AAV-encoding shRNA against the mechanistic target of rapamycin (mTOR) exhibited an anti-angiogenic effect during oxygen-induced retinopathy (OIR) in rats (Lee et al. 2020). AAV encoding shRNA against dynamin 2 (DNM2) was shown to prevent myotubular myopathy in mice (Tasfaout et al. 2018). There are several other studies of the AAV-, adenovirus (Ad)-, retrovirus- and lentiviral-mediated delivery of RNAi therapeutic molecules (Lundstrom 2020). However, delivery through viral vectors often poses a threat because of its cytopathogenicity and pre-existing immunity against the virus (Rooij and Kauppinen 2014).

20.5.2 Stable Nucleic Acid Lipid Particles (SNALPs)

SNALPs are a second generation of lipid nanoparticle (LNP). SNALPs are generally neutral in charge and are composed of three components: an ionizable cationic lipid, a neutral helper lipid and a PEG-derivatized lipid. These particles combine the potential of ionizable cationic lipid, which is important for the encapsulation of siRNA at an acidic pH. Helper lipids 1,2-distearoyl-sn-glycero-3-phosphocholine (DSPC) and cholesterol are used to stabilize the lipid bilayer, and PEG-derivatized lipids are used to provide an aqueous shield around the nanoparticle surface, which prolongs the presence of siRNA in circulation (Coutinho et al. 2020). SNALPs were employed in the development of Patisiran (Onpattro). Currently, multiple delivery systems are in different stages of clinical trial from Tekmira Pharmaceuticals (currently Arbutus Biopharma) and Alnylam Pharmaceuticals (Singh and Peer 2016).

20.5.3 N-Acetyl-D-Galactosamine (GalNAc)

GalNAc is conjugated with siRNA and miRNA mimics, resulting in their uptake by the cell through clathrin-mediated endocytosis. They do not require an additional delivery molecule such as lipids. GalNAc have a high affinity for Asialoglycoprotein receptor (ASGPR), which is generally expressed on hepatocytes. Once inside the blood, GalNAc-conjugated siRNA translocates to the liver where GalNAc ligand binds to an ASGPR receptor. This results in the hepatocyte-specific uptake of siRNA and thus results in gene silencing (Nair et al. 2014). Whether GalNAc-siRNA conjugates ALN-PCSsc (Alnylam Pharmaceuticals) and GalNAc-miRNA conjugate RG-101 (Regulus) are under clinical trials.

20.5.4 EnGeneIC Delivery Vehicle Nanocells

These are also known as TargomiRs which are minicells containing miRNA mimics. They comprise three components, specifically, a miR-16-based microRNA mimic, EVDs (non-living bacterial minicells) that are the drug delivery vehicle and a targeting moiety (EGFR-expressing cancer cells) (Reid et al. 2015). TargomiRs coated with antibodies specific to EGFR are currently in a Phase I trial (clinicaltrials.gov identifier: NCT02369198).

20.6 RNAi THERAPEUTICS IN CLINICAL TRIALS

siRNA, miRNA and shRNA are powerful RNAi therapeutic molecules that can be considered medications that treat different diseases by targeting the mRNA in cells (Bumcrot et al. 2006). Along with the FDA-approved drugs Patisiran and Givosiran, several other drug candidates are currently under different stages of clinical trial. The various siRNA-mediated therapeutics under clinical trial are summarized in Table 20.1. The miRNA-based therapeutics are mentioned in Table 20.2. The shRNA-mediated therapeutics are listed in Table 20.3. Different miRNA-, siRNA- and shRNA-mediated therapeutic strategies to treat different diseases are mentioned in Figure 20.3.

TABLE 20.1
siRNA-Based Therapeutics under Clinical Trial

Drug	Target/ Mode of delivery	Delivery vehicle	Disease	Clinical trial number	Phase	Status	Company/ Current status	Reference
ONPATTRO (Patisiran, ALN-TTR02)	Transthyretin (TTR)/ Intravenous	LNP (DLinMC3-DMA)	TTR-Mediated Amyloidosis	NCT01 559077	I	Completed	Alnylam/ Approved	(Adams et al. 2018; Suhr et al. 2015)
				NCT02 053454	I	Completed		
				NCT01 617967	II	Completed		
				NCT01 961921	II	Completed		
				NCT01 960348	III	Completed		
				NCT02 939820	III	Approved		
Givosiran (ALNAS1)	5'-Aminolevulinate Synthase 1 (ALAS1)/ Subcutaneous	GalNAc-siRNA	Acute Intermittent Porphyria	NCT02 452372	I	Completed	Alnylam/ Approved	(Scott 2020; Sardh et al. 2019; Balwani et al. 2020)
				NCT02 949830	I/II	On-going		
				NCT03 505853	I	Recruiting		
				NCT03 338816	III	On-going		
Inclisiran (ALNPCSsc)	Proprotein Convertase Subtilisin/Kexin Type 2 (PCSK2)/ Subcutaneous	GalNAc-siRNA	Hypercholesterolemia, Atherosclerotic cardiovascular disease (ASCVD)	NCT02 314442	I	Completed	Alnylam/ Phase III	(Kosmas et al. 2018)
				NCT02 597127	II	Completed		
				NCT03 159416	I	Completed		
				NCT02 963311	II	Completed		

				NCT03397121	III	Active, not recruiting		
				NCT03400800	III	Active, not recruiting		
				NCT03399370	III	Active, not recruiting		
				NCT03060577	II	Active, not recruiting		
				NCT03705234	III	Recruiting		
				NCT03851705	III	Recruiting		
Fitusiran (ALNAT3sc, ALNAPC, SAR439774)	Antithrombin (AT3)/ Subcutaneous	GalNAc-siRNA conjugate	Hemophilia A, B	NCT02035605	I	Completed	Alnylam/ Phase III	(Pasi et al. 2021, 2017)
				NCT02554773	I, II	Completed		
				NCT03549871	III	Recruiting		
				NCT03754790	III	Recruiting		
				NCT03417102	III	Recruiting		
				NCT03417245	III	Recruiting		
Lumasiran (ALNGO1)	Hydroxyacid Oxidase 1 (HAO1)/ Subcutaneous	GalNAc-siRNA conjugate	Primary Hyperoxaluria Type 1 (PH1)	NCT02706886	I, II	Active, not recruiting	Alnylam/ Phase III	(Liebow et al. 2017)
				NCT03350451	II	Enrolling by invitation		
				NCT03681184	III	Recruiting		
Vutrisiran (ALNTTRsc02)	Transthyretin (TTR)/ Subcutaneous	GalNAc-siRNA conjugate	TTR-Media ted Amyloidosis	NCT02797847	I	Completed	Alnylam/ Phase III	(Janas et al. 2019)
				NCT03759379	III	Recruiting		

(Continued)

TABLE 20.1 (*Continued*)
siRNA-Based Therapeutics under Clinical Trial

Drug	Target/ Mode of delivery	Delivery vehicle	Disease	Clinical trial number	Phase	Status	Company/ Current status	Reference
QPI-1002	p53/ Intravenous	Naked siRNA	Delayed Graft Function, Acute Kidney Injury	NCT00554359	I	Completed	Quark/ Phase III	(Demirjian et al. 2017)
				NCT00683553	I	Terminated		
				NCT00802347	I, II	Completed		
				NCT02610283	II	Completed		
				NCT02610296	III	Active, not recruiting		
				NCT03510897	III	Recruiting		
QPI-1007	Caspase 2/ Intravitreal	Naked siRNA	Optic Atrophy, Non-arteritis Anterior Ischemic Optic Neuropathy, Glaucoma, Angle closure	NCT01965106	II	Completed	Quark/ Suspended	(Solano et al. 2014)
				NCT01965106	II	Completed		
				NCT02341560	II/III	Recruiting		
				NCT01640912	I	Completed		
Tivanisiran (SYL1001)	Transient receptor potential cation channel subfamily V member 1 (TRPV1)/ Intravitreal	Naked siRNA	Ocular pain, dry eye syndrome	NCT01438281	I	Completed	Sylentis/ Phase III	(Moreno-Montañés, Bleau, and Jimenez 2018; Benitez-Del-Castillo et al. 2016)
				NCT01776658	I, II	Completed		
				NCT02455999	II	Completed		
				NCT03108664	III	Completed		

ARO-HBV	Hepatitis B virus (HBV) gene/ Subcutaneous	TRiM platform	Hepatitis B	NCT03365947	I, II	Recruiting	Arrowhead Pharmaceutical/ Phase II	(Wooddell et al. 2018)
TKM-080301	Polo Like Kinase 1 (PLK1)/ Intravenous	LNP	Adrenal Cortical Carcinoma, Hepatocellular Carcinoma, Neuroendocrine Tumor, Solid Tumors	NCT01437007	I	Completed	Arbutus Biopharma Corporation/ Phase II	(Demeure et al. 2016)
				NCT01262235	I, II	Completed		
				NCT02191878	I, II	Completed		
ALN-AAT02	Alpha-1 antitrypsin (AAT)/ Subcutaneous	GalNAc-siRNA conjugate	Antitrypsin Deficiency Liver Disease	NCT03767829	I, II	Recruiting	Alnylam/ Phase II	(Y. Huang 2017)
ALN-CC5	C5a Receptor/ Subcutaneous	GalNAc-siRNA conjugate	Paroxysmal Nocturnal Hemoglobinuria	NCT02352493	I, II	Completed	Alnylam/ Phase II	(Hill et al. 2015)
				NCT03841448	II	Recruiting		
DCR-PHXC	Lactate dehydrogenase A (LDHA)/ Subcutaneous	GalNAc-siRNA conjugate	Primary Hyperoxaluria	NCT03392896	I	Active, not recruiting	Dicerna Pharmaceuticals/ Phase III	(Dindo et al. 2019)
				NCT03847909	II	Recruiting		
Cotsiranib (STP705)	Transforming growth factor beta 1 (TGF-β1) and cyclooxygenase-2 (COX2)/ Topical	Histidine-Lysine Copolymer (HKP) peptide	Hypertrophic Scar	NCT02956317	I, II	Recruiting	Sirnaomics/ Phase II	(Zhou et al. 2017)
Atu027	Protein Kinase N3 (PKN3)/ Intravenous	AtuPlex Technology	Metastatic Pancreatic Cancer, Solid Tumors	NCT00938574	I	Completed	Silence Therapeutics GmbH/ Phase II	(Schultheis et al. 2020)
				NCT01808638	I, II	Completed		
siG12D LODER	Kirsten rat sarcoma virus (KRAS)/ Implant	LODER polymer	Pancreatic Ductal Adenocarcinoma, Pancreatic cancer	NCT01188785	I	Completed	Silenseed/ Phase II	(Golan et al. 2015)
				NCT01676259	II	Recruiting		

(Continued)

TABLE 20.1 (*Continued*)
siRNA-Based Therapeutics under Clinical Trial

Drug	Target/ Mode of delivery	Delivery vehicle	Disease	Clinical trial number	Phase	Status	Company/ Current status	Reference
AMG890	Lipopotein(a)/ Subcutaneous	TRiM platform	Cardiovascular diseases	NCT03 626662	I	Recruiting	Amgen/ Phase I	(Borrelli et al. 2019)
NU-0129	B-cell lymphoma 2 Like 12 (BCL2L12)/ Intravenous	Infusion	Gliosarcoma	NCT03 020017	I	Active, not recruiting	Northwestern University/ Phase I	(Jensen et al. 2013)
siRNA-EphA2	Ephrin type A receptor 2 (EphA2)/ Intravenous	LNP	Advanced cancers	NCT01 591356	I	Recruiting	MD Anderson/ Phase I	(Landen et al. 2005; Nishimura et al. 2013)
ARO-ANG3	Angiopoietin Like 3 (ANGPTL3)/ Subcutaneous	GalNAc-siRNA Conjugate	Dislipidemia, HypercholesTerolemia	NCT03 747224	I	Recruiting	Arrowhead/ Phase I	(Hamilton 2018)
iPsiRNA (Proteasome siRNA and tumor antigen RNA-transfected DC)	Latent Membrane Protein 1 (LMP1), Latent Membrane Protein 7 (LMP7), MECL1, also known as PSMB10, proteasome subunit beta 10/ Intradermal	Ex vivo transfection	Metastatic melanoma, Absence of CNS Metastases	NCT00 672542	I	Completed	Scott Pruitt, Duke University/ Phase I	(Dannull et al. 2013)
APN401 (siRNA-transfected PBMC)	Cbl proto-oncogene/ Dendritic cell (DC) cancer vaccine/ Intravenous infusion	Ex vivo siRNA electroporated PBMCs	Solid Tumors That Are Metastatic or Cannot Be Removed By Surgery	NCT02 166255	I	Completed	Wake Forest University Health Sciences/ Phase I	(Loibner et al. 2018; Triozzi et al. 2015)
				NCT03 087591	I	Recruiting		
				NCT01 808638	I, II	Completed		

TABLE 20.2
miRNA-Based Therapeutics under Clinical Studies

Drug	Target/ Therapeutic agent	Delivery vehicle	Disease	Clinical trial number	Phase	Status	Company/ Current status	Reference
Miravirsen (SPC3649)	miR-122/ AntimiR-122	Locked nucleic acid (LNA)-modified antisense inhibitor	Hepatitis C virus (HCV) infection	NCT01 646489	I	Completed	Santaris Pharma and Hoffmann-La Roche/Suspended	(Gebert et al. 2014; Janssen et al. 2013; S. Zhang et al. 2021)
				NCT01 200420	II	Completed		
				NCT01 872936	II	Ongoing		
				NCT02 031133	II	Ongoing		
				NCT02 508090	II	Ongoing		
MRX34	miR-34/ miR-34 mimic	Lipid nanoparticles LNPs (Smarticles)	Multiple solid tumours	NCT01 829971	I	Terminated	Mirna Therapeutics/ Discontinued	(Beg et al. 2017; S. Zhang et al. 2021)
RG-125/ AZD4076	miR-103/107/ AntimiR-103/107	GalNAc-conjugated	Type 2 diabetes and non-alcoholic fatty liver diseases	NCT02 612662	I	Ongoing	Regulus Therapeutics/ Phase II	(S. Zhang et al. 2021)
				NCT02 826525	I/IIa	Ongoing		
RG-101	miR-122/ AntimiR-122	GalNAc-conjugated	Hepatitis C infection	Not available	I	Completed	Regulus Therapeutics/ Discontinued	(S. Zhang et al. 2021)
				Not available	II	Ongoing		
Cobomarsen MRG-106	miR-155/ AntimiR-155	LNA-modified antisense inhibitor	Lymphomas, leukemias	NCT02 580552	I	Ongoing	miRagen Therapeutics/ Phase I	(Seto et al. 2018)
				NCT03 713320	II	Ongoing		
Remlarsen MRG-201	miR-29/ miR-29 mimic	Cholesterol conjugated miRNA duplex	Scleroderma	NCT03 601052	II	Ongoing	miRagen Therapeutics/ Phase II	(Bonneau et al. 2019; Gallant-Behm et al. 2019),
MesomiR-1	miR-16/ miR-16 mimic	EnGeneIC delivery vehicle	Mesothelioma, non-small cell lung cancer	NCT02 369198	I	Completed	EnGeneIC/ Suspended	(van Zandwijk et al. 2017)

TABLE 20.3
shRNA-Based Therapeutics under Clinical Trial

Drug	Target/ Therapeutic agent	Delivery vehicle	Disease	Clinical trial number	Phase	Status	Company/ Current status	Reference
Vigil vaccine (FANG, vigil, vigil EATC)	Furin/ Intradermal	Ex vivo electroporation of shRNA	Breast cancer, ovarian cancer, colorectal cancer, Ewing's sarcoma, metastatic melanoma, metastatic non-small-cell lung cancer, solid tumors	NCT01 061840	I	Completed	Gradalis, Inc/ Phase III	(Ghisoli et al. 2015, 2016, 2017)
				NCT02 574533	I	Completed		
				NCT01 453361	II	Terminated		
				NCT01 309230	II	Active, not recruiting		
				NCT01 505166	II	Terminated		
				NCT01 551745	II	Completed		
				NCT01 867086	II	Completed		
				NCT02 511132	II	Active, not recruiting		
				NCT02 639234	II	Withdrawn		
				NCT02 346747	II	Active, not recruiting		
				NCT02 725489	II	Enrolling by invitation		
				NCT03 073525	II	Active, not recruiting		
				NCT03 495921	III	Recruiting		

Lentivirus vector CCR5 shRNA	C-C chemokine receptor type 5 (CCR5)/ Intravenous infusion	Lentivirus vector, Ex vivo transfection	AIDS-related lymphoma	NCT02 797470	I, II	Recruiting	AIDS Malignancy Consortium/ Phase I	(Walker et al. 2012)
Cal-1 (LVsh5/C46, Cal-1 modified HSPC, Cal-1 modified CD4 + T lymphocytes)	C-C chemokine receptor type 5 (CCR5)/ Intravenous infusion	Lentivirus vector, Benitect's DNA-directed RNAi technology	HIV/AIDS	NCT01 734850	I, II	Completed	Calimmune Inc./ Phase II	(Burke et al. 2015; Delville et al. 2019)
				NCT02 378922	I	Withdrawn		
				NCT02 390297	Not specified	Enrolling by invitation		
				NCT03 593187	I, II	Recruiting		
pbi-shRNA STMN1 LP	Stathmin1 (STMN1)/ Intratumoral	Lipoplex	Advanced and/or Metastatic Cancer	NCT01 505153	I	Completed	Gradalis, Inc/ Phase I	(Phadke et al. 2011)
pbi-shRNA EWS/ FLI1 type 1 LPX	EWS/friend leukemia integration 1 transcription factor (FLI 1)/ Intravenous influsion	Lipoplex	Ewing's Sarcoma	NCT02 736565	I	Recruiting	Gradalis, Inc/ Phase I	(S. Zhang et al. 2021)
Lentivirus vector rHIV7-shI-TAR-CCR5RZ-transduced hematopoietic progenitor cells	HIV-1 tat/rev(shI)-trans-active response element (TAR), CCR5 ribozyme/ Intravenous	Lentivirus vector, Ex vivo transfection	AIDS-Related Lymphoma	NCT00 569985	I	Active, not recruiting	City of Hope Medical Center/ Phase I	(DiGiusto et al. 2010)
				NCT01 961063	I	Active, not recruiting		
				NCT02 337985	I	Active, not recruiting		
shRNA	Exportin 1 (XPO1) Not available	Unknown	Chronic Lymphocytic Leukemia (CLL)	NCT02 757586	Unknown	Not yet recruiting	Peking University People's Hospital/ Unknown	(S. Zhang et al. 2021)
TT-034 (PF-05095808)	Hepatitis C virus (HCV)/ Intravenous	AAV Vector (shRNA)	Hepatitis C	NCT01 899092	I, II	Completed	Tacere Therapeutics/ Suspended	(Patel et al. 2016; S. Zhang et al. 2021)
				NCT02 315638	I, II	Terminated		

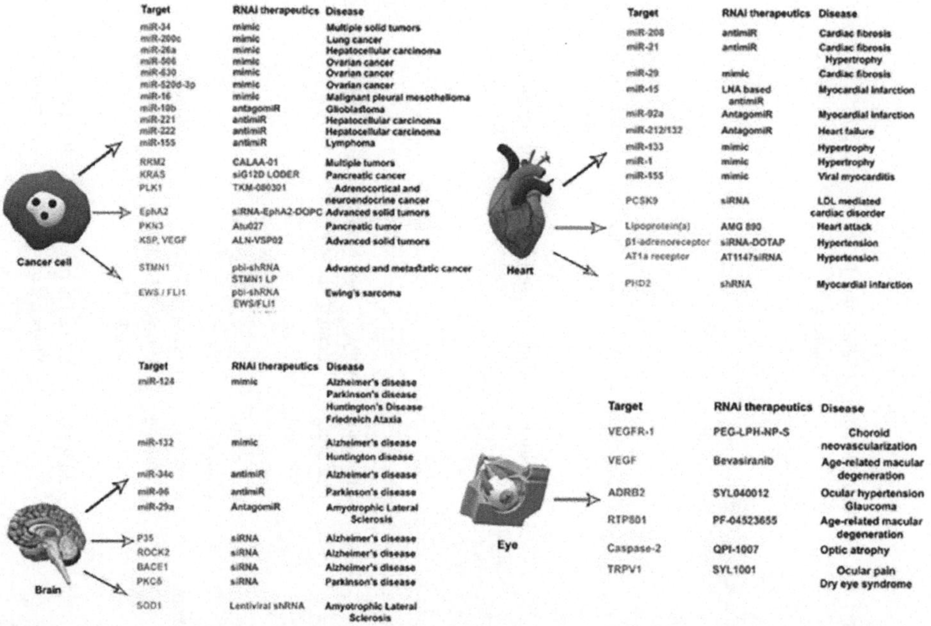

FIGURE 20.3 RNAi therapeutics for the treatment of diseases. Different miRNA-, siRNA- and shRNA-mediated treatments are available for the treatment of cancer (top left), cardiovascular diseases (top right), neurological disorders (bottom left) and ocular diseases (bottom left). The mimics against respective beneficial miRNAs are used to overexpress them for the treatment of diseases. AntimiR or antagomiR is used against the harmful miRNAs to downregulate them for the treatment of diseases. The following are the different tools used in the available treatments: RRM2-Ribonucleotide reductase, KRAS-Kirsten rat sarcoma virus, PLK1-Polo Like Kinase 1, EphA2-Ephrin type A receptor 2, KSP-Kinesin spindle protein, VEGF-Vascular endothelial growth factor, STMN1-Stathmin1, EWS/FL1-EWS/friend leukemia integration 1, a transcription factor, PCSK9-Proprotein convertase subtilisin/kexin type 9, AT1a-angiotensin II subtype 1a, PHD2-prolyl hydroxylase-2, ROCK2-Rho-associated coiled-coil-containing protein kinase 2, BACE1-Beta-Secretase 1, PKCδ-Protein kinase C-δ, SOD1-Superoxide dismutase type 1, VEGFR-1-Vascular endothelial growth factor receptor-1, ADRB2-Adrenoceptor Beta 2, RTP801-Hypoxia-inducible factor 1 responsive gene, TRPV1-Transient receptor potential cation channel, subfamily V member 1.

20.7 RNAi THERAPEUTICS AND CANCER

Cancer is one of the leading causes of death worldwide. Almost one-third of the RNAi therapeutics are targeted at cancer and cancer-related genes. miRNA-based therapeutics generally target an array of genes, thereby regulating distinct biological processes relating to malignancies (Iorio and Croce 2012). To the contrary, siRNA-based therapeutics are highly specific, which makes them efficient agents for targeted personalised cancer treatments (Daka and Peer 2012).

The first-in-human, Phase I study of MRX34, a liposomal miR-34a mimic, was conducted on people with advanced solid tumors (Beg et al. 2017). MRX34 treatment with dexamethasone premedication showed evidence of anti-tumor activity. The trial of MRX34 was suspended, however, because of immune-mediated toxicities (Hong et al. 2020). miR-34 mimic encapsulated in lipid nanoparticles was reported to be effective against liver, prostrate and lung cancer in a murine model system (Wiggins et al. 2010; C. Liu et al. 2011; Trang et al. 2011). Systemic intravenous administration of miR-34a via lipid emulsion nanoparticles was reported to impede pancreatic cancer growth by decreasing $CD44^{+}$ cell counts, decreasing tumor growth and increasing tumor cell apoptosis (Pramanik et al. 2011). Therapeutic transport of miR-200c mimic was shown to increase radiosensitivity in lung cancer and resulted in longer survival (Cortez et al. 2014). miR-200c mimic directly regulated oxygen response proteins such as peroxiredoxin 2 (PRDX2), NF-E2-related factor 2 (NRF2; also known as NFE2L2) and sestrin 1 (SESN1), thereby increasing ROS generation and causing the apoptosis of cancer cells. Adeno-associated virus (AAV)-associated delivery of miR-26a in a mouse model of hepatocellular carcinoma was shown to inhibit cancer cell progression and increase tumor cell apoptosis (Kota et al. 2009). Dioleoylphosphatidylcholine (DOPC) liposome-mediated delivery of miR-506 mimic was shown to increase E-cadherin expression and prevent TGFβ-induced epithelial-mesenchymal transition (EMT) by targeting SNAI2 in ovarian cancer (D. Yang et al. 2013). DOPC-mediated antimir delivery against miR-630 was shown to significantly reduce the growth of tumors and metastasis during ovarian cancer (Rupaimoole et al. 2016). Delivery of miR-16 mimic in a bacterially derived EGFR-targeted EnGeneIC Delivery Vehicle (EDV) nanocell delivery system was shown to reduce tumors in malignant pleural mesothelioma and Non-small cell lung cancer (NSCLC) xeno-graft mouse models (Reid et al. 2015). Moreover, therapeutic silencing of miR-10b using ASO antagomiRs was shown to augment a functionally important miR-10b target, Hoxd10, and it was also shown to suppress lung metastasis (Ma et al. 2010; Zhong et al. 2021). AntagomiR against miR-10b was reported to reduce the growth of tumors in malignant brain glioblastoma (Gabriely et al. 2011). Targeting miR-10b with nanoparticles conjugated to LNA was used to regress metastatic breast cancer (Yoo et al. 2015).

Inhibition of ribonucleotide reductase (RRM2) through siRNA was shown to reduce the proliferation of tumor cells (Heidel et al. 2007). CALAA-01 is one such therapeutic targeting RRM2. It is delivered by a cyclodextrin nanoparticle delivery system. However, due to its toxicity issues in a Phase I trial, further trials of CALAA-01 have been suspended (Zuckerman et al. 2014). Another siRNA therapeutic siG12D LODER, which targets the Kirsten rat sarcoma virus (KRAS) gene, is currently under a Phase II trial as a medication for pancreatic cancer (Golan et al. 2015). TKM-080301 is another siRNA therapeutic used to target Polo Like Kinase 1 (PLK1) that causes adrenocortical and neuroendocrine cancer (Demeure et al. 2016). siRNA-EphA2-DOPC is a siRNA therapeutic targeting the Ephrin type A receptor 2 (EphA2) gene for treatment of advanced solid tumors (Landen et al. 2005). Numerous other siRNA-based therapeutics are discussed in Table 20.1.

20.8 RNAi THERAPEUTICS AND CARDIOVASCULAR DISEASES

Several RNAi therapeutics were shown to have major roles in cardiovascular diseases such as myocardial infarction, hypertrophy and fibrosis (Romaine et al. 2015; Van Rooij and Olson 2012). miR-208, a cardiac-exclusive miRNA is encoded by the α-myosin heavy chain (αMHC) gene. miR-208 degrades thyroid hormone receptor-associated protein 1 (THRAP1), which was shown to control cardiac homeostasis. Another microRNA, miR-21, was shown to cause cardiac fibrosis and hypertrophy during pathological stress. miR-21 was shown to target Sprouty 1 (SPRY1), an inhibitor of mitogen-activated protein kinase (MAPK) signalling. The inhibition of miR-21 by antimiR-21 treatment was shown to increase the expression of SPRY1, thereby decreasing cardiac hypertrophy and fibrosis (Thum et al. 2008). Members of the miR-15 family such as miR-15a, miR-15b, miR-16–1, miR-16–2, miR-195 and miR-497 were found to be upregulated during cardiac stress and myocardial infarction. Tiny 8-mer LNA inhibitors that target the seed region of the miR-15 family were reported to target numerous members of this family and have similar seed sequences (Hullinger et al. 2012). The inhibition of the miR-15 family augmented postnatal cardiac regeneration (Porrello et al. 2013). miR-29 down-regulation was shown to augment collagen deposition resulting in cardiac fibrosis. Overexpression of miR-29 resulted in decreased cardiac fibrosis by targeting proteins such as collagens, fibrillins and elastin (Van Rooij et al. 2008). miR-92a controls neo-angiogenesis after ischemic injury. miR-92a antagomiR was shown to enhance angiogenesis and resulted in a revival of the damaged tissue (Bonauer et al. 2009).

siRNA-mediated silencing of proprotein convertase subtilisin/kexin type 9 (PCSK9) was shown to lower the levels of plasma low-density lipoprotein (LDL), thereby decreasing the chances of cardiac disorders (Watts and Ockene 2020). Use of siRNA (AMG 890) was shown to reduce lipoprotein(a) production, which reduced the risk of heart attack (Borrelli et al. 2019). siRNA-DOTAP lipoplexes were reported to target β1-adrenoreceptor expression to thus reduce blood pressure during hypertension (Arnold et al. 2007). AT1147siRNA was shown to target the AT1a receptor, thus decreasing the binding of angiotensin II (Ang II) to the cells and resulting in decreased hypertension (Vázquez et al. 2005). shRNA-mediated targeting of the prolyl hydroxylase-2 (PHD2) protein was shown to increase angiogenesis and contractility following myocardial infarction (M. Huang et al. 2008).

20.9 RNAi THERAPEUTICS AND NEURODEGENERATIVE DISORDERS

A multitude of miRNA and siRNA therapeutics have been discovered that are shown to target genes that cause various neurodegenerative disorders. miR-124 mimics were shown to reduce Aβ amyloid production by targeting the Beta-Secretase 1 (BACE1) gene during Alzheimer's disease (Fang et al. 2012). Similarly, miR-124 mimics were shown to target Bcl-2-like 11 (BIM), SRY-Box Transcription Factor 9 (SOX9) and Jagged 1 genes in Parkinson's disease (Wang et al. 2016; Saraiva et al. 2016), SOX9 in Huntington's Disease (Lee et al. 2017) and the Frataxin (FXN) gene in Friedreich Ataxia (Bandiera et al. 2013). miR-132 mimic was reported to silence Inositol-Trisphosphate 3-Kinase B (ITPKB), which caused decreased amyloidosis and Tau hyperphosphorylation during Alzheimer's disease (Salta et al. 2016). miR-132 mimic also silenced methyl CpG binding protein 2 (MeCP2), resulting in increased life extension in Huntington's disease (Fukuoka et al. 2018). AntimiR-34c was shown target miR-34c, which increased expression of Sirtuin 1 (SIRT1) and Synaptotagmin 1 (SYT1) during Alzheimer's disease (Zovoilis et al. 2011; Shi et al. 2020). This results in enhanced memory and recovery of dendritic spine density.

siRNA-mediated allele-specific silencing of amyloid precursor protein (APP) was shown to be a potential therapy for Alzheimer's disease (Feng et al. 2006). siRNA-mediated knockdown of P35, which is a cyclin-dependent protein kinase 5 (CDK5) activator was shown to treat Alzheimer's disease by decreasing the hyperphosphorylation of the Tau protein (López-Tobón et al. 2011; Ubeda, Kemp, and Habener 2004). siRNA-mediated silencing of Rho-associated coiled-coil-containing protein kinase 2 (ROCK2) augmented the expression of LR11 protein, thereby decreasing the formation of Aβ plaques (Y. Liu et al. 2013). siRNA-mediated silencing of BACE1 was reported to ameliorate APP cleavage and Aβ formation (Peng and Masliah 2010). siRNA targeted against Protein kinase C-δ (PKCδ) was shown to reduce Parkinson toxin MPP^+, induce dopaminergic degeneration and ameliorate neuronal loss in Parkinson's disease (Yang et al. 2004). Lentiviral shRNA-targeting superoxide dismutase type 1 (SOD1) was shown to increase the survival rate and delay the onset of Amyotrophic Lateral Sclerosis (ALS) (Raoul et al. 2005). Several other siRNA and shRNA therapeutics apart from these mentioned here are used for the treatment of neurodegenerative disorders (Zahir-Jouzdani et al. 2018).

20.10 RNAi THERAPEUTICS AND OCULAR DISORDERS

Several siRNAs have been shown to play prominent roles in ocular diseases. siRNA (PEG-LPH-NP-S) targets Vascular endothelial growth factor receptor-1 (VEGFR-1), thus reducing choroid neovascularization (CMV) (H. Liu et al. 2011). Bevasiranib (Cand5), an FDA-approved siRNA curative, was shown to target Vascular endothelial growth factor (VEGF), thus ameliorating diabetic macular edema and wet Age-related macular degeneration (AMD) (Barakat and Kaiser 2009). Bamosiran (SYL040012) is another therapeutic that has been shown to target Adrenoceptor Beta 2 (ADRB2) for treatment of Ocular hypertension and glaucoma. PF-04523655 was reported to target RTP801 (hypoxia-inducible factor 1 responsive) and thus reduce AMD. QPI-1007 was shown to reduce optic atrophy by targeting Caspase-2 gene (Solano et al. 2014).

20.11 RNAi THERAPEUTICS AND OTHER DISEASES

Several other RNAi therapeutics that are currently under clinical trial have also been indicated to impart their protective role in other diseases. ND-L02-s0201, a lipid nanoparticle encasing heat shock protein 47 (HSP47) siRNA, was shown to ameliorate hepatic fibrosis by silencing HSP47, a collagen-binding ER chaperone (Liu et al. 2021). QPI-1002, which is currently under Phase III trial, was shown to prevent major adverse kidney events by targeting p53 (Demirjian et al. 2017). Secreted Protein Acidic And Cysteine Rich (SPARC) gene inhibition by SPRAC siRNA was shown to attenuate the fibrosis of the skin and lungs (Wang et al. 2010). ALN-RSV01 was indicated to attenuate bronchiolitis obliterans by targeting Respiratory Syncytial Virus (RSV) nucleocapsid following RSV infection (Gottlieb et al. 2016). ARC-520 was shown to attenuate Hepatitis B virus infection by targeting cccDNA-derived viral mRNA transcripts, thus reducing the production of viral proteins and HBV DNA (Schluep et al. 2017).

20.12 CONCLUSION

The field of RNAi is expanding robustly with new discoveries appearing on a daily basis. It provides an efficient alternative to traditional pharmaceuticals, as medications based on RNAi are easy to synthesize and highly sequence-specific, and the suppression of a specified target molecule is possible. RNAi therapeutics have made significant progress in a comparatively short time span. However, challenges that remain are the effective transport of the therapeutics in vivo, efficient cellular uptake, renal and reticuloendothelial clearance and effective drug release to the target organ.

With the approval of Patisiran and Givosiran, the problem of systemic delivery is solved to a certain extent. Several other candidates such as Inclisiran, Fitusiran, Lumasiran, Vutrisiran, etc. are currently under Phase III clinal trials. Ongoing research on the safety and potency of these and several other drug candidates under clinal trial may elevate a potent, major class of drug molecules that could cure a myriad of diseases and revolutionize medical biotechnology.

REFERENCES

Adams, David, Alejandra Gonzalez-Duarte, William D. O'Riordan, Chih-Chao Yang, Mitsuharu Ueda, Arnt V. Kristen, Ivailo Tournev, et al. 2018. "Patisiran, an RNAi Therapeutic, for Hereditary Transthyretin Amyloidosis." *The New England Journal of Medicine* 379 (1). N Engl J Med: 11–21. doi:10.1056/NEJMOA1716153.

Amarzguioui, Mohammed, Torgeir Holen, Eshrat Babaie, and Hans Prydz. 2003. "Tolerance for Mutations and Chemical Modifications in a SiRNA." *Nucleic Acids Research* 31 (2). Nucleic Acids Res: 589–95. doi:10.1093/NAR/GKG147.

Arnold, Anne Sophie, Yao Liang Tang, Keping Qian, Leping Shen, Valery Valencia, Michael Ian Phillips, and Yuan Clare Zhang. 2007. "Specific Beta1-Adrenergic Receptor Silencing with Small Interfering RNA Lowers High Blood Pressure and Improves Cardiac Function in Myocardial Ischemia." *Journal of Hypertension* 25 (1). J Hypertens: 197–205. doi:10.1097/01.HJH.0000254374.73241.AB.

Balwani, Manisha, Eliane Sardh, Paolo Ventura, Paula Aguilera Peiró, David C. Rees, Ulrich Stölzel, D. Montgomery Bissell, et al. 2020. "Phase 3 Trial of RNAi Therapeutic Givosiran for Acute Intermittent Porphyria." *The New England Journal of Medicine* 382 (24). N Engl J Med: 2289–2301. doi:10.1056/NEJMOA1913147.

Bandiera, Simonetta, François Cartault, Anne Sophie Jannot, Elie Hatem, Muriel Girard, Laila Rifai, Clemence Loiseau, et al. 2013. "Genetic Variations Creating MicroRNA Target Sites in the FXN 3'-UTR Affect Frataxin Expression in Friedreich Ataxia." *PLoS One* 8 (1). PLoS One. doi:10.1371/JOURNAL.PONE.0054791.

Barakat, Mark Rami, and P. K. Kaiser. 2009. "VEGF Inhibitors for the Treatment of Neovascular Age-Related Macular Degeneration." *Expert Opinion on Investigational Drugs* 18 (5). Expert Opin Investig Drugs: 637–46. doi:10.1517/13543780902855316.

Beg, Muhammad S., Andrew J. Brenner, Jasgit Sachdev, Mitesh Borad, Yoon Koo Kang, Jay Stoudemire, Susan Smith, et al. 2017. "Phase I Study of MRX34, a Liposomal MiR-34a Mimic, Administered Twice Weekly in Patients with Advanced Solid Tumors." *Investigational New Drugs* 35 (2). Invest New Drugs: 180–88. doi:10.1007/S10637-016-0407-Y.

Benitez-Del-Castillo, José Manuel, Javier Moreno-Montañés, Ignacio Jiménez-Alfaro, Francisco José Muñoz-Negrete, Krista Turman, Kadi Palumaa, Belén Sádaba, et al. 2016. "Safety and Efficacy Clinical Trials for SYL1001, a Novel Short Interfering RNA for the Treatment of Dry Eye Disease." *Investigative Ophthalmology & Visual Science* 57 (14). Invest Ophthalmol Vis Sci: 6447–54. doi:10.1167/IOVS.16-20303.

Bernstein, Emily, Amy A. Caudy, Scott M. Hammond, and Gregory J. Hannon. 2001. "Role for a Bidentate Ribonuclease in the Initiation Step of RNA Interference." *Nature* 409 (6818). Nature: 363–66. doi:10.1038/35053110.

Bonauer, Angelika, Guillaume Carmona, Masayoshi Iwasaki, Marina Mione, Masamichi Koyanagi, Ariane Fischer, Jana Burchfield, et al. 2009. "MicroRNA-92a Controls Angiogenesis and Functional Recovery of Ischemic Tissues in Mice." *Science (New York, N.Y.)* 324 (5935). Science: 1710–13. doi:10.1126/SCIENCE.1174381.

Bonneau, E., B. Neveu, E. Kostantin, G. J. Tsongalis, and V. De Guire. 2019. "How Close Are MiRNAs from Clinical Practice? A Perspective on the Diagnostic and Therapeutic Market." *EJIFCC* 30 (2). International Federation of Clinical Chemistry and Laboratory Medicine: 114. /pmc/articles/PMC6599191/.

Boone, Deborah R., Jeanna M. Leek, Michael T. Falduto, Karen E. O. Torres, Stacy L. Sell, Margaret A. Parsley, Jeremy C. Cowart, et al. 2017. "Effects of AAV-Mediated Knockdown of NNOS and GPx-1 Gene Expression in Rat Hippocampus after Traumatic Brain Injury." *PLoS One* 12 (10). PLoS One. doi:10.1371/JOURNAL.PONE.0185943.

Borrelli, Matthew J., Amer Youssef, Michael B. Boffa, and Marlys L. Koschinsky. 2019. "New Frontiers in Lp(a)-Targeted Therapies." *Trends in Pharmacological Sciences* 40 (3). Trends Pharmacol Sci: 212–25. doi:10.1016/J.TIPS.2019.01.004.

Braglia, Priscilla, Riccardo Percudani, and Giorgio Dieci. 2005. "Sequence Context Effects on Oligo(DT) Termination Signal Recognition by Saccharomyces Cerevisiae RNA Polymerase III." *The Journal of Biological Chemistry* 280 (20). J Biol Chem: 19551–62. doi:10.1074/JBC.M412238200.

Brummelkamp, Thijn R., René Bernards, and Reuven Agami. 2002. "A System for Stable Expression of Short Interfering RNAs in Mammalian Cells." *Science (New York, N.Y.)* 296 (5567). Science: 550–53. doi:10.1126/SCIENCE.1068999.

Bumcrot, David, Muthiah Manoharan, Victor Koteliansky, and Dinah W. Y. Sah. 2006. "RNAi Therapeutics: A Potential New Class of Pharmaceutical Drugs." *Nature Chemical Biology* 2 (12). Nat Chem Biol: 711–19. doi:10.1038/NCHEMBIO839.

Burke, Bryan P., Bernard R. Levin, Jane Zhang, Anna Sahakyan, Joshua Boyer, Maria V. Carroll, Joanna Camba Colón, et al. 2015. "Engineering Cellular Resistance to HIV-1 Infection in Vivo Using a Dual Therapeutic Lentiviral Vector." *Molecular Therapy. Nucleic Acids* 4 (4). Mol Ther Nucleic Acids: e236. doi:10.1038/MTNA.2015.10.

Chipman, Laura B., and Amy E. Pasquinelli. 2019. "MiRNA Targeting: Growing Beyond the Seed." *Trends in Genetics* 35 (3). Trends Genet: 215–22. doi:10.1016/J.TIG.2018.12.005.

Choung, Sorim, Young Joo Kim, Seonhoe Kim, Han Oh Park, and Young Chul Choi. 2006. "Chemical Modification of SiRNAs to Improve Serum Stability without Loss of Efficacy." *Biochemical and Biophysical Research Communications* 342 (3). Biochem Biophys Res Commun: 919–27. doi:10.1016/J.BBRC.2006.02.049.

Cogoni, Carlo, and Giuseppe Macino. 2000. "Post-Transcriptional Gene Silencing across Kingdoms." *Current Opinion in Genetics & Development* 10 (6). Curr Opin Genet Dev: 638–43. doi:10.1016/S0959-437X(00)00134-9.

Cortez, Maria Angelica, David Valdecanas, Xiaochun Zhang, Yanai Zhan, Vikas Bhardwaj, George A. Calin, Ritsuko Komaki, et al. 2014. "Therapeutic Delivery of MiR-200c Enhances Radiosensitivity in Lung Cancer." *Molecular Therapy : The Journal of the American Society of Gene Therapy* 22 (8). Mol Ther: 1494–1503. doi:10.1038/MT.2014.79.

Coutinho, Maria Francisca, Juliana Inês Santos, Liliana S. Mendonça, Liliana Matos, Maria João Prata, Amália S. Jurado, Maria C. Pedroso de Lima, et al. 2020. "Lysosomal Storage Disease-Associated Neuropathy: Targeting Stable Nucleic Acid Lipid Particle (SNALP)-Formulated SiRNAs to the Brain as a Therapeutic Approach." *International Journal of Molecular Sciences* 21 (16). Int J Mol Sci: 1–34. doi:10.3390/IJMS21165732.

Daka, Ala, and Dan Peer. 2012. "RNAi-Based Nanomedicines for Targeted Personalized Therapy." *Advanced Drug Delivery Reviews* 64 (13). Adv Drug Deliv Rev: 1508–21. doi:10.1016/J.ADDR.2012.08.014.

Dannull, Jens, N. Rebecca Haley, Gary Archer, Smita Nair, David Boczkowski, Mark Harper, Nicole De Rosa, et al. 2013. "Melanoma Immunotherapy Using Mature DCs Expressing the Constitutive Proteasome." *The Journal of Clinical Investigation* 123 (7). J Clin Invest: 3135–45. doi:10.1172/JCI67544.

Davis, Mark E. 2009. "The First Targeted Delivery of SiRNA in Humans via a Self-Assembling, Cyclodextrin Polymer-Based Nanoparticle: From Concept to Clinic." *Molecular Pharmaceutics* 6 (3). Mol Pharm: 659–68. doi:10.1021/MP900015Y.

Delville, Marianne, Fabien Touzot, Chloé Couzin, Isabelle Hmitou, Lounes Djerroudi, Amani Ouedrani, François Lefrère, et al. 2019. "Safety of CD34 + Hematopoietic Stem Cells and CD4 + T Lymphocytes Transduced with LVsh5/C46 in HIV-1 Infected Patients with High-Risk Lymphoma." *Molecular Therapy. Methods & Clinical Development* 13 (June). Mol Ther Methods Clin Dev: 303–9. doi:10.1016/J.OMTM.2019.02.006.

Demeure, Michael J., Tannaz Armaghany, Samuel Ejadi, Ramesh K. Ramanathan, Aymen Elfiky, Jonathan R. Strosberg, David C. Smith, et al. 2016. "A Phase I/II Study of TKM-080301, a PLK1-Targeted RNAi in Patients with Adrenocortical Cancer (ACC)." *Journal of Clinical Oncology* 34 (15_suppl). American Society of Clinical Oncology (ASCO): 2547–2547. doi:10.1200/JCO.2016.34.15_SUPPL.2547.

Demirjian, Segav, Gorav Ailawadi, Martin Polinsky, Dani Bitran, Shuli Silberman, Stanton Keith Shernan, Michel Burnier, et al. 2017. "Safety and Tolerability Study of an Intravenously Administered Small Interfering Ribonucleic Acid (SiRNA) Post on-Pump Cardiothoracic Surgery in Patients at Risk of Acute Kidney Injury." *Kidney International Reports* 2 (5). Kidney Int Rep: 836–43. doi:10.1016/J.EKIR.2017.03.016.

Denli, Ahmet M., Bastiaan B. J. Tops, Ronald H. A. Plasterk, René F. Ketting, and Gregory J. Hannon. 2004. "Processing of Primary MicroRNAs by the Microprocessor Complex." *Nature* 432 (7014). Nature: 231–35. doi:10.1038/NATURE03049.

DiGiusto, David L., Amrita Krishnan, Lijing Li, Haitang Li, Shirley Li, Anitha Rao, Shu Mi, et al. 2010. "RNA-Based Gene Therapy for HIV with Lentiviral Vector-Modified CD34(+) Cells in Patients Undergoing Transplantation for AIDS-Related Lymphoma." *Science Translational Medicine* 2 (36). Sci Transl Med. doi:10.1126/SCITRANSLMED.3000931.

Dindo, Mirco, Carolina Conter, Elisa Oppici, Veronica Ceccarelli, Lorella Marinucci, and Barbara Cellini. 2019. "Molecular Basis of Primary Hyperoxaluria: Clues to Innovative Treatments." *Urolithiasis* 47 (1). Urolithiasis: 67–78. doi:10.1007/S00240-018-1089-Z.

Elbashir, Sayda M., Jens Harborth, Winfried Lendeckel, Abdullah Yalcin, Klaus Weber, and Thomas Tuschl. 2001. "Duplexes of 21-Nucleotide RNAs Mediate RNA Interference in Cultured Mammalian Cells." *Nature* 411 (6836). Nature: 494–98. doi:10.1038/35078107.

Fang, Marong, Jing Wang, Xiaobing Zhang, Yu Geng, Zhiying Hu, John A. Rudd, Shucai Ling, et al. 2012. "The MiR-124 Regulates the Expression of BACE1/β-Secretase Correlated with Cell Death in Alzheimer's Disease." *Toxicology Letters* 209 (1). Toxicol Lett: 94–105. doi:10.1016/J.TOXLET.2011.11.032.

Feng, Xiaorong, Ping Zhao, Yunxiu He, and Zhiyi Zuo. 2006. "Allele-Specific Silencing of Alzheimer's Disease Genes: The Amyloid Precursor Protein Genes with Swedish or London Mutations." *Gene* 371 (1). Gene: 68–74. doi:10.1016/J.GENE.2005.11.006.

Fire, A., S. Xu, M. K. Montgomery, S. A. Kostas, S. E. Driver, and C. C. Mello. 1998. "Potent and Specific Genetic Interference by Double-Stranded RNA in Caenorhabditis Elegans." *Nature* 391 (6669). Nature: 806–11. doi:10.1038/35888.

Forsbach, Alexandra, Jean-Guy Nemorin, Carmen Montino, Christian Müller, Ulrike Samulowitz, Alain P. Vicari, Marion Jurk, et al. 2008. "Identification of RNA Sequence Motifs Stimulating Sequence-Specific TLR8-Dependent Immune Responses." *Journal of Immunology (Baltimore, Md. : 1950)* 180 (6). J Immunol: 3729–38. doi:10.4049/JIMMUNOL.180.6.3729.

Fougerolles, Antonin De, Muthiah Manoharan, Rachel Meyers, and Hans Peter Vornlocher. 2005. "RNA Interference in Vivo: Toward Synthetic Small Inhibitory RNA-Based Therapeutics." *Methods in Enzymology* 392. Methods Enzymol: 278–96. doi:10.1016/S0076-6879(04)92016-2.

Fougerolles, Antonin de, Hans Peter Vornlocher, John Maraganore, and Judy Lieberman. 2007. "Interfering with Disease: A Progress Report on SiRNA-Based Therapeutics." *Nature Reviews. Drug Discovery* 6 (6). Nat Rev Drug Discov: 443–53. doi:10.1038/NRD2310.

Frank, Filipp, Nahum Sonenberg, and Bhushan Nagar. 2010. "Structural Basis for 5'-Nucleotide Base-Specific Recognition of Guide RNA by Human AGO2." *Nature* 465 (7299). Nature: 818–22. doi:10.1038/NATURE09039.

Fukuoka, Masashi, Masaki Takahashi, Hiromi Fujita, Tomoko Chiyo, H. Akiko Popiel, Shoko Watanabe, Hirokazu Furuya, et al. 2018. "Supplemental Treatment for Huntington's Disease with MiR-132 That Is Deficient in Huntington's Disease Brain." *Molecular Therapy. Nucleic Acids* 11 (June). Mol Ther Nucleic Acids: 79–90. doi:10.1016/J.OMTN.2018.01.007.

Gabriely, Galina, Ming Yi, Ravi S. Narayan, Johanna M. Niers, Thomas Wurdinger, Jaime Imitola, Keith L. Ligon, et al. 2011. "Human Glioma Growth Is Controlled by MicroRNA-10b." *Cancer Research* 71 (10). Cancer Res: 3563–72. doi:10.1158/0008-5472.CAN-10-3568.

Gallant-Behm, Corrie L., Joseph Piper, Joshua M. Lynch, Anita G. Seto, Seok Jong Hong, Thomas A. Mustoe, Catherine Maari, et al. 2019. "A MicroRNA-29 Mimic (Remlarsen) Represses Extracellular Matrix Expression and Fibroplasia in the Skin." *The Journal of Investigative Dermatology* 139 (5). J Invest Dermatol: 1073–81. doi:10.1016/J.JID.2018.11.007.

Gantier, Michael P., Stephen Tong, Mark A. Behlke, Dakang Xu, Simon Phipps, Paul S. Foster, and Bryan R. G. Williams. 2008. "TLR7 Is Involved in Sequence-Specific Sensing of Single-Stranded RNAs in Human Macrophages." *Journal of Immunology (Baltimore, Md. : 1950)* 180 (4). J Immunol: 2117–24. doi:10.4049/JIMMUNOL.180.4.2117.

Gebert, Luca F. R., Mario A. E. Rebhan, Silvia E. M. Crivelli, Rémy Denzler, Markus Stoffel, and Jonathan Hall. 2014. "Miravirsen (SPC3649) Can Inhibit the Biogenesis of MiR-122." *Nucleic Acids Research* 42 (1). Nucleic Acids Res: 609–21. doi:10.1093/NAR/GKT852.

Ghisoli, Maurizio, Minal Barve, Robert Mennel, Carl Lenarsky, Staci Horvath, Gladice Wallraven, Beena O. Pappen, et al. 2016. "Three-Year Follow Up of GMCSF/Bi-ShRNA(Furin) DNA-Transfected Autologous Tumor Immunotherapy (Vigil) in Metastatic Advanced Ewing's Sarcoma." *Molecular Therapy: The Journal of the American Society of Gene Therapy* 24 (8). Mol Ther: 1478–83. doi:10.1038/MT.2016.86.

Ghisoli, Maurizio, Minal Barve, Reva Schneider, Robert Mennel, Carl Lenarsky, Gladice Wallraven, Beena O. Pappen, et al. 2015. "Pilot Trial of FANG Immunotherapy in Ewing's Sarcoma." *Molecular Therapy: The Journal of the American Society of Gene Therapy* 23 (6). Mol Ther: 1103–9. doi:10.1038/MT.2015.43.

Ghisoli, Maurizio, McCarley Rutledge, Philip J. Stephens, Robert Mennel, Minal Barve, Meghan Manley, Bahram R. Oliai, et al. 2017. "Case Report: Immune-Mediated Complete Response in a Patient with Recurrent Advanced Ewing Sarcoma (EWS) After Vigil Immunotherapy." *Journal of Pediatric Hematology/Oncology* 39 (4). J Pediatr Hematol Oncol: e183–86. doi:10.1097/MPH.0000000000000822.

Gilmore, Ian, Stephen Fox, Andrew Hollins, and Saghir Akhtar. 2006. "Delivery Strategies for SiRNA-Mediated Gene Silencing." *Current Drug Delivery* 3 (2). Curr Drug Deliv: 147–55. doi:10.2174/156720106776359159.

Golan, Talia, Elina Zorde Khvalevsky, Ayala Hubert, Rachel Malka Gabai, Naama Hen, Amiel Segal, Abraham Domb, et al. 2015. "RNAi Therapy Targeting KRAS in Combination with Chemotherapy for Locally Advanced Pancreatic Cancer Patients." *Oncotarget* 6 (27). Oncotarget: 24560–70. doi:10.18632/ONCOTARGET.4183.

Gottlieb, Jens, Martin R. Zamora, Tony Hodges, A. W. Musk, Urte Sommerwerk, Daniel Dilling, Selim Arcasoy, et al. 2016. "ALN-RSV01 for Prevention of Bronchiolitis Obliterans Syndrome after Respiratory Syncytial Virus Infection in Lung Transplant Recipients." *The Journal of Heart and Lung Transplantation: The Official Publication of the International Society for Heart Transplantation* 35 (2). J Heart Lung Transplant: 213–21. doi:10.1016/J.HEALUN.2015.08.012.

Grün, Dominic, Yi Lu Wang, David Langenberger, Kristin C. Gunsalus, and Nikolaus Rajewsky. 2005. "MicroRNA Target Predictions across Seven Drosophila Species and Comparison to Mammalian Targets." *PLoS Computational Biology* 1 (1). PLoS Comput Biol: 0051–0066. doi:10.1371/JOURNAL.PCBI.0010013.

Gu, Shuo, Lan Jin, Yue Zhang, Yong Huang, Feijie Zhang, Paul N. Valdmanis, and Mark A. Kay. 2012. "The Loop Position of ShRNAs and Pre-MiRNAs Is Critical for the Accuracy of Dicer Processing in Vivo." *Cell* 151 (4). Cell: 900–911. doi:10.1016/J.CELL.2012.09.042.

Ha, Minju, and V. Narry Kim. 2014. "Regulation of MicroRNA Biogenesis." *Nature Reviews Molecular Cell Biology* 15 (8). Nat Rev Mol Cell Biol: 509–24. doi:10.1038/nrm3838.

Haley, Benjamin, and Phillip D. Zamore. 2004. "Kinetic Analysis of the RNAi Enzyme Complex." *Nature Structural & Molecular Biology* 11 (7). Nat Struct Mol Biol: 599–606. doi:10.1038/NSMB780.

Hamilton, James. 2018. "Overcoming the Challenges of RNAi-Based Therapy: An Interview with James Hamilton." *Therapeutic Delivery* 9 (7). Ther Deliv: 511–13. doi:10.4155/TDE-2018-0030.

Hammond, Scott M., Emily Bernstein, David Beach, and Gregory J. Hannon. 2000. "An RNA-Directed Nuclease Mediates Post-Transcriptional Gene Silencing in Drosophila Cells." *Nature* 404 (6775). Nature: 293–96. doi:10.1038/35005107.

Hammond, Scott M., Amy A. Caudy, and Gregory J. Hannon. 2001. "Post-Transcriptional Gene Silencing by Double-Stranded RNA." *Nature Reviews Genetics* 2 (2). Nat Rev Genet: 110–19. doi:10.1038/35052556.

Heidel, Jeremy D., Joanna Yi Ching Liu, Yun Yen, Bingsen Zhou, Bret S. E. Heale, John J. Rossi, Derek W. Bartlett, and Mark E. Davis. 2007. "Potent SiRNA Inhibitors of Ribonucleotide Reductase Subunit RRM2 Reduce Cell Proliferation in Vitro and in Vivo." *Clinical Cancer Research: An Official Journal of the American Association for Cancer Research* 13 (7). Clin Cancer Res: 2207–15. doi:10.1158/1078-0432.CCR-06-2218.

Hill, Anita, Jorg Taubel, Jim Bush, Anna Borodovsky, Noriyuki Kawahata, Helen Mclean, Christine Powell, et al. 2015. "A Subcutaneously Administered Investigational RNAi Therapeutic (ALN-CC5) Targeting Complement C5 for Treatment of PNH and Complement-Mediated Diseases: Interim Phase 1 Study Results." *Blood* 126 (23). Blood: 2413–2413. doi:10.1182/BLOOD.V126.23.2413.2413.

Hong, David S., Yoon Koo Kang, Mitesh Borad, Jasgit Sachdev, Samuel Ejadi, Ho Yeong Lim, Andrew J. Brenner, et al. 2020. "Phase 1 Study of MRX34, a Liposomal MiR-34a Mimic, in Patients with Advanced Solid Tumours." *British Journal of Cancer* 122 (11). Br J Cancer: 1630–37. doi:10.1038/s41416-020-0802-1.

Hornung, Veit, Margit Guenthner-Biller, Carole Bourquin, Andrea Ablasser, Martin Schlee, Satoshi Uematsu, Anne Noronha, et al. 2005. "Sequence-Specific Potent Induction of IFN-Alpha by Short Interfering RNA in Plasmacytoid Dendritic Cells Through TLR7." *Nature Medicine* 11 (3). Nat Med: 263–70. doi:10.1038/NM1191.

Hu, Hai Y., Zheng Yan, Ying Xu, Hao Hu, Corinna Menzel, Yan H. Zhou, Wei Chen, and Philipp Khaitovich. 2009. "Sequence Features Associated with MicroRNA Strand Selection in Humans and Flies." *BMC Genomics* 10 (September). BMC Genom: 413. doi:10.1186/1471-2164-10-413.

Huang, Mei, Denise A. Chan, Fangjun Jia, Xiaoyan Xie, Zongjin Li, Grant Hoyt, Robert C. Robbins, et al. 2008. "Short Hairpin RNA Interference Therapy for Ischemic Heart Disease." *Circulation* 118 (14). Circulation: S226. doi:10.1161/CIRCULATIONAHA.107.760785.

Huang, Yuanyu. 2017. "Preclinical and Clinical Advances of GalNAc-Decorated Nucleic Acid Therapeutics." *Molecular Therapy Nucleic Acids* 6 (March). Mol Ther Nucleic Acids: 116–32. doi:10.1016/J.OMTN.2016.12.003.

Hullinger, Thomas G., Rusty L. Montgomery, Anita G. Seto, Brent A. Dickinson, Hillary M. Semus, Joshua M. Lynch, Christina M. Dalby, et al. 2012. "Inhibition of MiR-15 Protects against Cardiac Ischemic Injury." *Circulation Research* 110 (1). Circ Res: 71–81. doi:10.1161/CIRCRESAHA.111.244442.

Huntzinger, Eric, and Elisa Izaurralde. 2011. "Gene Silencing by MicroRNAs: Contributions of Translational Repression and MRNA Decay." *Nature Reviews Genetics* 12 (2). Nat Rev Genet: 99–110. doi:10.1038/NRG2936.

Iorio, Marilena V., and Carlo M. Croce. 2012. "MicroRNA Dysregulation in Cancer: Diagnostics, Monitoring and Therapeutics. A Comprehensive Review." *EMBO Molecular Medicine* 4 (3). EMBO Mol Med: 143–59. doi:10.1002/EMMM.201100209.

Jackson, Aimee L., and Peter S. Linsley. 2010. "Recognizing and Avoiding SiRNA Off-Target Effects for Target Identification and Therapeutic Application." *Nature Reviews Drug Discovery* 9 (1). Nat Rev Drug Discov: 57–67. doi:10.1038/NRD3010.

Jain, Rakesh K., and Triantafyllos Stylianopoulos. 2010. "Delivering Nanomedicine to Solid Tumors." *Nature Reviews: Clinical Oncology* 7 (11). Nat Rev Clin Oncol: 653–64. doi:10.1038/NRCLINONC.2010.139.

Janas, Maja M., Ivan Zlatev, Ju Liu, Yongfeng Jiang, Scott A. Barros, Jessica E. Sutherland, Wendell P. Davis, et al. 2019. "Safety Evaluation of 2'-Deoxy-2'-Fluoro Nucleotides in GalNAc-SiRNA Conjugates." *Nucleic Acids Research* 47 (7). Nucleic Acids Res: 3306–20. doi:10.1093/NAR/GKZ140.

Janssen, Harry L. A., Hendrik W. Reesink, Eric J. Lawitz, Stefan Zeuzem, Maribel Rodriguez-Torres, Keyur Patel, Adriaan J. van der Meer, et al. 2013. "Treatment of HCV Infection by Targeting MicroRNA." *The New England Journal of Medicine* 368 (18). N Engl J Med: 1685–94. doi:10.1056/NEJMOA1209026.

Jensen, Samuel A., Emily S. Day, Caroline H. Ko, Lisa A. Hurley, Janina P. Luciano, Fotini M. Kouri, Timothy J. Merkel, et al. 2013. "Spherical Nucleic Acid Nanoparticle Conjugates as an RNAi-Based Therapy for Glioblastoma." *Science Translational Medicine* 5 (209). Sci Transl Med. doi:10.1126/SCITRANSLMED.3006839.

Judge, Adam D., Vandana Sood, Janet R. Shaw, Dianne Fang, Kevin McClintock, and Ian MacLachlan. 2005. "Sequence-Dependent Stimulation of the Mammalian Innate Immune Response by Synthetic SiRNA." *Nature Biotechnology* 23 (4). Nat Biotechnol: 457–62. doi:10.1038/NBT1081.

Kawamata, Tomoko, Mayuko Yoda, and Yukihide Tomari. 2011. "Multilayer Checkpoints for MicroRNA Authenticity During RISC Assembly." *EMBO Reports* 12 (9). EMBO Rep: 944–49. doi:10.1038/EMBOR.2011.128.

Ketting, R. F., S. E. J. Fischer, E. Bernstein, T. Sijen, G. J. Hannon, and R. H. A. Plasterk. 2001. “Dicer Functions in RNA Interference and in Synthesis of Small RNA Involved in Developmental Timing in C. Elegans.” *Genes & Development* 15 (20). Genes Dev: 2654–59. doi:10.1101/GAD.927801.

Kosmas, Constantine, Alba Muñoz Estrella, Andreas Sourlas, Delia Silverio, Elizabeth Hilario, Peter Montan, and Eliscer Guzman. 2018. “Inclisiran: A New Promising Agent in the Management of Hypercholesterolemia.” *Diseases (Basel, Switzerland)* 6 (3). Diseases: 63. doi:10.3390/DISEASES6030063.

Kota, Janaiah, Raghu R. Chivukula, Kathryn A. O’Donnell, Erik A. Wentzel, Chrystal L. Montgomery, Hun Way Hwang, Tsung Cheng Chang, et al. 2009. “Therapeutic MicroRNA Delivery Suppresses Tumorigenesis in a Murine Liver Cancer Model.” *Cell* 137 (6). Cell: 1005–17. doi:10.1016/J.CELL.2009.04.021.

Kumari, Avnesh, Vineet Kumar, and Sudesh Kumar Yadav. 2011. “Nanocarriers: A Tool to Overcome Biological Barriers in SiRNA Delivery.” *Expert Opinion on Biological Therapy* 11 (10). Expert Opin Biol Ther: 1327–39. doi:10.1517/14712598.2011.587801.

Kuninty, Praneeth R., Jonas Schnittert, Gert Storm, and Jai Prakash. 2016. “MicroRNA Targeting to Modulate Tumor Microenvironment.” *Frontiers in Oncology* 6 (January). Front Oncol. doi:10.3389/FONC.2016.00003.

Landen, Charles N., Arturo Chavez-Reyes, Corazon Bucana, Rosemarie Schmandt, Michael T. Deavers, Gabriel Lopez-Berestein, and Anil K. Sood. 2005. “Therapeutic EphA2 Gene Targeting in Vivo Using Neutral Liposomal Small Interfering RNA Delivery.” *Cancer Research* 65 (15). Cancer Res: 6910–18. doi:10.1158/0008-5472.CAN-05-0530.

Lee, Soon-Tae, Wooseok Im, Jae-Jun Ban, Mijung Lee, Keun-Hwa Jung, Sang Kun Lee, Kon Chu, et al. 2017. “Exosome-Based Delivery of MiR-124 in a Huntington’s Disease Model.” *Journal of Movement Disorders* 10 (1). J Mov Disord: 45–52. doi:10.14802/JMD.16054.

Lee, Steven Hyun Seung, Hee Soon Chang, Ji Hyun Kim, Hee Jong Kim, Jun Sub Choi, Sunho Chung, Ha Na Woo, et al. 2020. “Inhibition of MTOR via an AAV-Delivered ShRNA Tested in a Rat OIR Model as a Potential Antiangiogenic Gene Therapy.” *Investigative Ophthalmology & Visual Science* 61 (2). Invest Ophthalmol Vis Sci. doi:10.1167/IOVS.61.2.45.

Lee, Yoontae, Chiyoung Ahn, Jinju Han, Hyounjeong Choi, Jaekwang Kim, Jeongbin Yim, Junho Lee, et al. 2003. “The Nuclear RNase III Drosha Initiates MicroRNA Processing.” *Nature* 425 (6956). Nature: 415–19. doi:10.1038/NATURE01957.

Lee, Yoontae, Kipyoung Jeon, Jun Tae Lee, Sunyoung Kim, and V. Narry Kim. 2002. “MicroRNA Maturation: Stepwise Processing and Subcellular Localization.” *The EMBO Journal* 21 (17). EMBO J: 4663–70. doi:10.1093/EMBOJ/CDF476.

Liebow, Abigail, Xingsheng Li, Timothy Racie, Julia Hettinger, Brian R. Bettencourt, Nader Najafian, Patrick Haslett, et al. 2017. “An Investigational RNAi Therapeutic Targeting Glycolate Oxidase Reduces Oxalate Production in Models of Primary Hyperoxaluria.” *Journal of the American Society of Nephrology: JASN* 28 (2). J Am Soc Nephrol: 494–503. doi:10.1681/ASN.2016030338.

Lin, Xiaoyu, Xiaoan Ruan, Mark G. Anderson, Jeffrey A. McDowell, Paul E. Kroeger, Stephen W. Fesik, and Yu Shen. 2005. “SiRNA-Mediated Off-Target Gene Silencing Triggered by a 7 Nt Complementation.” *Nucleic Acids Research* 33 (14). Nucleic Acids Res: 4527–35. doi:10.1093/NAR/GKI762.

Liu, Can, Kevin Kelnar, Bigang Liu, Xin Chen, Tammy Calhoun-Davis, Hangwen Li, Lubna Patrawala, et al. 2011. “The MicroRNA MiR-34a Inhibits Prostate Cancer Stem Cells and Metastasis by Directly Repressing CD44.” *Nature Medicine* 17 (2). Nat Med: 211–16. doi:10.1038/NM.2284.

Liu, Hong An, Yu ling Liu, Zhi Zhong Ma, Jian Cheng Wang, and Qiang Zhang. 2011. “A Lipid Nanoparticle System Improves SiRNA Efficacy in RPE Cells and a Laser-Induced Murine CNV Model.” *Investigative Ophthalmology & Visual Science* 52 (7). Invest Ophthalmol Vis Sci: 4789–94. doi:10.1167/IOVS.10-5891.

Liu, Yun, Jian Liu, Alistair Quimbo, Fengcheng Xia, Jiping Yao, Jean Pierre Clamme, Sonya Zabludoff, et al. 2021. “Anti-HSP47 SiRNA Lipid Nanoparticle ND-L02-S0201 Reverses Interstitial Pulmonary Fibrosis in Preclinical Rat Models.” *ERJ Open Research* 7 (2). ERJ Open Res. doi:10.1183/23120541.00733-2020.

Liu, Yunyun, Zhonglin Liu, Yong Wang, Yan Ran Liang, Xiaojun Wen, Jingyang Hu, Xingyi Yang, et al. 2013. “Investigation of the Performance of PEG-PEI/ROCK-II-SiRNA Complexes for Alzheimer’s Disease in Vitro.” *Brain Research* 1490 (January). Brain Res: 43–51. doi:10.1016/J.BRAINRES.2012.10.039.

Loibner, Hans, Guenther Lametschwandtner, Kerstin Westritschnig, Oliver Mutschlechner, Alexander Dohnal, Marc Oliver Salzberg, and Pierre L. Triozzi. 2018. “Adoptive Cellular Immunotherapy with APN401, Autologous Cbl-b Silenced Peripheral Blood Mononuclear Cells: Data from a Phase I Study in Patients with Solid Tumors.” *American Society of Clinical Oncology* 36 (15_suppl). Am Soc Clin Oncol: 3055–3055. doi:10.1200/JCO.2018.36.15_SUPPL.3055.

López-Tobón, Alejandro, John Fredy Castro-Álvarez, Diego Piedrahita, Ryan L. Boudreau, Juan Carlos Gallego-Gómez, and Gloria Patricia Cardona-Gómez. 2011. "Silencing of CDK5 as Potential Therapy for Alzheimer's Disease." *Reviews in the Neurosciences* 22 (2). Rev Neurosci: 143–52. doi:10.1515/RNS.2011.015.

Lundstrom, Kenneth. 2020. "Viral Vectors Applied for RNAi-Based Antiviral Therapy." *Viruses* 12 (9). Viruses. doi:10.3390/V12090924.

Ma, Li, Ferenc Reinhardt, Elizabeth Pan, Jürgen Soutschek, Balkrishen Bhat, Eric G. Marcusson, Julie Teruya-Feldstein, et al. 2010. "Therapeutic Silencing of MiR-10b Inhibits Metastasis in a Mouse Mammary Tumor Model." *Nature Biotechnology* 28 (4). Nat Biotechnol: 341–47. doi:10.1038/NBT.1618.

Marques, Joao Trindade, Thalie Devosse, Die Wang, Maryam Zamanian-Daryoush, Paul Serbinowski, Rune Hartmann, Takashi Fujita, et al. 2006. "A Structural Basis for Discriminating between Self and Nonself Double-Stranded RNAs in Mammalian Cells." *Nature Biotechnology* 24 (5). Nat Biotechnol: 559–65. doi:10.1038/NBT1205.

Martin, Raquel, Peter Smibert, Abdullah Yalcin, David M. Tyler, Ulrich Schäfer, Thomas Tuschl, and Eric C. Lai. 2009. "A Drosophila Pasha Mutant Distinguishes the Canonical MicroRNA and Mirtron Pathways." *Molecular and Cellular Biology* 29 (3). Mol Cell Biol: 861–70. doi:10.1128/MCB.01524-08.

Martinez, Javier, Agnieszka Patkaniowska, Henning Urlaub, Reinhard Lührmann, and Thomas Tuschl. 2002. "Single-Stranded Antisense SiRNAs Guide Target RNA Cleavage in RNAi." *Cell* 110 (5). Cell: 563–74. doi:10.1016/S0092-8674(02)00908-X.

Meng, Zhongji, and Mengji Lu. 2017. "RNA Interference-Induced Innate Immunity, Off-Target Effect, or Immune Adjuvant?" *Frontiers in Immunology* 8 (March). Frontiers Media SA: 331. doi:10.3389/FIMMU.2017.00331.

Miyazaki, Yu, Hiroaki Adachi, Masahisa Katsuno, Makoto Minamiyama, Yue Mei Jiang, Zhe Huang, Hideki Doi, et al. 2012. "Viral Delivery of MiR-196a Ameliorates the SBMA Phenotype via the Silencing of CELF2." *Nature Medicine* 18 (7). Nat Med: 1136–41. doi:10.1038/NM.2791.

Moreno-Montañés, Javier, Anne Marie Bleau, and Ana Isabel Jimenez. 2018. "Tivanisiran, a Novel SiRNA for the Treatment of Dry Eye Disease." *Expert Opinion on Investigational Drugs* 27 (4). Expert Opin Investig Drugs: 421–26. doi:10.1080/13543784.2018.1457647.

Nair, Jayaprakash K., Jennifer L. S. Willoughby, Amy Chan, Klaus Charisse, Md Rowshon Alam, Qianfan Wang, Menno Hoekstra, et al. 2014. "Multivalent N-Acetylgalactosamine-Conjugated SiRNA Localizes in Hepatocytes and Elicits Robust RNAi-Mediated Gene Silencing." *Journal of the American Chemical Society* 136 (49). J Am Chem Soc: 16958–61. doi:10.1021/JA505986A.

Napoli, Carolyn, Christine Lemieux, and Richard Jorgensen. 1990. "Lntroduction of a Chimeric Chalcone Synthase Gene into Petunia Results in Reversible Co-Suppression of Homologous Genes in Trans." *The Plant Cell* 2. Plant Cell: 279–89. https://academic.oup.com/plcell/article/2/4/279/5983193.

Neumeier, Julia, and Gunter Meister. 2021. "SiRNA Specificity: RNAi Mechanisms and Strategies to Reduce Off-Target Effects." *Frontiers in Plant Science* 11 (January). Front Plant Sci. doi:10.3389/FPLS.2020.526455.

Niaz, Saife. 2018. "The AGO Proteins: An Overview." *Biological Chemistry* 399 (6). Biol Chem: 525–47. doi:10.1515/HSZ-2017-0329.

Nishimura, Masato, Eun Jung Seife, Maitri Y. Shah, Chunhua Lu, Riccardo Spizzo, Masayoshi Shimizu, Hee Dong Han, et al. 2013. "Therapeutic Synergy between MicroRNA and SiRNA in Ovarian Cancer Treatment." *Cancer Discovery* 3 (11). Cancer Discov: 1302–15. doi:10.1158/2159-8290.CD-13-0159.

Nykänen, Antti, Benjamin Haley, and Phillip D. Zamore. 2001. "ATP Requirements and Small Interfering RNA Structure in the RNA Interference Pathway." *Cell* 107 (3). Cell: 309–21. doi:10.1016/S0092-8674(01)00547-5.

Park, Jong Eun, Inha Heo, Yuan Tian, Dhirendra K. Simanshu, Hyeshik Chang, David Jee, Dinshaw J. Patel, et al. 2011. "Dicer Recognizes the 5′ End of RNA for Efficient and Accurate Processing." *Nature* 475 (7355). Nature: 201–5. doi:10.1038/nature10198.

Pasi, K. John, Toshko Lissitchkov, Vasily Mamonov, Tim Mant, Margarita Timofeeva, Catherine Bagot, Pratima Chowdary, et al. 2021. "Targeting of Antithrombin in Hemophilia A or B with Investigational SiRNA Therapeutic Fitusiran-Results of the Phase 1 Inhibitor Cohort." *Journal of Thrombosis and Haemostasis* 19 (6). J Thromb Haemost: 1436–46. doi:10.1111/JTH.15270.

Pasi, K. John, Savita Rangarajan, Pencho Georgiev, Tim Mant, Michael D. Creagh, Toshko Lissitchkov, David Bevan, et al. 2017. "Targeting of Antithrombin in Hemophilia A or B with RNAi Therapy." *The New England Journal of Medicine* 377 (9). N Engl J Med: 819–28. doi:10.1056/NEJMOA1616569.

Patel, Keyur, Georgina Kilfoil, David L. Wyles, Susanna Naggie, Eric Lawitz, Stephen Bradley, Per Lindell, et al. 2016. "258. Phase I/IIa Study of TT-034, a DNA-Directed RNA Interference (DdRNAi) Agent

Delivered as a Single Administration for the Treatment of Subjects with Chronic Hepatitis C Virus (HCV).” *Molecular Therapy* 24 (May). Mol Ther: S102. doi:10.1016/S1525-0016(16)33067-2.

Peng, Kevin A., and Eliezer Masliah. 2010. “Lentivirus-Expressed SiRNA Vectors against Alzheimer Disease.” *Methods in Molecular Biology (Clifton, N.J.)* 614. Methods Mol Biol: 215–24. doi:10.1007/978-1-60761-533-0_15.

Phadke, Anagha P., Chris M. Jay, Zhaohui Wang, Salina Chen, Shengnan Liu, Courtney Haddock, Padmasini Kumar, et al. 2011. “In Vivo Safety and Antitumor Efficacy of Bifunctional Small Hairpin RNAs Specific for the Human Stathmin 1 Oncoprotein.” *DNA and Cell Biology* 30 (9). DNA Cell Biol: 715–26. doi:10.1089/DNA.2011.1240.

Porrello, Enzo R., Ahmed I. Mahmoud, Emma Simpson, Brett A. Johnson, David Grinsfelder, Diana Canseco, Pradeep P. Mammen, et al. 2013. “Regulation of Neonatal and Adult Mammalian Heart Regeneration by the MiR-15 Family.” *Proceedings of the National Academy of Sciences of the United States of America* 110 (1). Proc Natl Acad Sci U S A: 187–92. doi:10.1073/PNAS.1208863110.

Pramanik, Dipankar, Nathaniel R. Campbell, Collins Karikari, Raghu Chivukula, Oliver A. Kent, Joshua T. Mendell, and Anirban Maitra. 2011. “Restitution of Tumor Suppressor MicroRNAs Using a Systemic Nanovector Inhibits Pancreatic Cancer Growth in Mice.” *Molecular Cancer Therapeutics* 10 (8). Mol Cancer Ther: 1470–80. doi:10.1158/1535-7163.MCT-11-0152.

Rajewsky, Nikolaus, and Nicholas D. Socci. 2004. “Computational Identification of MicroRNA Targets.” *Developmental Biology* 267 (2). Dev Biol: 529–35. doi:10.1016/J.YDBIO.2003.12.003.

Raoul, Cédric, Toufik Abbas-Terki, Jean Charles Bensadoun, Sandrine Guillot, Georg Haase, Jolanta Szulc, Christopher E. Henderson, et al. 2005. “Lentiviral-Mediated Silencing of SOD1 Through RNA Interference Retards Disease Onset and Progression in a Mouse Model of ALS.” *Nature Medicine* 11 (4). Nat Med: 423–28. doi:10.1038/NM1207.

Reid, Glen, Marissa Williams, Michaela B. Kirschner, Nancy Mugridge, Jocelyn Weiss, Himanshu Brahmbhatt, Jennifer MacDiarmid, et al. 2015. “Abstract 3976: Targeted Delivery of a Synthetic MicroRNA-Based Mimic as an Approach to Cancer Therapy.” *Cancer Research* 75 (15_Supplement). Cancer Res: 3976. doi:10.1158/1538-7445.AM2015-3976.

Romaine, Simon P. R., Maciej Tomaszewski, Gianluigi Condorelli, and Nilesh J. Samani. 2015. “MicroRNAs in Cardiovascular Disease: An Introduction for Clinicians.” *Heart* 101 (12). Heart: 921. doi:10.1136/HEARTJNL-2013-305402.

Romano, Nicoletta, and Giuseppe Macino. 1992. “Quelling: Transient Inactivation of Gene Expression in Neurospora Crassa by Transformation with Homologous Sequences.” *Molecular Microbiology* 6 (22). Mol Microbiol: 3343–53. doi:10.1111/J.1365-2958.1992.TB02202.X.

Rooij, Eva, and Sakari Kauppinen. 2014. “Development of MicroRNA Therapeutics Is Coming of Age.” *EMBO Molecular Medicine* 6 (7). EMBO Mol Med: 851–64. doi:10.15252/EMMM.201100899.

Rooij, Eva Van, and Eric N. Olson. 2012. “MicroRNA Therapeutics for Cardiovascular Disease: Opportunities and Obstacles.” *Nature Reviews: Drug Discovery* 11 (11). Nat Rev Drug Discov: 860. doi:10.1038/NRD3864.

Rooij, Eva Van, Lillian B. Sutherland, Jeffrey E. Thatcher, J. Michael DiMaio, R. Haris Naseem, William S. Marshall, Joseph A. Hill, et al. 2008. “Dysregulation of MicroRNAs after Myocardial Infarction Reveals a Role of MiR-29 in Cardiac Fibrosis.” *Proceedings of the National Academy of Sciences of the United States of America* 105 (35). Proc Natl Acad Sci U S A: 13027–32. doi:10.1073/PNAS.0805038105.

Rupaimoole, R., C. Ivan, D. Yang, K. M. Gharpure, S. Y. Wu, C. V. Pecot, R. A. Previs, et al. 2016. “Hypoxia-Upregulated MicroRNA-630 Targets Dicer, Leading to Increased Tumor Progression.” *Oncogene* 35 (33). Oncogene: 4312–20. doi:10.1038/ONC.2015.492.

Salta, Evgenia, Annerieke Sierksma, Elke Vanden Eynden, and Bart De Strooper. 2016. “MiR-132 Loss de-Represses ITPKB and Aggravates Amyloid and TAU Pathology in Alzheimer's Brain.” *EMBO Molecular Medicine* 8 (9). EMBO Mol Med: 1005–18. doi:10.15252/EMMM.201606520.

Saraiva, C., J. Paiva, T. Santos, L. Ferreira, and L. Bernardino. 2016. “MicroRNA-124 Loaded Nanoparticles Enhance Brain Repair in Parkinson's Disease.” *Journal of Controlled Release : Official Journal of the Controlled Release Society* 235 (August). J Control Release: 291–305. doi:10.1016/J.JCONREL.2016.06.005.

Sardh, Eliane, Pauline Harper, Manisha Balwani, Penelope Stein, David Rees, D. Montgomery Bissell, Robert Desnick, et al. 2019. “Phase 1 Trial of an RNA Interference Therapy for Acute Intermittent Porphyria.” *The New England Journal of Medicine* 380 (6). N Engl J Med: 549–58. doi:10.1056/NEJMOA1807838.

Schluep, Thomas, Jason Lickliter, James Hamilton, David L. Lewis, Ching Lung Lai, Johnson Y. N. Lau, Stephen A. Locarnini, et al. 2017. “Safety, Tolerability, and Pharmacokinetics of ARC-520 Injection, an RNA Interference-Based Therapeutic for the Treatment of Chronic Hepatitis B Virus Infection, in Healthy Volunteers." *Clinical Pharmacology in Drug Development* 6 (4). Clin Pharmacol Drug Dev: 350–62. doi:10.1002/CPDD.318.

Schultheis, Beate, Dirk Strumberg, Jan Kuhlmann, Martin Wolf, Karin Link, Thomas Seufferlein, Joerg Kaufmann, et al. 2020. "Safety, Efficacy and Pharcacokinetics of Targeted Therapy with the Liposomal RNA Interference Therapeutic Atu027 Combined with Gemcitabine in Patients with Pancreatic Adenocarcinoma. A Randomized Phase Ib/IIa Study." *Cancers* 12 (11). Cancers (Basel): 1–13. doi:10.3390/CANCERS12113130.

Schwarz, Dianne S., György Hutvágner, Tingting Du, Zuoshang Xu, Neil Aronin, and Phillip D. Zamore. 2003. "Asymmetry in the Assembly of the RNAi Enzyme Complex." *Cell* 115 (2). Cell: 199–208. doi:10.1016/S0092-8674(03)00759-1.

Scott, Lesley J. 2020. "Givosiran: First Approval." *Drugs* 80 (3). Drugs: 335–39. doi:10.1007/S40265-020-01269-0.

Seok, Heeyoung, Haejeong Lee, Eun Sook Jang, and Sung Wook Chi. 2018. "Evaluation and Control of MiRNA-like off-Target Repression for RNA Interference." *Cellular and Molecular Life Sciences: CMLS* 75 (5). Cell Mol Life Sci: 797–814. doi:10.1007/S00018-017-2656-0.

Seto, Anita G., Xuan Beatty, Joshua M. Lynch, Melanie Hermreck, Michael Tetzlaff, Madeleine Duvic, and Aimee L. Jackson. 2018. "Cobomarsen, an Oligonucleotide Inhibitor of MiR-155, Co-Ordinately Regulates Multiple Survival Pathways to Reduce Cellular Proliferation and Survival in Cutaneous T-Cell Lymphoma." *British Journal of Haematology* 183 (3). Br J Haematol: 428–44. doi:10.1111/BJH.15547.

Shi, Zhongli, Kaixia Zhang, Huimin Zhou, Lei Jiang, Bing Xie, Ruiyuan Wang, Wenzhen Xia, et al. 2020. "Increased MiR-34c Mediates Synaptic Deficits by Targeting Synaptotagmin 1 Through ROS-JNK-P53 Pathway in Alzheimer's Disease." *Aging Cell* 19 (3). Aging Cell. doi:10.1111/ACEL.13125.

Singh, Manu Smriti, and Dan Peer. 2016. "SiRNA Delivery: Current Trends and Future Perspectives." *Therapeutic Delivery* 7 (2). Ther Deliv: 51–53. doi:10.4155/TDE.15.88.

Sioud, Mouldy. 2005. "Induction of Inflammatory Cytokines and Interferon Responses by Double-Stranded and Single-Stranded SiRNAs Is Sequence-Dependent and Requires Endosomal Localization." *Journal of Molecular Biology* 348 (5). J Mol Biol: 1079–90. doi:10.1016/J.JMB.2005.03.013.

Solano, Elisabeth C. R., Douglas J. Kornbrust, Amber Beaudry, Jeffrey W. D. Foy, David J. Schneider, and James D. Thompson. 2014. "Toxicological and Pharmacokinetic Properties of QPI-1007, a Chemically Modified Synthetic SiRNA Targeting Caspase 2 MRNA, Following Intravitreal Injection." *Nucleic Acid Therapeutics* 24 (4). Nucleic Acid Ther: 258–66. doi:10.1089/NAT.2014.0489.

Song, Erwei, Sang Kyung Lee, Jie Wang, Nedim Ince, Nengtai Ouyang, Jun Min, Jisheng Chen, et al. 2003. "RNA Interference Targeting Fas Protects Mice from Fulminant Hepatitis." *Nature Medicine* 9 (3). Nat Med: 347–51. doi:10.1038/NM828.

Suhr, Ole B., Teresa Coelho, Juan Buades, Jean Pouget, Isabel Conceicao, John Berk, Hartmut Schmidt, et al. 2015. "Efficacy and Safety of Patisiran for Familial Amyloidotic Polyneuropathy: A Phase II Multi-Dose Study." *Orphanet Journal of Rare Diseases* 10 (1). Orphanet J Rare Dis. doi:10.1186/S13023-015-0326-6.

Sui, Guangchao, Christina Soohoo, El Bachir Affar, Frédérique Gay, Yujiang Shi, William C. Forrester, and Yang Shi. 2002. "A DNA Vector-Based RNAi Technology to Suppress Gene Expression in Mammalian Cells." *Proceedings of the National Academy of Sciences of the United States of America* 99 (8). Proc Natl Acad Sci U S A: 5515–20. doi:10.1073/PNAS.082117599.

Tang, Guiliang. 2005. "SiRNA and MiRNA: An Insight into RISCs." *Trends in Biochemical Sciences* 30 (2). Trends Biochem Sci: 106–14. doi:10.1016/J.TIBS.2004.12.007.

Tasfaout, Hichem, Valentina M. Lionello, Christine Kretz, Pascale Koebel, Nadia Messaddeq, Deborah Bitz, Jocelyn Laporte, and Belinda S. Cowling. 2018. "Single Intramuscular Injection of AAV-ShRNA Reduces DNM2 and Prevents Myotubular Myopathy in Mice." *Molecular Therapy : The Journal of the American Society of Gene Therapy* 26 (4). Mol Ther: 1082–92. doi:10.1016/J.YMTHE.2018.02.008.

Thum, Thomas, Carina Gross, Jan Fiedler, Thomas Fischer, Stephan Kissler, Markus Bussen, Paolo Galuppo, et al. 2008. "MicroRNA-21 Contributes to Myocardial Disease by Stimulating MAP Kinase Signalling in Fibroblasts." *Nature* 456 (7224). Nature: 980–84. doi:10.1038/NATURE07511.

Trang, Phong, Jason F. Wiggins, Christopher L. Daige, Chris Cho, Michael Omotola, David Brown, Joanne B. Weidhaas, et al. 2011. "Systemic Delivery of Tumor Suppressor MicroRNA Mimics Using a Neutral Lipid Emulsion Inhibits Lung Tumors in Mice." *Molecular Therapy* 19 (6). Mol Ther: 1116. doi:10.1038/MT.2011.48.

Triozzi, Pierre, Mitra Kooshki, Angela Alistar, Rhonda Bitting, Amy Neal, Guenther Lametschwandtner, and Hans Loibner. 2015. "Phase I Clinical Trial of Adoptive Cellular Immunotherapy with APN401 in

Patients with Solid Tumors." *Journal for Immunotherapy of Cancer* 3 (Suppl 2). J Immunother Cancer: P175. doi:10.1186/2051-1426-3-S2-P175.

Ubeda, Mariano, Daniel M. Kemp, and Joel F. Habener. 2004. "Glucose-Induced Expression of the Cyclin-Dependent Protein Kinase 5 Activator P35 Involved in Alzheimer's Disease Regulates Insulin Gene Transcription in Pancreatic Beta-Cells." *Endocrinology* 145 (6). Endocrinology: 3023–31. doi:10.1210/EN.2003-1522.

Vázquez, Jorge, María F. Correa De Adjounian, Colin Sumners, Aaron González, Carlos Diez-Freire, and Mohan K. Raizada. 2005. "Selective Silencing of Angiotensin Receptor Subtype 1a (AT1aR) by RNA Interference." *Hypertension (Dallas, Tex. : 1979)* 45 (1). Hypertension: 115–19. doi:10.1161/01.HYP.0000150161.78556.C6.

Walker, Jon E., Rachel X. Chen, Jeannine McGee, Catherine Nacey, Richard B. Pollard, Mehrdad Abedi, Gerhard Bauer, et al. 2012. "Generation of an HIV-1-Resistant Immune System with CD34(+) Hematopoietic Stem Cells Transduced with a Triple-Combination Anti-HIV Lentiviral Vector." *Journal of Virology* 86 (10). J Virol: 5719–29. doi:10.1128/JVI.06300-11.

Wang, Huiqing, Yongyi Ye, Zhiyuan Zhu, Liqian Mo, Chunnan Lin, Qifu Wang, Haoyuan Wang, et al. 2016. "MiR-124 Regulates Apoptosis and Autophagy Process in MPTP Model of Parkinson's Disease by Targeting to Bim." *Brain Pathology (Zurich, Switzerland)* 26 (2). Brain Pathol: 167–76. doi:10.1111/BPA.12267.

Wang, Jiu Cun, Syeling Lai, Xinjian Guo, Xuefeng Zhang, Benoit de Crombrugghe, Sonali Sonnylal, Frank C. Arnett, and Xiaodong Zhou. 2010. "Attenuation of Fibrosis in Vitro and in Vivo with SPARC SiRNA." *Arthritis Research & Therapy* 12 (2). Arthritis Res Ther: R60. doi:10.1186/AR2973.

Warf, M. Bryan, W. Evan Johnson, and Brenda L. Bass. 2011. "Improved Annotation of C. Elegans MicroRNAs by Deep Sequencing Reveals Structures Associated with Processing by Drosha and Dicer." *RNA (New York, N.Y.)* 17 (4). RNA: 563–77. doi:10.1261/RNA.2432311.

Watts, Jonathan K., and Ira S. Ockene. 2020. "RNA Interference for the Masses? SiRNA Targeting PCSK9 Promises Prevention of Cardiovascular Disease." 30 (1). Mary Ann Liebert: 1–3. doi:10.1089/NAT.2019.0835. https://home.liebertpub.com/Nat

Wiggins, Jason F., Lynnsie Ruffino, Kevin Kelnar, Michael Omotola, Lubna Patrawala, David Brown, and Andreas G. Bader. 2010. "Development of a Lung Cancer Therapeutic Based on the Tumor Suppressor MicroRNA-34." *Cancer Research* 70 (14). Cancer Res: 5923–30. doi:10.1158/0008-5472.CAN-10-0655.

Wooddell, C., R. Zhu, H. Hamilton, Q. Chu, H. Sternard, J. Schumacher, T. Schluep, M. Seefeld, Z. Li, and B. Given. 2018. "Development of Subcutaneously Administered RNAi Therapeutic ARO-HBV for Chronic Hepatitis B Virus Infection." *Journal of Hepatology* 68 (April). J Hepatol: S18–19. doi:10.1016/S0168-8278(18)30255-1.

Wu, Ke, Juan He, Wenchen Pu, and Yong Peng. 2018. "The Role of Exportin-5 in MicroRNA Biogenesis and Cancer." *Genomics, Proteomics & Bioinformatics* 16 (2). Genom Proteom Bioinform: 120. doi:10.1016/J.GPB.2017.09.004.

Yang, Da, Yan Sun, Limei Hu, Hong Zheng, Ping Ji, Chad V. Pecot, Yanrui Zhao, et al. 2013. "Integrated Analyses Identify a Master MicroRNA Regulatory Network for the Mesenchymal Subtype in Serous Ovarian Cancer." *Cancer Cell* 23 (2). Cancer Cell: 186–99. doi:10.1016/J.CCR.2012.12.020.

Yang, Yongjie, Siddharth Kaul, Danhui Zhang, Vellareddy Anantharam, and Anumantha G. Kanthasamy. 2004. "Suppression of Caspase-3-Dependent Proteolytic Activation of Protein Kinase C Delta by Small Interfering RNA Prevents MPP+-Induced Dopaminergic Degeneration." *Molecular and Cellular Neurosciences* 25 (3). Mol Cell Neurosci: 406–21. doi:10.1016/J.MCN.2003.11.011.

Yoo, Byunghee, Amol Kavishwar, Alana Ross, Ping Wang, Doris P. Tabassum, Kornelia Polyak, Natalia Barteneva, et al. 2015. "Combining MiR-10b-Targeted Nanotherapy with Low-Dose Doxorubicin Elicits Durable Regressions of Metastatic Breast Cancer." *Cancer Research* 75 (20). Cancer Res: 4407–15. doi:10.1158/0008-5472.CAN-15-0888.

Zahir-Jouzdani, Forouhe, Fatemeh Mottaghitalab, Meshkat Dinarvand, and Fatemeh Atyabi. 2018. "SiRNA Delivery for Treatment of Degenerative Diseases, New Hopes and Challenges." *Journal of Drug Delivery Science and Technology* 45 (June). J Drug Deliv Sci Technol: 428–41. doi:10.1016/J.JDDST.2018.04.001.

Zandwijk, Nico van, Nick Pavlakis, Steven C. Kao, Anthony Linton, Michael J. Boyer, Stephen Clarke, Yennie Huynh, et al. 2017. "Safety and Activity of MicroRNA-Loaded Minicells in Patients with Recurrent Malignant Pleural Mesothelioma: A First-in-Man, Phase 1, Open-Label, Dose-Escalation Study." *The Lancet. Oncology* 18 (10). Lancet Oncol: 1386–96. doi:10.1016/S1470-2045(17)30621-6.

Zhang, Song, Zhujun Cheng, Yanan Wang, and Tianyu Han. 2021. "The Risks of MiRNA Therapeutics: In a Drug Target Perspective." *Drug Design, Development and Therapy* 15. Drug Des Devel Ther: 721. doi:10.2147/DDDT.S288859.

Zhong, Shen, Heiko Golpon, Patrick Zardo, and Jürgen Borlak. 2021. "MiRNAs in Lung Cancer. A Systematic Review Identifies Predictive and Prognostic MiRNA Candidates for Precision Medicine in Lung Cancer." *Translational Research: The Journal of Laboratory and Clinical Medicine* 230 (April). Transl Res: 164–96. doi:10.1016/J.TRSL.2020.11.012.

Zhou, Jia, Yixuan Zhao, Vera Simonenko, John J. Xu, Kai Liu, Deling Wang, Jingli Shi, et al. 2017. "Simultaneous Silencing of TGF-Β1 and COX-2 Reduces Human Skin Hypertrophic Scar through Activation of Fibroblast Apoptosis." *Oncotarget* 8 (46). Oncotarget: 80651–65. doi:10.18632/ONCOTARGET.20869.

Zovoilis, Athanasios, Hope Y. Agbemenyah, Roberto C. Agis-Balboa, Roman M. Stilling, Dieter Edbauer, Pooja Rao, Laurent Farinelli, et al. 2011. "MicroRNA-34c Is a Novel Target to Treat Dementias." *The EMBO Journal* 30 (20). EMBO J: 4299–4308. doi:10.1038/EMBOJ.2011.327.

Zuckerman, Jonathan E., Ismael Gritli, Anthony Tolcher, Jeremy D. Heidel, Dean Lim, Robert Morgan, Bartosz Chmielowski, Antoni Ribas, Mark E. Davis, and Yun Yen. 2014. "Correlating Animal and Human Phase Ia/Ib Clinical Data with CALAA-01, a Targeted, Polymer-Based Nanoparticle Containing SiRNA." *Proceedings of the National Academy of Sciences of the United States of America* 111 (31). Proc Natl Acad Sci U S A: 11449–54. doi:10.1073/PNAS.1411393111.

21 Nanocarriers for Targeted Drug Delivery

Rajeswari Tiwari[1] *and Sumit Kumar Pramanik*[1,*]
[1]CSIR-Central Salt & Marine Chemicals Research Institute, G.B. Marg, Bhavnagar, Gujarat, India
*Correspondence to: sumitpramanik@csmcri.res.in

21.1 INTRODUCTION

Materials that are nanoscale in size (1 nm or 10^{-9} m or one billionth of a metre) in any of its dimensions come under the category of nanomaterial, and the study and application of the substances and phenomenon with one or more dimensions between 1 nm to 100 nm is known as nanotechnology (Maynard 2007; Nikalje 2015). Nanotechnology in combination with biotechnology results in the outstandingly innovative area of nanobiotechnology (Banerji, Pramanik, Pal, et al. 2013; Maity et al. 2013; Pramanik et al. 2017; Pramanik, Sreedharan, et al. 2018; Banerji et al. 2012; H. Singh et al. 2019; Fakruddin, Hossain, and Afroz 2012). The current scenario of therapeutic (Banerji, Pramanik, Sanphui, et al. 2013; H. Singh et al. 2018; Hu et al. 2017; Mukherjee and Patra 2016), diagnostic (Agarwalla et al. 2018; Ashoka et al. 2020; H. Wang et al. 2015; Digesu et al. 2016) and theranostics (R. Tiwari, Shinde, et al. 2021; Gobbo et al. 2015; Swierczewska et al. 2016) has been majorly influenced by the field of nanobiotechnology (H. Singh et al. 2020). The increasing burden on human health with various chronic diseases increases the need for innovative research in the field of diagnosis, therapy and theranostics (Seneca et al. 2020; Kandoth et al. 2021; Sreedharan et al. 2021). This area of research in medicine mainly relies on nanotechnology and has great advantages over conventional techniques as they are small in size, helpful in the early detection of disease, repair tissue inside the body and deliver drugs. Drug delivery by classic methods has a lot of shortcomings such as nonspecific targeting, degradation in the gastrointestinal tract, low bioavailability, insolubility, etc. (Pramanik, Seneca, et al. 2018; Slenders et al. 2018). This draws the attention to the development of some novel drug delivery systems and nanocarriers that can be considered impeccable in this field (Pramanik and Das 2021). Nanocarriers are colloidal drug carrier systems used as a transportation module to deliver drugs to their site of action for therapy, diagnosis and theranostics. They have been extensively investigated in the past few decades. By virtue of their peculiar characteristics such as a high surface area-to-volume ratio, improved pharmacokinetics and biodistribution, improved stability (Raksha et al. 2022) and solubility (Kalepu and Nekkanti 2015), low toxicity (Dobrovolskaia and McNeil 2015; Choudhary et al. 2022) and controlled drug release, nanocarriers can be used as drug delivery systems (DDS). In the world of conventional medicine, nanocarriers mark an enormous boom for drug delivery (Jana et al. 2022; R. Tiwari, Banerjee, et al. 2021; K. Tiwari et al. 2022; Pasha et al. 2022). The problem of drug solubility can be resolved with the help of these nanocarriers; moreover, they increase bioavailability, target specific, protect sensitive drugs and minimize side effects without losing their activity (Dey et al. 2022; Fairbanks et al. 2021; Si et al. 2018). The critical function of the carrier is to perpetuate the drug action in vivo and preserve its optimal concentration at the targeted site (Si et al. 2019).

21.2 DIFFERENT NANOCARRIER PLATFORMS

Nanocarriers can be synthesised from a wide variety of organic and inorganic materials (Figure 21.1). Materials are being used to encapsulate and solubilize various chemotherapeutic agents

 DOI: 10.1201/9781003302131-22

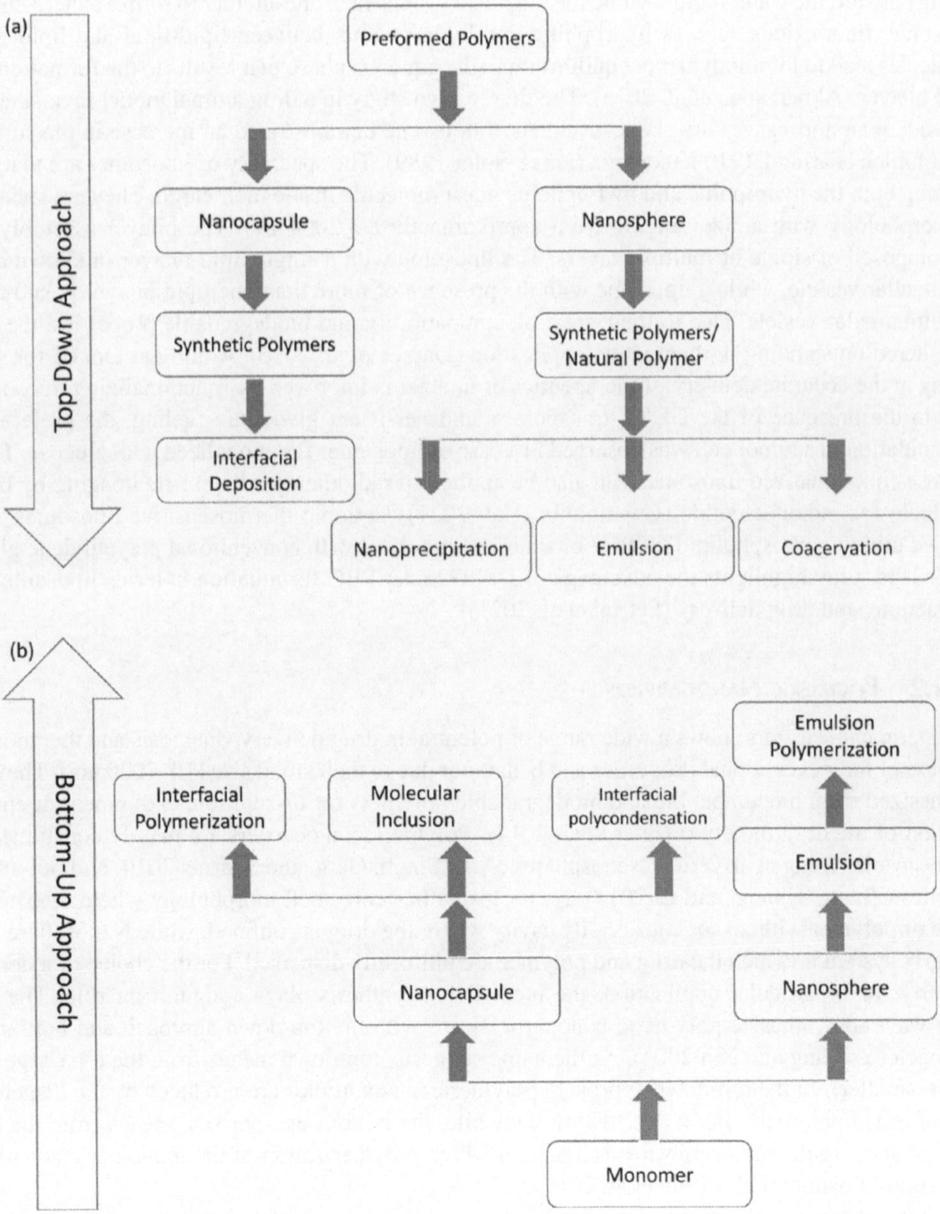

FIGURE 21.1 Method for the synthesis of polymeric nanomaterials (a) top-down approach (b) bottom-up approach.

for improved drug delivery in vivo. The various nanocarrier platforms used in the application of drug delivery are as follows.

21.2.1 Liposomes

Liposomes are microscopic vesicles used as a novel drug delivery system in which aqueous volume is entirely enclosed between phospholipid bilayers that form spontaneously when phospholipids are dispersed in water. When dispersed in water, they form lamellar sheets aligned by the polar head facing outward the water region, while the fatty acid groups face one another to form a sphere-shaped liposome. Interactions such as hydrophilic and hydrophobic, between lipid-lipid and lipid-water molecules lead to thermodynamic equilibrium in the aqueous phase that results in the formation of a lipid bilayer (Akbarzadeh et al. 2013). The first in vivo study in a drug animal model (cytosine arabinoside – an anti-cancer drug) encapsulated in liposome demonstrated an increase in the survival rate of mice bearing L1210 leukaemia (Perez-Soler 1989). The speciality of liposomes are that they can trap both the hydrophilic and hydrophobic guest molecule inside their cargo. They are spherical in morphology with a size ranging from approximately 80–200+ nm. The bilayer assembly can be composed of single or multiple layers. The liposome with a single lipid bilayer in known as an unilamellar vesicle, while a liposome with the presence of more than one lipid bilayer is known as a multilamellar vesicle. Due to their great biocompatibility and biodegradable properties, they are considered outstanding in therapeutic application (Daraee et al. 2016). Aiming at looked-for sites, aiding in the accurate delivery of therapeutics of interest is improved by functionalizing liposomes. Due to the presence of the GLUT transporters and persistent glycolytic fueling, the preferential accumulation in a tumor cell was observed in a case of liposomes functionalized with glucose. These glucose-functionalized liposomes can also be applied to radiotherapy/diagnostic imaging by being a vehicle for radionucleotide (Cvjetinović et al. 2021). Cationic thermosensitive liposomes were studied utilizing phospholipid DPPG2 or a surface modified with conventional polyethylene glycol PEG. This study highlights the advantage of DPPG2 over PEG formulation in terms liposome-cell interactions and drug delivery (Petrini et al. 2021).

21.2.2 Polymeric Nanocarriers

Polymeric nanocarriers shows a wide range of potential in drug delivery, diagnosis and theranostics because of their exceptional properties and behaviour due to their small size (10–1000 nm). They are synthesized from biocompatible and biodegradable polymers for dissolution, entrapment or encapsulation of the drug moiety (Daglar et al. 2014). Polymeric nanocarriers are mainly constituted of Nanocapsule (Deng et al. 2020), Nanospheres (A.K. Singh, Garg, and Sharma 2010) and polymeric micelles (Zhang, Huang, and Li 2014). Nanocapsule has core shell morphology where the shell is made of polymer with an aqueous or oily cavity where the drug is confined, while Nanospheres are a matrix system in which the drug and polymer are uniformly dispersed. For the choice of a desired property for a particular application, the method for synthesis plays a significant role. The two main ways to synthesize polymeric nanoparticles are with the top-down approach and bottom-up approach (Y. Wang and Xia 2004). As the name suggests, top-down means from the top (larger) to down (smaller). In the top-down approach, polymeric nanoparticles are produced by the dispersion of preformed polymers (Fu et al. 2018). Meanwhile, the bottom-up approach means from the bottom (smaller) to the top or upwards (larger), in which polymerization of the monomer leads to the formation of nanoparticles (Sun et al. 2017).

Cinnamaldehyde release was reported from a pH responsive dual-targeting polymeric nanocarrier to deliver ROS-activable prodrug and phenylboronic pinacol ester-caged doxorubicin. Surface functionalization was done with folic acid and mitochondria-targeting TPP (Xi et al. 2021). Similarly, Doxorubicin delivery was also reported by deep eutectic solvents grafted chitosan with a biotin nanocarrier (Pradeepkumar et al. 2021). For the active delivery, synthesis of a strong and steady targeting polymeric nanocarrier was conducted through a conjugation monoclonal antibody with biofunctionalized polymeric nanoparticles to enable the delivery of the fluorophore and the drug (Cano-Cortes et al. 2021).

Polymeric micelles are the self-assembled core shell nanostructure that belong to a set of nanosized colloids designed by an aqueous milieu of amphiphilic block copolymer with a size ranging from 10–100 nm. Due to their amphiphilic nature, block copolymers have both hydrophilic and hydrophobic counterparts. These block copolymers, when exposed to an aqueous solution above a certain concentration, form micelle. This concentration is called critical aggregation concentration

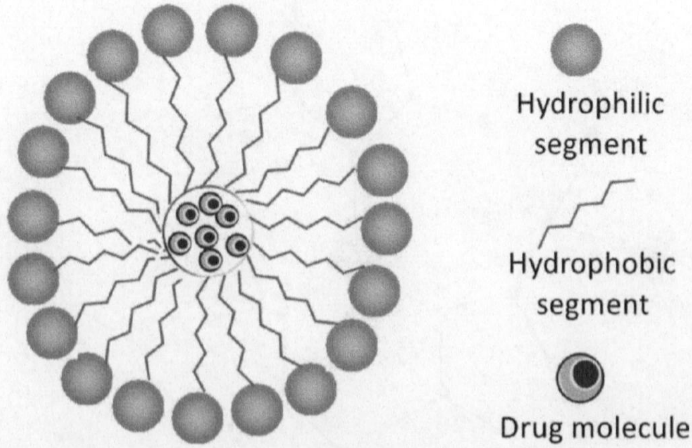

FIGURE 21.2 Schematic structures of polymeric micelles.

(CAC)/critical micelle concentration (CMC). Copolymer concentration should be over its CMC. At CMC, the hydrophobic segment of amphiphilic block copolymer is arranged in such a way as to decrease the contact with water molecules, which causes the formation of a core shell micellar morphology where the hydrophobic part of the amphiphilic block copolymer constitutes the core of the micelle, while the hydrophilic part forms the shell of the micelles. Thus, the hydrophobic segment facilitates the encapsulation of the hydrophobic drug, thus improving its stability and bioavailability. These drugs can be entrapped just by physical encapsulation or chemical attachment. Meanwhile, the hydrophilic shell region stabilizes the hydrophobic core, which makes the micelle the best candidate for in vivo administration. On dilution under CMC, the micelles disassemble into free unimers. Polymeric micelles show low CMC values, which makes them comparatively unresponsive to dilution; therefore, this results in enhanced circulation times (Croy and Kwon 2006).

Several stimuli-responsive polymeric micelles are reported in the literature. Glutathione Reductase-Sensitive polymeric micelle is synthesised using methoxypolyethylene glycol amine-glutathione-palmitic acid (mPEG-GSHn-PA). A loading capacity of 65% for dexamethasone was achieved. In the presence of a glutathione reductase enzyme, the disulphide bond formed by intermolecular oxidation of the thiol groups from glutathione is reduced to regenerate glutathione, thus resulting in the release of the encapsulated drug (Lima et al. 2021). Another enzyme-responsive disassembly of polymeric micelles was reported where NAD(P)H:quinone oxidoreductase-1 (NQO1), through intramolecular cyclization, produces a lactone-based group by reacting with trimethyl-locked quinone propionic acid (QPA) (Park et al. 2021).

21.2.3 Dendrimers

The name dendrimer is derived from the Greek work "Dendron" which resembles the meaning of a tree. A dendrimers is a three-dimensional, frequently branched and very systematic globular nanoscopic macromolecule with an easily modifiable surface and an average diameter of 1.5–14.5 nm (Madaan et al. 2014). They are also known as cascade molecules, arborols, dendritic molecules or nanometric architectures. Commonly, they are produced by using sugars, nucleotides and amino acids (Abbasi et al. 2014). The salient features of dendrimers that stand out from others are their excellent control over the molecular structure, low polydispersity index, cavities in the interior, various functionality at the surface, multivalency, a branched pattern, distinct molecular weight, etc. The layer between each cascade point is known as a generation. A typical dendrimer molecule comprises numerous generations that possess a chemical functional group at the exterior attached to an initiator core. There is an amplification in molecular size and surface functionality with the

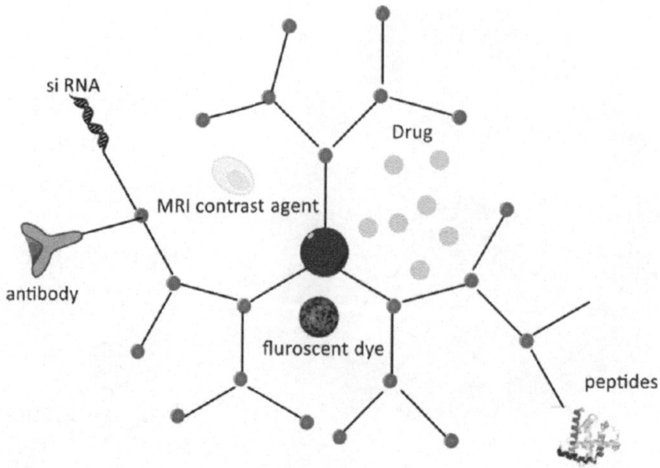

FIGURE 21.3 Schematic structures of dendrimers depicting various components.

rise in each generation. Drug delivery using a dendrimer platform can be performed in two ways. First, with the help of a hydrogen bond, chemical linkages and hydrophobic interactions, drugs can be loaded inside the cavities of the dendrimeric core. Second, to fulfil the required objectives, the drug or the targeting moiety get attached on the surface to transform functionality. When a drug is covalently bonded to the dendrimer core or the terminal groups, dendrimer-drug conjugates are formed. Trastuzumab was covalently attached to fluorinated fifth-generation G5 PAMAM. This can be utilized as a substitute to innate antibodies for tumor-targeting applications (Bartusik-Aebisher et al. 2021). Third-generation poly(amidoamine) (PAMAM) is modified with poly(2-methacryloyloxyethyl phosphorylcholine) (PMPC) to improve brain tumor-targeted drug delivery capability and to decrease PAMAM toxicity (Ban et al. 2021). Asialoglycoprotein receptor 1 (ASGP-R) on hepatocytes was targeted by using a biocompatible galactose-based generation 2 glycodendrimer built from GRAS galactose and tetraethylene glycol building blocks, which exhibited significant accumulation inside the liver (Sharma et al. 2021).

21.2.4 Carbon-Based Nanocarriers (CBN)

The essential constituent of CBNs is carbon atoms. The potential use of CBN in the imaging and diagnosis of cells and tissues is by virtue of their unique intrinsic properties. These intrinsic properties include an adjustable narrow emission spectrum and high photostability. Moreover, the properties of CBN's can be easily tuned by proper surface modification strategies. Furthermore, CBN's are considered excellent candidates for theranostic application due to their high surface area and mechanical and electrical properties. According to the shape and geometrical structures, carbon-based nanocarriers are classified as follows: nanotubes, graphene, nanodiamonds, fullerenes, porous carbon and carbon dots (Patel, Singh, and Kim 2019).

Carbon nanotubes (CNTs) primarily have tube or cylindrical morphology. They differ in length, diameters, number of layers, and chirality. CNTs can be characterized as single, double or multi-walled based on their structural configurations. Various methods are implemented for their preparation, which include chemical vapor deposition, arc discharge, laser ablation and high-pressure carbon monoxide disproportionation. Single-walled CNT constitutes an elongated enfolded graphene sheet that has a length to diameter ratio of about 1,000:1. They are 1D hollow and cylindrical shapes made with sp^2 hybridization. Meanwhile, double-walled CNT consists of a double-layer structure that includes two graphene sheets folded one on top of another along with two carbon nanotubes.

Multiwalled CNT comprises numerous layers of graphene sheets with a diameter varying between 5 and 50 nm. They possess more complicated electronic properties (Rathinavel, Priyadharshini, and Panda 2021). The interaction and bonding properties of armchair single-walled carbon nanotube (SWCNT), hydroxyl- a carboxyl-functionalized SWCNT (*f*-SWCNT) and doxorubicin were examined theoretically via a DFT study. Intermolecular interactions at critical bonding points in the aqueous and gas phases was conducted by the quantum theory of atoms in molecules (Karimzadeh, Safaei, and Jen 2021). Another study using Quantum chemical analysis-density functional methods was performed to analyse the ability to load and carry the 5-aminolevulinic acid anticancer drug of carbon nanotube, –COOH functionalized carbon nanotube and iron oxide nanoparticles (Rezaei et al. 2021).

Graphene consists of partially filled sp^2-orbits with a solo layer of carbon under and above the plane. Graphene is then further classified into two-class graphene nanoribbons and quantum dots. Graphene nanoribbons are synthesised by cutting graphene sheets in fine tiles that have a higher aspect ratio. They are prepared as finite, planar, and quasi-one-dimensional graphene structures. Graphene nanoribbons have great possible usage in the health field such as drug delivery, oncotherapy and DNA applications when functionalized with the oxide (Dutta and Pati 2010). Similarly, quantum dots are prepared by cutting the graphene into a piece in the range of 2–20 nm that has zero-dimensional geometry. Because of their nontoxic nature, ultra-small size, outstanding photostability and high-water solubility, they are used in several biomedical applications. Additionally, they have strong photoluminescence and photo-stability for their successful usage in bio-imaging (Tajik et al. 2020). GE11, an EGFR, is an antagonist peptide that is used to target nasopharyngeal carcinoma. To deliver cisplatin (CDDP) and doxorubicin (DOX), GE11-modified graphene quantum dots (GQDs@GE11) have been reported in the literature. FRET is very well observed here as an emission spectrum of GQDs (460 nm) that overlaps with the excitation spectrum of DOX (470 nm) (Yu et al. 2022). Likewise, to target cancer cells, hyaluronic acid was decorated on the surface of graphene quantum dots (Vahedi, Tabandeh, and Mahmoudifard 2022).

Nanodiamonds are CBN that are present in form of a 3D cubic lattice containing tetrahedrally bonded carbon atoms. On the basis of the synthesis process and size, they are divided into two types, specifically, detonation nanodiamonds (DNDs) and fluorescent nanodiamonds. DNDs are prepared by shocks with trinitrotoluene-like explosives and hexogen and have a size of around 5 nm. Fluorescent nanodiamonds are synthesised from high pressure and temperature and have a wider size distribution (Mochalin et al. 2012). Anti-HIV drug delivery to the brain was performed using nanodiamonds. One cART drug, efavirenz, was modified with unmodified nanodiamonds, -COOH surface modified ND and -NH2 surface-modified ND, thus showing the ability to transport anti-HIV-1 drugs across the blood-brain barrier (Roy et al. 2018).

Fullerenes are often called the molecular form of carbon, which comprises several atomic C_n clusters ($n > 20$) and constitutes carbon atoms on a spherical surface. Here, carbon atoms on spherical surface vertices form pentagons and hexagons. Fullerenes possess sp^2 hybridization where carbon atoms are linked by a covalent bond. C_{60} fullerene, also known as Buckminsterfullerene, is the most investigated type. Their capability to act as sensitizers for the photoproduction of singlet oxygen ($^1O_2$) ROS, which are exploited for blood sterilization and photodynamic cancer therapy, are well reported (Kazemzadeh and Mozafari 2019). The interaction between the drug hydroxyurea (HU) and the pure C70 and heterofullerene MC69 has been studied through DST calculations. It was found that hydroxyurea drugs were chemisorbed on heterofullerene SiC_{69}, BC_{69} and AlC_{69} and exhibited charge transfer and moderate adsorption energy (Li and Zhao 2021).

Carbon dots represent a class of CBN whose particle size is less than 10 nm and are quasi-spherical 0D carbon nanomaterials that possess fluorescence as their intrinsic property. They are used in a wide variety of applications such as explosives, chemical sensing, food safety, bioimaging, drug delivery and photocatalysis (M.L. Liu et al. 2019). Carbon dots were first discovered accidently from the purification of single-walled CNTs and were reported in 2004 (Xu et al. 2004). By tuning the shapes and sizes of CD and by doping heteroatoms such as oxygen, nitrogen, phosphorus, sulfur

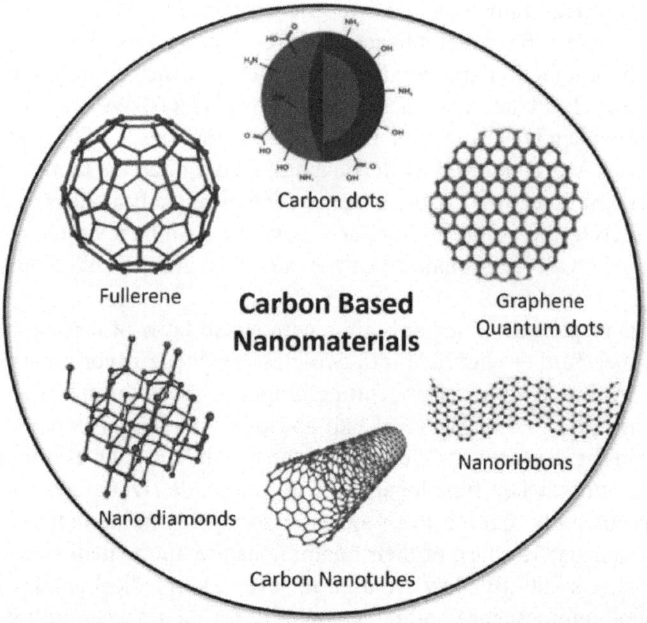

FIGURE 21.4 Schematic representation of various carbon-based nanomaterials used in drug delivery.

and boron, their photophysical and chemical properties can be dramatically altered. Compared to quantum dots, carbon dots have an upper hand in terms of better light stability, higher quantum yield, lower toxicity, abundant low-cost sources and excellent biocompatibility (J. Liu, Li, and Yang 2020). Glucose carbon dots conjugated with fluorescein were reported to cross the blood-brain barrier without any external targeting agent (Seven et al. 2021). Honeycomb-like nanoassemblies of carbon dots for the delivery of multiple therapeutic agents were synthesised, which are responsive against cancer-associated fibroblasts (Hou et al. 2021).

REFERENCES

Abbasi, Elham, Sedigheh Fekri Aval, Abolfazl Akbarzadeh, Morteza Milani, Hamid Tayefi Nasrabadi, Sang Woo Joo, Younes Hanifehpour, Kazem Nejati-Koshki, and Roghiyeh Pashaei-Asl. 2014. "Dendrimers: synthesis, applications, and properties." *Nanoscale Research Letters* 9 (1): 247. https://doi.org/10.1186/1556-276X-9-247.

Agarwalla, Hridesh, H. A. Anila, Firoj Ali, Swaraj Rashmi Pradhan, Bishwajit Ganguly, Sumit Kumar Pramanik, and Amitava Das. 2018. "Fluorescent chemodosimeter for quantification of cystathionine-γ-synthase activity in plant extracts and imaging of endogenous biothiols." *Chemical Communications* 54 (65): 9079–9082. https://doi.org/10.1039/C8CC04296A.

Akbarzadeh, Abolfazl, Rogaie Rezaei-Sadabady, Soodabeh Davaran, Sang Woo Joo, Nosratollah Zarghami, Younes Hanifehpour, Mohammad Samiei, Mohammad Kouhi, and Kazem Nejati-Koshki. 2013. "Liposome: classification, preparation, and applications." *Nanoscale Research Letters* 8 (1): 102. https://doi.org/10.1186/1556-276X-8-102.

Ashoka, Anila Hoskere, Firoj Ali, Rajeshwari Tiwari, Rina Kumari, Sumit Kumar Pramanik, and Amitava Das. 2020. "Recent Advances in Fluorescent Probes for Detection of HOCl and HNO." *ACS Omega* 5 (4): 1730–1742. https://doi.org/10.1021/acsomega.9b03420.

Ban, Jiamin, Sidi Li, Qi Zhan, Xuepin Li, Huike Xing, Ning Chen, Lixia Long, Xin Hou, Jin Zhao, and Xubo Yuan. 2021. "PMPC Modified PAMAM Dendrimer Enhances Brain Tumor-Targeted Drug Delivery." *Macromolecular Bioscience* 21 (4): 2000392. https://doi.org/10.1002/mabi.202000392.

Banerji, Biswadip, Sumit Kumar Pramanik, Samir Mandal, Nakul Chandra Maiti, and Keya Chaudhuri. 2012. "Synthesis, characterization and cytotoxicity study of magnetic (Fe_3O_4) nanoparticles and their drug conjugate." *RSC Advances* 2 (6): 2493–2497. https://doi.org/10.1039/C2RA01118B.

Banerji, Biswadip, Sumit Kumar Pramanik, Uttam Pal, and Nakul Chandra Maiti. 2013. "Dipeptide derived from benzylcystine forms unbranched nanotubes in aqueous solution." *Journal of Nanostructure in Chemistry* 3 (1): 12. https://doi.org/10.1186/2193-8865-3-12.

Banerji, Biswadip, Sumit Kumar Pramanik, Priyankar Sanphui, Sameer Nikhar, and Subhas C. Biswas. 2013. "Synthesis and cytotoxicity studies of novel triazolo-benzoxazepine as new anticancer agents." *Chemical Biology & Drug Design* 82 (4): 401–409. https://doi.org/10.1111/cbdd.12164.

Bartusik-Aebisher, Dorota, Grzegorz Chrzanowski, Zuzanna Bober, and David Aebisher. 2021. "An analytical study of trastuzumab-dendrimer-fluorine drug delivery system in breast cancer therapy in vitro." *Biomedicine & Pharmacotherapy* 133: 111053. https://doi.org/10.1016/j.biopha.2020.111053. www.sciencedirect.com/science/article/pii/S0753332220312452.

Cano-Cortes, Maria Victoria, Patricia Altea-Manzano, Jose Antonio Laz-Ruiz, Juan Diego Unciti-Broceta, Francisco Javier Lopez-Delgado, Jose Manuel Espejo-Roman, Juan Jose Diaz-Mochon, and Rosario M. Sanchez-Martin. 2021. "An effective polymeric nanocarrier that allows for active targeting and selective drug delivery in cell coculture systems." *Nanoscale* 13 (6): 3500–3511. https://doi.org/10.1039/D0NR07145E.

Choudhary, Preety, Sushama Biswas, Noufal Kandoth, Deepak Tayde, Abhishek Chatterjee, Samit Chattopadhyay, Amitava Das, Snehasikta Swarnakar, and Sumit Kumar Pramanik. 2022. "Graphene quantum dots alleviate ROS-mediated gastric damage." *iScience* 25 (4): 104062. https://doi.org/10.1016/j.isci.2022.104062. www.sciencedirect.com/science/article/pii/S2589004222003327.

Croy, S. R., and G. S. Kwon. 2006. "Polymeric Micelles for Drug Delivery." *Current Pharmaceutical Design* 12 (36): 4669–4684. http://dx.doi.org/10.2174/138161206779026245. www.eurekaselect.com/node/58294/article.

Cvjetinović, Đorđe, Željko Prijović, Drina Janković, Magdalena Radović, Marija Mirković, Zorana Milanović, Miloš Mojović, Đani Škalamera, and Sanja Vranješ-Đurić. 2021. "Bioevaluation of glucose-modified liposomes as a potential drug delivery system for cancer treatment using 177-Lu radiotracking." *Journal of Controlled Release* 332: 301–311. https://doi.org/10.1016/j.jconrel.2021.03.006. www.sciencedirect.com/science/article/pii/S0168365921001176.

Daglar, B., E. Ozgur, M. E. Corman, L. Uzun, and G. B. Demirel. 2014. "Polymeric nanocarriers for expected nanomedicine: current challenges and future prospects." *RSC Advances* 4 (89): 48639–48659. https://doi.org/10.1039/C4RA06406B.

Daraee, Hadis, Ali Etemadi, Mohammad Kouhi, Samira Alimirzalu, and Abolfazl Akbarzadeh. 2016. "Application of liposomes in medicine and drug delivery." *Artificial Cells, Nanomedicine, and Biotechnology* 44 (1): 381–391. https://doi.org/10.3109/21691401.2014.953633.

Deng, Siyuan, Maria R. Gigliobianco, Roberta Censi, and Piera Di Martino. 2020. "Polymeric nanocapsules as nanotechnological alternative for drug delivery system: current status, challenges and opportunities." *Nanomaterials* 10 (5). https://doi.org/10.3390/nano10050847.

Dey, Ananta, Manisha Yadav, Deepak Kumar, Anik Kumar Dey, Sweety Samal, Subhash Tanwar, Debrupa Sarkar, Sumit Kumar Pramanik, Susmita Chaudhuri, and Amitava Das. 2022. "A combination therapy strategy for treating antibiotic resistant biofilm infection using a guanidinium derivative and nanoparticulate Ag(0) derived hybrid gel conjugate." *Chemical Science* 13 (34): 10103–10118. https://doi.org/10.1039/D2SC02980D.

Digesu, Christopher S., Sophie C. Hofferberth, Mark W. Grinstaff, and Yolonda L. Colson. 2016. "From diagnosis to treatment: clinical applications of nanotechnology in thoracic surgery." *Thoracic Surgery Clinics* 26 (2): 215–228. https://doi.org/10.1016/j.thorsurg.2015.12.009. https://pubmed.ncbi.nlm.nih.gov/27112260 www.ncbi.nlm.nih.gov/pmc/articles/PMC4851727/.

Dobrovolskaia, Marina A., and Scott E. McNeil. 2015. "Strategy for selecting nanotechnology carriers to overcome immunological and hematological toxicities challenging clinical translation of nucleic acid-based therapeutics." *Expert Opinion on Drug Delivery* 12 (7): 1163–1175. https://doi.org/10.1517/17425247.2015.1042857.

Dutta, Sudipta, and Swapan K. Pati. 2010. "Novel properties of graphene nanoribbons: a review." *Journal of Materials Chemistry* 20 (38): 8207–8223. https://doi.org/10.1039/C0JM00261E.

Fairbanks, Simon D., Sumit Kumar Pramanik, Jim A. Thomas, Amitiva Das, and Nicolas Martin. 2021. "The management of mercury from dental amalgam in wastewater effluent." *Environmental Technology Reviews* 10 (1): 213–223. https://doi.org/10.1080/21622515.2021.1960642.

Fakruddin, Md, Zakir Hossain, and Hafsa Afroz. 2012. "Prospects and applications of nanobiotechnology: a medical perspective." *Journal of Nanobiotechnology* 10 (1): 31. https://doi.org/10.1186/1477-3155-10-31.

Fu, Xinxin, Jingxuan Cai, Xiang Zhang, Wen-Di Li, Haixiong Ge, and Yong Hu. 2018. "Top-down fabrication of shape-controlled, monodisperse nanoparticles for biomedical applications." *Advanced Drug*

Delivery Reviews 132: 169–187. https://doi.org/10.1016/j.addr.2018.07.006. www.sciencedirect.com/science/article/pii/S0169409X18301704.

Gobbo, Oliviero L., Kristine Sjaastad, Marek W. Radomski, Yuri Volkov, and Adriele Prina-Mello. 2015. "Magnetic nanoparticles in cancer theranostics." *Theranostics* 5 (11): 1249–1263. https://doi.org/10.7150/thno.11544. https://pubmed.ncbi.nlm.nih.gov/26379790 www.ncbi.nlm.nih.gov/pmc/articles/PMC4568452/.

Hou, Lin, Dandan Chen, Ruiting Wang, Ruibing Wang, Huijuan Zhang, Zhenzhong Zhang, Zhihong Nie, and Siyu Lu. 2021. "Transformable honeycomb-like nanoassemblies of carbon dots for regulated multisite delivery and enhanced antitumor chemoimmunotherapy." *Angewandte Chemie International Edition* 60 (12): 6581–6592. https://doi.org/10.1002/anie.202014397.

Hu, Xiaomeng, Tingting Wu, Yuling Bao, and Zhiping Zhang. 2017. "Nanotechnology based therapeutic modality to boost anti-tumor immunity and collapse tumor defense." *Journal of Controlled Release* 256: 26–45. https://doi.org/10.1016/j.jconrel.2017.04.026. www.sciencedirect.com/science/article/pii/S016836591730545X.

Jana, Biswajit, Aroni Chatterjee, Debsopan Roy, Shubhankar Ghorai, Dipika Pan, Sumit Kumar Pramanik, Nilanjan Chakraborty, and Jhuma Ganguly. 2022. "Chitosan/benzyloxy-benzaldehyde modified ZnO nano template having optimized and distinct antiviral potency to human cytomegalovirus." *Carbohydrate Polymers* 278: 118965. https://doi.org/10.1016/j.carbpol.2021.118965. www.sciencedirect.com/science/article/pii/S0144861721013527.

Kalepu, Sandeep, and Vijaykumar Nekkanti. 2015. "Insoluble drug delivery strategies: review of recent advances and business prospects." *Acta Pharmaceutica Sinica B* 5 (5): 442–453. https://doi.org/10.1016/j.apsb.2015.07.003. www.sciencedirect.com/science/article/pii/S2211383515001069.

Kandoth, Noufal, Subhadip Barman, Atin Chatterjee, Sandip Sarkar, Anik Kumar Dey, Sumit Kumar Pramanik, and Amitava Das. 2021. "Photoactive lanthanide-based upconverting nanoclusters for antimicrobial applications." *Advanced Functional Materials* 31 (43): 2104480. https://doi.org/10.1002/adfm.202104480. https://onlinelibrary.wiley.com/doi/abs/10.1002/adfm.202104480.

Karimzadeh, Sina, Babak Safaei, and Tien-Chien Jen. 2021. "Theorical investigation of adsorption mechanism of doxorubicin anticancer drug on the pristine and functionalized single-walled carbon nanotube surface as a drug delivery vehicle: A DFT study." *Journal of Molecular Liquids* 322: 114890. https://doi.org/10.1016/j.molliq.2020.114890. www.sciencedirect.com/science/article/pii/S0167732220371324.

Kazemzadeh, Houman, and Masoud Mozafari. 2019. "Fullerene-based delivery systems." *Drug Discovery Today* 24 (3): 898–905. https://doi.org/10.1016/j.drudis.2019.01.013. www.sciencedirect.com/science/article/pii/S1359644618302964.

Li, Wei, and Tinghu Zhao. 2021. "Hydroxyurea anticancer drug adsorption on the pristine and doped C70 fullerene as potential carriers for drug delivery." *Journal of Molecular Liquids* 340: 117226. https://doi.org/10.1016/j.molliq.2021.117226. www.sciencedirect.com/science/article/pii/S0167732221019504.

Lima, Ana Cláudia, Rui L. Reis, Helena Ferreira, and Nuno M. Neves. 2021. "Glutathione reductase-sensitive polymeric micelles for controlled drug delivery on arthritic diseases." *ACS Biomaterials Science & Engineering* 7 (7): 3229–3241. https://doi.org/10.1021/acsbiomaterials.1c00412.

Liu, Junjun, Rui Li, and Bai Yang. 2020. "Carbon dots: a new type of carbon-based nanomaterial with wide applications." *ACS Central Science* 6 (12): 2179–2195. https://doi.org/10.1021/acscentsci.0c01306.

Liu, Meng Li, Bin Tingyu Chen, Chun Mei Li, and Cheng Zhi Huang. 2019. "Carbon dots: synthesis, formation mechanism, fluorescence origin and sensing applications." *Green Chemistry* 21 (3): 449–471. https://doi.org/10.1039/C8GC02736F.

Madaan, Kanika, Sandeep Kumar, Neelam Poonia, Viney Lather, and Deepti Pandita. 2014. "Dendrimers in drug delivery and targeting: drug-dendrimer interactions and toxicity issues." *Journal of Pharmacy & Bioallied Sciences* 6 (3): 139–150. https://doi.org/10.4103/0975-7406.130965. https://pubmed.ncbi.nlm.nih.gov/25035633. www.ncbi.nlm.nih.gov/pmc/articles/PMC4097927/.

Maity, Mritunjoy, Sumit Kumar Pramanik, Uttam Pal, Biswadip Banerji, and Nakul Chandra Maiti. 2013. "Copper(I) oxide nanoparticle and tryptophan as its biological conjugate: a modulation of cytotoxic effects." *Journal of Nanoparticle Research* 16 (1): 2179. https://doi.org/10.1007/s11051-013-2179-z.

Maynard, Andrew D. 2007. "Nanotechnology: the next big thing, or much ado about nothing?" *The Annals of Occupational Hygiene* 51 (1): 1–12. https://doi.org/10.1093/annhyg/mel071.

Mochalin, Vadym N., Olga Shenderova, Dean Ho, and Yury Gogotsi. 2012. "The properties and applications of nanodiamonds." *Nature Nanotechnology* 7 (1): 11–23. https://doi.org/10.1038/nnano.2011.209.

Mukherjee, Sudip, and Chitta Ranjan Patra. 2016. "Therapeutic application of anti-angiogenic nanomaterials in cancers." *Nanoscale* 8 (25): 12444–12470. https://doi.org/10.1039/C5NR07887C.

Nikalje, Anna. 2015. "Nanotechnology and its applications in medicine." *Medicinal Chemistry* 5: 5: 81–89. https://doi.org/10.4172/2161-0444.1000247.

Park, Jaehyun, Seokhee Jo, Yeong Mi Lee, Gurusamy Saravanakumar, Junseok Lee, Dongsik Park, and Won Jong Kim. 2021. "Enzyme-triggered disassembly of polymeric micelles by controlled depolymerization via cascade cyclization for anticancer drug delivery." *ACS Applied Materials & Interfaces* 13 (7): 8060–8070. https://doi.org/10.1021/acsami.0c22644.

Pasha, Sheik Saleem, Amrita Banerjee, Sreejesh Sreedharan, Soumendra Singh, Noufal Kandoth, Katherine A. Vallis, Samir Kumar Pal, Sumit Kumar Pramanik, and Amitava Das. 2022. "Ultrasensitive reagent for ratiometric detection and detoxification of iAsIII in water and mitochondria." *Inorganic Chemistry* 61 (33): 13115–13124. https://doi.org/10.1021/acs.inorgchem.2c01761.

Patel, Kapil D., Rajendra K. Singh, and Hae-Won Kim. 2019. "Carbon-based nanomaterials as an emerging platform for theranostics." *Materials Horizons* 6 (3): 434–469. https://doi.org/10.1039/C8MH00966J.

Perez-Soler, R. 1989. "Liposomes as carriers of antitumor agents: toward a clinical reality." *Cancer Treatment Reviews* 16 (2): 67–82. https://doi.org/10.1016/0305-7372(89)90011-X. www.sciencedirect.com/science/article/pii/030573728990011X.

Petrini, Matteo, Wouter J. M. Lokerse, Agnieszka Mach, Martin Hossann, Olivia M. Merkel, and Lars H. Lindner. 2021. "Effects of surface charge, PEGylation and functionalization with dipalmitoylphosphatidyldiglycerol on liposome-cell interactions and local drug delivery to solid tumors via thermosensitive liposomes." *International Journal of Nanomedicine* 16: 4045–4061. https://doi.org/10.2147/IJN.S305106. https://pubmed.ncbi.nlm.nih.gov/34163158. www.ncbi.nlm.nih.gov/pmc/articles/PMC8214027/.

Pradeepkumar, Periyakaruppan, Mariappan Rajan, Hesham S. Almoallim, and Sulaiman Ali Alharbi. 2021. "Targeted delivery of doxorubicin in HeLa cells using self-assembled polymeric nanocarriers guided by deep eutectic solvents." *ChemistrySelect* 6 (28): 7232–7241. https://doi.org/10.1002/slct.202000036.

Pramanik, Sumit Kumar, and Amitva Das. 2021. "Fluorescent probes for imaging bioactive species in subcellular organelles." *Chemical Communications* 57 (91): 12058–12073. https://doi.org/10.1039/D1CC04273D.

Pramanik, Sumit Kumar, Senne Seneca, Martijn Peters, Lien D'Olieslaeger, Gunter Reekmans, Dirk Vanderzande, Peter Adriaensens, and Anitha Ethirajan. 2018. "Morphology-dependent pH-responsive release of hydrophilic payloads using biodegradable nanocarriers." *RSC Advances* 8 (64): 36869–36878. https://doi.org/10.1039/C8RA07066K.

Pramanik, Sumit Kumar, Sreejesh Sreedharan, Harwinder Singh, Nicola H. Green, Carl Smythe, Jim A. Thomas, and Amitava Das. 2017. "Imaging cellular trafficking processes in real time using lysosome targeted up-conversion nanoparticles." *Chemical Communications* 53 (94): 12672–12675. https://doi.org/10.1039/C7CC08185E.

Pramanik, Sumit Kumar, Sreejesh Sreedharan, Harwinder Singh, Mohsina Khan, Karishma Tiwari, Anjali Shiras, Carl Smythe, Jim A. Thomas, and Amitava Das. 2018. "Mitochondria targeting non-isocyanate-based polyurethane nanocapsules for enzyme-triggered drug release." *Bioconjugate Chemistry* 29 (11): 3532–3543. https://doi.org/10.1021/acs.bioconjchem.8b00460.

Raksha, Kumari, Noufal Kandoth, Shresth Gupta, Subhadeep Gupta, Sumit Kumar Pramanik, and Amitava Das. 2022. "Modulating resonance energy transfer with supramolecular control in a layered hybrid perovskite and chromium photosensitizer assembly." *ACS Applied Materials & Interfaces*. https://doi.org/10.1021/acsami.2c09281.

Rathinavel, S., K. Priyadharshini, and Dhananjaya Panda. 2021. "A review on carbon nanotube: an overview of synthesis, properties, functionalization, characterization, and the application." *Materials Science and Engineering: B* 268: 115095. https://doi.org/10.1016/j.mseb.2021.115095. www.sciencedirect.com/science/article/pii/S0921510721000556.

Rezaei, Azadeh, Ali Morsali, Mohammad Reza Bozorgmehr, and Marjan Nasrabadi. 2021. "Quantum chemical analysis of 5-aminolevulinic acid anticancer drug delivery systems: carbon nanotube, –COOH functionalized carbon nanotube and iron oxide nanoparticle." *Journal of Molecular Liquids* 340: 117182. https://doi.org/10.1016/j.molliq.2021.117182. www.sciencedirect.com/science/article/pii/S0167732221019061.

Roy, Upal, Vadym Drozd, Andriy Durygin, Jesse Rodriguez, Paul Barber, Venkata Atluri, Xiaohua Liu, Thomas G. Voss, Surendra Saxena, and Madhavan Nair. 2018. "Characterization of nanodiamond-based anti-HIV drug delivery to the brain." *Scientific Reports* 8 (1): 1603. https://doi.org/10.1038/s41598-017-16703-9.

Seneca, Senne, Sumit Kumar Pramanik, Lien D'Olieslaeger, Gunter Reekmans, Dirk Vanderzande, Peter Adriaensens, and Anitha Ethirajan. 2020. "Nanocapsules with stimuli-responsive moieties for controlled release employing light and enzymatic triggers." *Materials Chemistry Frontiers* 4 (7): 2103–2112. https://doi.org/10.1039/D0QM00244E.

Seven, Elif S., Yasin B. Seven, Yiqun Zhou, Sijan Poudel-Sharma, Juan J. Diaz-Rucco, Emel Kirbas Cilingir, Gordon S. Mitchell, J. David Van Dyken, and Roger M. Leblanc. 2021. "Crossing the blood–brain barrier with carbon dots: uptake mechanism and in vivo cargo delivery." *Nanoscale Advances* 3 (13): 3942–3953. https://doi.org/10.1039/D1NA00145K.

Sharma, Rishi, Joshua E. Porterfield, Hyoung-Tae An, Ambar Scarlet Jimenez, Seulki Lee, Sujatha Kannan, Anjali Sharma, and Rangaramanujam M. Kannan. 2021. "Rationally designed galactose dendrimer for hepatocyte-specific targeting and intracellular drug delivery for the treatment of liver disorders." *Biomacromolecules* 22 (8): 3574–3589. https://doi.org/10.1021/acs.biomac.1c00649.

Si, Mrinal Kanti, Sumit Kumar Pramanik, Vinayak Hingu, and Bishwajit Ganguly. 2018. "Computational predictions turning the isomers of alanine to generate distinct morphs of free-flowing salt crystals." *Physical Chemistry Chemical Physics* 20 (25): 17125–17131. https://doi.org/10.1039/C8CP01697F.

Si, Mrinal Kanti, Sumit Kumar Pramanik, Vinayak Hingu, and Bishwajit Ganguly. 2019. "2-Amino-pentanoic acid and 1-Pyridiniumylacetate are new additives to generate distinct morphs of salt crystals: a combined computational and experimental study." *Journal of Molecular Structure* 1175: 728–733. https://doi.org/10.1016/j.molstruc.2018.08.024. www.sciencedirect.com/science/article/pii/S0022286018309670.

Singh, Amit K., G. Garg, and P. Sharma. 2010. "Nanospheres: a novel approach for targeted drug delivery system." *International Journal of Pharmaceutical Sciences Review and Research* 5 (3): 84–88.

Singh, Harwinder, Sreejesh Sreedharan, Esteban Oyarzabal, Tufan Singha Mahapatra, Nicola Green, Yen-Yu Ian Shih, Manasmita Das, Jim A. Thomas, Sumit Kumar Pramanik, and Amitava Das. 2020. "Mitochondriotropic lanthanide nanorods: implications for multimodal imaging." *Chemical Communications* 56 (57): 7945–7948. https://doi.org/10.1039/D0CC02698K.

Singh, Harwinder, Sreejesh Sreedharan, Karishma Tiwari, Nicola H. Green, Carl Smythe, Sumit Kumar Pramanik, Jim A. Thomas, and Amitava Das. 2019. "Two photon excitable graphene quantum dots for structured illumination microscopy and imaging applications: lysosome specificity and tissue-dependent imaging." *Chemical Communications* 55 (4): 521–524. https://doi.org/10.1039/C8CC08610A.

Singh, Harwinder, Sreejesh Sreedharan, Rajeshwari Tiwari, Christa Walther, Carl Smythe, Sumit Kumar Pramanik, Jim A. Thomas, and Amitava Das. 2018. "A fluorescent chemodosimeter for organelle-specific imaging of nucleoside polyphosphate dynamics in living cells." *Crystal Growth & Design* 18 (11): 7199–7206. https://doi.org/10.1021/acs.cgd.8b01409.

Slenders, Eli, Senne Seneca, Sumit Kumar Pramanik, Nick Smisdom, Peter Adriaensens, Martin vandeVen, Anitha Ethirajan, and Marcel Ameloot. 2018. "Dynamics of the phospholipid shell of microbubbles: a fluorescence photoselection and spectral phasor approach." *Chemical Communications* 54 (38): 4854–4857. https://doi.org/10.1039/C8CC01012A.

Sreedharan, Sreejesh, Rajeshwari Tiwari, Deepak Tyde, Stephen O. Aderinto, Sumit Kumar Pramanik, Amitava Das, and James A. Thomas. 2021. "Nanocarriers used as probes for super-resolution microscopy." *Materials Chemistry Frontiers* 5 (3): 1268–1282. https://doi.org/10.1039/D0QM00684J.

Sun, Pei, Dong Chen, Hongping Deng, Nan Wang, Ping Huang, Xin Jin, and Xinyuan Zhu. 2017. " 'Bottom-up' construction of multi-polyprodrug-arm hyperbranched amphiphiles for cancer therapy." *Bioconjugate Chemistry* 28. https://doi.org/10.1021/acs.bioconjchem.7b00146.

Swierczewska, M., H. S. Han, K. Kim, J. H. Park, and S. Lee. 2016. "Polysaccharide-based nanoparticles for theranostic nanomedicine." *Advanced Drug Delivery Reviews* 99 (Pt A): 70–84. https://doi.org/10.1016/j.addr.2015.11.015. https://pubmed.ncbi.nlm.nih.gov/26639578 www.ncbi.nlm.nih.gov/pmc/articles/PMC4798864/.

Tajik, Somayeh, Zahra Dourandish, Kaiqiang Zhang, Hadi Beitollahi, Quyet Van Le, Ho Won Jang, and Mohammadreza Shokouhimehr. 2020. "Carbon and graphene quantum dots: a review on syntheses, characterization, biological and sensing applications for neurotransmitter determination." *RSC Advances* 10 (26): 15406–15429. https://doi.org/10.1039/D0RA00799D.

Tiwari, Karishma, Solagna Modak, Pulak Sarkar, Santanu Ray, Vasista Adupa, K. Anki Reddy, Sumit Kumar Pramanik, Amitava Das, and Santanu Karan. 2022. "Interfacial synthesis of large-area ultrathin polyimine nanofilms as molecular separation membrane." *iScience* 25 (4): 104027. https://doi.org/10.1016/j.isci.2022.104027. www.sciencedirect.com/science/article/pii/S2589004222002978.

Tiwari, Rajeshwari, Saswati Banerjee, Deepak Tyde, Krishna Das Saha, Anitha Ethirajan, Niladri Mukherjee, Samit Chattopadhy, Sumit Kumar Pramanik, and Amitava Das. 2021. "Redox-responsive nanocapsules for the spatiotemporal release of miltefosine in lysosome: protection against leishmania." *Bioconjugate Chemistry* 32 (2): 245–253. https://doi.org/10.1021/acs.bioconjchem.0c00667.

Tiwari, Rajeshwari, Prashant S. Shinde, Sreejesh Sreedharan, Anik Kumar Dey, Katherine A. Vallis, Santosh B. Mhaske, Sumit Kumar Pramanik, and Amitava Das. 2021. "Photoactivatable prodrug for simultaneous release of mertansine and CO along with a BODIPY derivative as a luminescent marker in mitochondria: a proof of concept for NIR image-guided cancer therapy." *Chemical Science* 12 (7): 2667–2673. https://doi.org/10.1039/D0SC06270G.

Vahedi, Nasrin, Fatemeh Tabandeh, and Matin Mahmoudifard. 2022. "Hyaluronic acid–graphene quantum dot nanocomposite: potential target drug delivery and cancer cell imaging." *Biotechnology and Applied Biochemistry* 69 (3): 1068–1079. https://doi.org/10.1002/bab.2178.

Wang, Haolu, Camilla A. Thorling, Xiaowen Liang, Kim R. Bridle, Jeffrey E. Grice, Yian Zhu, Darrell H. G. Crawford, Zhi Ping Xu, Xin Liu, and Michael S. Roberts. 2015. "Diagnostic imaging and therapeutic application of nanoparticles targeting the liver." *Journal of Materials Chemistry B* 3 (6): 939–958. https://doi.org/10.1039/C4TB01611D.

Wang, Yuliang, and Younan Xia. 2004. "Bottom-up and top-down approaches to the synthesis of monodispersed spherical colloids of low melting-point metals." *Nano Letters* 4 (10): 2047–2050. https://doi.org/10.1021/nl048689j.

Xi, Longchang, Jingbo Wang, Yuheng Wang, and Zhishen Ge. 2021. "Dual-targeting polymeric nanocarriers to deliver ROS-responsive prodrugs and combat multidrug resistance of cancer cells." *Macromolecular Bioscience* 21 (9): 2100091. https://doi.org/10.1002/mabi.202100091.

Xu, Xiaoyou, Robert Ray, Yunlong Gu, Harry J. Ploehn, Latha Gearheart, Kyle Raker, and Walter A. Scrivens. 2004. "Electrophoretic analysis and purification of fluorescent single-walled carbon nanotube fragments." *Journal of the American Chemical Society* 126 (40): 12736–12737. https://doi.org/10.1021/ja040082h.

Yu, Chaosheng, Zhen Long, Qianhui Qiu, Fang Liu, Yiming Xu, Tao Zhang, Rui Guo, Wen Zhong, Shuixian Huang, and Shuaijun Chen. 2022. "Graphene quantum dots-based targeted nanoprobes detecting drug delivery, imaging, and enhanced chemotherapy of nasopharyngeal carcinoma." *Bioengineering & Translational Medicine* 7 (2): e10270. https://doi.org/10.1002/btm2.10270.

Zhang, Yifei, Yixian Huang, and Song Li. 2014. "Polymeric micelles: nanocarriers for cancer-targeted drug delivery." *AAPS PharmSciTech* 15 (4): 862–871. https://doi.org/10.1208/s12249-014-0113-z. https://pubmed.ncbi.nlm.nih.gov/24700296. www.ncbi.nlm.nih.gov/pmc/articles/PMC4113619/.

22 Theranostic Application of Nanomaterial in Human Health and Disease

Anupam Maity[1], Rajdip Misra[1], Shubham Kundu[1], Kaushik Bera[1], Pranga Prava Mandal[1], Uttam Pal[1,] and Nakul C. Maiti[1,*]*
[1]CSIR – Indian Institute of Chemical Biology, Kolkata, India
*Correpondence to: uttam2707@gmail.com (U.P.) and ncmaiti@iicb.res.in (N.C.M.)

22.1 INTRODUCTION

Nanomaterials are usually nanosized (1–100 nm) clusters of atoms or molecules and show physical and chemical behaviors that are much different from the bulk content of the material from which they are made. Due to the presence of a large functionalized surface-to-volume ratio, they possess unique physical properties. Furthermore, size-dependent clustering of the electron environment provides nanomaterials certain chemical abilities and optical behavior. These properties make them capable of hosting different molecules for cellular binding and as a therapeutic agent for drug delivery and diagnostic agent for imaging. Drug molecules can be covalently attached on nanoparticle (NP) surfaces and act as a drug carrier vehicle (Figure 22.1). Nanomaterials can be used to utilize precision medicine for cancer treatment, viral diseases, neurodegenerative diseases, etc. (Pal and Pramanik 2018). In this chapter, we discuss the application of nanomaterials in drug delivery, imaging and therapeutics to human diseases.

For therapeutic uses, one important expected property is that NPs should have the ability to cross the biological (blood-brain) barrier. Therefore, the NPs for biomedical applications are engineered with enhanced capability to cross biological barriers and increase disease-specific responsiveness. For this purpose, several types of NPs are designed, and the three major categories are lipid-based, polymeric and inorganic nanoparticles, which show potential in biomedical applications.

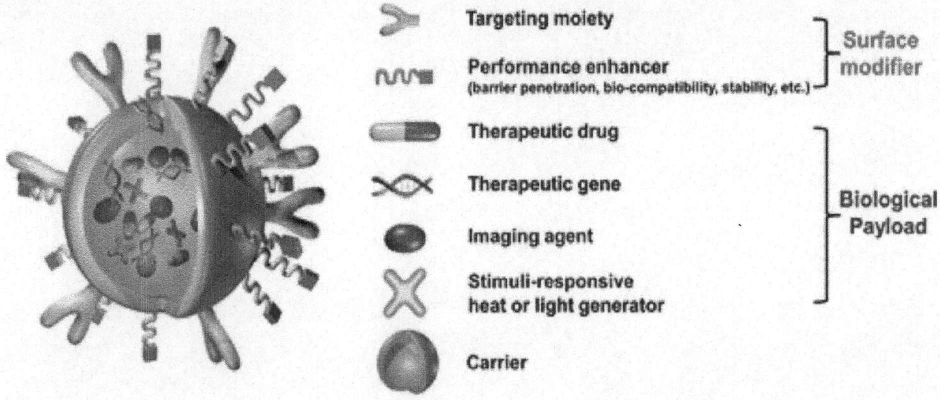

FIGURE 22.1 Schematic representation of a multifunctional nanocomposite carrying several biological payloads. It may be a hollow metal/metal-oxide/carbon nanostructure. Surface can be modified to enhance performance and for precise targeting.

Source: Reprinted from Lim et al. (2015) with permission © 2014 American Chemical Society.

DOI: 10.1201/9781003302131-23

22.1.1 Lipid-Based Nanoparticles

Lipid-based NPs are emerging as new therapeutic agents in nanomedicine. They are composed of a lipid bilayer and an aqueous compartment for high bioavailability and biocompatibility for drug delivery. They have the efficiency to carry large amounts of payloads to regulate biological properties (Figure 22.1). The liposome vesicular structure can be multilamellar or unilamellar. For this reason, it can carry hydrophilic, hydrophobic and lipophilic drugs (Sarfraz et al. 2018). Their stability can be altered by surface modification, NP size, surface charge, etc. (Sedighi et al. 2019). Surface modifications by ligand or by polymers are conducted for the enhancement of the rate of drug delivery and circulation efficiency (Alyautdin et al. 2014). Datta et al. discussed various synthetic and cell-derived lipid-based nanostructures and their biomedical applications (B. Datta et al. 2021).

22.1.2 Polymeric Nanoparticles

Polymeric NP is formed by natural or synthetic compounds. Some natural biopolymers such as chitosan are used for pulmonary (Al-Qadi et al. 2012) and intestine (Artursson et al. 1994) treatment and drug delivery. Xanthan gum is widely used in the pharmaceutical industry due to its good bioadhesive properties, low toxicity and low irritation effect (Goswami and Naik 2014). Cellulose is also one type of bio-polysaccharide used for drug control release by the alteration of solubility and gelation of the drug. It is used as an efficient drug delivery carrier. Polymeric NP can be synthesized by nanoprecipitation (C. Zhang et al. 2019), emulsification (Brown et al. 2020), ionic gelation, etc. NP cores with polymer matrices are chemically conjugate with polymer or nanosurface that have a great ability to carry both hydrophilic and hydrophobic cargo molecules such as protein,

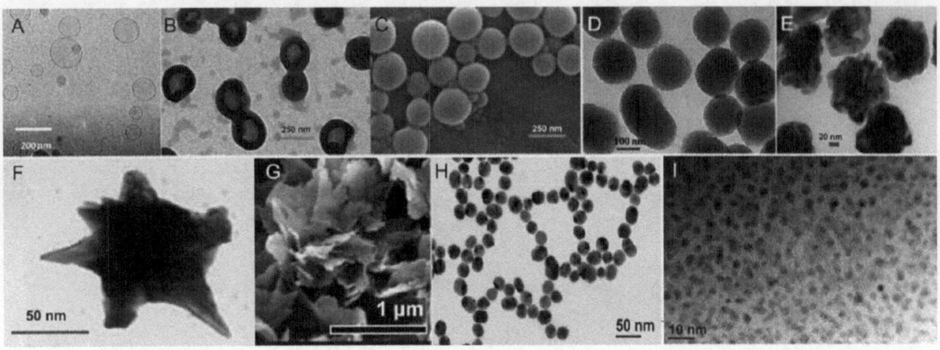

FIGURE 22.2 Schematic representation of different types of nanoplatforms. (A) Typical cryo-TEM of unilamellar liposomes. Reprinted from (Omarova et al. 2021) under Creative Commons Attribution 4.0 International (CC BY 4.0) license. (B and C) Transmission and scanning electron micrograph of pH-sensitive polymeric nanocapsules for stimuli-responsive drug spatiotemporal drug delivery. Reprinted from Pramanik et al. 2019 with permission © 2019 American Chemical Society. (D and E) Transmission electron micrographs of mesoporous silica and porous gold (inorganic nanoparticles) that show reduced insulin fibrillization *in vitro* and Huntingtin aggregation in neuronal cell cultures. Reprinted from Bardhan et al. 2021 under a Creative Commons Attribution-NonCommercial 3.0 Unported (CC BY-NC 3.0) license. (F) Gold nanoflower that can be used by exploiting the surface properties for the detection of ascorbic acid with nanomolar sensitivity. Reprinted from De et al. 2018 under author permission © 2018 The Royal Society of Chemistry. (G) ZnO nanoflower grown on cotton cloth can potentially trap SARS-CoV-2 and prevent the spread of COVID19. Reprinted from Adhikari, Pal, et al. 2021 with permission © 2021 American Chemical Society. (H) Riboflavin-decorated AuNPs for targeted delivery to cancer cells that overexpress riboflavin receptors, with potential photodynamic therapy applications. Reprinted from Sau, Sanyal, et al. 2018 with permission © 2018 American Chemical Society. (I) Fluorescent carbon quantum dots with potential applications in imaging.

Source: Reprinted from Sau, Bera, et al. 2018 with permission © 2018 American Chemical Society.

small molecules and vaccines (Knight et al. 2019). Polymersome and dendrimer are examples of polymeric nanoparticles (Figure 22.2). Polymersome is liposomes such as artificial vesicle structures, but are generally applicable in local regions. These enhance the stability and cargo retention efficiency and are effective as drug delivery vehicles in cytosol (Zelmer et al. 2020). Dendrimers are used for the delivery of nucleic acid and small molecules (Xu et al. 2014; Palmerston Mendes et al. 2017) by cargo formation while retaining the ability to change their mass, shape and surface structure. Pramanik et al. demonstrated a synthesis strategy of polymeric nanocapsules that have polyurethane, poly-thiourethane and biodegradable polyester linkages (Figure 22.3). At a low pH of the target organ/tissue, the degradation ester linkages of polylactide leads to the release of the cargo (Pramanik et al. 2019). Thus, the stimuli-responsive spatiotemporal release of drugs can be achieved with such polymeric nanoparticles. Polymeric NPs have good advantages for drug delivery such as biocompatibility, their biodegradable property and water solubility. In addition, surface modification (Valcourt et al. 2020) can be done easily, making them very useful for cancer treatment, gene therapy and diagnostics. However, notably, despite all the advantages mentioned above, the use of polymeric NP may be risky when particles are prone to aggregation (Mitchell et al. 2021).

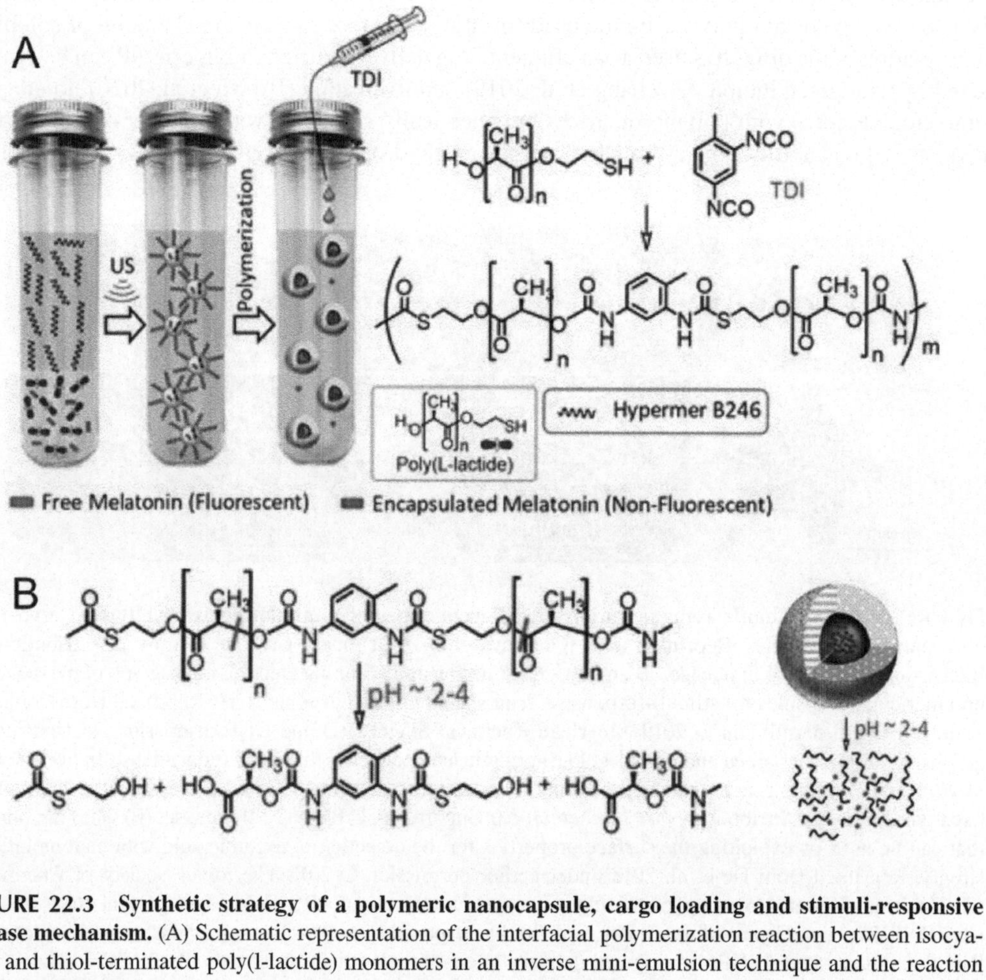

FIGURE 22.3 Synthetic strategy of a polymeric nanocapsule, cargo loading and stimuli-responsive release mechanism. (A) Schematic representation of the interfacial polymerization reaction between isocyanate and thiol-terminated poly(l-lactide) monomers in an inverse mini-emulsion technique and the reaction between thiol-terminated poly(l-lactide) and toluene diisocyanate (TDI). (B) Scheme for the degradation ester linkages of polylactide.

Source: Adopted from Pramanik et al. 2019 with permission © 2019 American Chemical Society.

22.1.3 Inorganic Nanoparticles

The use of inorganic nanoparticles in drug delivery is another emerging field for the treatment of some diseases. Some of the well-established inorganic nanomaterials for therapeutics are gold, silver, zinc, silica, etc. (Figures 22.2 and 22.4). Surface modification is often used to reduce cytotoxicity and improve the biocompatibility of nanoparticles (Maity et al. 2014). A large number of such nanoparticles are under different stages of clinical trials (Patra et al. 2018). Nanostructures are synthesized from these metals and used for drug delivery and imaging application. Different shapes and various types of nanostructures are formulated for biological application, for example, gold nanoparticles are mostly investigated and have the ability to form nanotubes, nanocages, nanospheres, nanorods, etc. (W. Yang et al. 2019). Inorganic NPs have physical, magnetic, electrical and optical properties that can be exploited for precision drug delivery, therapy and diagnosis (W. Yang et al. 2019; S. Datta et al. 2021; Chakraborty et al. 2018). AuNPs have free electrons on the surface that are continuously oscillating at a frequency based on their size and shape and give photothermal properties (Wang et al. 2020). Gold and silver nanoparticles have great surface plasmon resonance effects that are not shown in the cases of liposome, micell and dendrimer (Patra et al. 2018; S. Datta et al. 2021). From further study, we know that nanoparticles are delivered through paracellular transport and transcytosis. For surface functionalization, silver and gold NPs show high biocompatibility with in vivo experiments (Patra et al. 2018). Porous silica NPs are used for gene and drug uptake (K.-W. Huang et al. 2020) and delivery (Bardhan et al. 2018). Quantum dots made up of silicon semiconductor also have application for in vitro imaging (Wagner et al. 2019).

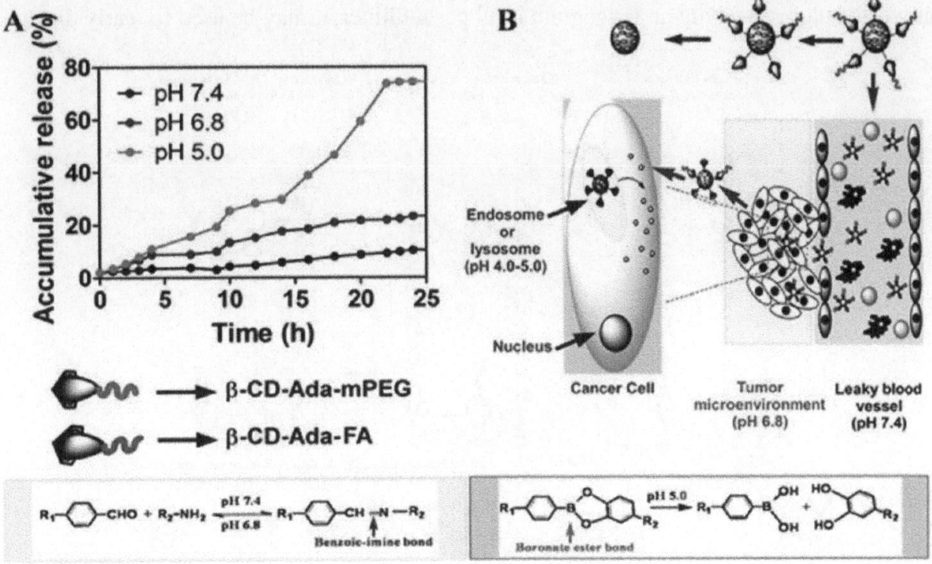

FIGURE 22.4 Targeted pH-sensitive cascade release of drug from mesoporous silica nanoplatform. (A) Time-dependent release of drugs at different pH. (B) Schematics of the cascade release mechanism. β-CD, β-cyclodextrin; Ada, 1-Adamantanemethylamine; FA, folic acid; mPEG, polyethylene glycol monomethyl ether. The benzoic-imine bond breaks at ~pH 6.8 in the tumor environment, releasing the gatekeeper only in the tumor microenvironment. Folate helps in the targeted uptake of the composite by the cancer cells. The boronate ester bond breaks at ~pH 5.0 in the endosome, releasing the cargo.

Source: Reprinted from Bhandary et al. 2017 under a Creative Commons Attribution 4.0 (CC BY 4.0) license.

22.2 NANOPARTICLES IN DIAGNOSTICS, DETECTION AND USE AS SENSORS

Nanoparticles not only are the emerging alternatives for small molecule drugs but also can be used for diagnostic purposes (Pal and Pramanik 2018). Nanotechnology-based diagnostics is a very modern technique for the detection of diseases. It could be a platform for the detection of the DNA, metabolites and protein markers related to diseases (Quesada-González and Merkoçi 2018). Various nanomaterials such as metal and semiconductor nanoparticles are used for diagnostics. Among these, AuNP has been shown to have great advantages in the detection of disease biomarkers (Oh et al. 2011). The localized surface plasmon resonance (LSPR), fluorescence resonance energy transfer (FRET) and surface-enhanced Raman scattering (SERS) of nanomaterials are sensitive tools for both detection and imaging. The photodynamic and electronic behavior of nanomaterials and scattering efficiency of visible and near infrared light (NIR) provide them compatibility with microscope techniques (electron, dark field scattering and confocal) (J.-L. Li et al. 2009), heterodyne imaging, UV-Vis, Raman spectroscopy, etc. (Yeh et al. 2012). De et al. recently demonstrated that crystal-defect-induced facet-dependent electrocatalytic activity of gold nanoflowers can be exploited for the selective nanomolar detection of ascorbic acid (De et al. 2018). Nanoparticles in nanocluster form can be used as a fluorescence agent due to their very small size, which is usually less than 10 nm in diameter (Sau, Bera, et al. 2018). Among them, gold nanoclusters are mostly studied and used as highly fluorescence probes for the detection of some ions and different disease biomarkers, etc. In the next few sections, the detection of viral infections, cancer diagnosis and early diagnosis of neurodegenerative diseases using the nanoplatform are discussed.

22.2.1 Virus Detection

Gold nanoclusters have a vital role in the detection of viral diseases. Streptavidin-coated gold nanoclusters (AuNC-SA) can be used for the detection of the HIV-1 p24 antigen by immunoassay (Figure 22.5). Because of its high sensitivity at a picogram level per milliliter, it may be used for early diagnosis of

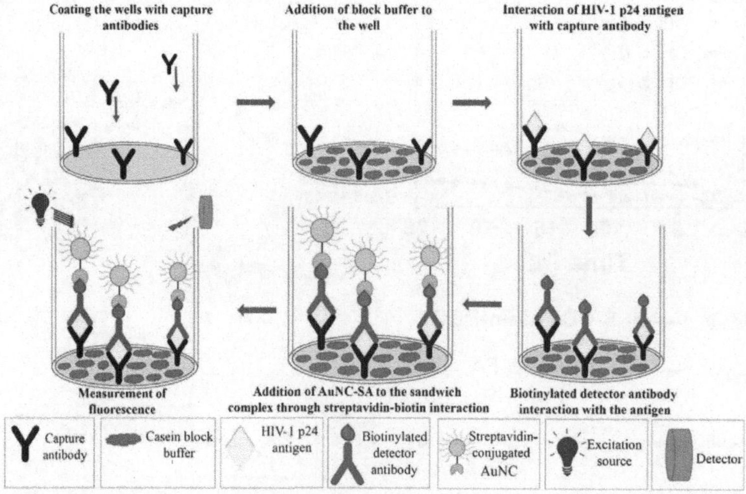

FIGURE 22.5 Schematic representation of the detection of HIV-1 P24 antigen by gold nanocluster immunoassay (GNCIA) using streptavidin-capped gold nanoclusters (AuNC-SA). Capture antibody is coated on the microtiter well. HIV P24 antigen binds with the capture antibody. Afterwards, biotinylated antibodies attach with captured antigens and form sandwich complexes. Subsequently, AuNC-SA binds with the sandwich complexes and gives fluorescence. The HIV P24 antigen is used as a target antigen of viral protein and is detected by the formation of antibody-antigen-antibody sandwich complexes bound to AuNC-SA fluorophore.

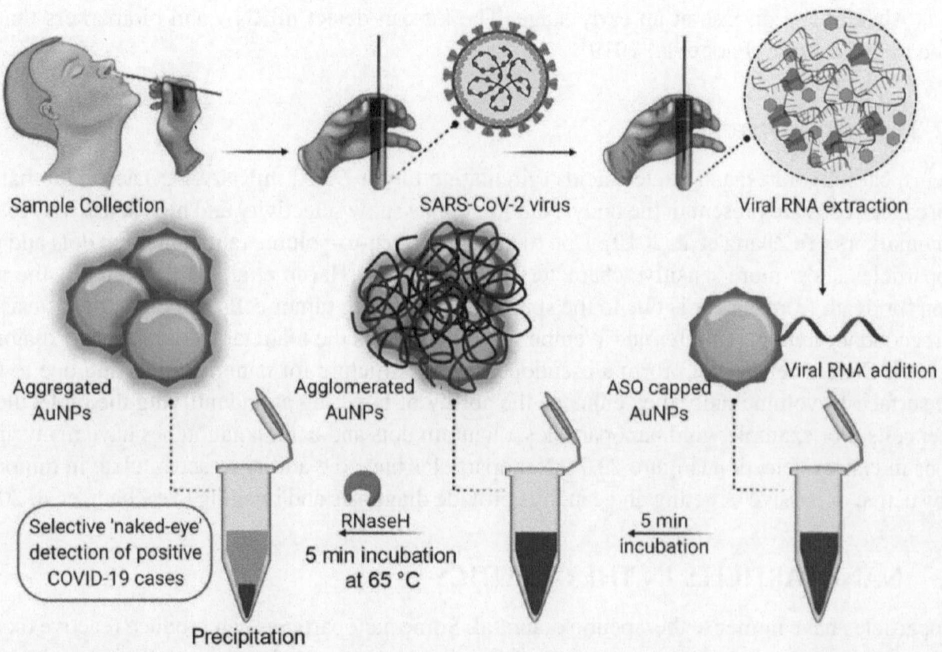

FIGURE 22.6 Schematic representation for the selective naked-eye/colorimetric detection of SARS-CoV-2 mediated by the suitably designed antisense oligonucleotide-capped AuNP (ASO-AuNP). ASO-capped AuNP is specific for the nucleocapsid phosphoprotein gene of SARS-CoV. Thiol-capped ASO-AuNP binds with the specific target RNA. This AuNP agglomerates only in the presence of the target RNA sequence of SARS-CoV-2. Incubation with RNaseH cleaves the RNA strand from the RNA-DNA hybrid that leads to the formation of a visually detectable precipitate indicating COVID positivity.

Source: Reprinted from Moitra et al. 2020 with permission © 2020 American Chemical Society.

HIV infection and to prevent the development of AIDS. Gold nanocluster immunoassay (GNCIA) can improve the efficiency and sensitivity of ELISA. In these investigations, the authors saw that emission intensity is directly proportional to the amount of p24 antigen (Kurdekar et al. 2018). A nanoparticle-enabled smartphone (NES) is a very recent development and an important method for the rapid detection of zika, hepatitis b and hepatitis c viruses. Platinum nanoprobes in NES help to detect these viruses by bubble formation in H_2O_2 (Draz et al. 2020). Infectious coronavirus disease (COVID19) can be detected by a colorimetric assay of gold nanoparticles with a higher accuracy than the conventional rapid antigen test (Figure 22.6). AuNP tagged with antisense oligonucleotide is a novel discovery for the rapid diagnosis of COVID19 (Moitra et al. 2020). It also provides a method for the naked eye detection of SARS-CoV-2 without any help of advanced instruments. The nucleocapsid phosphoprotein gene of SARS-CoV-2 is mainly used as the target gene for AuNP-based thiol-capped antisense oligonucleotide. Agglomeration of nanoparticles are found in the presence of the specific target RNA of SARS-CoV-2, and the surface plasmon resonance of the AuNP changes producing visual cues.

22.2.2 Neurodegenerative Disease Detection

Aβ amyloids have a major role in Alzheimer's disease development. Aggregation and misfolding of Aβ amyloid has been associated with this type of neurodegenerative disease. Gold nanoparticles have the ability to transport through the blood-brain barrier (Sela et al. 2015). AuNP-mediated colorimetric assay for the detection of Aβ peptide aggregation exhibits good selectivity and sensitivity (Khan and Park 2022). A novel kit with tagged fluorescent nanoparticles has also been invented that detects Alzheimer's disease at an early stage. The kit can detect miRNA and biomarkers that are related to this disease (Park et al. 2019)

22.2.3 Cancer Detection

Proteins, carbohydrates, and nucleic acids (circulating tumor DNA, miRNA, etc.) act as biomarkers when cancer cells are present in the body. Nanomaterials show selectivity and high sensitivity to cancer biomarkers (Ye Zhang et al. 2019). Due to a large surface-to-volume ratio, quantum dots and gold nanoparticles show more sensitive characters in diagnostics (Harun et al. 2013). Usually, the main reason for death from cancer is due to the spread of circulating tumor cells (CTC) by metastasis and from secondary tumors (Chaffer and Weinberg 2011). CTC is the main target for the early diagnosis of cancer. Cancer cells (CTC) form a pseudopodia-like structure for nanomaterials and due to their large surface-to-volume ratio, they enhance the ability of targeting and identifying the molecules of cancer cells. For example, gold nanoparticles, quantum dots and carbon nanotubes have many applications in cancer detection (Figure 22.7). Nanoparticles have the ability to accumulate in tumor tissue by active or passive targeting and can thus provide diagnosis and imaging (Ye Zhang et al. 2019).

22.3 NANOPARTICLES IN THERAPEUTICS

Nanoparticles have immense therapeutic potential. Some nanoparticles can produce reactive oxygen species (ROS), whereas some can scavenge ROS. Some nanoparticles show antimicrobial properties, and some show anti-cancer activities. Nanoparticles showing catalytic activities can be used to replenish deficient enzymes in metabolic disorders or in neurodegenerative diseases. However, it should be noted that uncontrolled interaction of nanoparticles with biomacromolecules can pose potential health hazards (Banerji et al. 2014). Nanoparticles can be made less toxic and more target-specific through surface modifications. The surface of NPs can be tweaked to have an effective binding affinity to specific drug molecules. The following sections give an overview of recent advances in the application of nanotherapeutics.

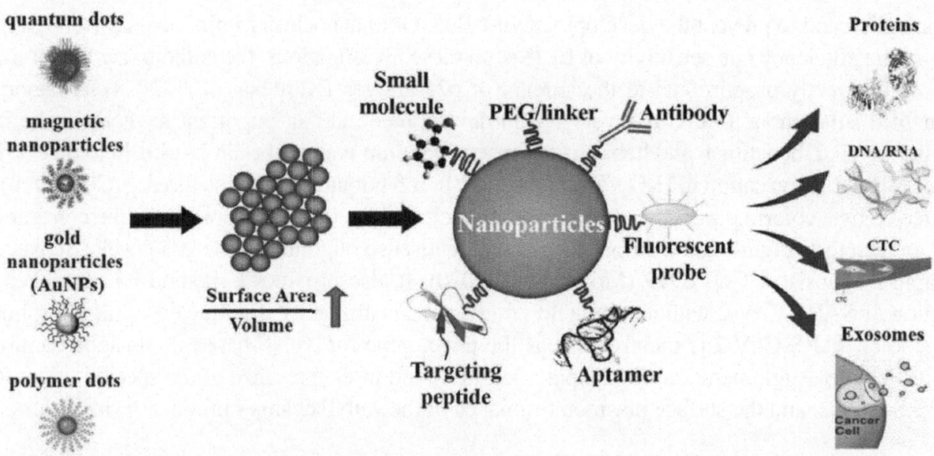

FIGURE 22.7 Schematic representation of nanotechnology-based improvement of cancer detection and diagnosis. Surface modification can be done to the nanoparticles by adding small molecules, peptides, antibodies, linker molecules, polyethylene glycol, fluorescent probes, etc. It can detect disease-specific proteins and DNA/RNA. It also can detect circulatory tumor cells.

Source: Reproduced from Ye Zhang et al. 2019 under a Creative Commons Attribution 4.0 (CC-BY 4.0) license.

22.3.1 Antimicrobial Agent

Nanomaterials such as copper, TiO_2 and Zn may act as antiviral or antibacterial agents by inhibiting transcription and translation or by producing ROS that causes damage to the lipid membrane of the virus or bacteria. A modern approach has also been discovered to suppress the spreading of COVID19 by coating surfaces with nanomaterials with antimicrobial properties (Imani et al. 2020; Adhikari, Pal, et al. 2021). Ultrasmall gold nanoclusters produce ROS, damage membranes and disrupt metabolism of the gram-positive and gram-negative bacteria (Zheng et al. 2017). Graphene materials, silver and ZnO nanomaterials kill the bacteria by damaging the membrane or producing oxidative stress. Mondal et al. showed antimicrobial activities of organic folic acid (FA) functionalized Mn_3O_4 NPs, which also offers significant X-ray contrast in *in vitro* studies, indicating the FA–Mn_3O_4 NPs to be a potential theranostic agent against bacteria (Mondal et al. 2021). Hasan et al. demonstrated the use of a wide bandgap semiconductor-based (ZnO) nanohybrid for potential antibacterial activity and photodynamic therapy (Hasan et al. 2020). Multi-drug-resistant bacteria such as methicillin-resistant *Staphylococcus aureus* (MRSA) have also been shown to die by the treatment of surface-modified AuNP (X. Li et al. 2014).

The antifungal activity of nanoparticles, especially the graphene-based nanoparticles (GBNs), is yet another interesting aspect. After the development of nanomaterials, there has been tremendous development over the past decades. It has been seen that many researchers are showing interest in using GBNs to develop new antifungal nanomaterials because of their unique germicidal activities with relatively low cytotoxicity towards mammalian cells (Alimardani et al. 2019). Fungi can colonize any organic surface and quickly spread fungal spores. Thus, humans are vastly affected by them. For this reason, suitable materials against fungi are very much in demand. GBNs have been proven to show antifungal activity (Sawangphruk et al. 2012) and studied by synthesizing different types of derivatives of graphene, especially graphene oxide and reduced graphene oxide.

22.3.2 In Cancer Treatment

The treatment of cancer is one of the challenging fields in medical science all over the world. Usually surgery, chemotherapy and radiotherapy are used for cancer disease recovery, but radiotherapy and chemotherapy have severe side effects on normal cells. Therefore, some new therapies utilizing nanoparticles are emerging for better treatment such as photodynamic therapy (PDT) (Sun et al. 2018), photothermal therapy (PTT), immunotherapy, gene therapy, etc. The surface coating of nanoparticles with biomolecules often makes them non-toxic toward normal cells while preserving anticancer activities (Maity et al. 2014). Nanostructures form a self-assembly system that works like a multifunctional building block (Ma and Zhao 2015), and in addition to the diagnosis of cancer, this can be used for its treatment. Enhanced permeability and retention effect (EPR) in the tumor cell microenvironment are very important properties of nanomaterials for PDT. Due to small changes in parameters such as pH and temperature, there is a major change in nanomaterials that can be utilized to specifically target tumor cells (Y. Hu et al. 2016). Light, photosensitizer (PS) and oxygen are major factors in PDT. PS acts as a light-absorbing dye that produces radical ions or converts the triplet oxygen to a singlet, causing destruction of the tumor cell (J. Hu et al. 2015). Coordination with nanomaterials is shown in drug delivery and target specificity to tumor cells. PTT is a method where light is transformed into heat with the help of a photothermal transducing agent (PTA). Cancer cells are killed through the heating and raising of temperature in a tumor cell; localized delivery can avoid any side effect on normal healthy cells (X. Huang et al. 2006). Sau et al. showed targeted delivery by taking advantage of the overexpression of a tumor biomarker receptor protein in malignant cells of human breast and prostate cancer and by conjugating the ligand of this receptor on the nanosurface. They have further shown that the engineered thiol-modified riboflavin-gold nano assemblies (RfS@AuNPs) accumulate selectively in the nucleus of cancer cells even without the nuclear localization signal peptide (Sau, Sen, et al. 2018; Sau, Sanyal, et al. 2018). Bera et al. showed that the fabrication of nanohybrids by attaching NIR-sensitive dye indocyanine green on ZnO nanoparticles improved its photodynamic therapeutic potential (A. Bera et al. 2022).

Among various cancers, brain cancer is a serious disease, as it is very difficult to detect and treat brain tumors. Moreover, drugs are not directly transported through the blood-brain barrier. In such a scenario, nanoparticles could be used as carriers for drug delivery. The recent discovery of nanoparticle probes encapsulated by biologically localized embedding (PEBBLE) makes drug delivery possible through surface modification (Sumner and Kopelman 2005). Three types of functionalizations are shown: first, the target molecule for guiding the PEBBLE; second, modification is for imaging; and third, it is used for delivering drugs near cancer cells. Doxorubicin, an anti-cancer drug, has been targeted by nanosystem-mediated delivery in brain cancer cells. Super paramagnetic iron oxide has a role in detecting the location of brain cancer cells (Yong Zhang et al. 2004).

22.3.3 Neurodegenerative Disease

Neurodegenerative diseases such as Alzheimer's, Parkinson's, Huntington's, etc. are a heterogeneous group of diseases due to the development of the slow degeneration of structure and function of the central nervous system (CNS). In these diseases, severe behavioral disorders and abnormality in movement have been seen (Dugger and Dickson 2017). This type of abnormality of neurons is caused by oxidative stress, neuroinflammation, protein misfolding and toxic stress. The proteins that are mainly responsible for this disease are β-amyloid, α-Synuclein and tau protein (Dugger and Dickson 2017). Alzheimer's disease is caused by a deposition of β-amyloid in the extracellular portion and induced toxicity by neuro-fibrillation. The small-sized nanoparticles capped with peptide (KLVFF) have been shown to inhibit the self-aggregation and prevent the fibrillization of β-amyloids (Zhao et al. 2019). This nanocomposite also reduces the oligomeric Aβ by microglial phagocytosis (Zhao et al. 2019) (Figure 22.8). Parkinson's disease (PD) is associated with

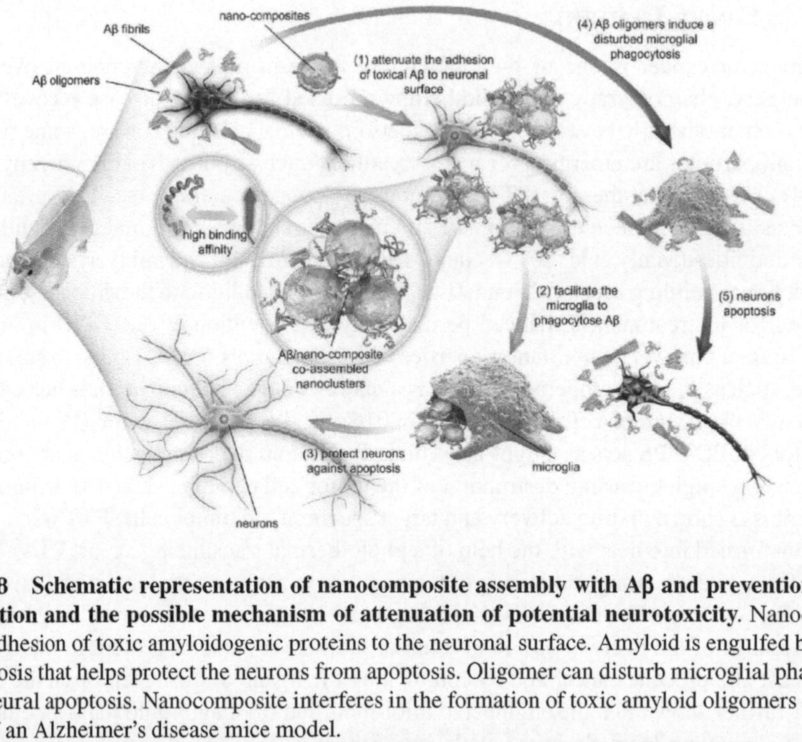

FIGURE 22.8 Schematic representation of nanocomposite assembly with Aβ and prevention of amyloid aggregation and the possible mechanism of attenuation of potential neurotoxicity. Nanocomposite inhibits the adhesion of toxic amyloidogenic proteins to the neuronal surface. Amyloid is engulfed by microglial phagocytosis that helps protect the neurons from apoptosis. Oligomer can disturb microglial phagocytosis and causes neural apoptosis. Nanocomposite interferes in the formation of toxic amyloid oligomers and fibrils in the case of an Alzheimer's disease mice model.

Source: Reprinted from Zhao et al. 2019 with permission © 2018 American Chemical Society.

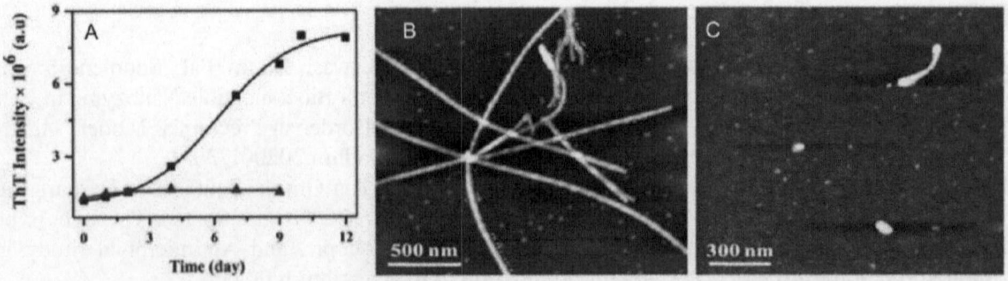

FIGURE 22.9 Fibril formation of α-synuclein protein is arrested by porphyrin loaded gold nanospheres. (A) ThT fluorescence shows the sigmoid growth curve of fibril formation; in the presence of porphyrin-capped gold nanoparticle (TPPS@AuNP), fibril formation was hindered. (B) AFM images of α-Synuclein fibril in the absence of TPPS@AuNP. (C) Fibril is usually absent in the presence of nanoparticles.

Source: Reprinted from K. Bera et al. 2020 with permission © 2020 American Chemical Society.

less secretion of dopamine from the substantia nigra of the brain stem due to neural degradation (Toulouse and Sullivan 2008). Some symptoms of PD include tremors, walking difficulties and slowness of movement (Jagaran and Singh 2021). Overexpression and fibrillization of α-synuclein causes toxicity and may lead to PD. Smaller nanoparticles can often transport through the blood-brain barrier (BBB) that leads to target-specific drug delivery with surface functionalization. Gene therapy has a great role in controlling neurodegenerative diseases such as AD and PD where NP is a good vehicle for gene delivery. Nanocarrier-mediated siRNA gene silencing is one of the best methods to regulate the disease. The main target of gene therapy is the misfolded protein accumulation and aggregation of β-amyloid and α-Synuclein that causes endoplasmic reticulum stress (L. Yang et al. 2014). Inorganic nanomaterials such as AuNP prevent fibril formation of β-amyloid and α-synuclein and play a very important role in controlling these diseases. Porphyrin-loaded gold nanospheres have been shown to hinder the fibril formation of α-synuclein protein in vitro (K. Bera et al. 2020) (Figure 22.9). Porous nanosilica and nano gold even without any surface modifications have been shown to reduce Huntingtin (Htt) aggregation in SH-SY5Y cells (Bardhan et al. 2021). Nanoparticles can also be administered to replenish some deficient enzymes in neurodegenerative disease. For example, Adhikari et al. demonstrated the incorporation of a biocompatible citrate-capped Mn_3O_4 nanozyme in a cellular antioxidant enzyme cascade to mimic the activity of glutathione peroxidase leading to the reversal of Huntington's-like symptoms in a preclinical model (Adhikari, Mondal, et al. 2021). Thus, nanotechnology has shown considerable therapeutic potential for neurodegenerative diseases.

22.4 CONCLUSION

Nanomedicine is a modern discovery in medical and scientific research. Nanoparticles have a high surface-to-volume ratio, which enhances the effective drug delivery rate and specificity. Nanomaterials such as gold silver nanoparticles and silica porous particles have unique physical and chemical properties that make them useful in theranostic application. Due to less side effects, surface functionalized nanomaterials are being directly or indirectly tested for various critical diseases such as cancer, neurodegenerative diseases, viral infections, etc. Nanomaterials can also detect and neutralize some dangerous infectious agents such as the HIV and SARS-CoV-2 viruses, some virulent bacteria, etc. Several nanomaterials have passed clinical trials. Thus, nanoparticles, quantum dots, porous silica, etc. will soon be available for targeted drug delivery and in theranostic applications. Nanoformulations are becoming indispensable in modern medicine and emerging directions for the treatment of diseases in the future.

REFERENCES

Adhikari, Aniruddha, Susmita Mondal, Monojit Das, Pritam Biswas, Uttam Pal, Soumendra Darbar, Siddhartha Sankar Bhattacharya, et al. 2021. "Incorporation of a Biocompatible Nanozyme in Cellular Antioxidant Enzyme Cascade Reverses Huntington's Like Disorder in Preclinical Model." *Advanced Healthcare Materials* 10 (7): 2001736. https://doi.org/10.1002/adhm.202001736.

Adhikari, Aniruddha, Uttam Pal, Sayan Bayan, Susmita Mondal, Ria Ghosh, Soumendra Darbar, Tanusri Saha-Dasgupta, et al. 2021. "Nanoceutical Fabric Prevents COVID-19 Spread Through Expelled Respiratory Droplets: A Combined Computational, Spectroscopic, and Antimicrobial Study." *ACS Applied Bio Materials* 4 (7): 5471–84. https://doi.org/10.1021/acsabm.1c00238.

Alimardani, Vahid, Samira Sadat Abolmaali, and Sedigheh Borandeh. 2019. "Antifungal and Antibacterial Properties of Graphene-Based Nanomaterials: A Mini-Review." *Journal of Nanostructures* 9 (3): 402–13. https://doi.org/10.22052/JNS.2019.03.002.

Al-Qadi, S., A. Grenha, D. Carrión-Recio, B. Seijo, and C. Remuñán-López. 2012. "Microencapsulated Chitosan Nanoparticles for Pulmonary Protein Delivery: In Vivo Evaluation of Insulin-Loaded Formulations." *Journal of Controlled Release: Official Journal of the Controlled Release Society* 157 (3): 383–90. https://doi.org/10.1016/j.jconrel.2011.08.008.

Alyautdin, Renad, Igor Khalin, Mohd Ismail Nafeeza, Muhammad Huzaimi Haron, and Dmitry Kuznetsov. 2014. "Nanoscale Drug Delivery Systems and the Blood-Brain Barrier." *International Journal of Nanomedicine* 9: 795–811. https://doi.org/10.2147/IJN.S52236.

Artursson, P., T. Lindmark, S. S. Davis, and L. Illum. 1994. "Effect of Chitosan on the Permeability of Monolayers of Intestinal Epithelial Cells (Caco-2)." *Pharmaceutical Research* 11 (9): 1358–61. https://doi.org/10.1023/a:1018967116988.

Banerji, Biswadip, Sumit Kumar Pramanik, Uttam Pal, and Nakul Chandra Maiti. 2014. "Binding of Hemoglobin to Ultrafine Carbon Nanoparticles: A Spectroscopic Insight into a Major Health Hazard." *RSC Advances* 4 (43): 22536–41. https://doi.org/10.1039/C4RA02569E.

Bardhan, Munmun, Sandip Dolui, Siddhi Chaudhuri, Uttam Paul, Gaurav Bhattacharjee, Manorama Ghosal, Nakul C. Maiti, Debashis Mukhopadhyay, and Dulal Senapati. 2021. "Impact of Porous Nanomaterials on Inhibiting Protein Aggregation Behaviour." *RSC Advances* 11 (6): 3354–62. https://doi.org/10.1039/D0RA10927D.

Bardhan, Munmun, Anupa Majumdar, Sayantan Jana, Tapas Ghosh, Uttam Pal, Snehasikta Swarnakar, and Dulal Senapati. 2018. "Mesoporous Silica for Drug Delivery: Interactions with Model Fluorescent Lipid Vesicles and Live Cells." *Journal of Photochemistry and Photobiology B: Biology* 178 (January): 19–26. https://doi.org/10.1016/j.jphotobiol.2017.10.023.

Bera, Arpan, Md. Nur Hasan, Uttam Pal, Damayanti Bagchi, Tuhin Kumar Maji, Tanusri Saha-Dasgupta, Ranjan Das, and Samir Kumar Pal. 2022. "Fabrication of Nanohybrids Toward Improving Therapeutic Potential of a NIR Photo-Sensitizer: An Optical Spectroscopic and Computational Study." *Journal of Photochemistry and Photobiology A: Chemistry* 424 (February): 113610. https://doi.org/10.1016/j.jphotochem.2021.113610.

Bera, Kaushik, Animesh Mondal, Uttam Pal, and Nakul C. Maiti. 2020. "Porphyrin-Armored Gold Nanospheres Modulate the Secondary Structure of α-Synuclein and Arrest Its Fibrillation." *The Journal of Physical Chemistry C* 124 (11): 6418–34. https://doi.org/10.1021/acs.jpcc.9b11503.

Bhandary, Suman, Arijit Bhowmik, Aparajita Ghosh, Suchandrima Saha, Uttam Pal, Nivedita Roy, Nilanjan Chakraborty, et al. 2017. "Targeting IL-6/IL-6R Signaling Axis in Triple-Negative Breast Cancer by a Novel Nifetepimine-Loaded Cascade PH Responsive Mesoporous Silica Based Nanoplatform." *Global Journal of Nanomedicine* 3 (2): 555609. https://doi.org/10.19080/GJN.2017.03.555609.

Brown, Shannon B., Lei Wang, Ryan R. Jungels, and Blanka Sharma. 2020. "Effects of Cartilage-Targeting Moieties on Nanoparticle Biodistribution in Healthy and Osteoarthritic Joints." *Acta Biomaterialia* 101 (January): 469–83. https://doi.org/10.1016/j.actbio.2019.10.003.

Chaffer, Christine L., and Robert A. Weinberg. 2011. "A Perspective on Cancer Cell Metastasis." *Science (New York, N.Y.)* 331 (6024): 1559–64. https://doi.org/10.1126/science.1203543.

Chakraborty, Sudeshna Das, Abhishek Sau, Arnab Maity, Uttam Pal, Chaitrali Sengupta, Maireyee Bhattacharya, Gourab Bhattacharjee, et al. 2018. "Low Magnetic Field Induced Surface Enhanced Transient Spin-Trajectory Modulation of a Prototype Anticancer Drug Sanguinarine on a Single Domain Superparamagnetic Nanosurface." *The Journal of Physical Chemistry C* 122 (36): 20619–31. https://doi.org/10.1021/acs.jpcc.8b06129.

Datta, Brateen, Debashish Paul, Uttam Pal, and Tatini Rakshit. 2021. "Intriguing Biomedical Applications of Synthetic and Natural Cell-Derived Vesicles: A Comparative Overview." *ACS Applied Bio Materials* 4 (4): 2863–85. https://doi.org/10.1021/acsabm.0c01480.

Datta, Soumendu, Uttam Pal, and Tanusri Saha-Dasgupta. 2021. "Effect of Mg Doping on Structural, Alloying, Electronic, Optical, and Bactericidal Properties of CunAg25–n Nanoclusters: A Computational Study." *The Journal of Physical Chemistry C* 125 (20): 11066–74. https://doi.org/10.1021/acs.jpcc.1c02292.

De, Sandip Kumar, Subrata Mondal, Pintu Sen, Uttam Pal, Biswarup Pathak, Kuber Singh Rawat, Munmun Bardhan, et al. 2018. "Crystal-Defect-Induced Facet-Dependent Electrocatalytic Activity of 3D Gold Nanoflowers for the Selective Nanomolar Detection of Ascorbic Acid." *Nanoscale* 10 (23): 11091–102. https://doi.org/10.1039/C8NR03087A.

Draz, Mohamed S., Anish Vasan, Aradana Muthupandian, Manoj Kumar Kanakasabapathy, Prudhvi Thirumalaraju, Aparna Sreeram, Sanchana Krishnakumar, et al. 2020. "Virus Detection Using Nanoparticles and Deep Neural Network–Enabled Smartphone System." *Science Advances* 6 (51): eabd5354. https://doi.org/10.1126/sciadv.abd5354.

Dugger, Brittany N., and Dennis W. Dickson. 2017. "Pathology of Neurodegenerative Diseases." *Cold Spring Harbor Perspectives in Biology* 9 (7): a028035. https://doi.org/10.1101/cshperspect.a028035.

Goswami, S., and Sonali D. Naik. 2014. "Natural Gums and Its Pharmaceutical Application." *Journal of Scientific and Innovative Research*. https://doi.org/10.31254/jsir.2014.3118.

Harun, Noor Aniza, Matthew J. Benning, Benjamin R. Horrocks, and David A. Fulton. 2013. "Gold Nanoparticle-Enhanced Luminescence of Silicon Quantum Dots Co-Encapsulated in Polymer Nanoparticles." *Nanoscale* 5 (9): 3817–27. https://doi.org/10.1039/C3NR00421J.

Hasan, Md Nur, Tuhin Kumar Maji, Uttam Pal, Arpan Bera, Damayanti Bagchi, Animesh Halder, Saleh A. Ahmed, et al. 2020. "Wide Bandgap Semiconductor-Based Novel Nanohybrid for Potential Antibacterial Activity: Ultrafast Spectroscopy and Computational Studies." *RSC Advances* 10 (64): 38890–99. https://doi.org/10.1039/D0RA07441A.

Hu, Jun, Yong'an Tang, Ahmed H. Elmenoufy, Huibi Xu, Zhen Cheng, and Xiangliang Yang. 2015. "Nanocomposite-Based Photodynamic Therapy Strategies for Deep Tumor Treatment." *Small (Weinheim an Der Bergstrasse, Germany)* 11 (44): 5860–87. https://doi.org/10.1002/smll.201501923.

Hu, Yang, Ran Lin, Pengcheng Zhang, Joshua Fern, Andrew G. Cheetham, Kunal Patel, Rebecca Schulman, et al. 2016. "Electrostatic-Driven Lamination and Untwisting of β-Sheet Assemblies." *ACS Nano* 10 (1): 880–88. https://doi.org/10.1021/acsnano.5b06011.

Huang, Kuan-Wei, Fu-Fei Hsu, Jiantai Timothy Qiu, Guann-Jen Chern, Yi-An Lee, Chih-Chun Chang, Yu-Ting Huang, et al. 2020. "Highly Efficient and Tumor-Selective Nanoparticles for Dual-Targeted Immunogene Therapy against Cancer." *Science Advances* 6 (3): eaax5032. https://doi.org/10.1126/sciadv.aax5032.

Huang, Xiaohua, Ivan H. El-Sayed, Wei Qian, and Mostafa A. El-Sayed. 2006. "Cancer Cell Imaging and Photothermal Therapy in the Near-Infrared Region by Using Gold Nanorods." *Journal of the American Chemical Society* 128 (6): 2115–20. https://doi.org/10.1021/ja057254a.

Imani, Sara M., Liane Ladouceur, Terrel Marshall, Roderick Maclachlan, Leyla Soleymani, and Tohid F. Didar. 2020. "Antimicrobial Nanomaterials and Coatings: Current Mechanisms and Future Perspectives to Control the Spread of Viruses Including SARS-CoV-2." *ACS Nano* 14 (10): 12341–69. https://doi.org/10.1021/acsnano.0c05937.

Jagaran, Keelan, and Moganavelli Singh. 2021. "Nanomedicine for Neurodegenerative Disorders: Focus on Alzheimer's and Parkinson's Diseases." *International Journal of Molecular Sciences* 22 (16): 9082. https://doi.org/10.3390/ijms22169082.

Khan, Zeeshan A., and Seungkyung Park. 2022. "AuNPs-Aβ-Ni-HRP Sandwich Assay: A New Sensitive Colorimetric Method for the Detection of Aβ 1–40." *Talanta* 237 (January): 122946. https://doi.org/10.1016/j.talanta.2021.122946.

Knight, Frances C., Pavlo Gilchuk, Amrendra Kumar, Kyle W. Becker, Sema Sevimli, Max E. Jacobson, Naveenchandra Suryadevara, et al. 2019. "Mucosal Immunization with a PH-Responsive Nanoparticle Vaccine Induces Protective CD8+ Lung-Resident Memory T Cells." *ACS Nano* 13 (10): 10939–60. https://doi.org/10.1021/acsnano.9b00326.

Kurdekar, Aditya Dileep, L. A. Avinash Chunduri, C. Sai Manohar, Mohan Kumar Haleyurgirisetty, Indira K. Hewlett, and Kamisetti Venkataramaniah. 2018. "Streptavidin-Conjugated Gold Nanoclusters as Ultrasensitive Fluorescent Sensors for Early Diagnosis of HIV Infection." *Science Advances* 4 (11): eaar6280. https://doi.org/10.1126/sciadv.aar6280.

Li, Jing-Liang, Lei Wang, Xiang-Yang Liu, Zhi-Ping Zhang, Hong-Chen Guo, Wei-Min Liu, and Sing-Hai Tang. 2009. "In Vitro Cancer Cell Imaging and Therapy Using Transferrin-Conjugated Gold Nanoparticles." *Cancer Letters* 274 (2): 319–26. https://doi.org/10.1016/j.canlet.2008.09.024.

Li, Xiaoning, Sandra M. Robinson, Akash Gupta, Krishnendu Saha, Ziwen Jiang, Daniel F. Moyano, Ali Sahar, et al. 2014. "Functional Gold Nanoparticles as Potent Antimicrobial Agents against Multi-Drug-Resistant Bacteria." *ACS Nano* 8 (10): 10682–86. https://doi.org/10.1021/nn5042625.

Lim, Eun-Kyung, Taekhoon Kim, Soonmyung Paik, Seungjoo Haam, Yong-Min Huh, and Kwangyeol Lee. 2015. "Nanomaterials for Theranostics: Recent Advances and Future Challenges." *Chemical Reviews* 115 (1): 327–94. https://doi.org/10.1021/cr300213b.

Ma, Xing, and Yanli Zhao. 2015. "Biomedical Applications of Supramolecular Systems Based on Host–Guest Interactions." *Chemical Reviews* 115 (15): 7794–839. https://doi.org/10.1021/cr500392w.

Maity, Mritunjoy, Sumit Kumar Pramanik, Uttam Pal, Biswadip Banerji, and Nakul Chandra Maiti. 2014. "Copper(I) Oxide Nanoparticle and Tryptophan as Its Biological Conjugate: A Modulation of Cytotoxic Effects." *Journal of Nanoparticle Research* 16 (1): 2179. https://doi.org/10.1007/s11051-013-2179-z.

Mitchell, Michael J., Margaret M. Billingsley, Rebecca M. Haley, Marissa E. Wechsler, Nicholas A. Peppas, and Robert Langer. 2021. "Engineering Precision Nanoparticles for Drug Delivery." *Nature Reviews Drug Discovery* 20 (2): 101–24. https://doi.org/10.1038/s41573-020-0090-8.

Moitra, Parikshit, Maha Alafeef, Ketan Dighe, Matthew B. Frieman, and Dipanjan Pan. 2020. "Selective Naked-Eye Detection of SARS-CoV-2 Mediated by N Gene Targeted Antisense Oligonucleotide Capped Plasmonic Nanoparticles." *ACS Nano* 14 (6): 7617–27. https://doi.org/10.1021/acsnano.0c03822.

Mondal, Susmita, Ria Ghosh, Aniruddha Adhikari, Uttam Pal, Dipanjan Mukherjee, Pritam Biswas, Soumendra Darbar, et al. 2021. "In Vitro and Microbiological Assay of Functionalized Hybrid Nanomaterials to Validate Their Efficacy in Nanotheranostics: A Combined Spectroscopic and Computational Study." *ChemMedChem* 16 (24): 3739–49. https://doi.org/10.1002/cmdc.202100494.

Oh, Eunkeu, James B. Delehanty, Kim E. Sapsford, Kimihiro Susumu, Ramasis Goswami, Juan B. Blanco-Canosa, Philip E. Dawson, et al. 2011. "Cellular Uptake and Fate of PEGylated Gold Nanoparticles Is Dependent on Both Cell-Penetration Peptides and Particle Size." *ACS Nano* 5 (8): 6434–48. https://doi.org/10.1021/nn201624c.

Omarova, Marzhana, Yueheng Zhang, Igor Kevin Mkam Tsengam, Jibao He, Tianyi Yu, Donghui Zhang, and Vijay John. 2021. "Hydrophobe Containing Polypeptoids Complex with Lipids and Induce Fusogenesis of Lipid Vesicles." *The Journal of Physical Chemistry B* 125 (12): 3145–52. https://doi.org/10.1021/acs.jpcb.0c11477.

Pal, Uttam, and Sumit Kumar Pramanik. 2018. "Advances in the Application of Nanomaterials and Nanoscaled Materials in Physiology or Medicine: Now and the Future." In *Nanomaterials in Bio-Medical Applications: A Novel Approach*. Materials Research Foundations. http://dx.doi.org/10.21741/9781945291739-7.

Palmerston Mendes, Livia, Jiayi Pan, and Vladimir P. Torchilin. 2017. "Dendrimers as Nanocarriers for Nucleic Acid and Drug Delivery in Cancer Therapy." *Molecules (Basel, Switzerland)* 22 (9): E1401. https://doi.org/10.3390/molecules22091401.

Park, Jun Sung, Sang Tae Kim, Sang Yun Kim, Min Gi Jo, Myeong Jun Choi, and Myeong Ok Kim. 2019. "A Novel Kit for Early Diagnosis of Alzheimer's Disease Using a Fluorescent Nanoparticle Imaging." *Scientific Reports* 9 (1): 13184. https://doi.org/10.1038/s41598-019-49711-y.

Patra, Jayanta Kumar, Gitishree Das, Leonardo Fernandes Fraceto, Estefania Vangelie Ramos Campos, Maria del Pilar Rodriguez-Torres, Laura Susana Acosta-Torres, Luis Armando Diaz-Torres, et al. 2018. "Nano Based Drug Delivery Systems: Recent Developments and Future Prospects." *Journal of Nanobiotechnology* 16 (1): 71. https://doi.org/10.1186/s12951-018-0392-8.

Pramanik, Sumit Kumar, Uttam Pal, Preety Choudhary, Harwinder Singh, Russel J. Reiter, Anitha Ethirajan, Snehasikta Swarnakar, and Amitava Das. 2019. "Stimuli-Responsive Nanocapsules for the Spatiotemporal Release of Melatonin: Protection against Gastric Inflammation." *ACS Applied Bio Materials* 2 (12): 5218–26. https://doi.org/10.1021/acsabm.9b00236.

Quesada-González, Daniel, and Arben Merkoçi. 2018. "Nanomaterial-Based Devices for Point-of-Care Diagnostic Applications." *Chemical Society Reviews* 47 (13): 4697–709. https://doi.org/10.1039/C7CS00837F.

Sarfraz, Muhammad, Attia Afzal, Tan Yang, Yongkang Gai, Shahid Masood Raza, Muhammad Waseem Khan, Yao Cheng, et al. 2018. "Development of Dual Drug Loaded Nanosized Liposomal Formulation by a Reengineered Ethanolic Injection Method and Its Pre-Clinical Pharmacokinetic Studies." *Pharmaceutics* 10 (3): E151. https://doi.org/10.3390/pharmaceutics10030151.

Sau, Abhishek, Kallol Bera, Uttam Pal, Arnab Maity, Pritiranjan Mondal, Samrat Basak, Alivia Mukherjee, et al. 2018. "Design and Synthesis of Fluorescent Carbon-Dot Polymer and Deciphering Its Electronic Structure." *The Journal of Physical Chemistry C* 122 (41): 23799–807. https://doi.org/10.1021/acs.jpcc.8b08322.

Sau, Abhishek, Sulagna Sanyal, Kallol Bera, Sabyasachi Sen, Amrit Krishna Mitra, Uttam Pal, Prabal Kumar Chakraborty, et al. 2018. "DNA Damage and Apoptosis Induction in Cancer Cells by Chemically Engineered Thiolated Riboflavin Gold Nanoassembly." *ACS Applied Materials & Interfaces* (January). https://doi.org/10.1021/acsami.7b18837.

Sau, Abhishek, Sabyasachi Sen, Kallol Bera, Uttam Pal, Biswarup Satpati, Chandrima Das, and Samita Basu. 2018. "Nuclear Uptake of Thiolated Riboflavin Gold Nanoassembly: DNA Damage and Apoptosis Induction in Cancer Cell." *Biophysical Journal* 114 (3): 360a. https://doi.org/10.1016/j.bpj.2017.11.2002.

Sawangphruk, Montree, Pattarachai Srimuk, Poramane Chiochan, Tanas Sangsri, and Patcharaporn Siwayaprahm. 2012. “Synthesis and Antifungal Activity of Reduced Graphene Oxide Nanosheets.” *Carbon* 50 (14): 5156–61. https://doi.org/10.1016/j.carbon.2012.06.056.

Sedighi, Mahsa, Sandro Sieber, Fereshteh Rahimi, Mohammad-Ali Shahbazi, Ali Hossein Rezayan, Jörg Huwyler, and Dominik Witzigmann. 2019. “Rapid Optimization of Liposome Characteristics Using a Combined Microfluidics and Design-of-Experiment Approach.” *Drug Delivery and Translational Research* 9 (1): 404–13. https://doi.org/10.1007/s13346-018-0587-4.

Sela, Hagit, Hagit Cohen, Paz Elia, Raya Zach, Zeev Karpas, and Yehuda Zeiri. 2015. “Spontaneous Penetration of Gold Nanoparticles Through the Blood Brain Barrier (BBB).” *Journal of Nanobiotechnology* 13 (1): 71. https://doi.org/10.1186/s12951-015-0133-1.

Sumner, James P., and Raoul Kopelman. 2005. “Alexa Fluor 488 as an Iron Sensing Molecule and Its Application in PEBBLE Nanosensors.” *The Analyst* 130 (4): 528–33. https://doi.org/10.1039/b414189j.

Sun, Haifeng, Shukun Li, Wei Qi, Ruirui Xing, Qianli Zou, and Xuehai Yan. 2018. “Stimuli-Responsive Nanoparticles Based on Co-Assembly of Naturally-Occurring Biomacromolecules for in Vitro Photodynamic Therapy.” *Colloids and Surfaces A: Physicochemical and Engineering Aspects* 538: 795–801. https://doi.org/10.1016/j.colsurfa.2017.11.072.

Toulouse, André, and Aideen M. Sullivan. 2008. “Progress in Parkinson’s Disease-Where Do We Stand?” *Progress in Neurobiology* 85 (4): 376–92. https://doi.org/10.1016/j.pneurobio.2008.05.003.

Valcourt, Danielle M., Megan N. Dang, Mackenzie A. Scully, and Emily S. Day. 2020. “Nanoparticle-Mediated Co-Delivery of Notch-1 Antibodies and ABT-737 as a Potent Treatment Strategy for Triple-Negative Breast Cancer.” *ACS Nano* 14 (3): 3378–88. https://doi.org/10.1021/acsnano.9b09263.

Wagner, Angela M., Jennifer M. Knipe, Gorka Orive, and Nicholas A. Peppas. 2019. “Quantum Dots in Biomedical Applications.” *Acta Biomaterialia* 94 (August): 44–63. https://doi.org/10.1016/j.actbio.2019.05.022.

Wang, Jianxin, Andrea M. Potocny, Joel Rosenthal, and Emily S. Day. 2020. “Gold Nanoshell-Linear Tetrapyrrole Conjugates for Near Infrared-Activated Dual Photodynamic and Photothermal Therapies.” *ACS Omega* 5 (1): 926–40. https://doi.org/10.1021/acsomega.9b04150.

Xu, Leyuan, Hao Zhang, and Yue Wu. 2014. “Dendrimer Advances for the Central Nervous System Delivery of Therapeutics.” *ACS Chemical Neuroscience* 5 (1): 2–13. https://doi.org/10.1021/cn400182z.

Yang, Liu, Linqing Miao, Feisi Liang, Haoliang Huang, Xiuyin Teng, Shaohua Li, Jaloliddin Nuriddinov, Michael E. Selzer, and Yang Hu. 2014. “The MTORC1 Effectors S6K1 and 4E-BP Play Different Roles in CNS Axon Regeneration.” *Nature Communications* 5 (1): 5416. https://doi.org/10.1038/ncomms6416.

Yang, Wenjie, Huazheng Liang, Songhua Ma, Devin Wang, and Jun Huang. 2019. “Gold Nanoparticle Based Photothermal Therapy: Development and Application for Effective Cancer Treatment.” *Sustainable Materials and Technologies* 22 (December): e00109. https://doi.org/10.1016/j.susmat.2019.e00109.

Yeh, Yi-Cheun, Brian Creran, and Vincent M. Rotello. 2012. “Gold Nanoparticles: Preparation, Properties, and Applications in Bionanotechnology.” *Nanoscale* 4 (6): 1871–80. https://doi.org/10.1039/c1nr11188d.

Zelmer, Christina, Ludovit P. Zweifel, Larisa E. Kapinos, Ioana Craciun, Zekiye P. Güven, Cornelia G. Palivan, and Roderick Y. H. Lim. 2020. “Organelle-Specific Targeting of Polymersomes into the Cell Nucleus.” *Proceedings of the National Academy of Sciences* 117 (6): 2770–78. https://doi.org/10.1073/pnas.1916395117.

Zhang, Chang-Xiong, Ying Cheng, Dao-Zhou Liu, Miao Liu, Han Cui, Bang-Le Zhang, Qi-Bing Mei, et al. 2019. “Mitochondria-Targeted Cyclosporin a Delivery System to Treat Myocardial Ischemia Reperfusion Injury of Rats.” *Journal of Nanobiotechnology* 17 (1): 18. https://doi.org/10.1186/s12951-019-0451-9.

Zhang, Ye, Maoyu Li, Xiaomei Gao, Yongheng Chen, and Ting Liu. 2019. “Nanotechnology in Cancer Diagnosis: Progress, Challenges and Opportunities.” *Journal of Hematology & Oncology* 12 (1): 137. https://doi.org/10.1186/s13045-019-0833-3.

Zhang, Yong, Conroy Sun, Nathan Kohler, and Miqin Zhang. 2004. “Self-Assembled Coatings on Individual Monodisperse Magnetite Nanoparticles for Efficient Intracellular Uptake.” *Biomedical Microdevices* 6 (1): 33–40. https://doi.org/10.1023/b:bmmd.0000013363.77466.63.

Zhao, Yu, Jinquan Cai, Zichen Liu, Yansheng Li, Chunxiong Zheng, Yadan Zheng, Qun Chen, et al. 2019. “Nanocomposites Inhibit the Formation, Mitigate the Neurotoxicity, and Facilitate the Removal of β-Amyloid Aggregates in Alzheimer’s Disease Mice.” *Nano Letters* 19 (2): 674–83. https://doi.org/10.1021/acs.nanolett.8b03644.

Zheng, Kaiyuan, Magdiel I. Setyawati, David Tai Leong, and Jianping Xie. 2017. “Antimicrobial Gold Nanoclusters.” *ACS Nano* 11 (7): 6904–10. https://doi.org/10.1021/acsnano.7b02035.

Index

F

G

H

I

K

L

M

N

O

P

Q

R

S

T

U

V

W

X

Z